Handbook of
CATHODIC CORROSION PROTECTION

Handbook of
CATHODIC CORROSION PROTECTION

Theory and Practice of Electrochemical Protection Processes

Third Edition

W. von Baeckmann, W. Schwenk, and W. Prinz, Editors

With contributions from:
W. von Baeckmann, H. Bohnes, G. Franke, D. Funk, C. Gey, H. Gräfen,
G. Heim, V. Heinzelmann, K. Horras, B. Isecke, H. Kampermann, B. Leutner,
H. -U. Paul, F. Paulekat, W. Prinz, B. Richter, G. Rieger, H. G. Schöneich, W. Schwenk

G|P **Gulf Professional Publishing**
an imprint of Elsevier Science

Gulf Professional Publishing is an imprint of Elsevier Science.

Copyright © 1997 by Elsevier Science (USA).

All rights reserved.

Originally published by Gulf Publishing Company, Houston, TX.

No part of this publication may be reproduced, stored in a retrieval system, or transmitted in any form or by any means, electronic, mechanical, photocopying, recording, or otherwise, without the prior written permission of the publisher.

Permissions may be sought directly from Elsevier's Science & Technology Rights Department in Oxford, UK: phone: (+44) 1865 843830, fax: (+44) 1865 853333, e-mail: permissions@elsevier.co.uk. You may also complete your request on-line via the Elsevier Science homepage (http://www.elsevier.com), by selecting 'Customer Support' and then 'Obtaining Permissions'.

 This book is printed on acid-free paper.

ISBN 0-88415-056-9

The publisher offers special discounts on bulk orders of this book.
For information, please contact:
Manager of Special Sales
Elsevier Science
200 Wheeler Road
Burlington, MA 01803
Tel: 781-313-4700
Fax: 781-313-4802

For information on all Gulf publications available, contact our World Wide Web homepage at http://www.bh.com/gulf

10 9 8 7 6 5 4 3

Printed in the United States of America.

Contents

Foreword to the Third Edition xix
Foreword to the First Edition xx
Preface xxii
Acknowledgments xxiii
Index of Authors xxiv
Commonly Used Quantities, Constants, and Symbols xxv
Frequently Used Indices xxx
  Chemical and Thermodynamic Quantities Y xxx
  Electrochemical Quantities Y xxx
  Electrical Quantities Y xxx
  General Symbols xxx

American and European Electrical Cable Sizes xxxii

1   **The History of Corrosion Protection** 1
        W. V. BAECKMANN
1.1   Corrosion Protection for Buried Pipelines 1
1.2   Corrosion Protection by Painting 8
1.3   History of Cathodic Protection 9
1.4   Development of Stray Current Protection 20
1.5   Corrosion Protection by Information 23
1.6   References 25

2   **Fundamentals and Concepts of Corrosion and Electrochemical Corrosion Protection** 27
        W. SCHWENK
2.1   Corrosion Processes, Corrosion Damage, and Protective Countermeasures 27
2.2   Electrochemical Corrosion 29
2.2.1   Metallic Materials 30

v

2.2.2	Aqueous Electrolytes	34
2.2.3	Electrochemical Phase Boundary Reactions	36
2.2.3.1	Basic Thermodynamics	37
2.2.3.2	Electrochemical Kinetics	40
2.2.4	Mixed Electrodes	44
2.2.4.1	Homogeneous Mixed Electrodes	44
2.2.4.2	Heterogeneous Mixed Electrodes and Cell Formation	46
2.2.5	Observations of Current Distribution	50
2.3	Potential Dependence of Corrosion Extent	52
2.3.1	Almost Uniform Weight Loss Corrosion	53
2.3.1.1	Weight Loss Corrosion of Active Metals	53
2.3.1.2	Weight Loss Corrosion of Passive Metals	59
2.3.2	Pitting Corrosion	62
2.3.3	Stress Corrosion	63
2.3.4	Hydrogen-Induced Corrosion	66
2.3.5	Corrosion Fatigue	70
2.3.6	Limits of Applicability of Electrochemical Protection Processes	70
2.4	Critical Protection Potentials and Ranges	71
2.5	References	76

3 Fundamentals and Practice of Electrical Measurements ... 79
W. v. Baeckmann and W. Schwenk

3.1	The Electrical Parameters: Current, Voltage, and Resistance	79
3.2	Reference Electrodes	85
3.3	Potential Measurement	88
3.3.1	Bases of Potential Measurement of Electrodes with Flowing Current	88
3.3.2	Application of Potential Measurement	96
3.3.2.1	Measuring Instruments and Their Properties	96
3.3.2.2	Potential Measurements on Pipelines and Storage Tanks	97
3.3.2.3	Potential Measurement under the Influence of Stray Currents	100
3.3.2.4	Potential Measurement under the Influence of Alternating Currents	102
3.3.3	Application of Protection Criteria	103
3.3.3.1	Pragmatic Protection Criteria for Nonalloyed Ferrous Materials	104
3.3.3.2	Potential Measurement with Potential Test Probes	106
3.4	Current Measurement	107
3.4.1	General Advice for Measurement of Current	107
3.4.2	Pipe Current Measurement	108

3.4.3	Measurement of Current Density and Coating Resistance	110
3.5	Resistivity Measurement	112
3.5.1	Resistivity Measuring Instruments	113
3.5.2	Measurement of Specific Soil Resistivity	114
3.5.3	Measurement of Grounding Resistance	118
3.6	Location of Faults	119
3.6.1	Measurement of Foreign Contacts	120
3.6.1.1	Fault Location using dc	120
3.6.1.2	Fault Location with ac	122
3.6.2	Location of Heterogeneous Surface Areas by Measurements of Field Strength	123
3.6.2.1	Location of Local Anodes	124
3.6.2.2	Location of Coating Defects	125
	(a) Circular damage	127
	(b) Porous coating	129
3.7	Intensive Measurement Technique	131
3.7.1	Quantities to be Measured and Objectives of Intensive Measurement Technique	132
3.7.2	Carrying Out an Intensive Measurement	135
	(a) Determination of the pipe/soil potential	135
	(b) Determination of ΔU values	136
3.8	References	137

4 Corrosion in Aqueous Solutions and Soil ... 139
G. HEIM AND W. SCHWENK

4.1	Action of Corrosion Products and Types of Corrosion	139
4.2	Determining the Corrosion Likelihood of Uncoated Metals	142
4.2.1	Corrosion in Soils	144
4.2.2	Corrosion in Aqueous Media	148
4.3	Enhancement of Anodic Corrosion by Cell Formation or Stray Currents from dc Installations	148
4.4	Corrosion Due to ac Interference	150
4.5	References	152

5 Coatings for Corrosion Protection ... 153
G. HEIM AND W. SCHWENK

| 5.1 | Objectives and Types of Corrosion Protection by Coatings | 153 |
| 5.1.1 | Organic Coatings | 153 |

5.1.2	Cement Mortar Coatings	154
5.1.3	Enamel Coatings	154
5.1.4	Metallic Coatings	154
5.2	Properties of Organic Coatings	155
5.2.1	Electrical and Electrochemical Properties	155
5.2.1.1	Review of the Types of Reactions	155
5.2.1.2	Coating Resistance and Protection Current Demand	156
5.2.1.3	Effectiveness of Cathodes and Cell Formation	162
5.2.1.4	Electrochemical Blistering	163
5.2.1.5	Cathodic Disbonding	166
5.2.2	Physicochemical Properties	169
5.2.3	Mechanical Properties	170
5.2.4	Corrosion of the Steel under the Coating	171
5.3	Properties of Cement Mortar and Concrete	173
5.3.1	Corrosion of Mortar	173
5.3.2	Corrosion of Steel in Mortar	174
5.4	Properties of Enamel Coatings	175
5.5	Properties of Metallic Coatings	176
5.6	References	177

6 Galvanic (Sacrificial) Anodes 179
H. Bohnes and G. Franke

6.1	General Information	179
6.1.1	Current Capacity of Galvanic Anodes	180
6.1.2	Current Discharge from Galvanic Anodes	183
6.2	Anode Materials	185
6.2.1	Iron	185
6.2.2	Zinc	185
6.2.3	Aluminum	188
6.2.4	Magnesium	191
6.3	Backfill Materials	196
6.4	Supports	198
6.5	Forms of Anodes	199
6.5.1	Rod Anodes	200
6.5.2	Plates and Compact Anodes	200
6.5.3	Anodes for Tanks	201
6.5.4	Offshore Anodes	202
6.5.5	Special Forms	202
6.6	Quality Control and Performance Testing	203

6.7	Advantages and Disadvantages of Galvanic Anodes	204
6.8	References	206

7   Impressed Current Anodes 207
H. Bohnes and D. Funk

7.1	General Comments	207
7.2	Anode Materials	208
7.2.1	Solid Anodes	208
7.2.2	Noble Metals and Valve Metals Coated with Noble Metals	213
7.2.3	Metal Oxide-Coated Valve Metals	216
7.2.4	Polymer Cable Anodes	217
7.3	Insulating Materials	217
7.4	Cables	218
7.5	Forms of Anode	219
7.5.1	Anodes Suitable for Soil	219
7.5.2	Anodes Suitable for Water	221
7.5.3	Anodes for Internal Application	222
7.6	References	224

8   Impressed Current Equipment and Transformer-Rectifiers 225
W. v. Baeckmann and H. Kamperman

8.1	Site and Electrical Protection Measures	226
8.2	Design and Circuitry of Impressed Current	228
8.3	Rectifier Circuit	229
8.4	Adjustable Transformer-Rectifiers	230
8.5	Rectifiers Resistant to High Voltage	232
8.6	Control Rectifiers	233
8.7	Transformer-Rectifiers without Mains Connections	237
8.8	Equipment and Control of Transformer-Rectifiers	237
8.9	References	242

9   Impressed Current Ground Beds and Interference Problems 243
W. von Baeckmann and W. Prinz

9.1	Impressed Current Ground Beds	244
9.1.1	Continuous Horizontal Anode Beds	244

9.1.2	Single Anode Installations	248
9.1.3	Deep Anode Beds	250
9.1.4	Design of Anodes	254
9.2	Interference with Foreign Pipelines and Cables	256
9.2.1	Interference from the Voltage Cone of Anodes	257
9.2.2	Interference from the Cathodic Voltage Cone of the Protected Object	259
9.2.3	Avoidance of Interference	261
9.3	References	264

10 Pipelines ... 265
W. Prinz

10.1	Electrical Properties of Steel Pipelines	265
10.2	Preconditions for Pipeline Protection	268
10.2.1	Measures for Achieving a Low Resistance Load	268
10.2.2	Measures for Achieving a Low Leakage Load	268
10.2.2.1	Pipeline Coating	268
10.2.2.2	Insulating Joints	269
10.2.2.3	Electrically Operated Valves and Regulators	271
10.2.2.4	Casings	272
10.2.2.5	Special Installations on the Pipeline	274
10.2.2.6	Prevention of Electrical Contact with Foreign Objects	274
10.3	Design of Cathodic Protection	276
10.3.1	Design Documents	276
10.3.2	Test Points	276
10.3.3	Determination of Current Demand	277
10.3.4	Choice of Protection Method	278
10.3.4.1	Galvanic Anodes	278
10.3.4.2	Impressed Current Anodes	279
10.3.5	Pipelines for Electrolytically Conducting Liquids	280
10.3.6	Distribution Networks	283
10.4	Commissioning the Cathodic Protection Station	285
10.5	Monitoring and Supervision	287
10.6	References	289

11 Storage Tanks and Tank Farms ... 290
K. Horras and G. Rieger

| 11.1 | Special Problems Relating to the Protection of Tanks | 290 |
| 11.2 | Preparatory Measures | 290 |

11.3	Storage Tanks	292
11.3.1	Determination of Current Demand, Evaluation, and Connections of the Protection Equipment	292
11.3.2	Choice of Protection Method	295
11.3.3	Examples of the Design of Protective Installations	296
11.3.3.1	Equipment Using Galvanic Anodes	296
11.3.3.2	Impressed Current Station	298
11.4	Tank Farms and Filling Stations	299
11.5	Special Problems in Cathodic Protection Near Railways	300
11.5.1	General Comments	300
11.5.2	Equipotential Bonding and Insulating Joints	301
11.5.3	Protective Grounding with Electrified Railways	302
11.5.4	Lightning Protection	302
11.5.5	Interference and Working in the Area of Railways	303
11.6	Measures in the Case of Dissimilar Metal Installations	304
11.7	Internal Protection of Fuel Tanks	304
11.8	Consideration of Other Protection Measures	306
11.9	Operation and Maintenance of Cathodic Protection Stations	307
11.10	References	308

12 Local Cathodic Protection 309
W. v. Baeckmann and W. Prinz

12.1	Range of Applications	309
12.2	Special Features of the Local Cathodic Protection	310
12.3	Power Stations	312
12.4	Oil Refineries	315
12.5	Installations with Small Steel-Reinforced Concrete Foundations	317
12.6	Tank Farms	318
12.7	References	322

13 Telephone Cables 323
C. Gey

13.1	Laying Cables	323
13.2	Passive Corrosion Protection	324
13.3	Cathodic Protection	326
13.3.1	Stray Current Protection	327
13.3.2	Cathodic Protection with Impressed Current Anodes	329
13.4	References	334

14 Power Cables 335
H.-U. PAUL AND W. PRINZ

14.1 Properties of Buried Power Cables 335
14.2 Cathodic Protection of the Steel Conduits for Power Cables 336
14.2.1 Requirements for dc Decoupling Devices (between Casing and Ground) 337
14.2.2 Types and Circuits of dc Decoupling Devices 338
14.2.2.1 Low Ohmic Resistances 338
14.2.2.2 Higher Ohmic Resistances 339
14.2.2.3 dc Coupling Devices with Nickel-Cadmium Cell 340
14.2.2.4 Polarization Cell 340
14.2.2.5 dc Decoupling Devices with Silicon Diodes 341
14.2.3 Installation of Cathodic Protection Station 343
14.2.4 Control and Maintenance of Cathodic Protection 343
14.3 Stray Current Protection 344
14.4 References 346

15 Stray Current Interference and Stray Current Protection 347
W. v. BAECKMANN AND W. PRINZ

15.1 Causes of Stray Current Interference 347
15.1.1 dc Equipment 347
15.1.2 General Measures at dc Equipment 348
15.2 Stray Currents from dc Railways 348
15.2.1 Regulations for dc Railways 348
15.2.2 Tunnels for dc Railways 352
15.3 Stray Currents from High-Voltage dc Power Lines 353
15.4 Stray Currents Due to Telluric Currents 355
15.5 Protective Measures 358
15.5.1 Stray Current Protection for Individual Pipelines 358
15.5.2 Combined Stray Current Protective Measures in Urban Areas 362
15.6 Stray Current Protection in Harbor Areas 364
15.7 References 366

16 Marine Structures and Offshore Pipelines 367
W. v. BAECKMANN AND B. RICHTER

16.1 Cathodic Protection Measures 367
16.1.1 Design Criteria 368

16.1.2	Protection with Galvanic Anodes	372
16.1.3	Impressed Current Protection	373
16.2	Platforms	373
16.2.1	Steel Structures	373
16.2.2	Concrete Structures	376
16.3	Harbor Structures	376
16.3.1	Impressed Current Equipment	377
16.3.2	Protection with Galvanic Anodes	379
16.4	Steel Sheet Piling	380
16.5	Piling Foundations	380
16.6	Offshore Pipelines	383
16.7	Control and Maintenance of Cathodic Protection	385
16.7.1	Production Platforms	385
16.7.2	Harbor Structures	387
16.7.3	Offshore Pipelines	388
16.8	References	390

17 Cathodic Protection of Ships ... 391
H. Bohnes and B. Richter

17.1	Water Parameters	391
17.1.1	Dissolved Salts and Solid Particles	391
17.1.2	Aeration and Oxygen Content	393
17.1.3	Flow Rate in the Case of a Moving Ship	394
17.1.4	Variations in Temperature and Concentration	394
17.2	Effect of Materials and Coating Parameters	395
17.3	Cathodic Protection Below the Waterline	397
17.3.1	Calculations of the Protection Current Requirement	398
17.3.2	Protection by Galvanic Anodes	399
17.3.2.1	Size and Number of Anodes	399
17.3.2.2	Arrangement of Anodes	401
17.3.2.3	Control and Maintenance of Cathodic Protection	402
17.3.3	Protection with Impressed Current	403
17.3.3.1	Current Supply and Rectifiers	404
17.3.3.2	Impressed Current Anodes and Reference Electrodes	405
17.3.3.3	Arrangement of Anodes and Reference Electrodes	408
17.4	Internal Cathodic Protection of Tanks and Containers	410
17.5	Cathodic Protection of Heat Exchangers, Condensers and Tubing	412
17.6	Cathodic Protection of Bilges	412

17.7	Cathodic Protection of Docks	413
17.8	References	414

18 Cathodic Protection of Well Casings 415
W. PRINZ AND B. LEUTNER

18.1	Description of the Object to be Protected	415
18.2	Causes of Corrosion Danger	415
18.2.1	Formation of Corrosion Cells	415
18.2.2	Free Corrosion in Different Soil Layers	417
18.2.3	Conditions for the Occurrence of Stress Corrosion	417
18.2.4	Corrosion by Anodic Interference (Cell Formation, Stray Currents)	417
18.3	Measurements for Assessing Corrosion Protection of Well Casings	418
18.3.1	Investigations for Corrosion Damage	418
18.3.2	Measurement of ΔU Profiles	418
18.3.3	Measurement of the Tafel Potential	421
18.4	Design and Construction of Cathodic Protection Stations	422
18.5	Commissioning, Maintenance and Control	425
18.6	References	426

19 Cathodic Protection of Reinforcing Steel in Concrete Structures 427
B. ISECKE

19.1	The Corrosion System Steel-Concrete	427
19.2	Causes of Corrosion of Steel in Concrete	428
19.3	Electrolytic Properties of Concrete	428
19.4	Criteria for Cathodic Protection	429
19.5	Application of Cathodic Protection to Reinforced Concrete Structures	431
19.5.1	Design and Installation	431
19.5.2	Determination of the State of Corrosion of the Reinforcing Steel	432
19.5.3	Reinforcement Continuity	433
19.5.4	Installation and Types of Anode System	434
19.5.5	Concrete Replacement Systems for Cathodic Protection	435
19.5.6	Commissioning, Maintenance and Control	436

| 19.6 | Stray Current Effects and Protective Measures | 438 |
| 19.7 | References | 439 |

20 Internal Cathodic Protection of Water Tanks and Boilers ... 441
G. Franke and U. Heinzelmann

20.1	Description and Function of Objects to be Protected	441
20.1.1	Materials for Objects to be Protected and Installation Components	442
20.1.2	Types of Linings and Coatings	443
20.1.3	Preconditions for Internal Cathodic Protection	443
20.1.4	Measures to Prevent Anodic Interference	444
20.1.5	Measures to Prevent Danger from Hydrogen Evolution	446
20.2	Protection with Galvanic Anodes	447
20.3	Protection with Impressed Current	448
20.3.1	Equipment with Potential Control	448
20.3.2	Equipment with Current Control Based on Water Consumption	450
20.4	Description of Objects to be Protected	450
20.4.1	Boilers with Enamel Linings	450
20.4.2	Boilers with Electrolytically Treated Water	456
20.4.3	Water Storage Tanks	458
20.4.4	Filter Tanks	461
20.5	Requirements for Drinking Water	462
20.6	References	463

21 Internal Electrochemical Corrosion Protection of Processing Equipment, Vessels, and Tubes ... 464
H. Gräfen and F. Paulekat

21.1	Special Features of Internal Protection	464
21.2	Cathodic Protection with Galvanic Anodes	466
21.3	Cathodic Protection with Impressed Current	467
21.3.1	Internal Cathodic Protection of Wet Oil Tanks	467
21.3.2	Internal Cathodic Protection of a Wet Gasometer	468
21.3.3	Internal Cathodic Protection of a Power Plant Condenser Cooled by Seawater	469
21.3.4	Internal Cathodic Protection of a Water Turbine	469
21.4	Anodic Protection of Chemical Plant	474
21.4.1	Special Features of Anodic Protection	474

21.4.2	Anodic Protection with Impressed Current	476
21.4.2.1	Preparatory Investigations	476
21.4.2.2	Protection Against Acids	478
21.4.2.3	Protection Against Media of Different Composition	480
21.4.2.4	Protection Against Alkaline Solutions	480
21.4.2.5	Combined Protection by Impressed Current and Inhibitors	483
21.4.3	Protective Effect of Local Cathodes due to Alloying	483
21.4.4	Protective Action of Inhibitors	484
21.5	Trends in the Application of Internal Electrochemical Protection	485
21.6	References	487

22 Safety and Economics 489
W. v. Baeckmann and W. Prinz

22.1	Safety	489
22.1.1	Statistics of Pipeline Failures	489
22.1.2	Measures for Control and Maintenance	490
22.2	General Comments on Economics	491
22.3	Costs of Cathodic Protection of Buried Pipelines	492
22.3.1	Galvanic Anodes	493
22.3.2	Impressed Current Anodes	495
22.3.3	Prolonging the Life of Pipelines	496
22.4	Corrosion Protection of Well Casings	499
22.5	Corrosion Protection in Seawater	500
22.6	Cost of Internal Protection	501
22.6.1	Internal Cathodic Protection	501
22.6.2	Internal Anodic Protection	502
22.7	References	504

23 Interference Effects of High-Voltage Transmission Lines on Pipelines 505
H.-U. Paul and H. G. Schöneich

23.1	Capacitive Interference	506
23.2	Ohmic Interference	507
23.2.1	Contact with a Conductor under High Voltage	507
23.2.2	Voltage Cone of a Pylon Grounding Electrode	508
23.3	Inductive Interference	510
23.3.1	Causes and Factors Involved	510

23.3.2	Calculation of Pipeline Potentials in the Case of Parallel Routing of a High-Voltage Transmission Line and a Pipeline	511
23.3.3	Obliquely Routed Sections of the Lines	516
23.3.4	Simplified Calculation Methods	517
23.3.4.1	Interference by Fault Currents and by Railway Operating Currents	517
23.3.4.2	Interference from a Three-Phase High-Voltage Transmission Line in Normal Operation	519
23.3.5	Representation of the Characteristics of a Pipeline	521
23.4	Limiting Lengths and Limiting Distances	524
23.4.1	Allowable Contact Voltages	525
23.4.1.1	Short-Term Interference	525
23.4.1.2	Long-Term Interference	526
23.4.2	Determination of Pipeline Potentials	526
23.5	Protection Measures against Unallowably High Pipeline Potentials	526
23.5.1	Short-Term Interference	526
23.5.2	Long-Term Interference	526
23.5.3	Protective Measures by Grounding	527
23.5.4	Grounding Electrodes and Cathodic Protection	528
23.5.4.1	Overvoltage Arresters for Short-Term Interference	528
23.5.4.2	Voltage Arresters for Long-Term and Short-Term Interference	529
23.5.4.2.1	Polarization Cells	529
23.5.4.2.2	Diodes	529
23.6	Measurement of Pipeline Potentials	530
23.6.1	Measurement of Short-Term Interference	531
23.6.1.1	Beat Method (by Superposition)	531
23.6.1.2	Reverse Polarity Method	531
23.6.2	Measurement of Long-Term Interference	532
23.6.3	Results of Pipeline Potential Measurement	533
23.7	References	534

24 Distribution of Current and Potential in a Stationary Electric Field 535

W. v. BAECKMANN AND W. SCHWENK

24.1	Grounding Resistance of Anodes and Grounds	536
24.2	Interference Factor with Several Anodes	544
24.3	Potential Distribution at Ground Level	545
24.3.1	Soil Resistance Formulas	545

24.3.2	Anodic Voltage Cone	546
24.3.3	Cathodic Voltage Cone in a Cylindrical Field	547
24.3.4	Interference from the Cathodic Voltage Cone	548
24.4	Calculation of Current and Potential Distribution	549
24.4.1	General Relationships for a Two-Conductor Model	549
24.4.2	Calculation of Ground Electrodes Having a Longitudinal Resistance	550
24.4.3	Range of Cathodic Protection and Current Requirement	552
24.4.4	Potential Distribution in the Case of Overprotection	555
24.4.5	Cathodic Protection in Narrow Gaps	556
24.4.6	Distribution of Current and Potential Inside a Pipe at Insulating Units	557
24.5	General Comments on Current Distribution	558
24.6	References	560

Index 561

Foreword to the Third Edition

The preparation of this third edition after about 10 years since publication of the second edition of this handbook has required a complete revision of the major part of the book. The reason is not only new developments in technology and application, but also the identification of vital factors in the protection system. Developments in standards and regulations also had to be taken into account.

Electrochemical corrosion and electrochemical corrosion protection have the same bases. These and the uniform terminology given in the DIN manual 219 form the scientific basis for this handbook. Descriptions of new corrosion systems and increased questions concerning the use of potential and current distribution as well as the influence of high tension have improved our understanding and favor new applications, even if less useful parts of the previous tests have had to be omitted. There are new developments in *IR*-free potential measurements, intensive measuring techniques, and computer-controlled evaluation data. Only interactions between corrosion and polarization are described in dealing with the properties of coatings for passive corrosion protection.

Considerable alterations have been made in the chapters concerned with technical applications which are the result of advances in electrochemical corrosion protection in general practice. Here also, abbreviation and omission of less relevant parts of the older editions have had to be made to create space for more recent information. Recent applications in the chemical industry have necessitated a complete rewriting of the industrial chapter. A new chapter is included on the cathodic protection of steel reinforcement in concrete.

The editors thank all the collaborators in this handbook for their effort as well as Ruhrgas AG and Mannesmannröhren-Werke AG for their generous support in editing the manuscript, and the publishers for their cooperation in shaping and publishing this handbook.

W. VON BAECKMANN
W. SCHWENK
and W. PRINZ
Essen and Duisberg
Summer 1988

Foreword to the First Edition

The discovery and use of metals at the end of the Stone Age was one of the most important steps in the development of modern technology. Most base metals are, unfortunately, not stable. In unfavorable environments they can be destroyed at variable rates by corrosion. The study of such corrosion reactions and the methods by which corrosion of metals can be fought is a task of great economic significance.

The processes of cathodic protection can be scientifically explained far more concisely than many other protective systems. Corrosion of metals in aqueous solutions or in the soil is principally an electrolytic process controlled by an electric tension, i.e., the potential of a metal in an electrolytic solution. According to the laws of electrochemistry, the reaction tendency and the rate of reaction will decrease with reducing potential. Although these relationships have been known for more than a century and although cathodic protection has been practiced in isolated cases for a long time, it required an extended period for its technical application on a wider scale. This may have been because cathodic protection used to appear curious and strange, and the electrical engineering requirements hindered its practical application. The practice of cathodic protection is indeed more complex than its theoretical base.

There are extensive publications on many individual problems together with practical instructions. However, it was difficult for the technologist in Germany to master the subject because no comprehensive up-to-date publication was available in German. The Subcommittee for Corrosion of DVGW instigated the publication of a handbook of cathodic protection, and a number of members offered their cooperation as authors of individual chapters.

This handbook deals mainly with the practice of cathodic protection, but the discussion includes fundamentals and related fields as far as these are necessary for a complete review of the subject. We thought it appropriate to include a historical introduction in order to explain the technological development of corrosion protection. The second chapter explains the theoretical basis of metal corrosion and corrosion protection. We have deliberately given practical examples of combinations of various materials and media in order to exemplify the numerous fields of application of electrochemical protection.

At present cathodic protection is only generally applied for materials in contact with natural waters and soil, but future applications are envisaged for industrial plants and containers. For this reason we have included a chapter on anodic protection that has been applied in isolated cases during the past 10 years. Cathodic and anodic protection are basically very similar systems and justify the description *electrochemical protection* in the subtitle of this book.

Most applications combine cathodic protection with a surface coating. The chapter on physical protection systems seemed appropriate because of the various interactions that must be taken into account. A chapter on general measuring technology has also been added since the practice of cathodic protection has repeatedly shown the importance of a careful study of measuring problems. It requires experience to account for possible sources of error in calculations, and it is always necessary to check unusual measured results by independent monitoring. Impressed current installations present particular measurement problems, keeping in mind that an installation with reversed polarity generates intensive corrosion. This is worse than an inoperative system or no corrosion protection at all.

Further chapters cover in detail the characteristics and applications of galvanic anodes and of cathodic protection rectifiers, including specialized instruments for stray current protection and impressed current anodes. The fields of application discussed are buried pipelines; storage tanks; tank farms; telephone, power and gas-pressurized cables; ships; harbor installations; and the internal protection of water tanks and industrial plants. A separate chapter deals with the problems of high-tension effects on pipelines and cables. A study of costs and economic factors concludes the discussion. The appendix contains those tables and mathematical derivations which appeared appropriate for practical purposes and for rounding off the subject.

The editors take the opportunity to thank all the contributors for their efforts; Ruhrgas AG and Mannesmann Research Institute GmbH for their kind assistance; and last but not least, the publishers Verlag Chemie for their generous help in publishing and designing the handbook.

W. V. BAECKMANN
and W. SCHWENK
Essen and Duisburg
Spring 1971

Preface

The editors of the German edition of the *Handbook of Cathodic Corrosion Protection* would like to express their cordial thanks to Gulf Publishing Company for their keen interest in the translation of this work. We are sure the English edition will promote a better exchange of experience in the field of corrosion protection, particularly with respect to problems of global safety and environment.

A serious problem in preparing this translation was that so many technical branches are involved with electrochemical corrosion and corrosion protection, and they often have their own "technical languages" and terms. A good translation required interdisciplinary teamwork. Consequently, the editors added to their team A. Baltes and J. Venkateswarlu, who are experts in their fields. Dipl.-Ing. A. Baltes, with Pipeline Engineering GmbH, Essen, is the German delegate to the European Committees for Standardisation in the field of pipeline protection and cathodic protection. In this capacity he took care of the proper English translation of the technical terms in this field. J. Venkateswarlu, B. Tech. (Met.), is a metallurgist and corrosion engineer with Mannesmann Research Institute, Duisburg. He was a great help in checking most of the metallurgical and technical terms as well as critical phrases and idioms.

We would also like to thank many of the authors of the German edition, namely, G. Franke, U. Heinzelmann, H. Gräfen, B. Isecke, B. Leutner, B. Richter, and H.G. Schöneich for their particular assistance. Thanks are due also to Pipeline Engineering GmbH, Essen, and Europipe Gmbh, Ratingen, for their keen interest in the English edition and for having supported the work of our team.

We hope the reader of this English edition will not be troubled with some of the symbols that have "German" letters and indices. Changing all of these symbols would have been an overwhelming task and would most likely have introduced unavoidable errors. To help our readers overcome this problem, a list of symbols is provided. However, we must caution the reader that a translation cannot recognize the local significance of both national standards and official regulations. In this book, one can only view these references as an example, keeping in mind that science and technology are international matters.

W. v. Baeckmann
W. Schwenk
Essen and Duisberg
Spring 1997

Acknowledgments

The publishers wish to acknowledge the contribution of Dr. Robert W. Waterhouse, who provided the initial draft of the translation from the German. The publishers are grateful to the authors for their diligent help with subsequent drafts of the translation.

A special note of thanks to Dirk van Oostendorp, whose reading of the final proofs was of great help rendering some of the most difficult aspects of the translation. We also wish to recognize the fine efforts of Ruth B. Haas in overseeing the editorial and production aspects of this project. We are also grateful for the contributions of Peter Dorn, who assisted Ms. Haas in rendering some of the most problematic passages.

Index of Authors

Dipl.-Phys. W. v. BAECKMANN
Ulmenstraße 12
4300 Essen 1

Dipl.-Chem. H. BOHNES
Gerrickstraße 23
4100 Duisburg 12

G. FRANKE
Norsk Hydro Magnesiumgesellschaft mbH
Scharnhölzstraße 350
4250 Bottrop

D. FUNK
Ruhrgas AG
Postfach 10 32 52
4300 Essen 1

Ing. C. CEY
Fernmeldetechnisches Zentralamt
Postfach 50 00
6100 Darmstadt

Prof. Dr. H. GRÄFEN
Ursulastraße 9
5010 Bergheim 8

Dr. G. HEIM
Korrosionstechnik
Rubensweg 1
4010 Hilden

Dipl.-Chem. U. HEINZELMANN
Guldager Electrolyse GmbH
Postfach 141
4660 Gelsenkirchen-Buer

Dipl.-Ing. K. HORRAS
Technische Akademie Wuppertal
Postfach 10 04 09
5600 Wuppertal 1

Dr. B. ISECKE
Bundesanstalt für Materialprüfung (BAM)
Uter den Eichen 87
1000 Berlin 45

Dipl.-Phys. H. KAMPERMANN
Quante Fernmeldetechnik GmbH
Uellendahler Straße 353
5600 Wuppertal 1

Dipl.-Ing. B. LEUTNER
BEB Erdgas und Erdöl GmbH
Postfach 51 03 60
3000 Hannover 1

Dipl.-Ing H.-U. PAUL
Rheinisch-Westfälische Elecktrizitätswerke AG
Postfach 10 31 65
4300 Essen 1

Ing. F. PAULEKAT
Starkstrom-und Signalbaugesellschaft mbH
Postfach 10 37 32
4300 Essen 1

Ing. W. PRINZ
Ruhrgas AG
Postfach 10 32 52
4300 Essen 1

Dr. B. RICHTER
Germanischer Lloyd AG, Hamburg
Postfach 11 16 06
2000 Hamburg 11

Dipl.-Ing. G. RIEGER
Technischer Überwachungsverein Rheinland e.V.
Postfach 10 17 50
5000 Köln 91

Dr. H. G. SCHÖNEICH
Ruhrgas AG
Postfach 10 32 52
4300 Essen 1

Prof. Dr. W. SCHWENK
Mannesmannröhren-Werke AG
Mannesmann Forschungsinstitut
Postfach D 25 11 67
4100 Duisburg 25

Commonly Used Quantities, Constants, and Symbols

Symbol	Meaning	Units
a	distance, length	cm, m
b	distance, length	cm, m
$b_{+/-}$	Tafel slope (log) (i.e., the logarithmic relation between current and applied voltage expressed as $\eta = a + b \log i$, where η is overvoltage, i is current, and a and b are constants	mV
B	mobility	cm^2 mol J^{-1} s^{-1}
B_0, B_1, B_E	total rating number (soil aggressiveness)	
$c(X_i)$	concentration of material X_i	mol cm^{-3}, mol L^{-1}
C	capacity	F = Ω^{-1} s
C	constant	
C_D	double layer capacity of an electrode	μF cm^{-2}
d	distance, diameter	mm, m
D_i	diffusion constant of material X_i	cm^2 s^{-1}
E	electric field strength	V cm^{-1}
f	frequency	Hz = s^{-1}
f_a	conversion factor	mm a^{-1}/(mA cm^{-2})
f_b	conversion factor	g m^{-2} h^{-1}/(mA cm^{-2})
f_c	conversion factor	mm a^{-1}/(g m^{-2} h^{-1})
f_v	conversion factor	L m^{-2} h^{-1}/(mA cm^{-2})
F	force	N
F	interference factor	
$\mathscr{F}$	Faraday constant = 96485 A s mol^{-1} = 26.8 A h mol^{-1}	

Symbol	Meaning	Units
g	limiting current density	$A\ m^{-2}$
G	limiting current	A
G	leakage	$S = \Omega^{-1}$
G'	leakage load (leakage per unit length)	$S\ m^{-1}$, $S\ km^{-1}$
ΔG	free enthalpy of formation	$J\ mol^{-1}$
h	height, earth covering	cm, m
i	run number	
I	current	A
I_s	current requirement protection current	A
I'	current supply, current load	$A\ km^{-1}$
j_H	H-permeation rate	$L\ cm^{-2}\ min^{-1}$
J_i	transport rate of material X_i	$mol\ cm^{-2}\ s^{-1}$
J	current density	$A\ m^{-2}$, $mA\ cm^{-2}$
J_{act}	activation current density	$A\ m^{-2}$, $mA\ cm^{-2}$
J_{max}	maximum current density of a sacrificial anode	$A\ m^{-2}$, $mA\ cm^{-2}$
J_{pass}, J_p	passivation current density	$A\ m^{-2}$, $mA\ cm^{-2}$
J_s	protection current density, lowest protection current density	$A\ m^{-2}$, $mA\ cm^{-2}$
J_0	exchange current density	$A\ m^{-2}$, $mA\ cm^{-2}$
k	polarization parameter	cm, m
k	specific cost (deutsche marks)	DM/unit
K	stress intensity	$N\ mm^{-3/2}$
K	equilibrium constant	$1\ (mol\ L^{-1})^{(\Sigma n_i)}$
K	cost (deutsche marks)	DM
K_{Sx}	acid capacity up to pH = x	$mol\ L^{-1}$
K_{Bx}	base capacity up to pH = x	$mol\ L^{-1}$
K_W	ionization constant for water ($10^{-14}\ mol^2\ L^{-2}$ at 25°C)	
K_w	reaction constant in oxygen corrosion	mm
l	length, distance	cm, m, km
l_i	ion mobility of material X_i	$S\ cm^2\ mol^{-1}$
l_k	characteristic length, nominal length	m, km

Symbol	Meaning	Units
L	protection range, length	m, km
L	inductivity	$H = \Omega\ s$
L_{Gr}	limiting length	m, km
m	mass	g, kg
m'	pipe mass per unit length	$kg\ m^{-1}$
M	atomic, molecular weight	$g\ mol^{-1}$
M'	mutual inductivity per unit length	$H\ km^{-1}$
n	number, number of cycles	
n'	number per unit length	m^{-1}, km^{-1}
n_i	stoichiometric coefficient, charge number of material X_i	
N	defect (holiday) density	m^{-2}
N	reciprocal slope of $\ln|J|–U$–curves	mV
p	pressure, gas pressure	bar
$p(X_i)$	partial pressure of component X_i	bar
P	permeation coefficient	$cm^2\ s^{-1}\ bar^{-1}$, $g\ cm^{-1}\ h^{-1}\ bar^{-1}$
Q	electric charge	A s, A h
Q'	current constant of sacrificial anodes per unit mass	$A\ h\ kg^{-1}$
Q''	current content of sacrificial anodes per unit volume	$A\ h\ dm^{-3}$
r	radius, distance	cm, m
r	reduction factor	
r_P	specific polarization resistance	$\Omega\ m^2$
r_u	specific coating resistance	$\Omega\ m^2$
R	electrical resistance, grounding resistance	Ω
R	gas constant = 8.31 $J\ mol^{-1}\ K^{-1}$	
R'	resistance per unit length, resistance load	$\Omega\ m^{-1}, \Omega\ km^{-1}$
R_m	ultimate tensile strength (UTS)	$N\ mm^{-2}$
R_p	polarization resistance	Ω
$R_{p0.2}$	0.2% proof stress	$N\ mm^{-2}$
R_u	coating resistance	Ω
s	distance, thickness, decrease in thickness	mm, cm

Symbol	Meaning	Units
S	surface, cross-section	m²
t	time	s, h, a
t	depth	cm, m
T	temperature	°C, K
u_i	electrochemical mobility of substance X_i	V⁻¹ cm² s⁻¹
U	voltage, potential	V
U_{off}	off potential	V
U_B''	potential difference between reference electrodes parallel over the pipeline	mV, V
$U_B^{\perp}$, ΔU_x	potential difference between reference electrodes perpendicular to the pipeline (distance x)	mV, V
$U_{\text{Cu-CuSO}_4}$	potential measured against the saturated Cu-CuSO₄ reference electrode	mV, V
U_{on}	on potential	V
U_H	potential measured against the standard hydrogen electrode	mV, V
U_{IR}	ohmic voltage drop	V
$U_{IR\text{-free}}$	IR-free potential	V
U_R	rest potential	V
U_s	protection potential	V
U_T	driving voltage	V
U_{over}	reverse switching potential	V
U_0	open circuit voltage (EMF)	V
v	weight loss per unit area and time	g m⁻² h⁻¹
v_{int}	mean value of v	
V	volume	cm³, dm³, L
$\mathbf{V}$	atomic, molecular volume	m³ mol⁻¹, L mol⁻¹
w, w_{int}	rate of reduction in thickness, mean value	mm, a⁻¹
w	degree of effectiveness	(%)
w	number of windings	
w_i	velocity of material X_i	cm s⁻¹
x	position coordinate	m, km

Symbol	Meaning	Units
Y'	admittance per unit area (admittance load)	S km^{-1}
Y_s	yield point	N mm^{-2}
z_i	charge number of material X_i	
Z	impedance	Ω
Z	characteristic resistance or impedance (of a line)	Ω
Z_i	rating number (soil aggressiveness)	

Greek Symbols

Symbol	Meaning	Units
α	symmetry factor	
α	path constant (dc)	km^{-1}
$\beta_{+/-}$	Tafel slope (Napieran loop)	mV
γ	transfer coefficient	km^{-1}
δ	diffusion layer thickness	cm
$\varepsilon, \varepsilon_r$	dielectric constant, relative	
ε_0	electric field constant = 8.85×10^{-14} F cm^{-1}	
η	overvoltage, polarization	mV, V
η_Ω	ohmic voltage drop, resistance polarization	mV, V
$\varkappa$	specific conductance, conductivity	S cm^{-1}
$\widetilde{\mu}_i$	electrochemical potential of material X_i	J mol^{-1}
μ_i	partial molar free enthalpy of material X_i	J mol^{-1}
μ_0	magnetic field constant = 1.26×10^{-8} H cm^{-1}	
μ_r	permeability number	
υ	relative number of cycles	
ρ	specific resistance, resistivity	Ω cm
ρ_{st}	specific resistance of steel (ca. 1.7×10^{-6} Ω m)	
ρ_s	density, specific weight	g cm^{-3}
σ	tensile strength	N mm^{-2}
τ	time constant	s
φ	electrical potential	V
φ	phase angle	
ω	cyclic frequency	s^{-1}

Frequently Used Indices

Chemical and Thermodynamic Quantities Y

Y° standard conditions

Y^* condition for thermodynamic equilibrium

Y_i quantity of component X_i

Electrochemical Quantities Y

$Y_{a,c}$ quantity of the anodic (a) or cathodic (c) region as well as the relevant total currents

$Y_{A,C}$ quantity of the anodic (A) or cathodic (C) partial reaction

Y_e quantity in cell formation

Electrical Quantities Y

Y' length-related quantity (Y-load)

Y_x Y at the point with coordinates x (e.g., r, 1, 0, ∞)

Y_X Y for a definite electrode or object X (B, reference electrode; Me, metal; E, ground; M, mast; R, pipe; S, rail; T, tunnel)

General Symbols

e^- electron

DN   nominal pressure

EP   epoxy resin

FI   failure current

FU   failure voltage

HV   Vickers hardness

HV_{dc} high-voltage dc transmission

General Symbols (*continued*)

IR	ohmic voltage drop
IT	protective system with isolated starpoint*
LCD	liquid crystal display
Me	metal
Ox	oxidizing agent
PE	polyethylene (HD-PE high pressure PE, ND-PE low pressure PE)
PEN	protective conductor with neutral conductor function
PN	nominal pressure
PUR	polyurethane
PVC	polyvinylchloride
Red	reducing agent or component
TN	protective system with PEN conductor*
TT	protective or grounding system
X_i	symbol for material *i*

---

* DIN VDE 0100, Pt 300, Beuth Publ., Berlin 1985

American and European Electrical Cable Sizes

Size (AWG)	Size (Metric)	Cross Sectional Area (mm^2)	Nominal Weight (lb / 1000 ft)	Resistance (ohms/1000 ft)	dc Current Rating (A)
# 12		3.308	20.16	1.65	20
	4 mm^2	3.9972	24.37	1.41	24
# 10		5.26	32.06	1.02	30
	6 mm^2	5.95	36.56	0.939	31
# 8		8.37	50.97	0.64	40
	10 mm^2	10.02	60.93	0.588	42
# 6		13.3	81.05	0.41	55
	16 mm^2	15.89	97.49	0.351	56
# 5		16.768	102.2	0.284	63
# 4		21.14	128.9	0.259	70
	25 mm^2	25.18	152.3	0.222	73
# 2		33.65	204.9	0.162	90
	35 mm^2	34.99	213.3	0.16	92
# 1		42.4	258.4	0.129	110
	50 mm^2	49.99	304.07	0.118	130
# 1/0		53.44	325.8	0.102	135
# 2/0		67.45	410.9	0.079	165
# 4/0		107.16	653.3	0.05	230

1

The History of Corrosion Protection

W. V. BAECKMANN

The works of Plato (427–347 B.C.) contained the first written description of corrosion. Plato defined rust as the earthy component separating out of the metal. Georgius Agricola held to the same opinion some 2000 years later in his great mineralogical work *De natura fossilium*: "Iron rust (lat. ferrugo or rubigo) is, so to speak, a secretion of metallic iron. Iron can be protected against this defect by various wrappings, such as red lead, white lead, gypsum, bitumen or tar." Gaius Secundus Pliny also mentioned bitumen, pitch, white lead, and gypsum as protecting iron and bronze against corrosion. He reported that Alexander the Great had constructed a pontoon bridge at Zeugmar on the Euphrates with the aid of an iron chain. Links that were inserted later suffered rust attacks, while the original ones remained immune. The opinion, sometimes expressed today, that modern iron is inferior and more corrosion-prone than old iron, was thus current even in ancient times [1].

The concept of the *corrosion* process, derived from the Latin *corrodere* (to eat away, to destroy), first appeared in the *Philosophical Transactions* in 1667 [2]. It was discussed in a German translation from the French on the manufacture of white lead in 1785 and was mentioned in 1836 in the translation of an English paper by Davy on the cathodic protection of iron in seawater [3]. However, almost until the present day, the term was used indiscriminately for *corrosion reaction*, *corrosion effects*, and *corrosion damage*. Only in DIN* 50900, Part I, were these terms distinguished and defined [4] (see Section 2.1).

1.1 Corrosion Protection for Buried Pipelines

The active and passive electrochemical processes on which present-day corrosion protection is based were already known in the 19th century, but reliable protection for pipelines only developed at the turn of the 20th century.

---

* All cable references contained in this text are based upon cable specifications in Germany as required by DIN (Deutsche Industrie Normen). We are aware that these specifications may or may not be applicable to the reader's specific requirements, and we therefore recommend the reader consult local standards and codes to ensure compliance with the necessary local codes. Some of the cables are defined in a list of American and European electrical cable sizes located in the front of this book.

Corrosion protection using bitumen coatings reaches back into antiquity. The most ancient occurrence of bitumen deposits was in Mesopotamia. Many writers of antiquity, such as Dido, Strabo, and Vitruvius, mention that asphalt was obtained for many years near Babylon. About 5000 years ago, the streets of Ur, capital of the Sumerians (north of present-day Kuwait), were lit at night with mineral oil. Natural gas was reported to be used for lighting in the Middle East and China.

Bitumen was used in ancient times as an adhesive for sealing hydraulic structures and as mortar for masonry [5]. The Bible mentions that Noah used pitch for caulking the Ark. Not unlike the Tower of Babylon, the houses of one of the most ancient cities in the world, Mohenjo-Daro in the upper Indus valley, were constructed with bricks of clay and bitumen mortar [6].

The earliest metal pipelines, made of copper, bronze, and lead, had no protection against corrosion. The pipes were often surrounded by lime and gypsum mortar for sealing, cohesion, and protection. These early metal pipes are rarely found today because the valuable metals were reused once the pipelines were abandoned. In 1907 the archaeologist Borchardt found the earliest metal pipe at a temple complex near the pyramid of King Sahu-re. It was part of a 250-m-long pipeline which was used to carry rain water from the temple courtyard. The 1-m-long sections, with a diameter of 47 mm, were made of 1.4-mm-thick beaten copper, curved, and the overlapping longitudinal edges hammered together. The pipes were set into a rock-hewn channel and covered with lime mortar. The only well-preserved pipe in its bedding is shown in Fig. 1-1. Its age can be taken as 4500 years since records indicate that King Sahu-re belonged to the fifth dynasty of Egyptian rulers [7].

The Phoenicians were building water ducts and pipelines of clay, stone, or bronze about 1000 B.C. and the construction of long-distance water pipelines flourished in imperial Roman times. The water supply lines of Rome had a total length of about 450 km, and consisted mainly of open or covered water ducts. The Roman writer Vitruvius gives a fairly accurate description of the manufacture of lead pipes [8]. The pipes were above ground and were often laid beside the roadway or in ducts inside houses [9].

Fig. 1-1 The world's oldest metal pipe from the temple of King Sahu-re (photo: Staatliches Museum, Berlin).

The History of Corrosion Protection   3

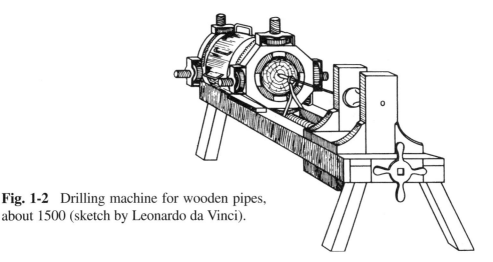

Fig. 1-2 Drilling machine for wooden pipes, about 1500 (sketch by Leonardo da Vinci).

With the fall of the Roman Empire, the ancient water supplies petered out. In early medieval times, people were content to conduct local water in wooden pipes to public cisterns. The first wooden pipelines for water were laid at Lübeck about 1293 and in 1365 at Nuremberg. In 1412 the Augsburg master builder Leopold Karg first used wrought-iron pipes in conjunction with wooden pipes to supply water. Because of their propensity to corrosion, they seem to have proved a failure and a few years later they were exchanged for wooden, lead, and cast-iron pipes.

The author of the first German natural history, *puoch von der Natur*, Chunrad von Megenberg, reported in 1349 that mainly larch and fir were used for water supply [11]. The trunks were often steeped in lime water or brine before they were bored through on a drilling machine like the one sketched by Leonardo da Vinci (Fig. 1-2). Wooden pipes with one end tapered and the other socketed were wedged into each other. Pipes with abutting ends were held together by wrought-iron ferrules (called "Tuchel" ferrules). Figure 1-3 shows wooden pipes with a cast-iron ring ferrule (laid before 1760). The sleeves were sealed with hemp, tallow, pitch, wax or resin, which also acted as protection for the iron rings. It is recorded that wooden pipes were painted with pitch or tar. Later on, wooden pipes with an internal coating of liquid tar were used in London and New York for town and natural

Fig. 1-3 Oak water pipe over 200 years old laid above ground in the Wadgassen Abbey (Saarland) with cast iron clamps (photo: Mannesmann Archives).

Table 1-1 Evolution of ferrous metals technology and corrosion protection

Year	4000	2000	0	1000	1500	1800	1900	2000
Age	Prehistoric and early days	Ancient times		Middle Ages	Age of rationalism		Modern Age	
Period of culture	Early Stone Age	Bronze Age		Iron Age				Plastics
Raw materials and fuels	Meteoric iron		Iron ore and charcoal				Hard coal, coke	Natural gas
Source of energy	Wind	Physical strength		Water power			Heat energy	Atomic energy
Smelting and production of wrought iron		Charcoal hearth and iron foundry (direct reduction from ore)			Blast furnace (indirect production of pig iron)			
Process for the production of wrought iron		Renn process		Oxidation		Puddling	Bessemer, Siemens and LD converter, open hearth	
Type of iron	Native iron		Weldable wrought iron and steel		Cast iron in Europe		Mild iron, steel and ductile cast iron	
Importance of iron	Disappearing of bronze	Predomination of bronze	Iron is generally used for the production of apparatus and weapons, wood is an important construction material besides stone				Most important construction material for machines, bridges, vehicles, ships	
Corrosion protection		Bitumen	Burnishing	Paint coatings	Metallic coatings	Tar	Thick organic coatings	Cathodic protection
Binder		Wax, bitumen, varnish	Pitch and resins	Asphalt lacquer, linseed oil, shellac		Turpentine oil	Synthetic resins	
Pigments		Iron oxide, ochre, gypsum, chalk		Red lead, cinnabar		White lead	Ferric oxide, zinc dust	

gas. The protection of wooden pipes against rot may be regarded as the precursor of corrosion protection of wrought-iron pipelines.

The Bavarian Duke Maximilian I commissioned the master builder Simon Reifenstuel in 1618 to lay the first pipeline for brine from Reichenhall to Traunstein. The 31-km-long line required 9000 wooden pipes. Two centuries later the King of Bavaria commissioned the extension of the pipeline from Reichenhall to Berchtesgaden. The noted Karlsruhe engineer, George Friedrich von Reichenbach, had iron pipes cast to his own specification for this first German high-pressure pipeline. The initially porous cast pipes had to be sealed with a mixture of linseed oil and finely ground quicklime to enable them to withstand a pressure of 43 atmospheres [12]. This treatment with linseed oil was apparently not intended to be a protection against internal corrosion, which was known by the end of the 17th century. So-called "calcination" of pipelines was understood to include not only the formation of iron rust nodules with wastage of pipelines but also internal corrosive attack.

It cannot be ascertained with accuracy when molten iron was first obtained in the European cultural sphere. The forge production of iron in Siegerland goes back to Roman times. Iron was made in the ancient world using charcoal-burning forges and only a small portion was converted into steel for weapons (see Table 1-1). Only in medieval times, when water wheels supplied the required air, were the temperatures necessary for iron smelting reached. We can assume that the first cast iron was obtained in Europe in about the year 1380 but a few decades elapsed before cast-iron pipes for water supply could be made, the impetus being given by the casting of gun barrels. The Master Christian Slanterer cast 30 small breech loaders in Siegen in 1445. Twelve years later Count Johann IV required a water supply for Dillenburg castle and the order for cast-iron pipes went to the same master. Figure 1-4 shows the socketed ends of a well-preserved pipe, 1.1 m long by 70 mm in diameter, with lead-sealed sleeves [13].

Forged and cast-iron pipes were painted with molten pitch or wood tar at the close of medieval times. A work in 1827 states that pipes had been protected by coal tar for a long time [14]. Before being buried, in 1847 cast-iron gas and water pipes were treated with tar in Hanover. In Germany the tarring of wooden roofs was known before 1770. Coal tar had been produced in quantity during production of lighting gas between 1792 and 1802 in England. William Murdoch constructed the first gas production plant in Soho and illuminated the factory of Boulton and Watt on the occasion of the peace of Amiens in 1802.

Cast-iron pipes were used for the mains in the early stages of town gas supply, their sockets being sealed with tarred rope, oakum, or lead. Originally the connection pipes were lead and later of galvanized or coal-tarred forged iron. After the defeat of Napoleon in 1815, there was a surplus of cheap musket barrels, and these

6 Handbook of Cathodic Corrosion Protection

Fig. 1-4 End socket of a cast-iron waterpipe sealed with lead laid in 1457 (photo: Rheinstahl, Gelsenkirchen).

were often used as house connections for town gas pipes. The term "barrel" is still used in England to describe gas connection pipes [15].

The Dresden and Leipzig gasworks were founded in 1828 and the red lead putty socket seals were changed to seals of tarred rope when considerable losses were experienced in the grid. An outer varnish coating was applied as corrosion protection for pipes and connections. The *Leipziger Regulator* of 1863 stresses external protection against destruction by oxidation in its instruction for safe pipe-laying. In England between 1830 and 1850 we find directions on the use of coal tar and asphalt tar together with other materials for pipe protection. An English company founded in 1884 for the manufacture of asphalt tar and mastic cladding was the first to use mineral filling materials.

After 1860 in the United States, water mains were only occasionally given coatings of tar. About 1896 the activities of English undertakings were extended to America, where chiefly bare metal pipelines had previously been laid. Water supply pipes were coated internally with bitumen in America after 1912. Vicat (1837) in France and J. Bull (1843) in America introduced the widely known cement mortar as a protective material for water pipes [16].

The best-known English pipe protection material was invented by Angus Smith, and consisted of a mixture of coal tar and linseed oil. Occasionally pipes were laid in sand or pitch-filled wooden ducts to protect them against especially aggressive soils. Bitumenized paper-wrapped pipes for gas lines that could withstand pressures of 20 atmospheres were first shown during the Paris Exhibition in 1867 though they had come into use for water supply shortly before then (Fig. 1-5). Zinc plating was reported to be an effective protection for wrought-iron pipes in 1864. F. Fischer mentioned cathodic protection for the first time in an exhaustive report. In 1875 there was a report on the use of mineral wool as insulation and of tarred or asphalted

Fig. 1-5 Asphalt paper pipe from a well water pipeline laid in Crailsheim in 1863.

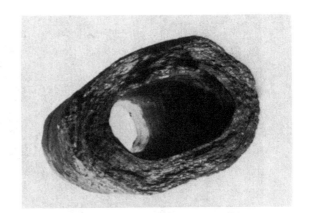

pipes. The first warning of the destruction of gas pipes by aggressive soils and stray currents dates from 1892.

Jute, soaked in molten coal tar pitch, was used after 1900 when it was found desirable to provide forged pipes with a thicker coating. After the First World War, this was further improved with mineral bitumen and a mixture of slate and lime powder or with powdered granite. The change from the use of tar to oxidized bitumen with much improved physical properties made possible the deposition of thick layers inside the water pipes. Wool felt was introduced at the end of the 1920s to eliminate the rot experienced with soaked jute, resulting in a considerable improvement in cladding quality. Figure 1-6 shows the application of bitumenized wool felt at a pipe works which achieved considerable improvement in the quality of the coating. A prognosis, repeated in the journal *Gas- und Wasserfach*, "that the external protection in its present form will not permit attack on pipes," proved to be too optimistic. Since 1953 pipe works have changed from wool felt to glass wool as a support material for bitumen for general use, mainly for mechanical reasons [17].

Field coating of welds has always presented problems. Straw and jute with a greasy material was employed in 1910, but this later saponified in the soil. By chance the pharmacist Schade of Berlin learned of this problem and recommended the use of petroleum jelly in a bandage-like application. Hot-applied bitumen bands, as used by pipe works since 1928, proved to be most durable. Since 1930, electrical measuring methods have played an important part in research into insulation bands and pipe coverings carried out by the Gas Institute in Karlsruhe, the present-day Engler-Bunte Institute [18].

As production increased, new coating materials for pipe protection evolved and a variety of synthetic materials became available. In 1950, continuous-laid pipelines in America and Italy were coated with polyvinyl chloride tapes applied

Fig. 1-6 Coating of a line pipe with bitumen in the pipemill in 1935 (photo: Deutsche Röhrenwerke, Düsseldorf).

by wrapping machines, but even multiple layers gave inadequate protection against mechanical damage. Better results were achieved from 1960 onward with a direct-ring, nozzle-extruded, polyethylene tube that was shrunk onto the pipe with the addition of an adhesive. A process was developed for large pipes by melting polyethylene powder onto the rotating preheated pipe [19]. In Germany, polyethylene coating is carried out and tested according to DIN 30670 [20]. Subsequent coating on the construction site is usually with plastic tapes according to DIN 30672 [21].

1.2 Corrosion Protection by Painting

Ancient iron structures sometimes show no sign of corrosion or at most, very little. The clean atmosphere of past centuries may be responsible in that it allowed a very thin adherent layer of oxide to develop on the surface [22]. This layer very often protects against even today's increasingly aggressive industrial pollutants. Very often the conditions of the initial corrosion are the ones that determine the lifespan of metals [23]. A well-known example is the sacred pillar of Kutub in Delhi, which was hand forged from large iron blooms in 410 A.D. In the pure dry air, the pillar remains free of rust traces but shows pitting corrosion of the iron

The History of Corrosion Protection 9

buried in the soil. However, a sample of this 99.7% pure iron brought to England corroded as fast as any other forged iron.

In Europe and India, iron blooms were made in small smelting furnaces using charcoal and air supplied by bellows; a single smelting yielded only 8 to 10 kilos of forged iron. The development of iron was different in China. The Chinese used anthracite as early as 200 B.C., and this enabled them to make cast iron mainly for utility purposes, e.g., ploughshares, cauldrons, or large vases. The technique of iron casting reached Europe only toward the end of the 14th century. Table 1-1 reviews the development of iron manufacturing technology and of corrosion [24] protection [25]-[28].

The necessity to protect steel and iron against corrosion was generally recognized during the 18th century [14]. The first modern reports on rust-protective paints appeared in 1822 in *Dinglers Polytechnischem Journal*. They proposed to use varnish, resin or vegetable oils for painting steel surfaces. The basic essential of good painting technique, the thorough cleaning of metallic surfaces before painting, seems to have been recognized by 1847. Red lead as a primary coating was recommended in 1885 [14]. Paints and varnishes made from coal tar were used in America after about 1860 to protect iron and steel in shipbuilding, but originally were used only for coating the interior surfaces of iron ships. Coal tar paint was first used in 1892 to paint a large floating dock. The locks, floodgates, and weirs of the Panama Canal were sprayed with tar paint in 1912.

A frequently cited example of protection from atmospheric corrosion is the Eiffel Tower. The narrow and, for that age, thin sections required a good priming of red lead for protection against corrosion. The top coat was linseed oil with white lead, and later coatings of ochre, iron oxide, and micaceous iron oxide were added. Since its construction the coating has been renewed several times [29]. Modern atmospheric corrosion protection uses quick-drying nitrocellulose, synthetic resins, and reaction resins (two-component mixes). The chemist Leo Baekeland discovered the synthetic material named after him, Bakelite, in 1907. Three years later the first synthetic resin (phenol formaldehyde) proved itself in a protective paint. A new materials era had dawned.

1.3 History of Cathodic Protection

In 1936, at Khuyut Rabuah near Baghdad, several clay jugs about 14 cm high were found. Inside was a narrow, pitch-sealed copper cylinder, containing a corroded iron kernel. Similar jugs were found in the ruins of Seleucis on the opposite bank of the Tigris. The assumption is that these objects originate from imperial Roman times (27 B.C. to 395 A.D.) and Wilhelm König, a past director of the Baghdad antiquities administration, believes that these objects are battery cells used for gilding small pieces of jewelry by electrolytic processes. He writes: "All these finds may prove that galvanic electricity was known a long time before Galvani (after whom it was named) experimented with frogs' legs (1789)" [30].

The attraction of rubbed amber and some other effects of electricity were known in ancient times. We know from finding nails in an old wreck that the Romans knew about contact corrosion combined with electric current flow. A skin of lead as a protection against boring worms covered the wooden planks of the ship and was nailed down with copper nails. Galvanic couples formed between the lead and the copper nails and the less noble lead sheets around the nails corroded in the seawater and fell off. The shipbuilders discovered a simple solution and covered the heads of the copper nails with lead as well. Galvanic current flow between the two metals was eliminated and corrosion was prevented [26].

It is not certain whether Sir Humphrey Davy (Fig. 1-7) knew of these considerations. He accepted a commission from the Admiralty for the protection of copper-clad wooden ships, which had been introduced in 1761. During his numerous laboratory experiments, he discovered the cathodic protection of copper by zinc or iron [3]. Davy had already put forward the hypothesis in 1812 that chemical and electrical changes are identical or at least arise from the same material property. He believed that chemical reaction forces could be reduced or increased by altering the electric state of the material. Materials can combine only if they have different electric charges. If an originally positive material can be artificially negatively

Fig. 1-7 Sir Humphrey Davy.

charged, the binding forces are destroyed and can no longer participate in corrosion reactions.

The beginnings of galvanic electricity and investigations on electrolytes were based on Galvani's experiments with frogs' legs in 1789. The Italian physicist Alessandro Volta discovered in Pavia in 1797 the so-called voltaic pillar. For the first time current was produced from an electric cell. The reverse process, electrolysis, had been discovered by Alexander von Humboldt in 1795 in an electrolytic cell with zinc and silver electrodes in an aqueous electrolyte. In 1798 Ritter noticed that the potential series of metals was identical with the ranking of metals according to their oxidizability.

Although these discoveries can hardly be called electrochemical, the explanations given by Davy are remarkable. Davy had established that copper was a metal which acted weakly positive in a galvanic potential series. He deduced from that, that the corrosive action of seawater on copper could be prevented if it were weakly negatively charged. If the copper surface became negative (i.e., a cathode) then all chemical reactions, including corrosion, would be prevented. To explain the process, Davy performed experiments in which polished copper coupons were immersed in weakly acidified seawater. A piece of tin was soldered onto one of the copper coupons. After 3 days the copper coupons without tin showed considerable corrosion whereas the specimen with the soldered tin bore no trace of any corrosion. Davy came to the conclusion that other non-noble metals such as zinc or iron could provide corrosion protection. Davy carried out further experiments with the help of his pupil Michael Faraday. From this work it became apparent that the location of the zinc was immaterial. On another copper coupon to which an iron coupon was soldered, and which was then connected to a piece of zinc, not only the copper but also the iron was protected against corrosion.

After Davy communicated these results to the Royal Society and the British Admiralty, he obtained permission in 1824 to begin practical experiments on the copper cladding of warships. These experiments were carried out at the Portsmouth naval base. Davy attached zinc and cast-iron plates to the copper-clad ships to protect against corrosion. He established that cast-iron was the most economical material. Cast-iron plates 5 cm thick and 60 cm long gave very satisfactory results on nine ships. On ships' hulls where rivets and nails were already rusted, the corrosion protection was only effective in the immediate vicinity of the anodes. To explain this, Davy carried out further experiments on the warship *Sammarang* (Fig. 1-8). The ship had been covered with new copper sheet in India in 1821. Cast-iron metal plates constituting 1.2% of the total copper surface of the ship's hull were fixed to the bow and the stern. The ship then made a voyage to Nova Scotia (Canada) and returned in January 1825. Apart from some attack at the stern which was attributed to water vortices, there was no corrosion damage to the rest of the ship. Equally good results were achieved with the Earl of Darnley's yacht *Elizabeth* and the 650-ton

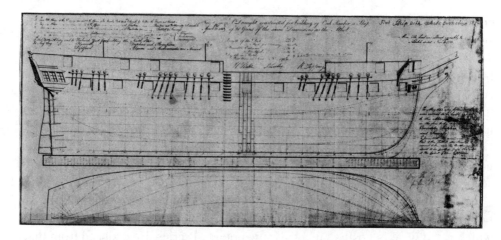

Fig. 1-8 Constructional drawing of the *Sammarang*, which was the first ship to make a sea voyage to Nova Scotia from March 1824 to January 1825 with cathodic protection of the copper sheathing.

freighter *Carnebra Castle*. Each ship was equipped on the stern and bow with two zinc plates amounting to 1% of the copper surface. The copper cladding looked as good as new after the freighter returned from Calcutta.

Some years after Davy's death, Faraday examined the corrosion of cast iron in sea water and found that it corrodes faster near the water surface than deeper down. In 1834 he discovered the quantitative connection between corrosion weight loss and electric current. With this discovery he laid the scientific foundation of electrolysis and the principles of cathodic protection.

Apparently without knowledge of Davy's experiments, the inspector of telegraphs in Germany, C. Frischen, reported to a meeting of the Architekten-und Ingenieur-Verein at Hanover in 1856 the results of a wider experimental enquiry, over a long period, "with particular regard to the protection of the most important and widely used metal, wrought iron, which constitutes the most important parts of the large structures, like bridges, locks, gates, etc." Frischen soldered or screwed pieces of zinc onto iron as a protection against seawater and concluded "that an effective protection of iron is doubtless due to the influence of galvanic electricity." However, achieving a successful and practical protective technique would have required many protracted, large-scale experiments [31] and [32].

It is little known that Thomas Alva Edison tried to achieve cathodic protection of ships with impressed current in 1890; however, the sources of current and anodic materials available to him were inadequate. In 1902, K. Cohen achieved practical cathodic protection using impressed direct current. The manager of urban works at

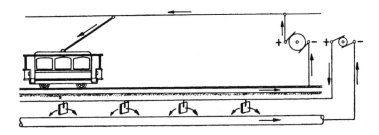

Fig. 1-9 Diagram of cathodic protection from the patent of H. Geppert of (DRP No. 211612).

Karlsruhe, Herbert Geppert, constructed the first cathodic protection installation for pipelines in 1906. This was a direct current generator of 10 V 12 A capacity protecting 300 m of gas and water pipelines within the electrical field of a tramline [33]. Figure 1-9 shows the principle for which H. Geppert obtained a German patent in 1908 [34]. Protection using consumable anodes was termed "electrochemical protection" at a Congress of the Institute of Metals in Geneva in the autumn of 1913.

To protect steam boilers and their tubes from corrosion, E. Cumberland used cathodic impressed current in America in 1905. Figure 1-10 has been taken from the corresponding German patent [35]. In 1924 several locomotives of the Chicago Railroad Company were provided with cathodic protection to prevent boiler corrosion. Where previously the heating tubes of steam boilers had to be renewed every 9 months, "the costs fell sharply after the introduction of the electrolytic

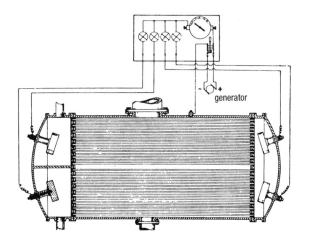

Fig. 1-10 Internal cathodic protection from the patent of E.G. Cumberland of (DRP No. 247544).

process." Aluminum anodes with applied dc were used by A. Guldager in Denmark for internal protection of hot water supply plants. Cathodic protection was thereby provided to the interior of the warm water tank and the connecting pipes by formation of a secondary surface film.

At the beginning of the 20th century, with the development of stainless steels, the passivity of metals became technically important in corrosion protection. In a presentation at the international exhibition of chemical engineering (ACHEMA) in Frankfurt in 1958, it was asserted that it was thanks to metal passivity that progress from the Stone Age to the age of metal technology had been possible [36]. Investigation of passivity phenomena in the 1930s and particularly after the Second World War led to electrochemical investigations and the knowledge that potential was an important variable in corrosion reactions. Great progress in measuring techniques occurred with the development of the potentiostat in the 1950s and systematic investigation of the dependence of corrosion parameters on potential began worldwide. The scientific basis for general electrochemical protection was laid. By determining the limiting potentials for the occurrence of certain corrosion phenomena, in particular local corrosion such as pitting and stress corrosion [37], this work led to the concept of protection potentials.

The passivating stainless steels presented a possibility for developing anodic protection. High-alloy steels, similar to carbon steels, are not capable of being cathodically protected in strong acids because hydrogen evolution prevents the necessary drop in potential. However, high-alloy steels can be passivated and maintained in the passive state by anodic protection. C. Edeleanu was the first to demonstrate in 1950 that anodic polarization of the pump housing and connecting pipework could protect a chromium-nickel steel pumping system against attack by concentrated sulfuric acid [38]. The unexpectedly wide range of anodic protection is due to the high polarization resistance of the passivated steel. Locke and Sudbury [39] investigated different metal/medium systems in which the application of anodic protection was relevant. Several anodically protected installations were in operation in the United States by 1960, e.g., storage tanks and reaction vessels for sulfonating and neutralization plants. Not only did the installations have a longer life but also a greater purity of the products was achieved. In 1961 anodic protection was first applied on a large scale to prevent stress corrosion cracking in a caustic soda electrolysis plant in Aswan [40]. Anodic protection for caustic soda tanks has been used on a large scale since the end of the 1960s and electrochemical corrosion protection methods have become of permanent importance for industrial plants (see Chapter 21).

During the previous century, the success of cathodic protection was often a matter of chance. In 1906 at the instigation of the DVGW,[1] F. Haber and L. Gold-

---

[1] The Deutscher Verein des Gas- und Wasserfachs, the current name of an earlier association of gas and waterworks engineers.

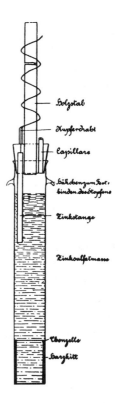

Fig. 1-11 Nonpolarizable Zn/ZnSO$_4$ electrode developed by Professor Haber in 1908.

schmidt studied the scientific fundamentals of cathodic protection. They recognized cathodic protection and stray current corrosion as electrochemical phenomena and Haber's well-known circuit for measuring current density, soil density, soil resistance, and pipe/soil potentials is described in the *Zeitschrift für Elektrochemie* [41]. Haber used nonpolarized zinc sulfate electrodes to measure potential (see Fig. 1-11). Two years later McCollum employed the first copper sulfate electrode, which since then has generally succeeded in measuring potential in buried installations. Between 1910 and 1918, O. Bauer and O. Vogel at the material testing station in Berlin determined the current density required for cathodic protection [42]. In 1920 the Rhineland cable near Hanover was damaged by corrosion as a result of geological factors in the soil, and for the first time in Germany zinc plates were built into the cable shafts to protect the metal sheathing [32]. The protection of iron by electric current was eventually the subject of a dissertation in 1927 [43].

By 1920 welding technology had reached a stage at which safe welds could be reliably produced and as a consequence continuously welded transmission pipelines could be constructed. Thus, there appeared to be nothing to hinder a common application of cathodic protection. Peterson [44] reported an electrical process for protecting pipelines against rust which had also given excellent protection against soil currents in a Galician oilfield. That this report did not lead to the use of cathodic protection was probably the fault of the engineering-oriented builders of pipe networks, who regarded electrochemical protection as a black art. Even electrical engineers overestimated the costs of the process and the danger to other pipelines from the applied currents. Instead, therefore, attempts were made to improve the resistance of pipe coating materials to aggressive soils and to reduce the danger of stray current corrosion by insulating joints.

Cathodic protection of pipelines did not develop in Germany but it was applied from 1928 onward in the United States. Figure 1-12 shows a medal with the head of Robert J. Kuhn, called the "Father of Cathodic Protection" in America. He installed the first cathodic protection rectifier in 1928 on a long-distance gas pipeline in New Orleans, and thus inaugurated the first practical application of cathodic protection of pipelines. As early as 1923, E.R. Shepard in New Orleans had diverted powerful tramline stray currents with an electrical drainage system. The protection range of plain cast pipes with poorly conducting joints did not extend to the end of the pipeline, so Kuhn put additional protective rectifiers in. He found by experiments that a protective potential of –0.85 V against a saturated copper/copper-sulfate electrode provided sufficient protection against any form of corrosion. At the Washington Conference for Corrosion Protection held by the National Bureau of Standards in 1928, Kuhn reported on the significant value of his experiments, on which the entire modern technology of cathodic protection is founded [45]. Considerable doubt existed in the minds of American scientists at that time over the causes of corrosion of buried pipelines. Kuhn's presentation was the only one that dealt with corrosion resulting from galvanic cell formation. It contained the description of a process that prevented corrosion by application of rectified dc, i.e., by cathodic protection. Kuhn wrote: "This method was not applied exclusively to prevent corrosion but to cut out electrolytic corrosion of the pipeline arising from streetcar stray currents by electric drainage." This application had shown that the pipes were not only protected against

Fig. 1-12 Gold medal with picture of Robert J. Kuhn. The medal was established by the Technical Committee of Cathodic Protection and first awarded in 1970 by the DVGW in Würzburg.

stray current electrolysis but also against galvanic cell currents and thus against soil corrosion. The experiment showed that on average 10 to 20 mA m^{-2} protective current density was sufficient to depress the potential of the pipeline to a value where pitting no longer occurred [46].

Some experts still regarded the experiments with skepticism. The American Petroleum Institute in Los Angeles declared in 1935 that galvanic protective currents from zinc anodes could no longer protect pipelines over the increasing distances being used, and that protection against chemical attack, such as by acids, would be definitely impossible. Since pipelines in America were mostly laid without any insulating coverings until well into the present century, the cathodic protection of such pipelines proved comparatively expensive and required considerable protective currents. It is therefore not surprising that even in the early 1930s only about 300 km of pipelines in America were protected by zinc anodes and 120 km by impressed current. Among them were pipelines in Houston, Texas, and Memphis, Tennessee, which had been cathodically protected by Kuhn from 1931 to 1934. In early 1954, I. Denison was awarded the Whitney prize by the National Association of Corrosion Engineers. Kuhn's discovery was given renewed publicity, for in his acceptance speech, Denison explained: "In the first corrosion protection conference in 1929 Kuhn described how the potential of a pipeline was depressed to −0.85 V against a saturated copper/copper sulfate electrode using a D.C. rectifier. I need not remind you that this value is the accepted potential criterion for cathodic protection used worldwide nowadays."

Smith reported in 1940 [47]:

"Cathodic protection applied in the United States has had mixed success and variable criticism. Failure is partly due to false evaluation of the active region and partly due to insufficient planning and operation of the plant to be protected. It is partly the failure to recognize that the process against corrosion due to high and widely ranging soil currents arising from stray currents from D.C. installations with correspondingly higher potentials to give protection against them, can only be successful if these currents are drained in addition. The cathodic protection process is only relevant for oil, gas, and water pipelines against soil attack in areas where there are no stray currents. Neither soil composition nor climate can impair the action of the protective medium if the anodic current is sufficient to reduce the potential of the pipeline to be protected against that of the soil by about 0.3 V."

The first anode installation for the cathodic protection of gas pipelines in New Orleans consisted of a 5-m-long horizontal cast-iron tube. Later old tramway lines were used. Since in downtown New Orleans there was no suitable place to install impressed current anodes and to avoid detrimental effects on other pipelines, Kuhn recommended the use of deep anodes which were first installed in 1952 at a depth of 90 m. The first deep anodes were installed in Hamburg in Germany in 1962 by F. Wolf [48] (Fig. 1-13).

Publications on the cathodic protection of pipelines became known in Europe at the end of the 1920s. In Belgium the drainage of streetcar stray currents was widely practiced. L. de Brouwer applied protection to gas supply lines in Brussels from 1932 onward and in 1939 the base plates of a gasholder were also protected with impressed current [49]. In Germany the following report was made in 1939 on the cathodic protection of pipelines [50]: "The following precautions against stray currents should first be taken to prevent the leakage of current from the rails into the surrounding earth. On the pipeline it is advisable to provide it with a double sheath and to choose electrically insulating connections to raise the insulation resistance for a distance of about 200 m either side of its crossing of the tracks. A conducting connection between the pipe and the track should only be made with the greatest caution so that the reverse action does not occur." From Soviet publications on cathodic protection, it appears that by about 1939 more than 500 cathodically protected installations existed in the USSR [51], [52]; judging by their numbers, these utilized sacrificial anodes. Cathodic protection of pipelines by sacrificial anodes appeared in Great Britain after 1940 [53]. In Germany in 1949 the water

Fig. 1-13 The first deep anode for cathodic protection in Germany.

The History of Corrosion Protection

supply network of the Brunswick brown coal mines were cathodically protected by zinc plates by W. Ufermann [54]. Figure 1-14 shows the cathodic protection installation in the Palatinate near Bogenheim, which was installed together with other protective measures against stray current by Saar-Ferngas in 1952. In 1953 in Duisburg-Hamborn [55] and in 1954 in Hamburg [56] particular cathodic impressed current systems had been installed to protect limited parts of old long-distance gas lines. After 1955 these were extended to all pipelines and in particular newly installed long-distance gas supply lines. However, in spite of the obvious advantages of cathodic protection, even for single-wall storage tanks buried in soil, it was only in 1972 that a directive was enacted [57].

In the past few years the technique has developed from local cathodic protection to the cathodic protection of the underground installations of whole power stations and industrial plants (see Chapter 12). Since 1974 cathodic protection has been compulsory in Germany for gas pipelines with pressures over 4 and 16 bars (DVGW instructions G462 and 463 respectively) and for oil pipelines according to the guidelines for transport of dangerous liquids by long-distance pipelines (TRbF* 301).

The basic standard for cathodic protection was laid down for the first time in DIN 30676 to which all the application areas of the different branches of protection can be referred. In this the most important point is the technique for accurately measuring the object/soil potential [58]. The usual off-potential measurement method for underground installations has been slowly implemented and enforced in Europe since the 1960s [59].

Fig. 1-14 Cathodic protection rectifier manufactured in France (1952) at Bogenheim in the Palatinate.

1.4 Development of Stray Current Protection

The first news of the destructive influence of stray currents reached Europe from America shortly before the turn of the century. Stray current electrolysis as a new corrosion hazard for buried pipes as a result of increasing domestic dc supplies and the construction of direct current railways also became evident in Germany. Werner von Siemens introduced at the 1879 Berlin Gewerbeausstellung (Trade Exhibition) the first direct current electric railway in the world. The first electric tram to Berlin-Lichterfelde ran 2 years later using a positive and a negative rail with an operating voltage of 140 V. In 1882 Siemens installed an experimental tram service with overhead supply, running from Westend to Spandauer Bock. As shown in Fig. 1-15, the line was initially equipped with two overhead wires so that the escape of stray currents into the soil was prevented [60]. Unfortunately, it was not possible to retain this system.

The first underground railway in the world was opened on January 10, 1863 in the City of London, operating with steam locomotives. The first line with electric traction and a three-rail system was built in 1890. The four-rail system, still in use today and consisting of two insulated conductor rails and two running rails, was introduced in 1903 in the course of electrification of the old steam tracks. The Metropolitan Company, responsible for a part of the track, used ac, while the District Railway preferred dc as a consequence of the connection with the American railways. This dispute came before a British arbitration tribunal in 1900. The problem of corrosion of public supply lines by the returning current from electric train

Fig. 1-15 Berlin streetcar in 1882 with double overhead conductor with rod contact.

locomotives was at that time already appreciated as a result of the considerable difficulties encountered in America, and the advantage of the ac system was that these inconveniences were eliminated. The arbitrators solved the problem by recommending the use of 600-V direct current and also the laying of a return-current rail insulated from the running rails [61].

The first serious electrically induced corrosion damage in the field of tram rails appeared in 1887 at Brooklyn on forged iron pipes, and in the summer of 1891 at Boston on the lead sheaths of telephone cables [62]. The first commission to investigate the phenomenon of stray currents was set up in America.[2] The commission recorded that considerable differences in potential exist between pipes and the track of electric railways, and that the pipes are endangered in positions "where the charge is positive and the current escapes into the electrolysis-prone surroundings." It was found experimentally by Flemming that iron surfaces buried in damp sand with a potential of +0.5 V and a current of 0.04 A corroded visibly after only a few days. The first direct stray current drainage was installed in 1895 by E. Thompson for the Brooklyn tramway in an attempt to move the stray current directly back to the rails without harmful effects [63]. However, this occasionally caused such a powerful increase in the stray currents that they melted the lead seals of the joints.

Deliberate stray current drainage was installed at a subrectifier in Germany as early as 1895 during the electrification of the Aachen tramway. The effective protection extended over a relatively small field since the comparatively large resistance of the pipe joints did not permit a greater extension of protection.

The Berlin City electrical engineer M. Kallmann reported in 1899 on a system for controlling stray currents of electric railways [64]. As early as 1894, the Board of Trade in London issued a safety regulation for the British electric railways which specified a potential differential of not more than 1.5 V where the pipeline was positive to the rails, but 4.5 V with the rails positive. Extensive research was undertaken on reducing the risk of stray current in the soil by metallic connections from pipes to rails. However, as one writer noted, "a procedure on these lines should definitely be discouraged as it carries the seed of its own destruction" [64].

The *Journal für Gasbeleuchtung* mentions electrolytic corrosion damage caused by direct current cables in Berlin in 1892, and a few years later damage by tramway currents was reported in 14 German towns. As early as 1894 the electrolytic processes of stray current corrosion were explained in detail in this journal by G.Rasch [65].

In 1910, a joint commission of the associations of German electrotechnicians and gas and waterworks engineers issued regulations for the protection of gas and

---

[2] Around 1890 about 100 tram lines were run by direct current in Germany and well over 500 in America. In 1981 there were only 33 such local trams in the Federal Republic and 25 in the German Democratic Republic.

water pipes against the harmful influence of currents from electric dc railways that used rails as conductors. These, however, forbade direct returning of stray currents to the rails. Attempts were made to reduce stray currents by incorporating insulating joints and improving the pipe coverings. This was mainly restricted to points where tram lines were crossed in order to economize on insulating flanges. This often resulted in new escape points for the stray current at the insulating flanges. To avoid the prohibited direct connection with tram rails, in regions of stray current, an exit connection was made to uncoated protective pipes or to iron girders buried alongside the rails. That the problem could not be solved in this manner was very soon recognized. Only in 1954 did a new version of VDE* 0150 establish a legal basis in the Federal Republic of Germany to sanction the stray current conductors and drains installed since 1950 [17]. In 1966 a joint group for corrosion problems, the Arbeitsgemeinschaft für Korrosionsfragen or Working Association for Corrosion Questions (AfK) and the Arbitration Board for Relevant Problems, decided on measures to protect pipelines against increasing high-tension effects by using better coatings [66].

The principles of stray current drainage by rectifier as practiced today depend on impressed current cathodic protection as described by H. Geppert. In his patent Geppert referred to the fact that stray currents escaping from a pipeline will be compensated, and he also mentions the possibility of a direct connection of the protective current source with the rails. A direct connection without additional impressed current between pipeline and rails is sufficient only with permanently negative rails, as for instance near rectifier installations. As early as 1930, 25 direct stray current drainages for post office cables existed in Milan and Turin. It is necessary to install return current blocks in the connections where rails become even temporarily positive. In 1934, L. de Brouwer, the Chief Engineer of Distrigaz, installed one of the first polarized relays in Fontaine-l'Eveane near Brussels [67,68]. By installing rectifiers into a temporary drainage connection in Berlin in 1942, reverse current was prevented. Figure 1-16 shows the first drainage relays made in Germany, which were installed for the protection of a long-distance pipeline near Immigrath in 1953.

A protective rectifier between pipeline and rails that was installed by Kuhn in 1928 was the forerunner of modern forced drainage. This form of controlled stray current drainage was developed further, particularly in France, in combination with stray current relays. The first such polarized drainage was installed at Bad Dürheim in 1975 and had an appearance similar to the cathodic protection rectifier in Bogenheim, shown in Fig. 1-14. Today mainly controlled potential rectifiers are used for stray current drainage. The first of these, in Wuppertal-Cronenberg, conducted peak currents of more than 200 A between 1961 and 1970.

Fig. 1-16 First polarized relay for stray current drainage in Germany at Immigrath in 1953.

1.5 Corrosion Protection by Information

The question of corrosion and corrosion protection has become even more important with the increased use of metallic materials as industry has grown. The concern over the great number of avoidable failures compelled an investigation of the causes for which in most cases organizational problems had been indicated [69]. It also showed that the science of corrosion protection had spread into diverse technical areas almost unsurveyed. The Arbeitsgemeinschaft Korrosion founded in 1931 was reconstituted in 1981 as a registered association with a scientific advisory committee. In the new statute its purpose was designated as the coordination of technical regulations as well as research and the exchange of experience. The information given in this handbook is largely contained in the DIN standards which were formulated and mutually agreed upon by the standard committees NAGas 5.2 and NMP 171. In this respect, the conceptual standards DIN 50900 and the basic standards for electrochemical protection, DIN 30676 and DIN 50927, are particularly relevant.

To provide a better survey, the Arbeitsgemeinschaft Korrosion undertook the sifting of the regulations collected by the Deutsche Informationszentrum für Technik and published them in an industry-related compilation with an advisory commentary [70]. Updates continue to be published in the journal *Werkstoffe und Korrosion*.

The state of the art in 1987 was published in "Survey of standards, technical rules and regulations in the area of corrosion, corrosion testing and corrosion protection," together with the most important DIN standards and some other regulations in the DIN 219 [71]. A commentary on these standards has also appeared [72].

1.6 References

[1] L. Beck, Geschichte des Eisens, 5, Bd., Vierweg & Sohn, Braunschweig 1884-1903.
[2] H. Leierzapf, Werkstoffe und Korrosion 36, 88 (1985).
[3] H. Davy, Phil. Transact. *144*, 151 (1824).
[4] DIN 50900. Teil 1, Beuth Verlag, Berlin, 1975 und 1982; Werkstoffe und Korrosion *32*, 33 (1981).
[5] R.J. Forbes, Bitumen *4*, 6 (1934).
[6] Vergessene Städte am Indus, Verlag Ph. v. Zabern, Mainz, 1987.
[7] L. Borchardt, Das Grabdenkmal des Königs Sahu-re, J.C. Hindrichs'sche Buchhandlung, Leipzig 1910.
[8] W. Schhrmann, Sanitär- u. Röhrenmarkt 7.21 (1952).
[9] Mannesmann AG, Rohre gab es immerschon, Dhsseldorf 1965.
[10] M. Kromer, Wasser in jedwedes Bürgers Haus, Ullstein GmbH, Frankfurt u. Berlin 1962.
[11] F. Feldhaus, Deutsche Industre B Deutsche Kultur *9*, 1 (1913).
[12] B. Gockel, gwf *108*, 191 (1967).
[13] A. Wittmoser, Gießerei *44*, 557 (1957).
[14] G. Seelmeyer, Werkstoffe u. Korrosion *4*, 14 (1953).
[15] J. Körting, Geschichte der deutschen Gasindustrie, Vulkan-Verlag, Essen 1963.
[16] F. Fischer, Journal für Gasbel. *19*, 304 (1876).
[17] W.v. Baeckmann, gwf *108*, 702 (1967).
[18] F. Eisenstecken, gwf *76*, 78 (1933).
[19] H. Klas, gwf *108*, 208 (1967).
[20] DIN 30670, Beuth Verlag, Berlin, 1980.
[21] G. Heim u. W.v. Baeckmann, 3R intern. *26*, 302 (1987).
[22] F. Tödt, Korrosion und Korrosionsschutz, Walter de Gruyter, Berlin 1961.
[23] U.R. Evans, Einführung in die Korrosion der Metalle. Historischer Überblick, S. 251, Verlag Chemie, Weinheim 1965.
[24] R. Johannsen, Geschichte des Eisens, 2. Bd., Verlag-Stahleisen, Düsseldorf 1935.
[25] G. Seelmeyer, Wichtige Daten aus der Geschichte der Korrosion bis 1900 in F. Tödt [16].
[26] Geschichte des Lackes, Dtsch. Farben-Z. *10*, 251 (1956).
[27] P. Baur, Farbe u. Lack *69*, 3 (1963).
[28] G.A. Walter, Die geschichtliche Entwicklung der rheinischen Mineralfarbenindustrie. G. Baedecker, Essen 1922.
[29] K.A. van Oeteren-Panhäuser, Werkstoffe u. Korrosion *10*, 422 (1959).
[30] W. König, Im verlorenen Paradies, R. Rohrer, Baden b. Wien 1940.
[31] C. Frischen, Zeitschrift des Architechten- und Ingenieur-Vereins fhr das Königreich von Hannover, *3*, 14 (1857).
[32] H. Steinrath, Zur geschichtlichen Entwicklung des kathodischen Schutzverfahrens, in Korrosion *11*, Verlag Chemia, Weinheim 1959.
[33] H. Geppert u. K. Liese, Journal für Gasbel. *53*, 953 (1910).
[34] H. Geppert, Patenschrift Nr.211 612 v. 27. 3. 1908.

[35] E.G. Cumberland, Patenschrift Nr. 247 544, 17 d, Gruppe 5 v. 28. 9. 1911.
[36] H. Gerischer, Angew, Chemie *70*, 285 (1958).
[37] W. Schwenk, Werkstoffe u. Korrosion *19*, 741 (1968).
[38] W.v. Baeckmann, Chemiker Ztg. *87*, 395 (1963).
[39] J.P. Sudbury, W.P. Banks u. C.E. Locke, Mat. Protection *4*, 81 (1965).
[40] H. Gräfen, E. Kahl, A. Rahmel, Korrosionum 1, Verlag Chemie, Weinheim (1974).
[41] F. Haber u. F. Goldschmidt, Zeitschrift für Elektrochemia *12*, 49 (1908).
[42] O. Vogel u. O. Bauer, gwf *63*, 172 (1920); *68*, 683 (1925).
[43] W. van Wüllen-Scholten, gwf *73*, 403 (1930).
[44] Perterson, gwf *71*, 848 (1928).
[45] R.J. Kuhn, Bureau of Standards 73B75 (1928).
[46] R.J. Kuhn, Corr. Prev. Control *5*, 46 (1958).
[47] W.T. Smith, Gas Age *85*, 49 (1940).
[48] F. Wolfe, gwf *97*, 104 (1956).
[49] L. de Brouwer, gwf *84*, 190 (1941).
[50] H. Steinrath, Röhren u. Armaturen Ztg. *4*, 180 (1939).
[51] V.H. Pritula, Kathodischer Schutz für Rohrleitungen IWF, London 1953.
[52] B.G. Volkov, N.I. Tesov, V.V. Suvanov, Korrosionsschutzhandbuch, Verlag Nedra, Leningrad 1975.
[53] J.H. Morgan, Cathodic Protection, Leonard Hill Books, London 1959.
[54] w. Ufermann, gwf *95*, 45 (1954).
[55] G. Reuter u. G. Schürmann, gwf *97*, 637 (1956).
[56] F. Wolf, gwf *97*, 100 (1956).
[57] TRbf 408, Carl Heymanns Verlag, Köln, 1972.
[58] DIN 30676, Beuth Verlag, Berlin 1984.
[59] W. Schwenk, 3R internal. *25*, 664 (1986).
[60] H. Dominik, Geballte Kraft (W.v. Siemens), W. Limpert, Berlin 1941.
[61] J.G. Bruce in T.H. Vogel, Jahrbuch des Eisenbahnwesens *17*, 76 (1966), Hestra-Verlag, Darmstadt.
[62] C. Michalke, Journal für Gasbel. *49*, 58 (1906).
[63] J.J. Meany, Materials Performance *10*, 22 (1974).
[64] M. Kallmann, ETZ *20*, 163 (1899).
[65] G. Rasch, Journal für Gasbel. *37*, 520 (1894); *38*, 313 (1895); *41*, 414 (1898).
[66] R. Buckel, Elektr, Wirtschaft *72*, 309 (1973).
[67] F. Besig, Korrosion u. Metallschutz *5*, 99 (1929); Journal für Gasbel. *77*, 37 (1934).
[68] A. Weiler, Werkstoffe und Korrosion *13*, 133 (1962).
[69] W. Schwenk, Werkstoffe u. Korrosion *37*, 297 (1986).
[70] Werkstoffe u. Korrosion *35*, 337 (1984); *37*, 277 (1986).
[71] DIN Taschenbuch Nr. 219 "Korrosionsverhalten – Korrosionsprüfung – Korrosionsschutz," Beuth Verlag, Berlin 1987.
[72] W. Fischer, Korrosionsschutz durch Information und Normung, Verlag I. Kuron, Bonn 1988.

2

Fundamentals and Concepts of Corrosion and Electrochemical Corrosion Protection

W. SCHWENK

2.1 Corrosion Processes, Corrosion Damage, and Protective Countermeasures

The term corrosion is used to describe the reaction of a material with its surroundings that produces measurable changes and can lead to damage. With metallic materials and aqueous solutions, the reactions are in general of an electrochemical nature. However, in addition, pure chemical reactions or entirely physical processes can also be occurring. Not every process necessarily leads to damage. This is a question of the extent of the reaction and the demands on the function of material or medium, which should always be considered together. Damage is said to occur when this function is impaired. Corrosion protection is designed to prevent such detrimental action [1-4].

The schematic diagram in Fig. 2-1 shows the various reactions that can occur sequentially and simultaneously in the corrosion process. Material transport and chemical reactions can supply or remove important reaction components. In addition to adsorption or desorption, a phase boundary reaction occurs which mostly involves electrochemical reactions and which can be affected by external currents. An example is formation of hydride on lead where the partner to the hydrogen reaction can arise from a cathodic process. Hydrogen can also penetrate the metal microstructure, where it can have a physical or chemical effect, causing a degradation of the mechanical properties of the material. In this type of corrosion cracks can form during mechanical testing, leading to fracture without the loss of any metal.

Types of damage can be classified as uniform or localized metal removal, corrosion cracking or detrimental effects to the environment from the corrosion products. Local attack can take the form of shallow pits, pitting, selective dissolution of small microstructure regions of the material or cracking. Detrimental effects are certainly not the case with buried pipelines, but have to be considered for environments in vessels and containers. It is usual, where different results of reactions lead

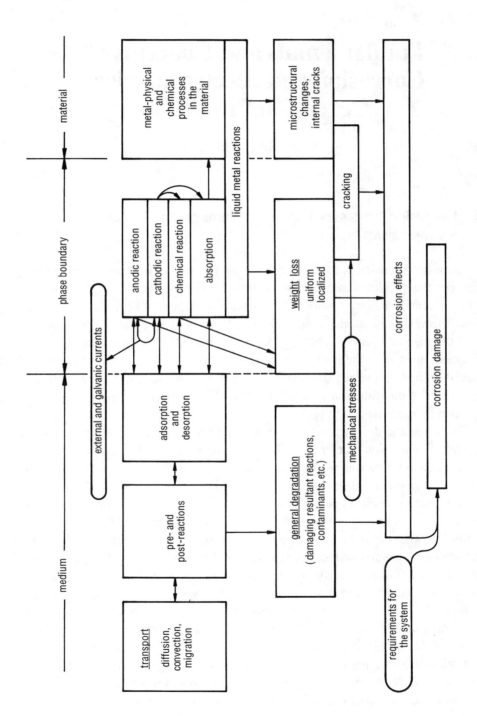

Fig. 2-1 Flow diagram of corrosion processes and types of damage (schematic).

Fundamentals of Corrosion and Electrochemical Corrosion Protection 29

to definite forms of corrosion effects, to classify them as particular types of corrosion. The most important are uniform corrosion, shallow pitting corrosion, pitting corrosion, crevice corrosion, intergranular corrosion, and those with accompanying mechanical action—stress corrosion and corrosion fatigue. The two latter are most to be feared. Most failures arise from pitting corrosion. Uniform corrosion almost always occurs in practice but seldom leads to failure.

2.2 Electrochemical Corrosion

This handbook deals only with systems involving metallic materials and electrolytes. Both partners to the reaction are conductors. In corrosion reactions a partial electrochemical step occurs that is influenced by electrical variables. These include the electric current I flowing through the metal/electrolyte phase boundary, and the potential difference $\Delta\phi = \phi_{Me} - \phi_{El}$ arising at the interface. ϕ_{Me} and ϕ_{El} represent the electric potentials of the partners to the reaction immediately at the interface. The potential difference $\Delta\phi$ is not directly measurable. Therefore, instead the voltage U of the cell Me'/metal/electrolyte/reference electrode/Me' is measured as the conventional electrode potential of the metal. The connection to the voltmeter is made of the same conductor metal Me'. The potential difference $\phi_{Me'} - \phi_{Me}$ is negligibly small; then since $\Delta\phi_B = \phi_B - \phi_{El}$:

$$U = \Delta\phi - \Delta\phi_B \tag{2-1}$$

ϕ_B is the potential of the reference electrode, without whose identification the potential U is undefined. Potentials are conveniently calculated against a standard reference value. Section 3.2 contains further details on reference electrodes and conversion factors. Section 3.3 describes practical methods for measuring potential in the case of flowing currents.

Electrochemical corrosion is understood to include all corrosion processes that can be influenced electrically. This is the case for *all* the types of corrosion described in this handbook and means that data on corrosion velocities (e.g., removal rate, penetration rate in pitting corrosion, or rate of pit formation, time to failure of stressed specimens in stress corrosion) are dependent on the potential U [5]. Potential can be altered by chemical action (influence of a redox system) or by electrical factors (electric currents), thereby reducing or enhancing the corrosion. *Thus exact knowledge of the dependence of corrosion on potential is the basic hypothesis for the concept of electrochemical corrosion protection processes.*

One must always bear in mind that potential dependence is not the same in different types of corrosion. Thus critical potential ranges for different kinds of corrosion can overlap or run counter to one another. This is particularly important to ascertain when dealing with electrochemical protection measures for unknown systems. Electrolytic corrosion is a subsidiary branch of electrochemical corrosion and often involves the removal of metal by an anodic phase boundary reaction (see Fig. 2-1). Such corrosion is often increased by positive currents. There are also systems with opposite effects. In general, hydrogen-induced types of corrosion are favored by negative currents.

Electrochemical corrosion protection has the objective of reducing corrosion damage or removing it altogether. Three different processes are discussed in this work:

(a) Suppressing damage by direct currents from cells through electrical separation (protective measures against cell formation).
(b) Suppressing damage by direct currents by means of drainage (stray current protection).
(c) Protective measures using the controlled action of direct currents (protection currents) to bring the object to be protected into a potential range (protection potential range) where corrosion damage cannot occur. Protection currents can be imposed by rectifier installations or by cell formation by contact with sacrificial metals (e.g., sacrificial anodes in cathodic protection) (see Chapters 6 to 8).

The measures in (a) are dealt with primarily in Chapter 10 but also in other chapters relating to applications. Chapter 15 deals with the measures in (b). Information on the measures in (c) can be found in Chapters 10 to 21 and also in Sections 2.3 and 2.4 of this chapter.

2.2.1 Metallic Materials

Metallic materials consist of one or more metallic phases, depending on their composition, and very small amounts of nonmetallic inclusions. In the metallic state, atoms donate some of their outer electrons to the electron gas that permeates the entire volume of the metal and is responsible for good electrical conductivity (10^5 S cm^{-1}). Pure elements do not react electrochemically as a single component. A mesomeric state can be approximately assumed

$$\text{Fe} \leftrightarrow \text{Fe}^{2+} + 2\,e^- \tag{2-2}$$

in which iron ions and electrons appear in the metal. Both components can react with electrolytes in which the dissolution of metal by the passage of Fe^{2+} results in a positive current I_A and metal loss of Δm, while the passage of electrons corresponds to a negative current I_C <u>without</u> removal of metal. In the first case an anodic reaction occurs, and in the second a cathodic reaction. The relations are reversed in

Fundamentals of Corrosion and Electrochemical Corrosion Protection

a transfer from the solution to the metal: in the anodic reaction, electrons are transferred to the metal, and in the cathodic reaction iron is deposited. Cathodic metal deposition is used in electroplating and is the reverse process of electrolytic corrosion. Faraday's Law applies to both:

$$\Delta m = \frac{MQ}{z\mathcal{F}} \qquad (2\text{-}3)$$

where Δm is the mass of dissolved metal, M is the atomic weight, Q is the transferred electric charge, z is the valence of the metal ions, and $\mathcal{F}$ is Faraday's constant. The following equations can be derived with the aid of the specific gravity ρ_s:

$$J_A = \frac{Q}{St} = \frac{\Delta m}{St}\frac{z\mathcal{F}}{M} = \frac{\Delta s}{t}\frac{z\mathcal{F}\rho_S}{M} \qquad (2\text{-}4)$$

where J_A is current density of the anodic partial reaction for the passage of metal ions; S is the surface area of the electrode; t is time; and Δs is the thickness of material removed. Equation (2-4) can be divided as follows:

$$w = f_a J_A = \frac{\Delta s}{t}; \quad f_a = \frac{M}{z\mathcal{F}\rho_S}$$

$$= \frac{3.2684}{z}\left(\frac{M}{\text{g mol}^{-1}}\right)\left(\frac{\text{g cm}^{-3}}{\rho_S}\right)\frac{\text{mm a}^{-1}}{\text{mA cm}^{-2}} \qquad (2\text{-}5)$$

$$\upsilon = f_b J_A = \frac{\Delta m}{St}; \quad f_b = \frac{M}{z\mathcal{F}}$$

$$= \frac{0.37311}{z}\left(\frac{M}{\text{g mol}^{-1}}\right)\frac{\text{g m}^{-2}\text{ h}^{-1}}{\text{mA cm}^{-2}} \qquad (2\text{-}6)$$

where w is the rate of decrease in thickness and υ is the rate of weight loss per unit surface area. The conversion factors f_a and f_b for some important metals are given in Table 2-1. A further relationship is given by:

$$w = f_c \upsilon; \quad f_c = \frac{f_a}{f_b} = \frac{1}{\rho_S} = 8.76\left(\frac{\text{g cm}^{-3}}{\rho_S}\right)\frac{\text{mm a}^{-1}}{\text{g m}^{-2}\text{ h}^{-1}} \qquad (2\text{-}7)$$

These corrosion parameters have to be modified for time- and place-related reaction velocities [6]. Different local removal rates are in general due to differences in composition or nonuniform surface films, where both thermodynamic and

Table 2-1 Conversion factors and standard potentials for electrochemical metal-metal ion reactions

Reaction from Eq. (2-21)	f_a Eq. (2-5) $\dfrac{\text{mm a}^{-1}}{\text{mA cm}^{-2}}$	f_b Eq. (2-6) $\dfrac{\text{g m}^{-2}\text{ h}^{-1}}{\text{mA cm}^{-2}}$	f_c Eq. (2-7) $\dfrac{\text{mm a}^{-1}}{\text{g m}^{-2}\text{ h}^{-1}}$	U°_H (25°C) Eq. (2-29) V
$Ag = Ag^+ + e^-$	33.6	40.2	0.83	+0.80
$2\,Hg = Hg^{2+} + 2\,e^-$	–	–	–	+0.80
$Cu = Cu^{2+} + 2\,e^-$	11.6	11.9	0.98	+0.34
$Pb = Pb^{2+} + 2\,e^-$	29.8	38.6	0.77	–0.13
$Mo = Mo^{3+} + 3\,e^-$	10.2	11.9	0.86	–0.20
$Ni = Ni^{2+} + 2\,e^-$	10.8	11.0	0.98	–0.24
$Tl = Tl^+ + e^-$	56.4	76.2	0.74	–0.34
$Cd = Cd^{2+} + e^-$	21.2	21.0	1.01	–0.40
$Fe = Fe^{2+} + 2\,e^-$	11.6	10.4	1.12	–0.44
$Cr = Cr^{3+} + 3\,e^-$	8.2	6.5	1.27	–0.74
$Zn = Zn^{2+} + 2\,e^-$	15.0	12.2	1.23	–0.76
$Cr = Cr^{2+} + 2\,e^-$	12.3	9.7	1.27	–0.91
$Mn = Mn^{2+} + 2\,e^-$	12.5	10.2	1.22	–1.18
$Al = Al^{3+} + 3\,e^-$	10.9	3.35	3.24	–1.66
$Mg = Mg^{2+} + 2\,e^-$	22.8	4.54	5.03	–2.38

kinetic effects have an influence. The tendency of metallic materials to local corrosion can be characterized as follows:

(a) uniform weight loss occurs mostly on active, almost single-phase metals;
(b) shallow pitting and pitting in general are only possible in the presence of surface films, particularly on passive metals;
(c) selective corrosion is only possible on multiphase alloys.

There are no films or protective surface films on active metals, e.g., mild steel in acid or saline solutions. Passive metals are protected by dense, less readily soluble surface films (see Section 2.3.1.2). These include, for example, high-alloy Cr steels and NiCr alloys as well as Al and Ti in neutral solutions. Selective corrosion of alloys is largely a result of local concentration differences of alloying elements which are important for corrosion resistance e.g., Cr [4].

Fundamentals of Corrosion and Electrochemical Corrosion Protection

The passage of electrons from the metal to the electrolyte is not directly related to metal removal, but has an indirect connection due to the electron neutrality law:

$$I_A = I_C \tag{2-8}$$

Electrons cannot be dissolved in aqueous solutions but react with oxidants in the following way:

$$Ox^{n+} + ze^- = Red^{(n-z)+} \tag{2-9}$$

Ox and Red are general symbols for oxidation and reduction media respectively, and n and $(n-z)$ indicate their numerical charge (see Section 2.2.2). Where there is no electrochemical redox reaction [Eq. (2-9)], the corrosion rate according to Eq. (2-4) is zero because of Eq. (2-8). This is roughly the case with passive metals whose surface films are electrical insulators (e.g., Al and Ti). Equation (2-8) does not take into account the possibility of electrons being diverted through a conductor. In this case the equilibrium

$$I = I_A - I_C \tag{2-10}$$

is valid instead of Eq. (2-8).

The current I is called the total current. In free corrosion, i.e., without the contribution of external currents (see Fig. 2-1), it is always zero, as given by Eq. (2-8). I_A and I_C are known as the anodic and cathodic partial currents. According to Eq. (2-10), generally in electrolytic corrosion anodic total currents and/or cathodic redox reactions are responsible.

All metallic materials can suffer electrolytic corrosion. Fractures caused by cathodic hydrogen only occur when the activity of the absorbed hydrogen and the level of the tensile stress, which can be external or internal, reach a critical value. In general, critical hydrogen absorption is achieved only in the presence of promoters. However, under very severe conditions such as at very low pH or very negative potential, critical hydrogen absorption can occur. Steels with a hardness greater than HV 350 are particularly susceptible.

Materials consisting of elements of subgroups 4 and 5 of the periodic table are prone to the formation of internal hydrides, leading to severe embrittlement and fracture. Titanium and zirconium are important examples. Materials consisting of elements in the main groups 4 to 6 of the periodic table suffer weight loss by corrosion due to the formation of volatile hydrides [7]. A typical example is lead. Types of corrosion arising from cathodic hydrogen can limit the application of cathodic protection and are dealt with in Refs. 8 and 9.

2.2.2 Aqueous Electrolytes

Anions and cations exist in water. They migrate in an electric field and thus carry a current. Ohm's Law is applicable:

$$\vec{J} = \varkappa \vec{E} = -\varkappa \operatorname{grad} \phi \qquad (2\text{-}11)$$

where $\varkappa$ is the specific conductivity and E is the electric field strength. In dilute electrolytes, the conductivity is the sum of the ion mobilities l_i:

$$\varkappa = \sum_i l_i \, c_i \, |z_i| \qquad (2\text{-}12)$$

The index i represents the type of ion and c is its concentration. In water, the ions have velocity wY_i, giving the relation:

$$\frac{l_i}{\mathscr{F}} = \left|\frac{\vec{w}_i}{\vec{E}}\right| = u_i \qquad (2\text{-}13)$$

The quotient u_i is called the electrochemical mobility and is tabulated along with ion mobilities. It is important to pay attention to the units because of possible confusion. Values of l_i are given in Table 2-2. Raising the temperature usually increases ion mobility, while increasing the concentration reduces the conductivity due to interactions:

$$l_i(c_i) = l_i(c_i = 0) - C\sqrt{c} \qquad (2\text{-}14)$$

Electrical conductivity is of interest in corrosion processes in cell formation (see Section 2.2.4.2), in stray currents, and in electrochemical protection methods. Conductivity is increased by dissolved salts even though they do not take part in the corrosion process. Similarly, the corrosion rate of carbon steels in brine, which is influenced by oxygen content according to Eq. (2-9), is not affected by the salt concentration [4]. Nevertheless, dissolved salts have a strong indirect influence on many local corrosion processes. For instance, chloride ions that accumulate at local anodes can stimulate dissolution of iron and prevent the formation of a film. Alkali ions are usually regarded as completely harmless, but as counterions to OH$^-$ ions in cathodic regions, they result in very high pH values and aid formation of films (see Section 2.2.4.2 and Chapter 4).

The pH value and thus the OH$^-$ ion concentration is important in the formation of surface films, since OH$^-$ ions generally form difficultly soluble compounds with metal ions (see Section 2.2.3.1). pH is an important parameter of the medium. One

Table 2-2 Ion mobilities l_i in S cm² mol⁻¹ for calculating specific conductivity with Eq. (2-12); between 10 and 25°C, conductivity increases between 2 and 3% per °C

$$\frac{l_i}{\text{S cm}^2 \text{ mol}^{-1}} = 96{,}487 \frac{u_i}{\text{V}^{-1} \text{ s}^{-1} \text{ cm}^2}$$

Cation	+z	l_i at 25°C	l_i at 100°C	Anion	−z	l_i at 25°C	l_i at 100°C
H_3O^+	1	350	637	OH^-	1	200	446
Na^+	1	50	150	Cl^-	1	76	207
K^+	1	73	200	NO_3^-	1	71	189
Mg^{2+}	2	53	170	HCO_2^-	1	44	–
Ca^{2+}	2	59	187	CO_3^{2-}	2	72	–
Fe^{2+}	2	53	–	SO_4^{2-}	2	79	256
Cu^{2+}	2	56	–				

has to remember, however, that considerable changes in the pH value can occur as a result of subsequent reactions on the metal surface. Generally the equilibrium is:

$$c(H^+)c(OH^-) = K_W; \quad K_W(25°C) = 10^{-14} \text{ mol}^2 \text{ L}^{-2} \tag{2-15}$$

$$\text{pH} = -\log\left[\frac{c(H^+)}{\text{mol L}^{-1}}\right] = -\log\left(\frac{K_W}{\text{mol}^2 \text{ L}^{-2}}\right) + \log\left[\frac{c(OH^-)}{\text{mol L}^{-1}}\right] \tag{2-16}$$

(Deviations from the ideal behavior that could be accounted for by the use of activity coefficients are neglected here.)

The concentration of oxidizing agents is essential for the course of reactions involving Eq. (2-9). These can be divided into two groups according to the type of oxidizing agent:

(a) oxygen corrosion (in all media)

$$O_2 + 2 H_2O + 4 e^- = 4 OH^- \tag{2-17}$$

(b) acid corrosion (particularly in strong acids)

$$2 H^+ + 2 e^- = 2 H \rightarrow H_2 \tag{2-18}$$

36 Handbook of Cathodic Corrosion Protection

The evolution of hydrogen from the acid molecule can also occur in slightly dissociated weak acids such as H_2CO_3 and H_2S. In the case of only slightly dissociated weak acids, such as H_2CO_3 and H_2S, production of hydrogen can also occur from the acid molecules. In this case, the acid concentration rather than the pH value is a measure of the aggressiveness of the corrosion. In the same way, hydrogen can be evolved from H_2O:

$$2\,H_2O + 2\,e^- = 2\,OH^- + 2\,H \rightarrow 2\,OH^- + H_2 \tag{2-19}$$

This reaction occurs with overall cathodic currents, i.e., with cathodic polarization. It can be practically ignored in the case of free corrosion of steel in a neutral solution. Other oxidizing media are of interest only in special cases.

In some cases the amount of gas evolved or consumed according to Eqs. (2-17) to (2-19) can be asked for. Then Eq. (2-6) is applicable. To obtain the volume, the following applies:

$$j_v = \frac{V}{St} = f_v\, J;\ f_v = \frac{\mathbf{V}}{z\mathcal{F}} = \frac{0.373}{z}\left(\frac{\mathbf{V}}{L\,mol^{-1}}\right) \frac{L\,m^{-2}h^{-1}}{mA\,cm^{-2}} \tag{2-20}$$

where j_v is the rate of evolution or consumption of gas; V is the volume of gas and $\mathbf{V}$ is the mol volume under standard conditions. For $\mathbf{V} = 22.4\,L$ it follows that:

$$f_v = \frac{8.36}{z}\,\frac{L\,m^{-2}h^{-1}}{mA\,cm^{-2}} \tag{2-20'}$$

Some data are given in Table 2-3.

2.2.3 Electrochemical Phase Boundary Reactions

In electrolytic corrosion, an anodic partial reaction takes place according to Eq. (2-3)

$$Me = Me^{z+} + z\,e^- \tag{2-21}$$

and a cathodic redox reaction according to Eq. (2-9) (see Fig. 2-1). The reaction rates can be expressed in general by using Eq. (2-6) with equivalent currents I_A and I_C. They are a function of the partners to the reaction and the potential U. For every partial reaction there is an equilibrium potential U^* in which the overall reaction is zero. The following section deals with the thermodynamic and kinetic fundamentals of these reactions.

Table 2-3 Conversion factors and standard potentials for electrochemical redox reactions

Reaction according to Eq. (2-9)	U_H° (25°C) Eq. (2-30) V	f_b Eq. (2-6) $\frac{g\ m^{-2}\ h^{-1}}{mA\ cm^{-2}}$	f_v Eq. (2-20) $\frac{L\ m^{-2}\ h^{-1}}{mA\ cm^{-2}}$
$2\ H^+ + 2\ e^- = H_2$	0.00	0.37	4.18
$O_2 + 2\ H_2O + 4\ e^- = 4\ OH^-$	+0.40	2.98	2.09
$Cl_2 + 2\ e^- = 2\ Cl^-$	+1.36	13.24	4.18
$Cr^{2+} + e^- = Cr^+$	−0.41		
$Cu^{2+} + e^- = Cu^+$	+0.16		
$Fe^{2+} + e^- = Fe^{2+}$	+0.77		

2.2.3.1 Basic Thermodynamics

The driving force for the transport of all particles is a change in the electrochemical potential $\tilde{\mu}_i$ which is related to the partial molar free enthalpy μ_i and the electric potential ϕ as follows:

$$\tilde{\mu}_i = \mu_i + z_i \mathcal{F} \phi \qquad (2\text{-}22)$$

For a homogeneous conductor and in the migration direction [see Eq. (3-1)]:

$$\vec{w}_i = -B\ \mathrm{grad}\ \tilde{\mu}_i = B\left(-\mathrm{grad}\ \mu_i + z_i \mathcal{F} \vec{E}\right) \qquad (2\text{-}23)$$

The factor $B = D/RT$ is the mobility and contains the diffusion coefficient D, the gas constant R, and the absolute temperature T. The equation includes a diffusion and a migration term. Correspondingly Eq. (2-23) gives the first diffusion law for $z_i = 0$ and Ohm's Law for $\mathrm{grad}\ \mu_i = 0$. For transfer across a phase boundary:

$$w_i = B(\Delta\mu_i + z_i \mathcal{F} \Delta\phi) \qquad (2\text{-}24)$$

Finally if $w_i = 0$, the equilibrium is given by:

$$0 = \Delta\tilde{\mu}_i^* = \Delta\mu_i^* + z_i \mathcal{F} \Delta\phi^* \qquad (2\text{-}25)$$

Application of Eq. (2-25) to the reaction under consideration, i.e., Eq. (2-21), and to the potential-determining reaction of the reference electrode, Eq. (2-18), leads with Eq. (2-1) to the Nernst potential equation:

$$-U^* = \Delta\phi^* - \Delta\phi_B = -\frac{\Delta\mu_i^* - \Delta\mu_B}{z_i \mathscr{F}} = -\frac{\Delta G}{z_i \mathscr{F}} \tag{2-26}$$

In this example, ΔG is the free reaction enthalpy of the chemical reaction

$$\mathrm{Me} + z\mathrm{H}^+ \rightarrow \mathrm{Me}^{z+} + \frac{z}{2}\mathrm{H}_2 \tag{2-27}$$

which corresponds to adding the electrochemical reactions in Eq. (2-18) and (2-21). The negative sign of U^* accounts for the fact that all $\Delta\phi$ contain potential differences in the reaction direction of Eq. (2-27) in the cell H_2/electrolyte/metal and ΔG is appropriately defined [10]. From the concentration dependence of μ_i it follows that

$$\mu_i = \mu_i^0 + RT \ln \frac{c_i}{\mathrm{mol\ L}^{-1}} \tag{2-28}$$

and for the standard state of the hydrogen electrode with a variable metal ion concentration $c(\mathrm{Me}^{z+})$, the equilibrium potential against the standard hydrogen electrode is

$$\begin{aligned}
U_\mathrm{H}^* &= \frac{\Delta G}{z\mathscr{F}} = \frac{\Delta G°}{z\mathscr{F}} + \frac{RT}{z\mathscr{F}} \ln \left[c\frac{(\mathrm{Me}^{z+})}{\mathrm{mol\ L}^{-1}} \right] \\
&= U_\mathrm{H}^o + \frac{RT}{z\mathscr{F}} \ln \left[\frac{c(\mathrm{Me}^{z+})}{\mathrm{mol\ L}^{-1}} \right]
\end{aligned} \tag{2-29}$$

and U_H^o is the standard potential which can be calculated from the free enthalpy of formation $\Delta G°$. Table 2-1 shows the more important values. The factor $RT/\mathscr{F}$ is 26 mV at 25°C. In the same way, a potential equilibrium can be derived for a simple redox reaction [Eq. (2-9)]:

$$U_\mathrm{H}^* = U_\mathrm{H}^o + \frac{RT}{z\mathscr{F}} \ln \frac{c_{\mathrm{Ox}}}{c_{\mathrm{Red}}} \tag{2-30}$$

Many redox reactions are more complicated than that given by Eq. (2-9). For a general redox reaction, with components X_i and their coefficients n_i written as

$$\sum_i n_i X_i = e^- \qquad (2\text{-}31)$$

the relation can be derived [4,10]:

$$U_H^* = U^\circ - \frac{RT}{z\mathscr{F}} \sum_i n_i \ln\left[\frac{c(X_i)}{\text{mol L}^{-1}}\right] \qquad (2\text{-}32)$$

A comprehensive list of standard potentials is found in Ref. 7. Table 2-3 gives a few values for redox reactions. Since most metal ions react with OH⁻ ions to form solid corrosion products giving protective surface films, it is appropriate to represent the corrosion behavior of metals in aqueous solutions in terms of pH and U_H. Figure 2-2 shows a Pourbaix diagram for the system Fe/H₂O. The boundary lines correspond to the equilibria:

line (1): Fe/Fe²⁺ corresponding to Eq. (2-21),
line (2): Fe/Fe(OH)₂ from Fe + 2 H₂O = Fe(OH)₂ + 2 H⁺ + 2 e⁻,
line (3): Fe²⁺/FeOOH from Fe²⁺ + 2 H₂O = FeOOH + 3 H⁺ + e⁻,
line (4): Fe³⁺/FeOOH from Fe³⁺ + 2 H₂O = FeOOH + 3 H⁺,
line (a): OH⁻/O₂ corresponding to Eq. (2-17),
line (b): H₂/H⁺ corresponding to Eq. (2-18).

Electrolytic corrosion occurs in regions I and IV with the formation of soluble iron ions. Solid corrosion products which can have a protective effect are formed in region II. This is the region of surface film passivity. Certain corrosive sub-

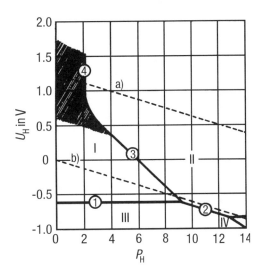

Fig. 2-2 Simplified potential-pH diagram for an iron/aqueous electrolyte system at 25°C; $c(\text{Fe}^{2+}) + c(\text{Fe}^{3+}) = 10^{-6}$ mol L⁻¹ (explanation in the text).

stances in the medium (e.g., chloride ions) and mechanical effects can destroy surface films locally, leading to intensive local corrosion such as pitting and stress corrosion. On the other hand, in certain passivating acids, such as HNO_3, H_2SO_4, and H_3PO_4, there is an area in region I (shown hatched), Fig. 2-2, where the material is covered with an extremely thin nonequilibrium film. This chemical passivity is not technically different from surface film passivity. For both cases, corrosion rates are extremely small but not zero, as in region III where the metal is thermodynamically stable. In addition, there is the latent danger of local corrosion [4].

Such pH–U_H diagrams are available for all metals [7]. They give an overview of corrosion behavior and the electrochemical protection that is possible when the potential is changed by means of impressed currents. Cathodic currents are required to reduce the potential in region III. Anodic currents are needed to raise the potential in region II or the hatched area in region I. This is the basis of cathodic and anodic protection. The region of H_2O stability between the straight lines (a) and (b) must be considered before a preliminary judgment can be made. Outside these lines, the possibility of changing the potential is limited by the electrolytic dissociation of water. From Fig. 2-2 it follows that cathodic protection in acid solutions is not practically possible but on the other hand anodic protection probably is possible.

Regions for soluble hydroxy complexes of the type $Fe(OH)^+$ are neglected in Fig. 2-2 and in corresponding diagrams in Ref. 7; compare the amended diagrams in the literature [11,12]. The regions in the pH-potential diagrams can be transposed to some extent by complex formation. This must be taken into account when dealing with unknown chemical solutions. In addition, when using the pH-potential diagrams in Ref. 7, care must be taken that, although regions of weight loss due to hydride formation are represented, regions where internal hydride formation occurs in metals of the 4 and 5 subgroups are not given.

2.2.3.2 Electrochemical Kinetics

The potential dependence of the velocity of an electrochemical phase boundary reaction is represented by a current-potential curve $I(U)$. It is convenient to relate such curves to the geometric electrode surface area S, i.e., to present them as current-density-potential curves $J(U)$. The determination of such curves is represented schematically in Fig. 2-3. A current is conducted to the counterelectrode E_3 in the electrolyte by means of an external circuit (voltage source U_0, ammeter, resistances R' and R'') and via the electrode E_1 to be measured, back to the external circuit. In the diagram, the current indicated ($\oplus$) is positive. The potential of E_1 is measured with a high-resistance voltmeter as the voltage difference of electrodes E_1 and E_2. To accomplish this, the reference electrode, E_2, must be equipped with a Haber-Luggin capillary whose probe end must be brought as close as possible to

Fundamentals of Corrosion and Electrochemical Corrosion Protection

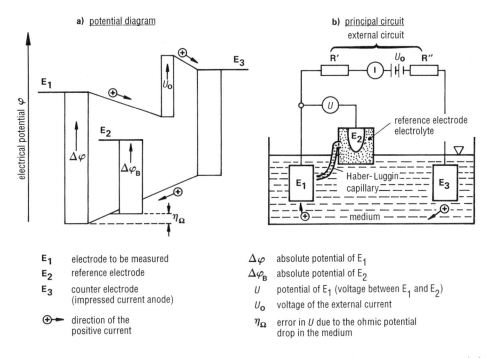

Fig. 2-3 Potential scheme (a) and circuit (b) for measuring a current-potential curve in cathodic polarization (explanation in the text).

the surface E_1 without shielding the current flow [13]. The potential diagram shows that the current in the electrolyte produces a voltage drop η_Ω that basically always results in an error in the potential measurement:

$$U = (\Delta\phi - \Delta\phi_B) - \eta_\Omega \tag{2-33}$$

By comparison with Eq. (2-1) the measured value in Fig. 2-3 is too negative by η_Ω according to Eq. (2-33) and correspondingly is too positive in the case of the anodic current. The error can be calculated for uniform current flow lines from Ohm's Law:

$$\eta_\Omega = J\frac{s}{\varkappa} \tag{2-34}$$

where s is the distance of the probe from the electrode surface. In the laboratory, potential measuring probes can be used and η_Ω from Eq. (2-34) can be kept very small. However, generally this is not possible for technical structures, and particularly not for buried objects. Possible ways to eliminate η_Ω errors (i.e., by *IR*-free potential measurements) are described in Section 3.3.1.

A simplifying assumption is made that only one electrochemical reaction occurs at E_1. Then the equilibrium potential U^* is present at $I = 0$. Positive (or negative) currents can only flow with positive (or negative) deviations of U^*. The difference $(U - U^*) = \eta$ is termed the overvoltage. The function $J(\eta)$ gives information on kinetics of the reaction and on the rate-determining step. If transport through the phase boundary itself is rate determining, then $J(\eta)$ is an exponential function (activation polarization). For this reason $J(U)$ curves are mostly plotted on a semilogarithmic scale. On the other hand, if a chemical reaction or diffusion in the medium is rate determining, then J is independent of potential, i.e., the curve $J(\eta)$ ends parallel to the potential axis (concentration overvoltage). A similar case can arise if less readily soluble surface films are formed which in the stationary state have a solubility rate equivalent to J. This is the case with passive metals (see Section 2.3.1.2). With poorly conducting surface films or in high-resistance media, the ohmic resistance controls the current so that $J(\eta)$ follows Ohm's Law. In this case η is not a genuine overvoltage, but corresponds to η_Ω in Eq. (2-34) and is thus in principle a measurement error, given that the potential is defined for the interface between the metal and film.

The literature [14] on electrochemical kinetics is extensive and specialized. Figure 2-4 shows a $J(\eta)$ curve of a redox reaction according to Eq. (2-9) with activation and diffusion polarization. It follows from theory [4, 10] for this example:

$$J = \frac{\exp\left(\dfrac{\eta}{\beta_+}\right) - \exp\left(\dfrac{-\eta}{\beta_-}\right)}{\dfrac{1}{J_0} + \dfrac{1}{G_A}\exp\left(\dfrac{\eta}{\beta_+}\right) + \dfrac{1}{G_C}\exp\left(\dfrac{-\eta}{\beta_-}\right)} \qquad (2\text{-}35)$$

where J_0 is exchange current density, and corresponds to the magnitude of the equally fast forward and reverse reactions in the equilibrium; G_A and G_C are the limiting diffusion current densities and are proportional to the concentration of the reactants concerned and increase according to the first law of diffusion with the flow velocity; and β_+ and β_- are the anodic and cathodic Tafel slopes (see definition in list of symbols) and are given by:

$$\beta_+ = \frac{RT}{\alpha z \mathcal{F}}; \quad \beta_- = \frac{RT}{(1-\alpha)z\mathcal{F}}; \quad \frac{1}{\beta_+} + \frac{1}{\beta_-} = \frac{z\mathcal{F}}{RT} \qquad (2\text{-}36)$$

The term α is a symmetry factor for the energy threshold for the passage of electrons and is approximately equal to 0.5. In Fig. 2-4, the value of α was chosen as $\tfrac{2}{3}$ for better distinction; integer exponents are chosen for J_0, G_A and G_C for clarity,

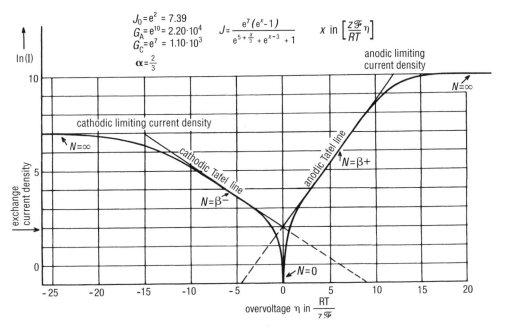

Fig. 2-4 Current-density-potential curves for an electrochemical partial reaction as in Eq. (2-35).

and η is plotted dimensionless. The slopes of the curves in semilogarithmic plots (natural logarithms)

$$N = \frac{\Delta \eta}{\Delta \ln |J|} \tag{2-37}$$

are given in the regions of Tafel lines β_- and β_+. The Tafel slopes in logs to the base 10 are obtained by multiplying by the factor $\ln 10 = 2.303$ for the system b_- and b_+. At equilibrium we obtain $N \to 0$ and in the limiting current region, $N \to \infty$. There are similar curves for electrolytic corrosion according to Eq. (2-21) as those in Fig. 2-4. An anodic limiting current is only possible with film formation.

In general, $J(U)$ curves for the anodic partial reaction follow a Tafel straight line. In neutral media, the cathodic partial reaction is mostly the reduction of oxygen, whose $J(U)$ curve ends in a limiting cathodic current density determined by the transport of oxygen. If at high overvoltages cathodic hydrogen evolves according to Eq. (2-19), the $J(U)$ curve bends with another constant slope.

The time dependence of the changes in the measured values is important in determining $J(U)$ curves. In the region of the Tafel lines, stationary states are reached relatively quickly. A time constant can be approximately calculated from the product of the double layer capacity $C_D \approx 10$ to 100 μF cm^{-2} and the polarization resistance $r_p = \Delta U/\Delta J \approx 1$ to 1000 Ω cm^2. From these volumes the time constant can be calculated to be in the range of 10^{-5} to 10^{-1} s. In contrast, diffusion and film formation are very time dependent. Stationary states are achieved very slowly in the region of limiting currents. This is very often the case in the field when surface films are present or have been developed.

2.2.4 Mixed Electrodes

In general, according to Eq. (2-10), two electrochemical reactions take place in electrolytic corrosion. In the experimental arrangement in Fig. 2-3, it is therefore not the $I(U)$ curve for one reaction that is being determined, but the total current-potential curve of the mixed electrode, E_I. Thus, according to Eq. (2-10), the total potential curve involves the superposition of both partial current-potential curves:

$$I(U) = I_A(U) + I_C(U) \tag{2-10'}$$

2.2.4.1 Homogeneous Mixed Electrodes

To simplify matters, it is assumed that the current densities for the partial reactions are independent of position on the electrode surface. Equation (2-10') can then be used to designate the current densities:

$$J(U) = J_A(U) + J_C(U) \tag{2-38}$$

Equation (2-38) is valid for every region of the surface. In this case only weight loss corrosion is possible and not localized corrosion. Figure 2-5 shows total and partial current densities of a mixed electrode. In free corrosion $J = 0$. The free corrosion potential U_R lies between the equilibrium potentials of the partial reactions U_A^* and U_C^*, and corresponds in this case to the rest potential. Deviations from the rest potential are called polarization voltage or polarization. At the rest potential $J_A = |J_C|$, which is the corrosion rate in free corrosion. With anodic polarization resulting from positive total current densities, the potential becomes more positive and the corrosion rate greater. This effect is known as anodic enhancement of corrosion. For a quantitative view, it is unfortunately often overlooked that neither the corrosion rate nor its increase corresponds to anodic total current density unless the cathodic partial current is negligibly small. Quantitative forecasts are possible only if the $J_C(U)$ curve is known.

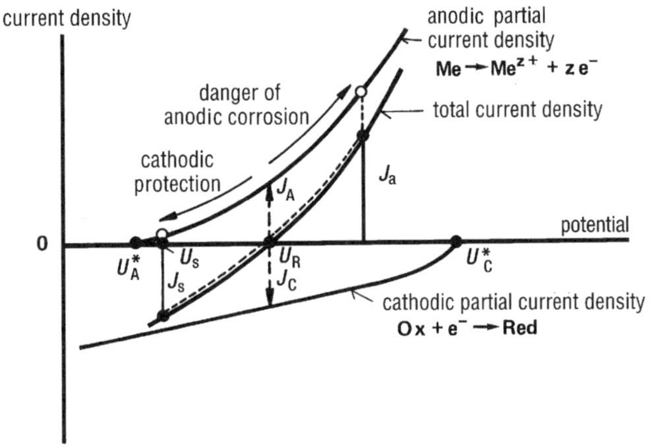

Fig. 2-5 Partial and total current densities in electrolytic corrosion of a homogeneous mixed electrode.

When cathodic polarization is a result of negative total current densities J_C, the potential becomes more negative and the corrosion rate lower. Finally, at the equilibrium potential U_A^* it becomes zero. In neutral water equilibrium potentials are undefined or not attainable. Instead, protective potentials U_S are quoted at which the corrosion rate is negligibly low. This is the case when $J_A \approx 1 \ \mu A \ cm^{-2} \ (w = 10 \ \mu m \ a^{-1})$ which is described by the following criteria for cathodic protection:

(a) potential criterion: $\quad U \leq U_s \quad$ (2-39)

(b) current criterion: $\quad |J_c| > |J_s| = J_c(U_s) \approx J_C(U_s) \quad$ (2-40)

where J_s is the lowest protection current density and corresponds to the overall total density at the protection potential U_s. In conjunction with Eq. (2-40), the following concepts are identified.

The terms "protection current" and "protection current densities" refer to any values of total cathodic currents that meet the criterion in Eq. (2-40). However, in the field, and for designing cathodic protection stations, another term is of interest, the protection current requirement. This term is concerned with the lowest value of the protection current that fulfills the criteria in Eqs. (2-39) or (2-40). Since with an extended object having a surface S the polarization varies locally, only the current density for the region with the most positive potential U_s has the value J_s. In other regions $|J_c| > J_s$. For this reason, the protection current requirement I_s is given by:

$$|I_s| > |J_s| S \quad (2-41)$$

where J_s is a constant for the system material/medium, whereas I_s is only defined for a given object.

46 Handbook of Cathodic Corrosion Protection

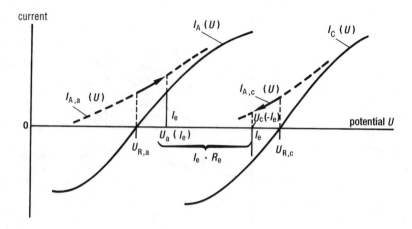

Fig. 2-6 Current-potential relationships for a heterogeneous mixed electrode or for cell formation (explanation in the text).

2.2.4.2 Heterogeneous Mixed Electrodes and Cell Formation

This is the general case where the current densities of the partial reactions vary over the electrode surface. Equation (2-10') and not Eq. (2-38) applies. As a simplification, a heterogeneous mixed electrode consisting of two homogeneous regions is considered in what follows. Figure 2-6 shows the total current-potential curves $I_a(U)$ and $I_c(U)$ (solid lines) and the relevant anodic partial current-potential curves $I_{A,a}(U)$ and $I_{A,c}(U)$ (dashed lines). The homogeneous regions have rest potentials $U_{R,a}$ and $U_{R,c}$. The index c represents the homogeneous region with the more positive rest potential because it represents the local cathode. The index a represents the corresponding quantities of the local anode. The metallic short circuit of the two homogeneous regions results in a heterogeneous mixed electrode and the cell current I_e flows as a result of the potential difference $(U_{R,c} - U_{R,a})$. In free corrosion, the potential of the local cathode is changed to $U_c(-I_e)$ and that of the local anode to $U_a(I_e)$ due to the internal polarization of the cell. The free corrosion potential of the heterogeneous mixed electrode is dependent on position. There is no rest potential because the local current densities are not zero. In the electrolyte the potential difference $(U_c - U_a)$ exists as the ohmic voltage drop $I_e \times R_e$. If the conductivity is sufficiently high, this difference can be very small so that a heterogeneous mixed electrode appears as a homogeneous electrode. The free corrosion potential that can be measured independent of position seems to be a rest potential of a homogeneous mixed electrode, but it is not.

There are no current-density-potential curves for mixed electrodes, only current-density-potential bands which can be represented in a three-dimensional J–U–x

Fundamentals of Corrosion and Electrochemical Corrosion Protection

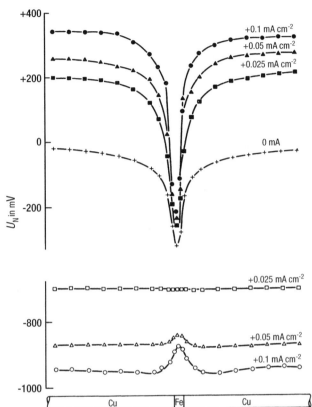

Fig. 2-7 Potential distribution curves for a cylindrical-shaped, mixed electrode of CuFeCu polarized in tapwater ($\varkappa \approx 10^{-3}$ S cm^{-1}). Spacing between the electrode and probe 1 mm.

diagram with x as the position coordinate [15]. Figure 2-7 shows as an example a U–x diagram for a cylindrical mixed electrode of copper with an iron ring in tap water. The parameter is the current related to the total area. In the case of free corrosion there exists a dip in the potential directly over the Fe anode. This dip becomes greater with increasing anodic polarization because the total current-potential curves of the homogeneous Cu and Fe regions become further apart at more positive currents. In contrast, the dip disappears with cathodic polarization. At high cathodic currents, it can even be reversed so that the Cu is at this stage more negative than iron. The reason for this is that the total current-potential curves of the homogeneous Cu and Fe regions intersect at negative currents. This unusual effect is known as potential reversal. Potential reversal is also possible with anodic polarization if, for example, as a result of differential film formation the total current-potential curves intersect at positive currents. This is possible, for example, for Fe and Zn in warm tapwater or in seawater [4]. Anodic potential reversal has to be taken into account in the cathodic protection of mixed installations (e.g., carbon steel and stainless steel) and sacrificial anodes should be used in order not to damage the less noble component of the object being protected.

Local corrosion is generally the result of the formation of heterogeneous mixed electrodes where the change in the local partial current-density-potential curves can result from the material or the medium. Where this is caused by the contact of different metals, it is known as a galvanic cell (see Fig. 2-7) [16,17]. Local differences in the composition of the medium result in concentration cells. These include the differential aeration cell which is characterized by subsequent chemical reactions that stabilize differences in pH values; chloride and alkali ions are important here [4]. Such corrosion cells can have very different sizes. In selective corrosion of multiphase alloys, anodes and cathodes can be separated by fractions of a millimeter. With objects having a large surface area (e.g., pipelines) the regions can cover several kilometers. It is immaterial whether the cathodic region is still part of the pipeline or an external component of the installation that is electrically connected to the pipeline. In the latter case this is referred to as a cathodic foreign structure. Examples of this are electrical grounds and reinforcing steel in concrete (see Section 4.3).

The ratio of the areas of cathodes to anodes is decisive for the potential damage resulting from cell formation [16,17]. Using the integral (mean) polarization resistances

$$R_a = \frac{U_a(I_e) - U_{R,a}}{I_e} \quad \text{and} \quad R_c = \frac{U_c(-I_e) - U_{R,c}}{-I_e} \tag{2-42}$$

$$I_e = \frac{U_{R,c} - U_{R,a}}{R_e + R_a + R_c} \tag{2-43}$$

The difference in rest potentials (see the practical potential series in Table 2-4) determines mostly the direction of the current and less of the level; for these the resistances are significant. In particular R_e can be neglected in the external corrosion of extended objects. In addition, the $I_a(U)$ curve is usually steeper than the $I_c(U)$ curve (i.e., $R_a \ll R_c$). By introducing the surface areas of anode and cathode S_a and S_c, it follows from Eq. (2-43) that:

$$J_a = \frac{I_e}{S_a} \approx \frac{U_{R,c} - U_{R,a}}{r_c} \times \frac{S_c}{S_a} \tag{2-44}$$

where $r_c = R_c \times S_c$ is the specific cathodic polarization resistance. At high cell currents, the cathodic partial current at the anode can be neglected, so that $J_{A,a} \approx J_a$ in

Fundamentals of Corrosion and Electrochemical Corrosion Protection

Table 2-4 Practical potential series

Metal	(a)	(b)
Titanium	(+181)	(–111)
Brass (SoMs) 70	+153	+28
Monel	(+148)	+12
Brass (Ms) 63	+145	+13
Copper	+140	+10
Nickel 99.6	+118	+46
CrNi steel 1.4301	(–84)	–45
AlMgSi	(–124)	–785
Aluminum 99.5	(–169)	–667
Hard Cr plating	(–249)	–291
Tin 98	(–275)	–809
Lead 99.9	(–283)	–259
Steel	–350	–335
Cadmium (anodes)	–574	–519
Electrolytic zinc coating	–794	–806
Zinc 98.5	–823	–284
Electron (AM) 503	–1460	–1355

Note: Rest potentials U_H in mV for common metals in (a) phthalate buffer at pH 6 and (b) artificial seawater [18], 25°C, air saturated and stirred. The rest potentials of the values in parentheses tend to become more positive with time due to film formation. (The values are dependent on the medium and operating conditions.)

Fig. 2-5. Equation (2-44) indicates that the activity of the cathode as well as the surface ratio and the potential difference also have an influence.

Sometimes difficulties arise because the cathode of the cell has a more positive potential than the anode (see Fig. 2-6). This is because the definition of anode and cathode is based on processes in the electrolyte, whereas potential measurement is based on events on the metal. This fact is illustrated in Fig. 2-8. If electrodes Pt and Fe are both in the same electrolyte with potential ϕ_{El}, then from the metal point of view, Pt is more positive than Fe. U is the electromagnetic force (emf) of the cell. When the switch S is closed, electrons flow from Fe(–) to Pt(+). If both electrodes are initially connected and immersed in separate electrolytes, they both have the potential ϕ_{Met}. The electrolyte at Fe is now more positive than that at Pt. The voltage U can be measured between two reference electrodes in the electrolytes. When the tap H is opened, a positive current flows in the electrolyte from Fe(+) to Pt(–).

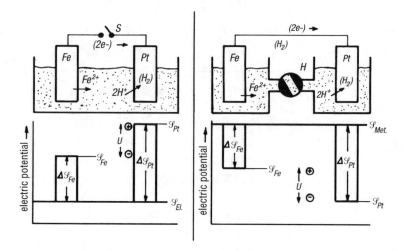

Fig. 2-8 Principle and potential diagram of a galvanic cell (explanation in the text).

This last process is the basis for the definition of anode and cathode. When current is flowing, the anode (Fe) is more negative on the metal side, but in the electrolyte it is more positive than the cathode (Pt).

2.2.5 Observations of Current Distribution

The distribution of current is of considerable interest in corrosion processes on heterogeneous mixed electrodes, particularly in the internal corrosion of tanks and complex shapes as well as generally in the application of electrochemical protection. A primary current distribution can be obtained from the laws of electrostatics by integration of the Laplace equation (div grad $\phi = 0$) [10,19]. Polarization resistances at the electrodes are neglected in this treatment. Current distribution is exclusively related to the geometry. When polarization resistance is taken into account, secondary and tertiary current distributions can be distinguished in which activation polarization alone, or together with concentration polarization, has an effect [20]. This is relevant in, for example, electroplating where uniform metal deposition is required. Current distribution is more uniform as a result of polarization resistance than in primary current distribution [4,10,19,20]. A polarization parameter

$$k = r_p / \varkappa \tag{2-45}$$

is introduced for comparability in which the ratio of polarization resistance to electrolyte resistance is taken into account. It corresponds to the thickness of a layer of electrolyte, whose ohmic resistance is equal to the polarization resistance. If in the primary current distribution this thickness is accounted for in the geometry, a better current distribution is obtained. The polarization parameter is also a measure of the range (see Section 24.5).

In electrochemical protection the necessary range of protection current is achieved by an appropriate arrangement of the electrodes. It follows that measures which raise the polarization resistance are beneficial. Coated objects have a coating resistance (see Section 5.2), which can be utilized in much the same way as the polarization resistance in Eq. (2-45). Therefore, the range in the medium can be extended almost at will by coatings for extended objects, even at low conductivity. However, the range is then limited by current supply to the object to be protected (see Section 24.4).

In many practical cases the question arises whether geometric hindrances may prevent sufficient protection current from being successfully supplied to the metal surface. Current could be shielded by stones, crevices, and in particular by poorly adhering tapes or disbonded coatings (see Section 5.2.4). Geometrically controlled resistance for the protection current exists in equal measure also for current from corrosion cells, for stray current, and for access of oxidizing agents to the cathode reaction according to Eq. (2-9). The current densities for electric conduction and for diffusion are given by the similar Eqs. (2-11) and

$$-J_C = -z\mathscr{F} D \operatorname{grad} c_{Ox} \tag{2-46}$$

A relation independent of geometry is given by the quotient:

$$\frac{-J_C}{J} = \frac{z\mathscr{F} D \Delta c_{Ox}}{\varkappa \Delta \phi} \tag{2-47}$$

For oxygen-controlled corrosion with $z = 4$, $D = 1$ cm^2 d^{-1}, and $\Delta c(O_2) = -10$ mg L^{-1}, it becomes:

$$\frac{-J_C}{J} = 1.4 \times 10^{-6} \left(\frac{\text{S cm}^{-1}}{\varkappa}\right)\left(\frac{\text{V}}{\Delta \phi}\right) \tag{2-47'}$$

Even for high-resistance media with $\varkappa = 10^{-4}$ S cm^{-1}, sufficient protection is obtained with only $\Delta\phi = 0.1$ V from the criterion of Eq. (2-40): $J_s \approx -J_c = 0.14\,J$ (i.e., the current supplied, J, is seven times greater than the current needed, J_s). Equation (2-47) applies exclusively to diffusion of the oxidizing component in stagnant medium and not for other possible types of transport, e.g., flow or aeration from the gas phase. Narrow crevices filled with stagnant water are less serious than stones that screen current (see Section 24.4.5).

2.3 Potential Dependence of Corrosion Extent

The principle of electrochemical corrosion protection processes is illustrated in Figs. 2-2 and 2-5. The necessary requirement for the protection process is the existence of a potential range in which corrosion reactions either do not occur or occur only at negligibly low rates. Unfortunately, it cannot be assumed that such a range always exists in electrochemical corrosion, since potential ranges for different types of corrosion overlap and because in addition theoretical protection ranges cannot be attained due to simultaneous disrupting reactions.

To discover the effective potential ranges for electrochemical protection, the dependence of the relevant corrosion quantities on the potential is ascertained in the laboratory. These include not only weight loss, but also the number and depth of pits, the penetration rate in selective corrosion, and service life as well as crack growth rate in mechanically stressed specimens, etc. Section 2.4 contains a summarized survey of the potential ranges for different systems and types of corrosion. Four groups can be distinguished:

I. The protection range lies at more negative potentials than the protection potential and is not limited:

$$U \leq U_s \tag{2-39}$$

II. The protection range lies at more negative potentials than the protection potential U_s, and is limited by a critical potential U_s':

$$U_s' \leq U \leq U_s \tag{2-48}$$

III. The protection range lies at more positive potentials than the protection potential and is not limited:

$$U \geq U_s \tag{2-49}$$

IV. The protection range lies at more positive potentials than the protection potential U_s and is limited by U_s'':

$$U_s \leq U \leq U_s'' \tag{2-50}$$

Cases I and II are examples of protection by cathodic polarization and III and IV of protection by anodic polarization. In cases I and III, the protection current can be uncontrolled, while in cases II and IV the potential has to be controlled. The development of operationally safe and widely available rectifiers (see Chapter 21)

Fundamentals of Corrosion and Electrochemical Corrosion Protection 53

has made it possible to meet the electrochemical protection requirements of many types of applications. The potential dependence of corrosion quantities for some typical systems and types of corrosion is described in the following section.

2.3.1 Almost Uniform Weight Loss Corrosion

Completely uniform removal of material occurs very seldom. In most cases it is only uniform on average over about 1 cm² in which irregularities with local depths and widths of up to about 1 mm can occur. It is practical to classify such systems as homogeneous although in these regions heterogeneous states can occur (see Fig. 4-3b). It is convenient to distinguish between weight loss corrosion of active metals in which no protective films are present, and that of passive metals.

2.3.1.1 Weight Loss Corrosion of Active Metals

In this type of corrosion, metal ions arising as a result of the process in Eq. (2-21) migrate into the medium. Solid corrosion products formed in subsequent reactions have little effect on the corrosion rate. The anodic partial current-density-potential curve is a constant straight line (see Fig. 2.4).

The protection potential can be evaluated kinetically for such cases [10,21]. It is assumed that the concentration of metal ions on the metal surface is c_0. The weight loss rate follows from the first law of diffusion:

$$v = \frac{D}{\delta} c_0 \quad (2\text{-}51)$$

and the relevant equilibrium potential from Eq. (2-29):

$$U_A^* = U° + \frac{RT}{z\mathscr{F}} \ln\left(\frac{c_0}{\text{mol L}^{-1}}\right) \quad (2\text{-}52)$$

If U_A^* is assumed to be the protection potential U_s, the velocity of the anodic partial reaction according to Eq. (2-21) is so fast that the concentration c_0 is maintained. At the same time, Eq. (2-51) represents the highest value for c_0. Therefore it follows from Eqs. (2-51) and (2-52) that:

$$U_S = U° + \frac{RT}{z\mathscr{F}} \ln\left(\frac{v\delta}{D} \times 10^3 \text{ cm}^3 \text{ mol}^{-1}\right) \quad (2\text{-}53)$$

with $RT/\mathscr{F} = 26$ mV, $D = 10^{-5}$ cm² s⁻¹, $\delta = 10^{-3}$ cm for rapid flow and assuming a low corrosion rate, $v = 10^{-12}$ mol cm⁻², s⁻¹ = 3.6 × 10⁻⁵ mol m⁻² h⁻¹ (for Fe: 2 mg m⁻² h⁻¹ ≃ 2 μa⁻¹). Equation (2-53) simplifies to:

$$U_s = U° - \frac{0.4 \text{ V}}{z} \tag{2-53'}$$

For Fe it follows from Eq. (2-53′) that $U_{Hs} = -0.64$ V. This value is 0.21 V more negative than the protection potential $U_{Cu\text{-}CuSO_4} = -0.85$ V which is adopted in practice (see Fig. 1-12) [22,23]. This more positive potential results because δ is actually greater and the cathodically generated surface films give additional protection [24-26].

Evaluation of the corrosion potential from Eq. (2-53) is inadmissible when the metal ion, for example, reacts further in strongly complexing solutions:

$$\text{Me}^{z+} + n\text{X} = \text{Me X}_n^{z+} \tag{2-54}$$

The concentration c (Me^{z+}) must be used in Eq. (2-52) and c (Me X$_n^{z+}$) in Eq. (2-51). A relation exists between the two concentrations from the law of mass action:

$$c(\text{Me}^{z+}) \times c^n(\text{X}) = K \times c(\text{Me X}_n^{z+}) \tag{2-55}$$

From Eqs. (2-51), (2-52), and (2-55), it follows in place of Eq. (2-53) that:

$$U_s = U° + \frac{RT}{z\mathcal{F}} \ln\left(\frac{v\delta}{D} \times 10^3 \text{ cm}^3 \text{ mol}^{-1}\right) - \frac{RT}{z\mathcal{F}} \ln\left[\frac{c^n(\text{X})}{K}\right] \tag{2-56}$$

The third term in Eq. (2-56) is negative and shows how the protection potential can deviate from Eq. (2-53) in such cases. Such effects can be observed in weakly alkali media as a result of hydroxo complex formation [27] and explain the unexpectedly high corrosion rates at relatively negative potentials [28]. All metals can to a lesser or greater extent form hydroxo complexes at sufficiently high pH values. This is of particular significance in cathodic protection in neutral and saline media since OH⁻ ions are formed cathodically by the reactions in Eqs. (2-17) and (2-19). The stronger the cathodic polarization to depress the potential below the protection potential U_s, the greater the rate of formation of OH⁻ ions, so that the third term in Eq. (2-56) depresses the protection potential to more negative potentials through the formation of a hydroxo complex. Therefore, thermodynamic balance can in general not be reached [3] and cathodic protection proves to be impossible with markedly amphoteric metals.

Protection potentials are usually determined experimentally because of the possibilities of error. Figure 2-9 shows experimental results for the potential dependence of weight loss rates for carbon steel [29,30]. Four curves are plotted at 25°C for the following media:

Fundamentals of Corrosion and Electrochemical Corrosion Protection

(a) neutral oxygen-saturated water; 0.5 M Na_3SO_4; artificial seawater; artificial soil solution according to [24];
(b) neutral oxygen-free water; as in (a);
(c) oxygen-free strong acid; bisulfate buffer at pH 1.7;
(d) oxygen-free weak acid; 0.5 M Na_2SO_4 saturated with CO_2 at pH 4.2.

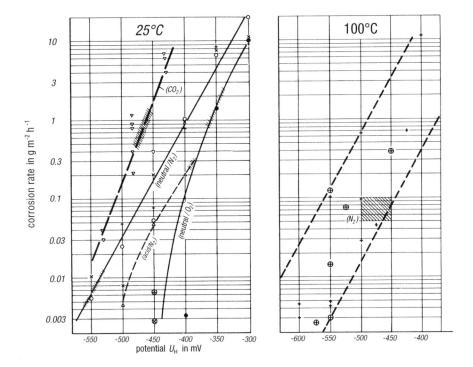

Fig. 2-9 Relation between potential and corrosion rate of a plain carbon steel in slowly circulating water. Definition of symbols:

Water composition	Purged with: N_2 [29]	O_2 [29]	CO_2 [30]
Artificial seawater	o-o	•-•	
Artificial soil solution [24] ($MgSO_4$, $CaCl_2$, $CaSO_4$)	x-x	⊗	
Bisulfate buffer pH 1.7	△-△		
0.5 M Na_2SO_4	+-+	⊕	▽-▽ (pH 4.2)

Range of potential under free corrosion: \\\\\

The hatched area gives the position of the rest potentials. Oxygen corrosion occurs in (a) and acid corrosion in (c) and (d). The corrosion rates are greater than 0.3 mm a^{-1} and can be reduced by cathodic polarization. In (b) the rest potential is at the protection potential. The barely measurable corrosion rate of 5 μm a^{-1} is due to the strongly inhibited cathodic partial reaction according to Eq. (2-19) with the evolution of H$_2$. Corrosion damage occurs in this medium only with anodic polarization. The curves show that a practical protection potential (e.g., for the condition $w < 10$ μm a^{-1}) is dependent on the medium but is assumed sufficiently safe at $U_H = -0.53$ V ($U_{Cu\text{-}CuSO_4} = -0.85$ V). The system Fe/water belongs to group I.

In oxygen-free neutral waters, the measured values lie on a Tafel straight line with $b_+ \approx 60$ mV. The curves for acidic water deviate markedly and in a distinctive manner owing to the complicated kinetics of Fe dissolution by the overall reaction in Eq. (2-21) [4]. In CO$_2$-containing waters, the steep slope with $b_+ \approx 40$ mV is of practical importance [30,31]. Here the protection potential lies apart from this at $U_H = -0.63$ V. A slight underprotection can lead to a large increase in the corrosion rate due to the steep $J_A(U)$ curve. Similar considerations apply to H$_2$S-containing media [32]. It can be assumed that in these acid media buffering has a crucial influence.

No Tafel straight line exists with oxygenated waters. The observed marked decrease in corrosion rate with cathodic polarization is due to oxide films in which cathodically formed OH$^-$ ions and O$_2$ are involved. Similar circumstances are found in high-resistance sandy soils where protection potentials at $U_H = -0.33$ V are possible [33,34].

Corrosion rates increase with rising temperature. At 60°C, the protection potential for unbuffered solutions containing CO$_2$ (pH about 4) is $U_H = -0.75$ V and for solutions buffered with CaCO$_3$ (pH about 6) $U_H = -0.63$ V. Figure 2-9 shows values for neutral Na$_2$SO$_4$ solutions at 100°C which are widely scattered. The protection potential is assumed to be $U_H = -0.63$ V. The increase in corrosion rate in free corrosion is attributable to a diminished restriction of the cathodic partial reaction according to Eq. (2-19) [35]. In the course of time, however, protective films of Fe$_3$O$_4$ can be formed. In media containing NaHCO$_3$ at $U_H = -0.53$ V, noticeable nonstationary corrosion can occur which rapidly diminishes with time so that special measures for cathodic protection are not necessary [28].

Figure 2-10 shows weight loss rate-potential curves for zinc in artificial soil solutions and tapwater [36,37]. Sufficiently low corrosion rates are only to be expected in water containing HCO$_3^-$. According to Eq. (2-53'), the protection potential is $U_H = -0.96$ V. Since zinc is an amphoteric metal, its susceptibility to corrosion increases in caustic solutions or in saline solutions with strong cathodic polarization. The system zinc/water belongs to group II, with a critical potential at $U'_{Hs} = -1.3$ V. The same protection potential can also be applied to zinc coatings.

Fundamentals of Corrosion and Electrochemical Corrosion Protection 57

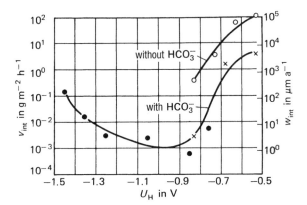

Fig. 2-10 Relation between corrosion rate of zinc and potential: x-x tapwater (pH 7.1, 4 mol m^{-3} HCO$_3^-$; 4 mol m^{-3} Ca^{2+}; 2.5 mol m^{-3} SO$_4^{2-}$, 2 mol m^{-3} Cl$^-$); o–o HCO$_3^-$ free soil solution according to [24]; •-• soil solution with HCO$_3^-$ according to [37].

With hot-dipped galvanized steel, hydrogen absorption with the formation of blisters can be observed in cathodic protection [38].

Figure 2-11 shows weight loss rate-potential curves for aluminum in neutral saline solution under cathodic protection [36,39]. Aluminum and its alloys are passive in neutral waters but can suffer pitting corrosion in the presence of chloride ions which can be prevented by cathodic protection [10, 40-42]. In alkaline media which arise by cathodic polarization according to Eq. (2-19), the passivating oxide films are soluble:

$$Al_2O_3 + 2\,OH^- + 3\,H_2O = 2\,Al(OH)_4^- \tag{2-57}$$

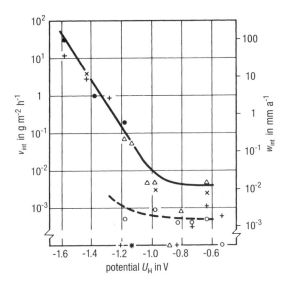

Fig. 2-11 Relation between corrosion rate of pure aluminum at 25°C; • 1 M Na$_2$SO$_4$ after [39]; △ 1.5 g L^{-1} NaCl; o tapwater (about 0.002 mol L^{-1} Na$^+$); + soil solution with 0.5 g L^{-1} NaCl after [24]; x soil solution with 1.5 g L^{-1} NaCl.

58 Handbook of Cathodic Corrosion Protection

Correspondingly, amphoteric aluminum materials suffer active corrosion by the anodic partial reaction:

$$Al + 4\,OH^- = Al(OH)_4^- + 3\,e^- \qquad (2\text{-}58)$$

Since OH⁻ ions are formed in neutral media in the cathodic partial reaction according to Eq. (2-19), the overall cathodic reaction appears by Eqs. (2-19) and (2-58):

$$Al + 4\,H_2O + e^- = Al(OH)_4^- + 2\,H_2 \qquad (2\text{-}59)$$

The system aluminum/water belongs to group II where U_s represents the pitting potential and U'_{Hs} lies between -0.8 and -1.0 V according to the material and the medium [22,23,36,39,42]. Since alkali ions are necessary as opposite ions to the OH⁻ ions in alkalization, the resistance increases with a decrease in alkali ion concentration (see Fig. 2-11). In principle, however, active aluminum cannot be protected cathodically [see the explanation of Eq. (2-56)].

Figure 2-12 shows weight loss rate-potential curves for lead in soil solution and clay soil [43]. This system also belongs to group II. The protection potential U_{Hs} is -0.33 V according to Eq. (2-53'). The $J_A(U)$ curve follows a Tafel straight line. At potentials more negative than $U'_{Hs} = -1.3$ V, chemical corrosion occurs with the evolution of atomic hydrogen according to Eq. (2-19) [7]:

$$Pb + 2\,H\,(PbH_2)\ Pb + H_2 \qquad (2\text{-}60)$$

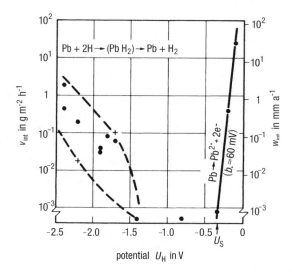

Fig. 2-12 Relation between corrosion rate of lead and potential: •-• soil solution containing NaHCO₃ according to [43]; +-+ loam soil.

Fig. 2-13 Anodic current-density-potential curve for Fe in 0.5 M H$_2$SO$_4$ at 25°C in the passive range (100 μA cm^{-2} ≃ 1 mm a^{-1}): activation potential: $U_{Hs} = 0.8$ V; breakdown potential: $U''_{Hs} = 1.6$ V; passive region: $U'_s < U < U''_s$.

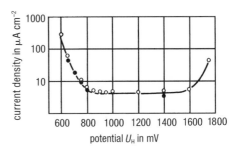

The primary corrosion product PbH$_2$ is unstable and decomposes in a subsequent reaction into lead powder and hydrogen gas. Figures 2-11 and 2-12 are typical examples of cathodic corrosion of amphoteric and hydride-forming metals.

2.3.1.2 Weight Loss Corrosion of Passive Metals

Two areas of passivity are located in Fig. 2-2 where Fe has a very low corrosion rate. In contrast to cathodically protected metals in groups I and II, the corrosion rate of anodically passivated metals in groups III and IV cannot in principle be zero. In most cases the systems belong to group IV where intensified weight loss corrosion or local corrosion occurs when $U > U''_s$. There are only a few metals belonging to group III: e.g., Ti, Zr [44] and Al in neutral waters free of halides.

Figure 2-13 shows the potential dependence of the corrosion rate in the passive region (Fig. 2-2) for the system Fe/H$_2$SO$_4$. At $U < U_{Hs} = 0.8$ V, the transition to active corrosion occurs, while at $U > U'_{Hs} = 1.6$ V, transpassive corrosion takes place [45]. These limiting potentials are strongly influenced by alloying elements in steels and the composition of the media [4]; in particular the corrosion rate is reduced in the passive range by Cr. Figure 2-14 shows as an example the potential dependence of the polarization current and corrosion rate for a CrNiMo stainless steel (AISI 316) in sulfuric acid for the potential region of active corrosion and for the transition into passivity. At $U < U_{Hs} = -0.15$ V, cathodic protection is indeed conceivable. However, this process is not applicable because of the very high protection current density of -300 A m^{-2}. There is also the danger of rapid corrosion in the active range with slight underprotection. The initial increase in corrosion rate upon cathodic polarization is due to the impairment of oxide films by hydrogen evolution. Such surface films can also be affected by chemical reagents in the medium, so that the corrosion rate-potential curve can be significantly displaced. Also, in comparison with carbon steel, the remarkable positive protection potential value of -0.15 V can be governed by secondary effects (e.g., enrichment in noble alloying elements in the surface). Such positive values are only observed with CrNi steels in pure sulfuric acid. The protection potentials in the presence of, for

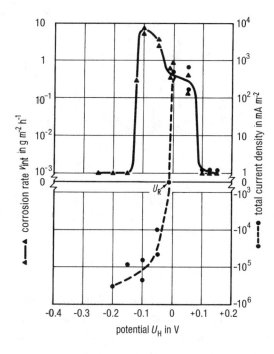

Fig. 2-14 Effect of potential on corrosion rate (▲) and on total current density (•), for No. 1.4401 CrNiMo steel (AISI 316) in 1 M H_2SO_4 at 60°C (purged with N_2).

example, SO_2 or H_2S [46] or for Cr steels [47], lie at such negative values that they are not capable of being achieved. Cathodic protection is therefore not possible for stainless steels in strong acids. On the other hand, anodic protection works very well on account of their ease of passivation. To achieve passivity $U_V > U_{Hs} = +0.15$ V, the passivating current density J_{pas} must exceed about 0.3 A m^{-2}. Only a few mA m^{-2} are sufficient to maintain the passive state. Since all the passivating steels belong to group IV, transpassive corrosion at $U > U_s''$ has to be heeded. For Cr and CrNi stainless steels in 1 M H_2SO_4, U_{Hs}' is 1.1 V at 25°C [48] and 0.8 V at 100°C [49].

The $I(U)$ curves for passivating metals have in general an N shape with several rest potentials. Figure 2-15 shows three characteristic cases, which are given by superimposing a constant value anodic partial current-potential curve on the equilibrium potential U_M^* and three cathodic partial current-potential curves with constant equilibrium potential U_L^* but with different slopes. In case I after passivating with $I_a > I_{pas}$, passivity can only be maintained by means of a constantly flowing protection current. If the protection current is absent, the material rapidly assumes the rest potential in the active corrosion range. The system is unstable-passive. In case II, after passivation the material remains at the rest potential within the pas-

Fundamentals of Corrosion and Electrochemical Corrosion Protection   61

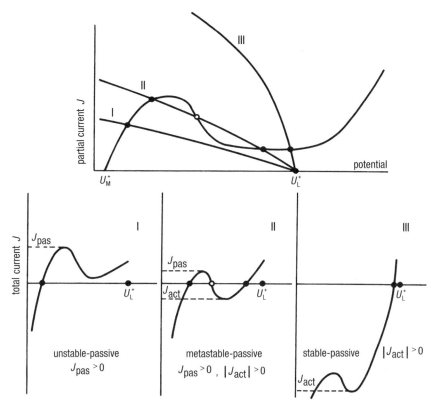

Fig. 2-15 Current-potential relation for passivatable metals (explanation in the text). I_{pas} = passivating current, I_{act} = activation current.

sive range. Activation is only possible if a cathodic disturbance $|I_c| > |I_{act}|$ occurs or if an oxidizing agent in the medium is consumed. The latter corresponds to a change of curve from II to I. In case II the system is metastable-passive. The anodic protection current does not need to flow continuously. Discontinuous potential control and regulation is sufficient. In case III the system is stable-passive. The active state can only be achieved by constant cathodic polarization with $|I_c| > |I_{act}|$.

These three passive systems are important in the technique of anodic protection (see Chapter 21). The kinetics of the cathodic partial reaction and therefore curves of type I, II or III depend on the material and the particular medium. Case III can be achieved by alloying additions of cathodically acting elements such as Pt, Pd, Ag, and Cu. In principle, this is a case of galvanic anodic protection by cathodic constituents of the microstructure [50].

2.3.2 Pitting Corrosion

Generally, pitting corrosion only occurs on passivated metals when the passive film is destroyed locally. In most cases chloride ions cause this local attack at potentials $U > U_{PC}$. Bromide ions also act in the same way [51]. The critical potential for pitting corrosion U_{PC} is called the pitting potential. It has the same significance as U_s in Eqs. (2-39) and (2-48).

Because of possible errors in determining pitting potentials from I(U) curves, it is safest to take them from pit density-potential diagrams which can be determined by chronopotentiostatic experiments. Figure 2-16 shows the results of experiments on 1.4031 CrNi stainless steel (AISI 304) in neutral waters [52].

In contrast to carbon steels and low-alloy steels, high-alloy Cr and CrNi stainless steels suffer no active corrosion in neutral waters [53] and belong to group I. The protection potential corresponds to the pitting potential and in general lies at more positive values than $U_H = 0.0$ V. Surface coatings and crevices [54,55] favor pitting corrosion. The pitting potential is displaced to more negative potentials by increases in temperature, an increasing concentration of chloride ions, and the presence of specific stimulators (H_2S and SO_2) [56] as well as material of poor purity (in particular sulfide inclusions) and a susceptibility to intergranular corrosion (according to the Strauss test such as DIN 50914) [4]. On the other hand, it is displaced to more positive values by other anions (e.g., sulfate [52]) and by high-alloying additions of Mo and Cr [57-59]. Information on the application of cathodic protection is collected in Refs. 60 and 61.

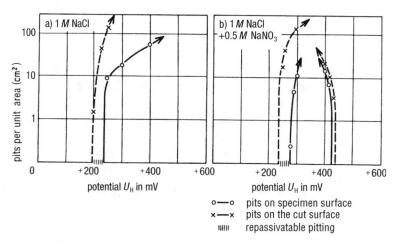

Fig. 2-16 Relation between pit density and potential for No. 1.4301 CrNi steel (AISI 304) at 25°C.

Fundamentals of Corrosion and Electrochemical Corrosion Protection

Nitrate ions have a special influence by inhibiting pitting corrosion in neutral and acid waters at $U > U_s$ [Eq. (2-50)] [48,52]. U_s corresponds to a second pitting potential and is designated the inhibition potential. The system belongs to group IV, with pitting corrosion at $U < U_s$ and transpassive corrosion at $U > U_s''$.

The system stainless steel/chloride-containing acid media belongs to group II. Active corrosion occurs at $U < U_s'$ and pitting corrosion at $U > U_{PC}$. In acid media containing chloride and nitrate ions, the following states can exist with increasing potential: cathodic protection – active corrosion – passivation – pitting corrosion – passivation – transpassive corrosion. The different types of corrosion are dependent on the potential, with the dependence varying greatly from type to type, which in a particular case can only be learned from chronopotentiostatic experiments.

Other passivating materials suffer pitting corrosion by chloride ions [62] in a way similar to stainless steels (e.g., Ti [63] and Cu [64]). The pitting potential for aluminum and its alloys lies between $U_H = -0.6$ and -0.3 V, depending on the material and concentration of chloride ions [10, 40-42].

2.3.3 Stress Corrosion

Similar to pitting, stress corrosion occurs predominantly on passive metals within a critical potential range. In addition to the specific properties of the material and the media, the type and level of mechanical loading influence the position of the limiting potential. In stress corrosion the loss of material can be vanishingly small or even zero. The fracture can run along the grain boundaries (intergranular)

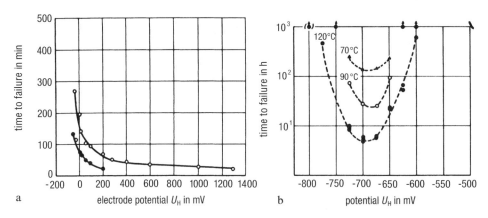

Fig. 2-17 Relation between the time to failure by intergranular stress corrosion cracking and potential for tensile specimens of soft iron: (a) boiling 55% $Ca(NO_3)_2$ solution, $\delta = 0.65\ R_m$, $\sigma = 0.90\ R_m$; (b) 33% NaOH, $\sigma = 300$ N mm^{-2}, at various temperatures.

or through the grains (transgranular). Lifetime-potential curves are used to assess corrosion damage or the applicability of corrosion protection methods. Figure 2-17 shows two examples for soft iron in nitrate solution [65] in (a), and in caustic soda [66] in (b). In both cases the specimens were cylindrical tensile specimens which were subjected to a constant load. There is a critical tensile stress below which stress corrosion does not occur; this can depend on the potential [67].

Stress corrosion systems can be divided into two categories depending on whether crack initiation occurs with constant static loading or whether it occurs only when a critical strain rate is exceeded in dynamic testing. These two types are known as "stress-induced" or "strain-induced" stress corrosion [68-70]. This differentiation is important in assessing the effects of damage and in experimental investigation. The critical potential ranges can thus depend on the strain rate.

Systems with lifetime-potential curves of type (a) in Fig. 2-17 can be cathodically protected against stress corrosion. The following metals belong to these systems:

(a) Plain carbon and low-alloy steels in nitrate solutions, particularly at elevated temperature. There are critical tensile stresses. The susceptibility range is widened under dynamic loading [71]. The systems belong to group I.
(b) Austenitic manganese steels in seawater [72,73]. The systems belong to group I.
(c) Austenitic stainless steels in Cl^- containing waters at elevated temperature [67,74,75]. The systems belong in general to group I, but at high tensile stresses and more negative potentials a new range of susceptibility (for stress corrosion cracking occurs) occurs (see j).
(d) Sensitized stainless steels in hot waters with a tendency to intergranular corrosion [76]. The systems belong to group I in neutral waters and to group Ii in acid waters.

Systems with lifetime-potential curves like type (b) in Fig. 2-17 can be protected anodically as well as cathodically against stress corrosion. The following metals belong to these systems:

(e) Plain carbon or low-alloy steels in caustic soda at elevated temperatures [66,77-80]. The systems belong to groups I or IV.
(f) Plain carbon or low-alloy steels in $NaHCO_3$ solutions [77,78,81]. The systems belong to groups I or IV.
(g) Plain carbon or low-alloy steels in $NaHCO_3$ solutions in general with a critical strain rate [77,78,80,82-84]. The systems belong to group I or IV.
(h) Plain carbon or low-alloy steels in Na_2CO_3 solutions at elevated temperature and with a critical strain rate [77]. The systems belong to groups I or IV.

Fundamentals of Corrosion and Electrochemical Corrosion Protection

(i) Plain carbon or low-alloy steels in $CO–CO_2–H_2O$ condensates [82,85], HCN [86], liquid NH_3 [87,111].

(j) Austenitic stainless steels in cold chloride-containing acids [88]. These systems belong to group IV on account of the anodic danger of pitting corrosion.

Cases (e), (g), and (h) are of interest in the cathodic protection of warm objects (e.g., district heating schemes [89] and high-pressure gas lines downstream from compressor stations [82]) because the media of concern can arise as products of cathodic polarization. The use of cathodic protection can be limited according to the temperature and the level of the mechanical stressing. The media in cases (a) and (f) are constituents of fertilizer salts in soil. Cathodic protection for group I is very effective [80].

In deciding on the type of protection for the systems shown in Fig. 2-17b, consideration has to be given to the level of protection current, current distribution according to Eq. (2-45), the by-products of electrolysis, and operational safety in conjunction with Fig. 2-15. Figure 2-18 shows the positions of nonstationary and quasi-stationary $J(U)$ curves relative to the critical range for stress corrosion. It is obvious that nonstationary $J(U)$ measurements lead to false conclusions and that anodic protection is more advantageous on account of the small difference among the protection range and the rest potential, the lower protection current density, and the higher polarization resistance [90].

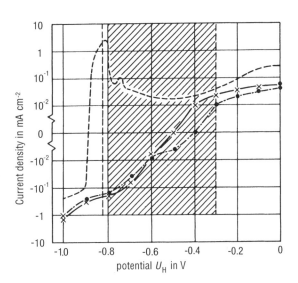

Fig. 2-18 $J(U)$ curves and critical potential range for intergranular stress corrosion (hatched) for a hardened 10 CrMo 9 10 steel (ASTM P21) in boiling 35% NaOH: --- potentiodynamically measured with +0.6 V h^{-1}; •--• potential change after every 0.5 h $\Delta U = +0.1$ V; x-x-x potential change after every 0.5 h $\Delta U = -0.1$ V.

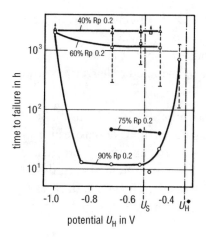

Fig. 2-19 Time to failure vs. potential for X70 pipeline steel in pH 5.5 buffer solution containing 150 mg L^{-1} sulfide ions at various loads, 15°C: $U_s = -0.53$ V (cathodic protection potential); $U_H^* = -0.32$ V (hydrogen equilibrium potential at 1 bar).

2.3.4 Hydrogen-Induced Corrosion

Adsorbed atomic hydrogen, which is formed in the cathodic partial reaction according to Eqs. (2-18) or (2-19), can be absorbed in the presence of promoters (e.g., H$_2$S), in acid media at pH < 3, or by strong cathodic polarization at $U_H < -0.8$ V as well as under high loading within the plastic region at a critical strain rate. As a consequence, blisters and internal cracks can be formed below the surface as well as fractures caused by stress corrosion under high loading [8,9,91,92]. Since the susceptibility to H-induced cracking increases at negative potentials, but on the other hand positive potentials favor anodic corrosion, electrochemical protection measures in these systems in general are not applicable. Protection measures that must be considered are choice of material, inhibition of the medium, and constructive measures to reduce stress. In general, with cathodic protection measures, it must be ascertained whether the conditions encourage H absorption.

If the technical regulations are adhered to for constructional steels in neutral waters, there are no conditions for H-induced corrosion. On the other hand, hardened and high-strength materials with hardnesses above HV 350 are very susceptible [60,82,92], since anodic polarization encourages crack formation in saline media and anodic pitting occurs with acid products of hydrolysis [93].

Figure 2-19 shows failure time-potential curves for a pipeline steel in weak acid buffer solution containing H$_2$S [94]. At potentials within the protection range for general corrosion, stress corrosion can occur with high-tensile stresses. At $U_H = -1.0$ V, the resistance increases because the promoter becomes inactive due to cathodic wall alkalinity according to Eq. (2-19). This effect does not occur in stronger acid solutions. The failure times become shorter at negative potentials

Fundamentals of Corrosion and Electrochemical Corrosion Protection 67

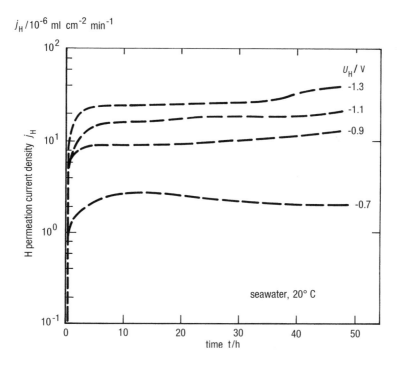

Fig. 2-20 Influence of potential on hydrogen permeation; 1-mm-thick steel sheet/aerated seawater.

[95]. Stress corrosion only occurs in high-strength steels with $R_m > 1200$ N mm^{-2} in the absence of promoters under static loading. Systems belonging to group II, such as stainless high-strength Cr steels, can experience H-induced stress corrosion cracking at $U < U_s$, and pitting or crevice corrosion at $U > U_s$ [96].

The activity of absorbed hydrogen is a measure of the damage for a given material. H absorption can be easily investigated in H-permeation experiments [9]. Figure 2-20 shows experimental results for steel/aerated seawater. Significant H absorption occurs only in the range of cathodic overprotection ($U_H < -0.8$ V).

Susceptibility is generally very high in the case of dynamic deformation in the plastic range where strain-induced stress corrosion occurs. The parameters that influence this type of corrosion can be investigated in a slow strain rate test at strain rates around 10^{-6} s^{-1} [91,92,97]. Figure 2-21 shows results for steel in seawater [92]. Included are the total current-density-potential curves and the reduction in area of the cracked specimens plotted as a function of potential. In addition, it is noted whether internal cracks are observed on longitudinal microsections. In aerated seawater, there is a transition at $U_H = -0.85$ V between the limiting current density for oxygen reduction according to Eq. (2-17) and the Tafel straight line for

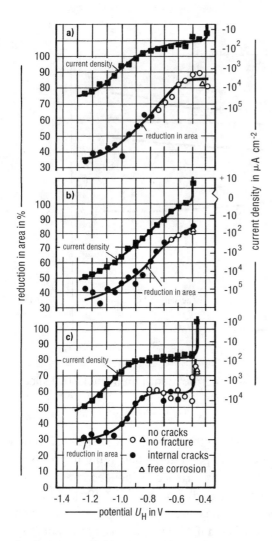

Fig. 2-21 Effect of potential on the current density and on the reduction in area for X65 steel in seawater (slow strain test, $\dot{\varepsilon} = 2 \times 10^{-6}$ s^{-1}). (a) Air purging, (b) nitrogen purging, (c) CO$_2$ purging.

hydrogen evolution according to Eq. (2-19) (see Fig. 2-21a). Since the current density increases exponentially with overvoltage, in practice the potential region of the Tafel straight line is not reached because of current distribution [see Eq. (2-45)]. This coincides with the range for cathodic overprotection (see also Section 24.4).

In oxygen-free seawater, the $J(U)$ curves, together with the Tafel straight lines for hydrogen evolution, correspond to Eq. (2-19) (see Fig. 2-21b). A limiting current density occurs with CO$_2$ flushing for which the reaction:

$$CO_2 + H_2O = H_2CO_3 = H^+ + HCO_3^- \tag{2-61}$$

is responsible [31] (see Fig. 2-21c). In Figs. 2-21a-c the Tafel line for the reaction according to Eq. (2-19) is unchanged.

Fundamentals of Corrosion and Electrochemical Corrosion Protection 69

The fall in reduction of area and the occurrence of internal cracks are a measure of the corrosion damage. There exists a clear correlation with cathodic current density in which a slight inhibition due to O_2 and stimulation by CO_2 can be recognized. The susceptibility is very high in the range of cathodic overprotection and is independent of the composition of the medium.

Strain-induced stress corrosion is to be expected in an increasing number of stressed structures (e.g., offshore rigs), especially in the region of notches. Here cathodic protection can certainly retard crack initiation, but it can accelerate crack propagation. It is understandable that the action of cathodic protection appears contradictory [98-102] (see also Chapter 16). Figure 2-22 shows crack propagation in notched steel specimens under cyclic loading for fatigue in air, for corrosion fatigue in NaCl solution, and for H-induced stress corrosion under cathodic polarization [98]. With increasing strain rate, i.e., increasing stress intensity range ΔK and frequency, all the curves merge with the Paris straight line for air [98] because the chemical corrosion processes do not have sufficient time to contribute [69]. Thus cathodic protection can be effective in the transition range.

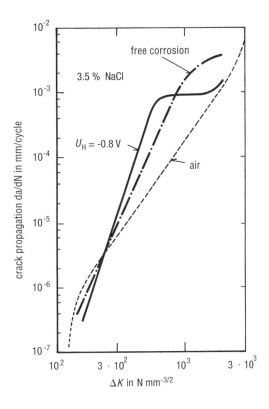

Fig. 2-22 Effect of stress intensity range ΔK on crack propagation for a notched specimen of X60 steel in 3.5% NaCl solution at 0.1 Hz.

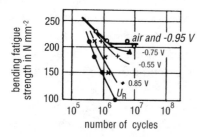

Fig. 2-23 S--N curves for plain carbon steel in 0.05 M potassium hydrogen benzoate (pH 4) at 30°C at various potentials U_H (U_R = rest potential).

2.3.5 Corrosion Fatigue

Under cyclic stressing, all metallic materials suffer crack formation well below their ultimate tensile strength (UTS). This process is called fatigue. There is a relationship between the stress amplitude and the number of cycles to failure which is represented by the Wöhler (S-N) curve. Figure 2-23 shows such a curve for a plain carbon steel with a fatigue limit in alternating bending of 210 N mm^{-2}. At this stress level, the curve becomes horizontal, i.e., at lower stress levels failure does not occur at any further number of cycles. In corrosion fatigue there is no fatigue limit. The S-N curve at the rest potential U_R falls steeply. Passivation by anodic protection at U_H = +0.85 V produces an insignificant increase in the fatigue life. On the other hand, cathodic protection at a much lower potential is very successful. At U_H = −0.95 V, the value reached is the same as that reached in air [103]. This favorable action of cathodic polarization relates to crack initiation in un-notched specimens and therefore does not contradict the opposite findings on crack propagation in Fig. 2-22. The boundary lines in the pH-potential diagram in Fig. 2-2 are not changed under static loading. With cyclic loading on the other hand, the passive regions disappear. In addition, line (1) is shifted with falling pH to more negative potentials and at pH = 4 bends downward.

Fatigue life can be slightly lengthened by anodic protection or by passivation. In acids even passive stainless CrNi steels suffer corrosion fatigue [104]. Resistance can occur if the passive film itself has a fatigue strength (e.g., in neutral waters [105]).

2.3.6 Limits of Applicability of Electrochemical Protection Processes

If the products of electrolysis favor other types of corrosion, electrochemical protection processes should not be applied or should be used only in a limited form. Hydrogen and OH$^-$ ions are produced in cathodic protection according to Eq. (2-19). The following possible corrosion danger must be heeded:

Fundamentals of Corrosion and Electrochemical Corrosion Protection 71

H-induced corrosion (see Section 2.2.1).

- Corrosion by hydride formation [e.g., Pb (Fig. 2-12)];
- Embrittlement by internal hydride formation (e.g., Ti, Zr, Nb, Ta) [These metals are important as valve metals for impressed current anodes (Section 7.1) and as materials in chemical plants (Section 21.4.3).]
- Stress corrosion (see Section 2.3.4), in particular accelerated crack propagation in fatigue (see Section 2.3.5).

OH⁻-induced corrosion

- Weight loss corrosion of amphoteric metals [e.g., Al (Fig. 2-11)];
- Weight loss corrosion of passive metals (e.g., CrNi stainless steels in acids) (cathodic protection in acids is not practicable; see Fig. 2-14);
- Stress corrosion of steels at elevated temperatures by NaOH.

There are limits to the use of anodic protection in the following cases:

- Too high a corrosion rate in the passive state;
- No possibility of protection in the case of current shielding (e.g., crevices under coatings or deposits);
- Too high fatigue stressing.

In general, in the case of coated materials, interaction with protection currents must be taken into account [106,107] (see Section 5.2.1). In internal protection, the effects of products of electrolysis must be checked [108] (see Chapters 20 and 21).

2.4 Critical Protection Potentials and Ranges

In this section a survey is given of the critical protection potentials as well as the critical potential ranges for a possible application of electrochemical protection. The compilation is divided into four groups for both cathodic and anodic protection with and without a limitation of the protection range to more negative or more positive potentials respectively.

Protection potentials and protection potential ranges for important systems

Group	System Material	System Medium	Protection potential/region (in volts) U_H	Protection potential/region (in volts) $U_{Cu-CuSO4}$	Notes/References
I $U < U_s$ Eq. (2-39)	Plain carbon and low-alloy ferrous materials	Neutral waters, saline and soil solutions (25°C)	<–0.53	<–0.85	Protection against weight loss corrosion Fig. 2-9 [29-34] (with film formation U_s is more positive)
		Boiling neutral waters	<–0.63	<–0.95	
		Weak acidic waters and anaerobic media (25°C)	<–0.63	<–0.95	
		High-resistance sandy soils	<–0.43 (–0.33)	<–0.75 (–0.65)	[34]
	High-alloy steels with >16% Cr[a] (e.g. 1.4301, AISI 304)	Neutral waters and soils (25°C)	<0.2	<–0.1	Protection against pitting and crevice corrosion
		Boiling neutral waters	<0.0	<–0.3	Heating surfaces are more susceptible than cooling surfaces [109] (see Fig. 2-16)
	CrNiMo stainless steels and Cr-rich special alloys	Seawater (25°C) Cl⁻ containing media	<0.0 (in general more positive; U_{PC} values determine)	<–0.3	U_s becomes more negative with increasing Cl⁻ concentration and temperature [4, 57-59]
	CrNi stainless steels	Cl⁻ containing hot water	About <0.0	About <–0.3	Protection against stress corrosion [74-76]

Protection potentials and protection potential ranges for important systems (continued)

Group	System Material	System Medium	Protection potential/region (in volts)		Notes/References
			U_H	$U_{Cu\text{-}CuSO_4}$	
	Plain carbon and low-alloy steels[b]	Warm solutions of			Protection against stress corrosion
		Nitrates	<–0.15	<–0.47	Fig. 2-17a [65,71]
		Caustic soda	<–0.98	<–1.30	Fig. 2-17b [66,79,89,90]
		Na_2CO_3	<–0.68	<–1.00 }	Strain induced [77,82]
		$NaHCO_3$	<–0.43	<–0.75	
		$(NH_4)_2CO_3$	<–0.35	<–0.67	[81]
	Cu, CuNi alloys	Neutral waters and soils (25°C)	<+0.14	<–0.18	Protection against weight loss corrosion [64]
	Sn	Neutral waters (25°C)	<–0.33	<–0.65	
II U'_s/U_s Eq. (2-48)	Plain carbon and low-alloy ferrous materials	Seawater	–0.53/ –0.78	–0.85/ –1.10	Protection for thin films and against stress corrosion at fluctuating loads (see Chapter 16)
		Cement, concrete	–0.43/–0.98	–0.75/–1.3	Section 5.3.2
	High-alloy heat-treated Cr steels ($R_m > 1200$ N mm^{-2})	Seawater (25°C)	–0.5/–0.0	–0.82/–0.32	Protection against H-induced stress corrosion and pitting corrosion [96, 110]
	Pb	Neutral waters and soils (25°C)	–1.4/–0.33	–1.7/–0.65	Protection against hydride formation and weight loss corrosion (see Fig. 2-12)

Protection potentials and protection potential ranges for important systems (continued)

Group	System Material	System Medium	Protection potential/region (in volts) U_H	Protection potential/region (in volts) $U_{Cu-CuSO4}$	Notes/References
	Zn	Neutral waters and soils (25°C)	−1.3/ −0.96	−1.6/ −1.3	Fig. 2-10 (with film formation U_s' is more positive)
	Al, Al alloys	Cold water			Protection against weight loss corrosion and pitting corrosion [36,39,42]
	Al Zn 4.5 Mg 1	Fresh water	−1.0/−0.3	−1.3/−0.62	(see Fig. 2-11)
		Seawater	−1.0/−0.5	−1.3/−0.82	U_s' becomes more negative with decreasing Na$^+$ concentration
		Seawater	−1.0/−0.7	−1.3/−1.02	
III $U > U_s$ Eq. (2.49)	Ti, Ti alloys	Halide-free acids	>0.0	>−0.32	Protection against weight loss corrosion [10,44]
		Increased concentration and temperature	U_s becomes more positive		See Chapter 21
IV $U_s'/U_s \leq$ Eq. (2-50)	Plain carbon and low-alloy steels ($R_p < 600$)[b] hardened region	Warm caustic soda ($R_m > 1000$ N mm^{-2})	−0.6/+0.2 −0.4/+0.2	−0.9/−0.1 −0.7/−0.1	Protection against stress corrosion and weight loss corrosion. Fig. 2-17b [66,79,89,90] [90]
		Warm Na$_2$CO$_3$ soln.	−0.6/?	−0.9/?	Strain induced [77,82]
		NaHCO$_3$ soln.	−0.3/?	−0.6/?	
		(NH$_4$)$_2$CO$_3$ soln.	−0.25/?	−0.57/?	[81]

Protection potentials and protection potential ranges for important systems (continued)

Group	System Material	System Medium	Protection potential/region (in volts)		Notes/References
			U_H	$U_{Cu-CuSO_4}$	
	Fe, plain carbon steels	0.5 M H_2SO_4 (25°C)	0.8/1.6	0.5/1.3	Protection against active and transpassive corrosion [45] (see Fig. 2-13)
	High-alloy steels with >16%Cr[c]	Halide-free cold acids	0.2/1.1	−0.1/0.8	Protection against active and transpassive corrosion
		Boiling conc. H_2SO_4	1.2/1.6	0.9/1.3	(See Chapter 21)
		Cl⁻ and NO_3^- containing waters (25°C)	0.5/1.1	0.2/0.8	Protection against pitting corrosion and transpassive corrosion (see Fig. 2-16)

[a] Rough values are given. Microstructure and composition of the material as well as temperature, composition, and flow rate of the medium can substantially influence U_s.

[b] With materials with high resistance to intergranular stress corrosion, U_s can be more positive in group I and more negative in group IV. Corrosion susceptibility increases with rising temperature (see Chapter 21).

[c] Certain chemical compounds in acids can influence U_s and U_s'' considerably. Mainly Ni, Mo, and Cu as alloying elements improve passivation. With increasing acid concentration, U_s becomes more positive. With increasing temperature U' becomes more negative (see Chapter 21).

2.5 References

[1] DIN 50900 Teil 1, Beuth-Verlag, Berlin 1982.
[2] Werkstoffe u. Korrosion, *32*, 33 (1981); Kommentar zu DIN 50900 in „Korrosion durch Information und Normung", Hrsg. W. Fischer, Verlag I. Kuron, Bonn 1988.
[3] H. Adrian u. C. L. Kruse, gwf wasser/abwasser *124*, 453 (1983).
[4] A. Rahmel u. W. Schwenk, Korrosion und Korrosionsschutz von Stählen, Verlag Chemie, Weinheim 1977.
[5] DIN 50900 Teil 2, Beuth-Verlag, Berlin 1984.
[6] DIN 50905 Teil 2 und 3, Beuth-Verlag, Berlin 1987.
[7] M. Pourbaix, Atlas d'Equilibres Électrochimiques, Gauthier-Villars & Cie., Paris 1963.
[8] D. Kuron, Wasserstoff und Korrosion, Bonner Studien-Reihe, Band 3, Verlag I. Kuron, Bonn 1986.
[9] W. Haumann, W. Heller, H.-A. Jungblut, H. Pircher, R. Pöpperling u. W. Schwenk, Stahl u. Eisen *107*, 585 (1987).
[10] H. Kaesche, Die Korrosion der Metalle, Springer-Verlag, Berlin–Heidelberg–New York 1966 u. 1979.
[11] T. Misawata, Corrosion Sci. *13*, 659 (1973).
[12] U. Rohlfs u. H. Kaesche, Der Maschinenschaden *57*, 11 (1984).
[13] DIN 50918, Beuth-Verlag, Berlin 1978.
[14] K.-J. Vetter, Elektrochemische Kinetik, Springer-Verlag, Berlin–Göttingen–Heidelberg 1961.
[15] H. Hildebrand u. W. Schwenk, Werkstoffe u. Korrosion *23*, 364 (1972).
[16] DIN 50919, Beuth-Verlag, Berlin 1984.
[17] W. Schwenk, Metalloberfläche *35*, 158 (1981).
[18] J. Elze u. G. Oelsner, Metalloberfläche *12*, 129 (1958).
[19] C. Wagner, Chemie-Ingenieur-Technik *32*, 1 (1960).
[20] E. Heitz u. G. Kreysa, Principles of Electrochemical Engineering, VCH-Verlag, Weinheim 1986.
[21] C. Wagner, J. electrochem. Soc. *99*, 1 (1952).
[22] DIN 30676, Beuth-Verlag, Berlin 1985.
[23] DIN 50927, Beuth-Verlag, Berlin 1985.
[24] H. Klas, Archiv Eisenhüttenwesen *29*, 321 (1958).
[25] H.-J. Engell u. P. Forchhammer, Corrosion Sci. *5*, 479 (1965).
[26] W. Fischer, Werkstoffe u. Korrosion *27*, 231 (1976).
[27] W. Schwenk, Werkstoffe und Korrosion *34*, 287 (1983).
[28] W. Schwenk, 3R international *23*, 188 (1984).
[29] G. Herbsleb u. W. Schwenk, Werkstoffe u. Korrosion *19*, 888 (1968).
[30] W. Schwenk, Werkstoffe u. Korrosion *25*, 643 (1974).
[31] G. Schmitt u. B. Rothmann, Werkstoffe u. Korrosion *29*, 98, 237 (1978).
[32] M. Solti u. J. Horvath, Werkstoffe u. Korrosion *9*, 283 (1958).
[33] W. Schwenk, 3R international *15*, 254 (1986).
[34] D. Funk, H. Hildebrand, W. Prinz u. W. Schwenk, Werkstoffe u. Korrosion *38*, 719 (1987).
[35] G. Resch, VGB-Sonderheft Speisewassertagung 1969, S. 17.
[36] W. Schwenk, 3R international *18*, 524 (1979).
[37] W. v. Baeckmann u. D. Funk, Werkstoffe u. Korrosion *33*, 542 (1982).
[38] W. Schwenk, Werkstoffe u. Korrosion *17*, 1033 (1966); ders., polytechn. tijdschr. process-techn. *26*, 345 (1971) (Mannesmann-Forschungsber. 548).
[39] H. Kaesche, Werkstoffe u. Korrosion *14*, 557 (1963).
[40] H. Kaesche, Z. phys. Chemie NF *26*, 138 (1960).
[41] H. Ginsberg u. W. Huppatz, Metall *26*, 565 (1972).
[42] W. Huppatz, Werkstoffe u. Korrosion *28*, 521 (1977); *30*, 673 (1979).

[43] W. von Baeckmann, Werkstoffe u. Korrosion 20, 578 (1969).
[44] O. Rüdiger u. W. R. Fischer, Z. Elektrochemie 62, 803 (1958).
[45] G. Herbsleb u. H.-J. Engell, Z. Elektrochemie 65, 881 (1961).
[46] G. Herbsleb u. W. Schwenk, Werkstoffe u. Korrosion 17, 745 (1966); 18, 521 (1967).
[47] G. Herbsleb, Werkstoffe u. Korrosion 20, 762 (1969).
[48] E. Brauns u. W. Schwenk, Arch. Eisenhüttenwes. 32, 387 (1961).
[49] P. Schwaab u. W. Schwenk, Z. Metallkde. 55, 321 (1964).
[50] H. Gräfen, Z. Werkstofftechnik 2, 406 (1971).
[51] G. Herbsleb, H. Hildebrand u. W. Schwenk, Werkstoffe u. Korrosion 27, 618 (1976).
[52] G. Herbsleb, Werkstoffe u. Korrosion 16, 929 (1965).
[53] G. Herbsleb u. W. Schwenk, gwf wasser/abwasser 119, 79 (1978).
[54] A. Kügler, K. Bohnenkamp, G. Lennartz u. K. Schäfer, Stahl u. Eisen 92, 1026 (1972).
[55] A. Kügler, G. Lennartz u. H.-E. Bock, Stahl u. Eisen 96, 21 (1976).
[56] G. Herbsleb, Werkstoffe u. Korrosion 33, 334 (1982).
[57] G. Herbsleb u. W. Schwenk, Werkstoffe u. Korrosion 26, 5 (1975).
[58] H. Kiesheyer, G. Lennartz u. H. Brandis, Werkstoffe u. Korrosion 27, 416 (1976).
[59] E.-M. Horn, D. Kuron u. H. Gräfen, Z. Werkstofftechnik 8, 37 (1977).
[60] DIN 50929 Teil 2, Beuth-Verlag, Berlin 1985.
[61] VG 81249 Teil 2 (Vornorm), Beuth-Verlag, Berlin 1983.
[62] G. Herbsleb, VDI-Bericht Nr. 243, S. 103 (1975).
[63] W. R. Fischer, Techn. Mitt. Krupp Forsch.-Ber. 22, 65 u. 125 (1964).
[64] K. D. Efird u. E. D. Verink, Corrosion 33, 328 (1977).
[65] A. Bäumel u. H.-J. Engell, Arch. Eisenhüttenwes. 32, 379 (1961).
[66] K. Bohnenkamp, Arch. Eisenhüttenwes. 39, 361 (1968).
[67] G. Herbsleb u. W. Schwenk, Werkstoffe u. Korrosion 21, 1 (1970).
[68] DIN 50922, Beuth-Verlag 1985.
[69] G. Herbsleb u. W. Schwenk, Corrosion 41, 431 (1985).
[70] Werkstoffe u. Korrosion 37, 45 (1986).
[71] W. Friehe, R. Pöpperling u. W. Schwenk, Stahl u. Eisen 95, 789 (1975).
[72] W. Prause u. H.-J. Engell, Werkstoffe u. Korrosion 22, 421 (1971).
[73] A. Bäumel, Werkstoffe u. Korrosion 20, 389 (1969).
[74] E. Brauns u. H. Ternes, Werkstoffe u. Korrosion 19, 1 (1968).
[75] G. Herbsleb u. H. Ternes, Werkstoffe u. Korrosion 20, 379 (1969).
[76] G. Herbsleb, VGB-Kraftwerkstechnik 55, 608 (1975).
[77] G. Herbsleb, B. Pfeiffer, R. Pöpperling u. W. Schwenk, Z. Werkstofftechnik 9, 1 (1978).
[78] H. Diekmann, P. Drodten, D. Kuron, G. Herbsleb, B. Pfeiffer u. E. Wendler-Kalsch, Stahl u. Eisen 103, 895 (1983).
[79] H. Gräfen u. D. Kuron, Arch. Eisenhüttenwes. 36, 285 (1965).
[80] W. Schwenk gwf gas/erdgas 123, 157 (1982).
[81] K. J. Kessler u. E. Wendler-Kalsch, Werkstoffe u. Korrosion 28, 78 (1977).
[82] Proc. 5th a. 6th Congr. Amer. Gas Assoc., Houston 1974, 1979.
[83] J. M. Sutcliffe, R. R. Fessler, W. K. Boyd u. R. N. Parkins, Corrosion 28, 313 (1972).
[84] G. Herbsleb, R. Pöpperling u. W. Schwenk, 3R international 20, 193 (1981).
[85] A. Brown, J. T. Harrison u. R. Wilkins, Corrosion Sci. 10, 547 (1970).
[86] H. Buchholtz u. R. Pusch, Stahl u. Eisen 62, 21 (1942).
[87] D. A. Jones u. B. E. Wilde, Corrosion 33, 46 (1977).
[88] G. Herbsleb, B. Pfeiffer u. H. Ternes, Werkstoffe u. Korrosion 30, 322 (1979).
[89] B. Poulson, L. C. Henrikson u. H. Arup, Brit. Corrosion J. 9, 91 (1974).
[90] H. Gräfen, G. Herbsleb, F. Paulekat u. W. Schwenk, Werkstoffe u. Korrosion 22, 16 (1971).
[91] H. Pircher u. R. Großerlinden, Werkstoffe u. Korrosion 38, 57 (1987).
[92] R. Pöpperling, W. Schwenk u. J. Venkateswarlu, Werkstoffe u. Korrosion 36, 389 (1985).
[93] B. E. Wilde, Corrosion 27, 326 (1971).

[94] G. Herbsleb, R. Pöpperling u. W. Schwenk, Werkstoffe u. Korrosion *31*, 97 (1980).
[95] F. K. Naumann u. W. Carius, Arch. Eisenhüttenwes. *30*, 283 (1959).
[96] G. Lennartz, Werkstoffe u. Korrosion *35*, 301 (1984).
[97] B. R. W. Hinton u. R. P. M. Procter, Corrosion Science *23*, 101 (1983).
[98] O. Vosikowski, Closed Loop *6*, 3 (1976).
[99] R. Pöpperling, W. Schwenk u. G. Vogt, Werkstoffe u. Korrosion *29*, 445 (1978).
[100] F. Schmelzer u. F. J. Schmitt, GKSS-Forschungszentrum Geesthacht, Bericht GKSS 85/E/29, Geesthacht 1985.
[101] R. Helms, H. Henke, G. Oelrich u. T. Saito, Bundesanstalt für Materialprüfung, Forschungsbericht Nr. 113, Berlin 1985.
[102] K. Nishiota, K. Hirakawa u. I. Kitaura, Sumitomo Search *16*, 40 (1976).
[103] K. Endo, K. Komai u. S. Oka, J. Soc. Mater. Sci. Japan, *19*, 36 (1970).
[104] H. Spähn, Z. phys. Chem. (Leipzig) *234*, 1 (1967).
[105] H. Tauscher u. H. Buchholz, Neue Hütte *18*, 484 (1973).
[106] DIN 50928, Beuth-Verlag, Berlin 1985.
[107] W. Schwenk, farbe+lack *90*, 350 (1985).
[108] DIN 50927, Beuth-Verlag, Berlin 1984.
[109] G. Herbsleb u. W. Schwenk, Werkstoffe u. Korrosion *26*, 93 (1975).
[110] D. L. Dull u. L. Raymond, Corrosion *29*, 205 (1973).
[111] H. Gräfen, H.Hennecken, E.-M. Horn, H. D. Kamphusmann u. D. Kuron, Werkstoffe u. Korrosion *36*, 203 (1985).

3

Fundamentals and Practice of Electrical Measurements

W. V. BAECKMANN AND W. SCHWENK

Practical measurements providing data on corrosion risk or cathodic protection are predominantly electrical in nature. In principle they concern the determination of the three principal parameters of electrical technology: voltage, current, and resistance. Also the measurement of the potential of metals in soil or in electrolytes is a high-resistance measurement of the voltage between the object and reference electrode and thus does not draw any current (see Table 3-1).

For measuring electrical corrosion, often several, synchronously registering measurement devices are necessary, which may entail considerable weight. In order to be able to transport the instruments quickly and safely to remote sites in the field, accommodation in a van designed for the purpose has stood the test of time. A station wagon is mostly suitable for maintenance and monitoring work, but for longer measurements of stray currents, a larger van in which one can stand up is used. Data on the most important measuring instruments are given in Section 3.3 and Table 3-2. The time to set up the connections can be reduced with a transverse distributor that is connected to terminals on the outside of the van and to the measuring instruments. A 12-V battery and a 220-V transformer are suitable as a current supply and for operating recording equipment. Information on measurement test points, the measurement range, and results as well as time and specific conditions are to be entered in a protocol. If no synchronous recordings are possible, mean and extreme values should be noted and compared with the tendencies of other varying parameters, e.g., potential and pipeline current.

The rapid development of microelectronics has enabled many similar measurements to be made with data collecting systems and then stored electronically. The raw data can then be downloaded to the data processing installation, where they can be plotted and evaluated at any time [1]. This applies particularly to monitoring measurements on pipelines; for intensive measurements, see Section 3.7. Figure 3-1 shows an example of a computer-aided data storage system.

3.1 The Electrical Parameters: Current, Voltage, and Resistance

In all electrical measurements, current and voltage measuring instruments with two terminals are employed. The object being measured similarly has two termi-

Table 3-1 Data and application range of important reference electrodes

Reference electrode	Me/Me$^+$ system	Electrolyte	Potential U_H at 25°C (V)	Temperature dependence (mV/°C)	Application
Cu-CuSO$_4$	Cu/Cu^{2+}	Sat. CuSO$_4$	+0.32	0.97	Soils, water
Ag-AgCl	Ag/Ag$^+$	Sat. KCl	+0.20	1.0	Saline and fresh water
Sat. calomel	Hg/Hg$_2^{2+}$	Sat. KCl	+0.24	0.65	Water, laboratory
1 M calomel	Hg/Hg$_2^{2+}$	1 M KCl	+0.29	0.24	Laboratory
Hg$_2$SO$_4$	Hg/Hg$_2^{2+}$	Sat. K$_2$SO$_4$	+0.71		Chloride-free water
Mercuric oxide	Hg/Hg$_2^{2+}$	0.1 M NaOH	+0.17		Dilute caustic soda
Mercuric oxide	Hg/Hg$_2^{2+}$	35% NaOH	+0.05		Concentrated caustic soda
Thalamid	Tl/Tl$^+$	3.5 M KCl	−0.57	<0.1	Warm media
Ag-saline	Ag	–	+0.25		Seawater and brine[a]
Pb-H$_2$SO$_4$	Pb/Pb^{2+}	–	−2.8		Concentrated sulfuric acid
Zn-saline	Rest potential	–	−0.79[b] −0.77±0.01		Seawater and brine
Zn-soil	Rest potential	–	−0.8±0.1		Soil
Fe-soil	Rest potential	–	−0.4±0.1		Soil
Stainless steel-soil	Rest potential	–	About −0.4 to +0.4		Soil

[a] The reference potential for other solutions containing Cl$^-$ ions must be determined.
[b] Activated with Hg.

nals which either correspond to both measurement connections (e.g., object and reference electrode) or the two ends of an open circuit. Every measuring instrument and every object to be measured is a two-terminal network, which is described by the characteristic $I(U)$.

In principle, when a measurement is made, the characteristic curves of the measuring instrument and those of the object being measured should intersect so that the coordinates of the intersection give the measured values. Figure 3-2 shows an $I(U)$ characteristic curve of the object being measured and the curves of two different measuring instruments, 1 and 2. The measured objects are in general two-terminal networks with a short-circuit current, I_0 at $U = 0$ and an electromagnetic force U_0 at $I = 0$. Such two-terminal networks are also called active two-terminal networks. Measuring instruments, on the other hand, are generally passive networks whose

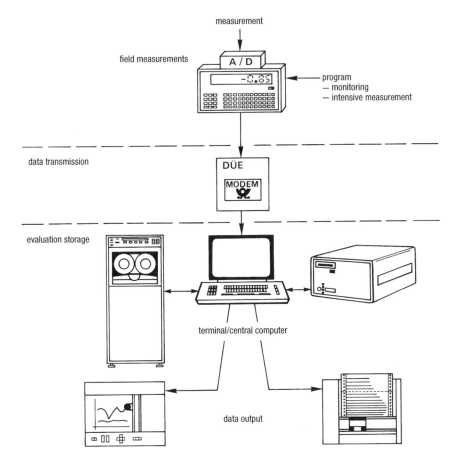

Fig. 3-1 Computer-aided data storage system for monitoring the cathodic protection of a long-distance pipeline.

Table 3-2 Survey of measuring instruments for corrosion protection measurements

	Measuring instrument	Type	Range	Input resistance/ voltage drop
1.	Electronic multimeter	dc and ac dc and ac	100 mV – 1000 V 10 μA – 10 A	10 MΩ
2.	Sensitive multimeter electronic	dc and ac dc and ac (potential, pipe current)	1 mV – 1000 V 1 μA – 3 A	1 MΩ/V; max 10 MΩ 1 – 500 mV
3.	Amplifier-voltmeter	dc	1 μA – 1000 V	1 MΩ – 100 MΩ
4.	Recording potentiometer	dc (potential)	10 mV – 500 mV	1 MΩ – 25 MΩ
5.	Multichannel recorder dc and ac (potential)	dc and ac potential	1 mV – 10 mV	1 MΩ – 10 MΩ
6.	Resistance measuring instrument dc and ac (potential)	Resistances (soil resistance, grounding resistance)	0 – 999 kΩ	1 kΩ
7.	Computerized measuring instrument for intensive measurement	dc (potential)	–5 V + 5 V	40 MΩ

Table 3-2 Survey of measuring instruments for corrosion protection measurements (*continued*)

Response	Scale length (mm)	Weight (kg)	Dimensions (mm)	Power demand (W)	Current supply	Remarks
—	101	0.45	146×118×44	0.004	9-V flat cell battery and mains instrument	Resistance measurement 1.0 Ω–20 $\mu\Omega$
0.8	110	1.5	205×128×100	0.015	V cells	Built-in filter for ac
0.3	2 × 35	3.0	205×160×170	2.0	4 × 9-V batteries or 220 V	1-V output for recorder, zero in center of scale
0.5	100	3.5	260×205×105	0.45	Unicell or 6-V and mains instrument	2 amplifiers, X-Y recorder
0.5	250	7.5	435×350×150	12	220-V or 12-V accumulator	2 recorders
—	LCD-indicator	1	185×70×170	2.1	6-V cells	108 Hz ± 4% automatic adjustment
—	LCD-indicator	3.6	420×300×100	—	NiCd 6-V cells	Damping 16 2/3 Hz: 40 dB 50 Hz: 45 dB

characteristic curves are straight lines through the origin. They are defined clearly by the internal resistance of the instrument. In Fig. 3-2, the internal resistance of instrument 1 is cot α and that of instrument 2, cot β. The measuring instrument should, if possible, be static with a rapid response, i.e., with a nonstationary pair of values (U,I). In addition, the stationary characteristic curve of the measuring instrument should be reached in a very short time. In contrast, two-terminal networks with capacities and inductances as well as electrochemical two-terminal networks are not static, but dynamic. Besides the measured values (U_1, I_1) and (U_2, I_2), there are nonstationary states of the measured object where the measured values all lie on lines 1 or 2. That makes it clear that measuring instruments with a static stationary characteristic curve are needed.

In measuring voltage, instrument 1 reads U_1 instead of U_0. The error in measurement becomes smaller with decreasing measured current I, and corresponding decreasing α, i.e., with increasing internal resistance. Voltmeters must be as high resistance as possible. The usual moving coil voltmeters have internal resistances of about 10 kΩ per volt ($I_1 = 0.1$ mA) and are not suitable for measuring potential. High-resistance instruments with about 1 MΩ per volt ($I_1 = 1$ μA) are usual in practice. Stationary potentials can be measured with them; their response times are, however, somewhat long (>1 s). For potential measurements, analog-reading electronic amplifier-voltmeters with resistances of 10^7 to 10^{12} Ω are generally used. The response times are <1 s, with electronic displays <1 ms.

In measuring current with instrument 2, a reading of I_2 instead of I_0 is obtained. Here the error is smaller with decreasing measured voltage U_2 and correspondingly increasing β, i.e., decreasing internal resistance. That means that in measuring current, the instrument must have as low an internal resistance as possible in order not to increase the total resistance of the circuit and thereby alter the measured values. The usual moving coil instruments have internal resistances of 100 Ω per mA^{-1}

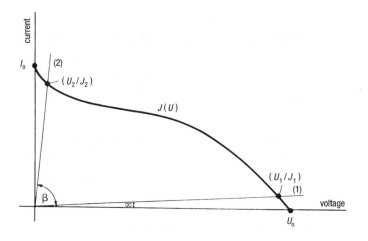

Fig. 3-2 Current and voltage measurement in a current-voltage diagram (explanation in the text).

($U_2 = 0.1$ V) and are suitable for current measurement. For smaller currents, sensitive instruments with 5 kΩ per μA^{-1} ($U_2 = 5$ mV) are used. Small currents are usually measured by the voltage drop across a fixed resistance (calibrated shunt) using an electronic amplifier-voltmeter. This method has the advantage that the circuit does not have to be interrupted to measure the current.

Resistance is measured either indirectly by separate current and voltage measurements or directly by comparison in a bridge. In both cases it involves two measurements. The instruments for measuring the current and voltage U_2 have to be chosen so as to make the deviation from U and I in Fig. 3-2 as small as possible.

In contrast to simple current and potential measurements, polarization measurements (i.e., current-potential curves) demand active systems with active external circuits with variable *characteristic curves* (see Fig. 2-3). These external circuits should be as static as possible so that all nonstationary values lie on the known curve, which is known as the resistance straight line of the external circuit. External circuits for electrochemical protection have steep straight lines in the $I(U)$ diagram (i.e., low internal resistance) because the potential can be well controlled independent of the current. The usual dc sources with high internal resistance are less suitable because changes in current requirements cause correspondingly large changes in voltage. For certain systems (e.g., groups II and IV in Section 2.4) only low-resistance protective rectifiers can be used (see Chapter 21).

3.2 Reference Electrodes

Reference electrodes are used in the measurement of potential [see the explanation to Eq. (2-1)]. A reference electrode is usually a metal/metal ion electrode. The electrolyte surrounding it is in electrolytically conducting contact via a diaphragm with the medium in which the object to be measured is situated. In most cases concentrated or saturated salt solutions are present in reference electrodes so that ions diffuse through the diaphragm into the medium. As a consequence, a diffusion potential arises at the diaphragm that is not taken into account in Eq. (2-1) and represents an error in the potential measurement. It is important that diffusion potentials be as small as possible or the same in the comparison of potential values. Table 3-1 provides information on reference electrodes.

Equations (2-13), (2-23), and (2-28) can be combined to give the following equation for the migration of ions:

$$\vec{w}_i = -\frac{u_i}{|z_i|}\left(\frac{RT}{c_i \mathcal{F}}\ \text{grad}\ c_i + z_i\ \text{grad}\ \phi\right) \tag{3-1}$$

The mobility u_i can be taken from Table 2-2 for dilute solutions and is proportional to the diffusion coefficient D_i. It follows, for the transport of anions and cations of an n–n valent salt, that

$$n = z_A = -z_C \text{ and } c = c_A = c_C \text{ from Eq. (3-1)}$$

$$w_A = -u_A \left(\frac{RT}{nc\mathscr{F}} \text{ grad } c + \text{ grad } \phi \right) \qquad (3\text{-}2a)$$

$$w_C = -u_C \left(\frac{RT}{nc\mathscr{F}} \text{ grad } c - \text{ grad } \phi \right) \qquad (3\text{-}2b)$$

The indices A and C correspond to anion and cation. In the stationary state w_A and w_C are equal because of the charge balance. It follows from Eqs. (3-2a) and (3-2b) and Eq. (2-13) that:

$$\frac{d\phi}{d \ln c} = \frac{l_C - l_A}{l_C + l_A} \frac{RT}{n\mathscr{F}} \qquad (3\text{-}3)$$

and after integration

$$\phi_1 - \phi_2 = \Delta\phi_{\text{Dif}} = \frac{l_C - l_A}{l_C + l_A} \frac{RT}{n\mathscr{F}} \ln \frac{c_1}{c_2} \qquad (3\text{-}4)$$

Instead of Eq. (2-1), it follows from Eq. (3-4) with $\Delta\phi_B = \phi_B - \phi_1$ and $\Delta\phi = \phi_{Me} - \phi_2$:

$$U = (\Delta\phi - \Delta\phi_B) - \Delta\phi_{\text{Dif}} \qquad (3\text{-}5)$$

The error due to diffusion potentials is small with similar electrolyte solutions ($c_1 \approx c_2$) and with ions of equal mobility ($l_A \approx l_C$) as in Eq. (3-4). This is the basis for the common use of electrolytic conductors (salt bridge) with saturated solutions of KCl or NH_4NO_3. The l-values in Table 2-2 are only applicable for dilute solutions. For concentrated solutions, Eq. (2-14) has to be used.

In the same way that potential differences can occur due to different mobility, they can also occur due to different adsorption of ions. There are therefore a large number of possibilities for potential errors in the field of reference electrodes [2], which, however, are generally less than 30 mV. Such potential errors can be neglected in the application of protection potential criteria, but they can lead to increased error in the evaluation of voltage cones (see Section 3.3.1). Equation (3-4) can be used for their evaluation in this case. It explains, for example, the increased potential difference that can occur between two reference electrodes when the reference electrode is immersed in strong acid. In general the errors are small in salt solutions and only to be expected in salt-free and cohesive soils [2].

Greater deviations which are occasionally observed between two reference electrodes in a medium are mostly due to stray electric fields or colloid chemical dielectric polarization effects of solid constituents of the medium (e.g., sand [3]) (see Section 3.3.1). Major changes in composition (e.g., in soils) do not lead to noticeable differences of diffusion potentials with reference electrodes in concentrated salt solutions. On the other hand, with simple metal electrodes which are sometimes used as probes for potential controlled rectifiers, certain changes are to be expected through the medium. In these cases the concern is not with reference electrodes, in principle, but metals that have a rest potential which is as constant as possible in the medium concerned. This is usually more constant the more active the metal is, which is the case, for example, for zinc but not stainless steel.

In addition, the temperature dependence of the diffusion potentials and the temperature dependence of the reference electrode potential itself must be considered. Also, the temperature dependence of the solubility of metal salts is important in Eq. (2-29). For these reasons reference electrodes with constant salt concentration are sometimes preferred to those with saturated solutions. For practical reasons, reference electrodes are often situated outside the system under investigation at room temperature and connected with the medium via a salt bridge in which pressure and temperature differences can be neglected. This is the case for all data on potentials given in this handbook unless otherwise stated.

In practice, different reference electrodes are used, depending on the medium and function. In particular, the following should be observed:

(a) stability of the reference electrode potential with time;
(b) grounding resistance and current loading ability;
(c) resistance against constituents of the medium and environmental influences as well as compatibility with the system being measured.

Point (a) only concerns simple metal electrodes and needs to be tested for each case. Point (b) is important for the measuring instrument being used. In this respect, polarization of the reference electrode leads to less error than an ohmic voltage drop at the diaphragm. Point (c) has to be tested for every system and can result in the exclusion of certain electrode systems for certain media and require special measures to be taken.

Cu-$CuSO_4$ electrodes with saturated $CuSO_4$ solution are recommended for potential measurements in soil. Their potential constancy is about 5 mV. Larger errors can be traced to chemical changes in the $CuSO_4$ solution. These electrodes have been developed for long-life applications in potential-controlled rectifiers and built-in potential measuring instruments on account of their robustness [4]. Their grounding resistance in the built-in arrangement in soil with $\rho = \leq 100\ \Omega$ m lies generally below 1000 Ω.

On the other hand, Cu-CuSO$_4$ should not be used as built-in electrodes for potential test probes (see Section 3.3.3.2) because there is a danger of copper precipitating on the steel electrode. Calomel electrodes with saturated KCl solution are preferred in this case and present no problems.

3.3 Potential Measurement

3.3.1 Bases of Potential Measurement of Electrodes with Flowing Current

According to the explanation of Fig. 2-3 and Eq. (2-33), there is always an ohmic voltage drop η_Ω in the potential measurement of polarized electrodes which is measured as an error. This can be kept small by using capillary probes on the reference electrode. This is used in laboratory experiments but in practice it is usually not feasible (e.g., with measurements in soils). Here one tries to separate electrochemical polarization from η_Ω by other methods. This is accomplished by making use of the different time dependencies [5]. It is assumed that the electrical properties of the phase boundary surface and of the medium can be approximately represented by a parallel circuit of resistance and capacity. These RC members are connected in series in an equivalent circuit with a voltage source U_R for the rest potential. With the specific parameters r_P and C_D for the phase boundary surface and r_M and C_M for the medium, the following expression can be derived for a galvanostatic switching experiment [6] from J_1 to J_2:

$$U(t) = U(J_2) + (J_1 - J_2)\left(r_P \exp\frac{-t}{\tau_P} + r_M \exp\frac{-t}{\tau_M}\right) \qquad (3\text{-}6)$$

where $U(J_2) = U_R + (r_P + r_M) J_2$ is the stationary potential. For the time constants:

$$\tau_P = r_P C_D \qquad (3\text{-}7)$$

$$r_M = \frac{s}{\varkappa},\ C_M = \frac{\varepsilon\varepsilon_0}{s},\ \tau_M = r_M C_M = \frac{\varepsilon\varepsilon_0}{\varkappa} \qquad (3\text{-}8)$$

The possible values for τ_P lie between 10^{-5} and 10^{-1} s according to the data in Section 2.2.3.2, but with a concentration polarization this can be much greater. In Eq. (3-8)

$\varepsilon_0 = 8.85 \times 10^{-14}\ \Omega^{-1}\ \text{s cm}^{-1}$, and ε is the relative dielectric constant. The thickness of the layer of medium s is cancelled out in τ_M. For aqueous media with $\varepsilon = 80$,

$$\left(\frac{\tau_M}{\mu s}\right) = \left(\frac{7.1}{\varkappa/\mu S\ \text{cm}^{-1}}\right) \tag{3-8'}$$

and consequently for water and soils where $\varkappa \gg 1\ \mu S\ \text{cm}^{-1}$, $\tau_M < 10^{-6}\ \text{s} \ll \tau_P$ [3, 5]. For a potential measurement immediately after the switch, $\exp[-t/\tau_P] \approx 1$ and $\exp[-t/\tau_M] = 0$ so that from Eq. (3-6) it follows:

$$U_{\text{over}} = U(J_2) + (J_1 - J_2)r_P = U(J_1) + (J_2 - J_1)r_M \tag{3-9}$$

U_{over} is the switching potential. The stationary potential before the switch $U(J_1)$ is the on potential U_{on}[1]:

$$U(J_1) = U_{\text{on}} = U_R + (r_P + r_M)J_1 \tag{3-10}$$

U_{on} contains the ohmic voltage drop $r_M J_1$ and the IR-free polarization potential

$$U_{IR\text{-free}} = U_R + r_P J_1 \tag{3-11}$$

It follows by rearranging Eqs. (3-9) to (3-11):

$$U_{\text{on}} = U_{IR\text{-free}}(J_1) + r_M J_2 \tag{3-12}$$

$$\frac{U_{\text{over}} - U_{\text{on}}}{J_2 - J_1} = r_M \tag{3-13}$$

$$U_{IR\text{-free}} = \frac{U_{\text{on}} J_2 - U_{\text{over}} J_1}{J_2 - J_1} \tag{3-14}$$

For $J_2 = 0$, it follows from Eqs. (3-9) to (3-14) that the off potential U_{off}:

$$U_{\text{off}} = U_{\text{on}} - r_M J_1 = U_{IR\text{-free}} \tag{3-15}$$

This result is immediately confirmed by the results given in Fig. 3-3. A steel electrode was cathodically polarized in a soil sludge. The potential was measured with a capillary probe IR-free as E_1 and without a probe as E_2. The difference directly

---

[1] For brevity, the terms "on potential" and "off potential" are used throughout to refer to a switched-on potential and a switched-off potential.

90 Handbook of Cathodic Corrosion Protection

gives the ohmic voltage drop. When the polarizing current is switched off, this difference momentarily disappears. Both values coincide and slowly reach the rest potential.

In the previous case a depolarization of 50 mV occurs within 0.1 s. The off potential measurement must accordingly take place within 0.1 s. Depending on the formation of surface films which give rise to an increase of polarization, the measuring time can be extended to about 1 s. This is generally the case for steel in neutral waters and soils after long operation. With unknown systems in a laboratory investigation, it should always be stated how quickly the electrochemical polarization changes on switching in order to determine measurement details for estimating IR-free potentials.

The validity of Eq. (3-15) further follows from the results given in Fig. 3-4. It represents the current-potential curve for a buried storage tank. The difference $U_\mathrm{on} - U_\mathrm{off}$ is proportional to the current I. The quotient R is equal to the grounding resistance of the tank. The $U_\mathrm{off}(I)$ curve corresponds to the true polarization curve.

An exact measurement of a true IR-free potential is only possible if one is dealing with a homogeneous mixed electrode, not a heterogeneous one (see Figs. 2-6 and 2-7). With heterogeneous mixed electrodes in free corrosion, separate surface regions can be polarized by the cell current, which also leads to an ohmic voltage drop in the medium. Since heterogeneous mixed electrodes are the normal case in practice, especially with extended objects such as pipelines, the action of cell currents is also of interest. In this respect it is assumed that isolated surface regions at defects in the coating have an IR-free potential U_n and a grounding resistance R_n. At all positions a uniform switching-on potential is measured:

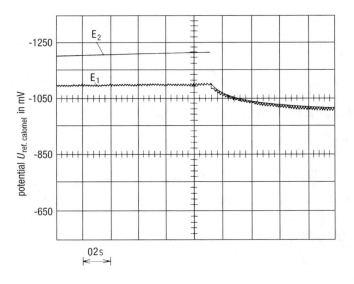

Fig. 3-3 Potential-time oscillogram for off potential measurement (explanation in the text).

$$U_{\text{on}} = U_n + I_n R_n \tag{3-16}$$

where I_n is the polarization current for an individual isolated region. The current for the entire object is given by:

$$I_1 = \sum_n I_n = \sum_n \frac{U_{\text{on}} - U_n}{R_n} \tag{3-17}$$

Correspondingly, for a switching experiment from I_1 to I_2:

$$I_2 = \sum_n \frac{U_{\text{over}} - U_n}{R_n} \tag{3-18}$$

Using Eqs. (3-13) and (3-15) and designating the currents by

$$U_{\text{over}} - U_{\text{on}} = (I_2 - I_1) R_M \tag{3-13'}$$

$$U_{IR\text{-free}} = U_{\text{on}} - R_M I_1 \tag{3-15'}$$

it follows from these expressions and Eqs. (3-17) and (3-18):

$$U_{IR\text{-free}} \times \sum_n \frac{1}{R_n} = \sum_n \frac{U_n}{R_n} \tag{3-19}$$

The *IR*-free potential is thus always the average of the single potentials U_n. The method of determination (i.e., the contribution of I_2) theoretically has no influence. In practice, however, with a small difference $I_2 - I_1$, a related inexact potential

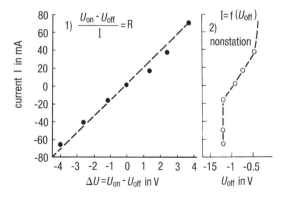

Fig. 3-4 Current-potential measurements on a buried steel storage tank with bitumen coating (surface area 4 m² with four defects 5 cm × 20 cm; soil resistivity $\rho \approx 30\ \Omega$ m).

difference $U_{over} - U_{on}$ causes an increase in the error. As an example, Eq. (3-19) is simplified for only two regions ($n = 1,2$):

$$U_{IR\text{-free}} = \frac{U_1 R_2 + U_2 R_1}{R_1 + R_2} \tag{3-19'}$$

For $U_1 = -0.7$ V, $U_2 = -0.9$ V, $R_1 = 10^4$ Ω and $R_2 = 10^3$ Ω; this gives $U_{IR\text{-free}} = -0.88$ V. This example shows that it is not practically possible to find relatively high ohmic places without extensive lowering of potential. Such places only occur in particularly unfavorable circumstances, e.g., in high-resistance stony soils. In the normal case, high resistances are coupled with sufficient polarization and vice versa. These can be represented as an example by a circular defect of radius r. From Eq. (3-10) with $r_M = R\pi r^2$ and R according to Eq. (24-17), it follows that:

$$U_{on} = U_R + J\left(r_P + \frac{\pi r^2}{4\varkappa r}\right) = U_R + \left(r_P + \frac{\pi r}{4\varkappa}\right) J \tag{3-20}$$

Replacing J from Eq. (3-20) in Eq. (3-11) gives:

$$U_{IR\text{-free}} = U_R + \frac{U_{on} - U_R}{1 + \frac{\pi r}{4\varkappa r_P}} \tag{3-21}$$

$U_{IR\text{-free}}$ reduces, for very large holidays ($r \to \infty$ and $R_n \to 0$), to U_R (no polarization!) and for very small holidays, ($r \to 0$ and $R_n \to \infty$), to U_{on} (maximal polarization!). The value of $U_{IR\text{-free}}$ is independent of J_2 both for homogeneous electrodes [Eqs. (3-11) or (3-14)] and also for heterogeneous electrodes [Eqs. (3-19)]; (i.e., it is the same in switching-off experiments and in reverse switching experiments). The reverse switching technique has, however, an advantage if τ_P is very small and consequently significant levels of electrochemical polarization are eliminated together with the ohmic voltage drop. This is possible for activation polarization but not for concentration polarization (see Section 2.2.3.2). In this the simplifying assumption is made that r_P is independent of J. According to Eq. (2-35) and Fig. 2-4, this is not the case. In the case of activation polarization, a different quotient can be derived from Eq. (2-35) with $G_A \to \infty$ and G_C for high cathodic overvoltages with $\exp[\eta/\beta_+] \approx 0$:

$$r'_P = \frac{\ln J_2 - \ln J_1}{J_2 - J_1} \beta_- = \beta_- \frac{f(x)}{J_1} \tag{3-22}$$

where $x = J_2 : J_1$ and $f(x) = (\ln x)/(x - 1)$; $f(x)$ has the following values:

x:	2	1.5	1	0.9	0.8	0.7	0.6	0.5	0.4
$f(x)$:	0.69	0.81	1.00	1.05	1.12	1.19	1.28	1.39	1.53

x:	0.3	0.25	0.2	0.15	0.1	0.05	0.02	0.01	0.001
$f(x)$:	1.72	1.85	2.01	2.23	2.56	3.15	3.99	4.65	6.91

For $1.5 < x < 0.5$, $f(x)$ is approximately 1. Such values, however, can only be measured through switching. For $x > 1$, one even has to switch J_2 to be larger than J_1. On switching off, x is very small where J_2 does not fall to zero but to the level of the rest potential. $f(x)$ can, however, become very large. The essential fact is that with increasing J_1, r'_P decreases for values of x that are not too small.

Equation (3-15) is the basis for determining IR-free potentials. On switching off, the activation polarization with resistance r_P, instead of Eq. (3-13), becomes:

$$U_{over} - U_{on} = (J_2 - J_1)(r_M + r'_P) \tag{3-23}$$

Then from Eqs. (3-15), (3-22), and (3-23), instead of Eq. (3-14):

$$U_{IR\text{-}free} - \beta_- f(x) = \frac{U_{on} J_2 - U_{over} J_1}{J_2 - J_1} = \frac{U_{on} x - U_{over}}{x - 1} \tag{3-24}$$

and after rearranging:

$$U_{over} = [U_{IR\text{-}free} - \beta_- f(x)] + x[U_{on} - U_{IR\text{-}free} + \beta_- f(x)] \tag{3-25}$$

$U_{IR\text{-}free}$ can thus be approximately correctly measured by extrapolating the $U_{over}(x)$ curve to $x \to 0$ from x values between 1.0 and 0.5. The error $\beta_- f(x)$ can be about 50 mV according to Eq. (2-36). Figure 3-5 shows results for lead and steel. U_{off} lies on the straight line for steel. In agreement with Fig. 3-2, electrochemical polarization is not eliminated. This is, however, the case with lead if $U(t)$ is not recorded very quickly ($\tau < 1$ ms) [7].

All methods of IR-free technique involve finally switching-off and reverse switching techniques. In these are included bridge methods [5, 8, 9], ac methods, and impulse methods [10, 11]. All of these have no practical significance. Finally, measurement methods are also derived from Eq. (3-25) with the help of varying potential gradients on the soil surface, so that x is replaced by:

$$x = \frac{J}{J_0} = \frac{\Delta U}{\Delta U_0} \tag{3-26}$$

In the normal case J_0 is the current density in the area of the measuring point, and U_0 is a potential difference at the ground above the pipeline (e.g., $U_B^\perp$ in Fig. 3-24).

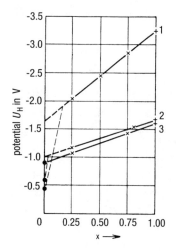

Fig. 3-5 Switching potential measurements for lead (1,2) and steel (3) in soil solution: + on potential; x switching-over potential; • off potential; $x = I_2/I_1$.

The data for J and ΔU in Eq. (3-26) change simultaneously with changes in the protection current or by the action of stray currents. Inserting them in Eq. (3-25) gives:

$$U = \left[U_{IR\text{-free}} - \beta_- f(x)\right] + \Delta U \, A \tag{3-25'}$$

where $A = [U_{on} - U_{IR\text{-free}} + \beta_- f(x)]/U_0$. Since A is a constant, $U_{IR\text{-free}}$ is obtained from Eq. (3-25') by extrapolation of the measured value of $U(\Delta U)$ to $\Delta U \to 0$. Equation (3-26) only applies when ΔU contains no foreign IR terms. For this reason ΔU should be obtained from symmetrically measured values [12,13]. Measurements have shown that this method is applicable for relatively poor coating, but not for well-coated pipelines with only occasional holidays, because obviously the proportionality of Eq. (3-26) is not always true [13]. In addition, errors can arise from differences in reference electrodes (see Section 3.2).

All techniques fail when Eq. (3-8) is not applicable with a small τ_M. It has been observed in a few cases with pure sandy soils that electrical polarization of the sand can occur, which falsifies a potential measurement made without probes [3]. The potential is consequently found to be noticeably too negative. Off potentials that are not realistic may be measured; they may have $U_{Cu\text{-}CuSO_4}$ values ranging from about −1.7 V to more negative values. This polarization effect of the soil does not occur when soluble salts are present or if the conductivity increases. The effect then only occurs when neither galvanic nor stray current corrosion due to high resistances can take place, or when the soil is not corrosive due to the absence of ions (see Chapter 4). Consequently an erroneous measurement limited by these effects does not lead to damage on account of an incorrect potential.

Ohmic voltage drops of the protection current were exclusively involved in the processes for *IR*-free potential measurements described above. Besides this, other foreign currents can cause potential drops and falsify the potential measurement (e.g., cell currents, equalizing currents and stray currents).

Cell currents and equalizing currents, which occur as a consequence of a local very nonuniform polarization of the object being protected when the protection current is switched off can be recognized by potential gradients on the soil surface with switching off of the protection current. The following paragraph shows a method of calculating *IR*-free potentials with the help of these potential gradients [2,13].

Application of this method or Eq. (3-25′) in the presence of stray currents is conceivable but would be very prone to error. It is particularly valid for good coating. Potential measurement is then only significant if stray currents are absent for a period, e.g., when the source of the stray current is not operating. In other cases only local direct measurements with the help of probes or test measurements at critical points can be considered. The potential test probes described in Section 3.3.3.2 have proved true in this respect.

For *IR*-free potential measurement with cell or equalizing currents it is assumed, analogous to Eq. (3-26), that current densities and potentials are proportional:

$$\frac{U_{IR\text{-free}} - U_{on}}{U_{IR\text{-free}} - U_{off}} = \frac{\Delta U_{on}}{\Delta U_{off}} = \frac{J_s}{J_e} \tag{3-27}$$

The first term in Eq. (3-27) represents the voltage drop between the reference electrode over the pipeline and the pipe surface. The second term represents the potential difference ΔU measured at the soil surface (ground level) perpendicular (directly above) to the pipeline. Average values of the $U_B^\perp$ values measured to the left and right of the pipeline are to be used (see Fig. 3-24) [2]. In this way stray *IR* components can be eliminated. The third term comprises the current densities where, in the switched-off state of the protection installation, there is a cell current J_e. In the normal case $J_e = 0$ and also correspondingly $\Delta U_{off} = 0$ as well as $U_{IR\text{-free}} = U_{off}$. On rearranging Eq. (3-27)

$$U_{IR\text{-free}} = U_{off} + \Delta U_{off} \frac{U_{off} - U_{on}}{\Delta U_{on} - \Delta U_{off}} \tag{3-28}$$

If ΔU_{on} and ΔU_{off} have the same sign, $U_{IR\text{-free}}$ is more positive than U_{off}. This arises when the measuring point has a more positive potential than the surroundings. This means that regions with a relatively positive pipe/soil potential have a still more positive potential than the measured U_{off} value! If ΔU_{off} and ΔU_{on} have different signs, then $U_{IR\text{-free}}$ is more negative than U_{off}. This occurs, for example, with local cathodic protection when there are foreign cathodic structures (see Chapter 12).

The second term of Eq. (3-28) is very prone to errors because the factor ΔU_{off} includes electrode errors (see Section 3.2). For small values of ΔU_{off}, Eq. (3-28) is not applicable in practice [2]. Applications are described in Section 3.7 along with intensive measurement.

3.3.2 Application of Potential Measurement

3.3.2.1 *Measuring Instruments and Their Properties*

As explained in Sections 3.1 and 3.2, potentials are measured with high-impedance electronically amplified voltmeters. The amplified voltmeters have high input resistances between 1 and 100 MΩ. The measured voltage is transformed electronically into an alternating voltage, amplified in two stages, rectified by a modulator, and fed via a direct voltage amplifier acting as integrator to a moving coil measuring instrument or digital display. Superimposed alternating voltages are attenuated with preconnected RC filters of 60 to 90 dB (which corresponds to a ratio of 1000 to 30,000:1) (see Section 3.3.2.4). For measurements in soil, analog measuring instruments are generally used. Digital display instruments are only suitable for relatively steady potential readings. The power rating of amplifiers is only a few tenths of a watt. They are usually supplied from nickel-cadmium cells. Their life is about 60 h.

Amplifying voltmeters generally have an output that can be connected to a recording and measuring instrument. Thus, recorders with low input impedance can also be used for recording results with high source resistance. High-impedance multimeter instruments used in electrotechnology for measuring voltage, current and resistance can also be employed in potential measurement. Multimeter instruments are mostly moving coils with tension-wire adjustment. They are robust, insensitive to temperature changes and have a linear scale. For response times of <1 s, which is necessary for potential measurement, these instruments have a maximum internal resistance of 100 kΩ per volt. Since the source resistance of reference electrodes with a large area is below 1 kΩ, sufficiently accurate potential measurements are possible with such instruments. With potential measurements in high-resistance sandy soils or in paved streets (small diaphragm), the source resistance of the reference electrode can, however, considerably exceed 1 kΩ. Table 3-2 includes data from potential measuring instruments, recorders as well as dc and ac multimeter instruments, which are frequently used in corrosion protection measurements.

The decay time of electrochemical depolarization depends not only on the prepolarization but also on the quality of the coating (see Fig. 3-6). With newly installed pipelines and storage tanks, it is generally only necessary to carry out a drainage test for a few hours. Figure 3-7 shows the potential decay of a polarized steel surface, which was recorded after switching off the protection current using a fast recorder with a response time of 2 ms for 10 cm at different chart speeds. The off potential at a chart speed of 1 cm/s corresponds to the value measured with an amplified voltmeter. It can be seen in Fig. 3-7 that the error arising from measuring instruments with response times of 1 s amounts to about 50 mV, because a small part of the polarization is measured as a part of the ohmic drop [10]. It is necessary in measuring off potentials that the measuring instrument have a response time of <1 s and aperiodic damping. The response time of multimeter instruments depends on the input resistance of the instrument and on the source resistance of the object to be measured, and in amplifying voltmeters on the amplifier circuit. The response time can be determined with the circuit shown in Fig. 3-8 [14]. The internal resistance of the current and voltage source which is being measured is demonstrated by a resistance R_p connected in parallel with the measurement instrument. R_v and R_p are suitable reverse switching decade resistances (20 to 50 kΩ). The potentiometer R_T (about 50 kΩ) serves as the input of the instrument to be tested on the full range. With aperiodic damped instruments, the response time of the indicator is stopped on reaching 1% of the end or the beginning of the scale. In instruments that overshoot, the time of the indicator movement, including the overshoot, is recorded and simultaneously the magnitude of the overshoot is determined as a percentage of the scale reading. Table 3-2 contains response times of some common instruments used in corrosion protection measurements (for recording instruments, see Section 3.3.2.3).

3.3.2.2 Potential Measurements on Pipelines and Storage Tanks

Errors can occur in potential measurements on pipelines and storage tanks in soil if foreign voltages such as, for example, ohmic voltage drops in soil, are not taken into account [15]. Figure 3-9 shows the potential curves for a single defect (spherical field) or for several statistically distributed defects (cylindrical field) in the pipeline coating. In general the on potential of the installation to be protected (e.g., a pipeline) is measured, with flowing protection current, against a reference

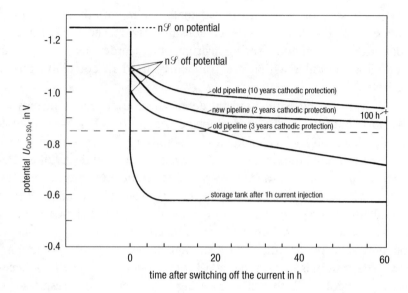

Fig. 3-6 Electrochemical depolarization after switching off the protection current as a function of the operating time of cathodic protection and the quality of the coating.

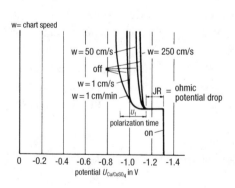

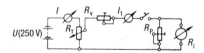

Fig. 3-8 Circuit for determining the response time of measuring instruments with internal resistance R_i (explanation in the text).

Fig. 3-7 Electrochemical depolarization after switching off the protection current for different recording speeds (polarization of steel in artificial soil solution for 200 h).

electrode installed above it on the ground. This on potential contains ohmic voltage components, according to the data in Section 3.3.1, which must be eliminated when determining the state of polarization. In the method used here, the current must be definitely changed, which is not possible with stray currents caused by dc railways.

For determining the off potentials of cathodically protected pipelines, time relays are built into the cathodic protection station to interrupt the protection current synchronously with neighboring protection stations for 3 s every 30 s. The synchronous on and off switching of the protection stations is achieved with a synchronous motor activated by a cam-operated switch. The synchronization of the protection station is achieved as follows: a time switch is built into the first protection station. An interruption of the protection current is detectable at the next protection station as a change in the pipe/soil potential. Since the switching time is known, the time switch of the second protection station can be activated synchronously. The switching of further protection stations can be synchronized in the same manner.

Other timing switch gear works with electronic digital clocks. The 1-s time interval derived from the mains frequency is counted in an integrating counter. The

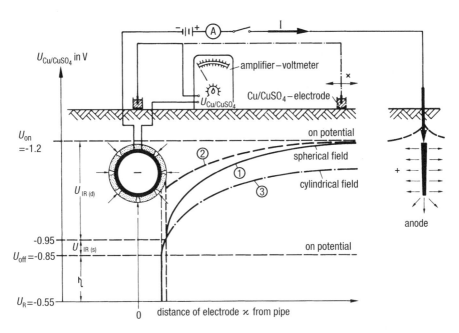

Fig. 3-9 Ohmic voltage drop at defects in the pipe coating. (1) spherical field; (2) voltage drop at the soil surface; (3) cylindrical field $U_{IR}(s)$ = voltage drop at the defect; $U_{IR}(d)$ = voltage drop at the grounding resistance.

switching on and switching off times can be set as desired. A completely synchronized switching of all protection systems can be effected by a transmitted impulse. For example, the Bundespost transmitter (Mainflingen) emits the frequency 77.5 kHz on the long wave band. On this frequency a time signal is transmitted every second, minute, etc. For identification the impulse is omitted on the 59th second. This can be used for synchronizing the switching of cathodic protection stations.

It is recommended that all instruments be provided with an additional time switch, so that the interruption of the protection current supply outside working hours (i.e., during the night) is not maintained. In this way the reduction in the protection current supply is kept as small as possible; this amounts to 10% with 27 s on and 3 s off. This can be significant in the measurement of a long pipeline, which can often take several weeks.

In analyzing the results on a cathodically protected pipeline, the protection current density and coating resistances should be calculated for individual sections of the pipeline in addition to the on and off potentials, the pipe current, and the resistances at insulating points and between the casing and the pipeline. The results should be shown by potential plots to give a good summary [15] (see Fig. 3-20).

In the cathodic protection of storage tanks, potentials should be measured in at least three places, i.e., at each end and at the top of the cover [16]. Widely different polarized areas arise due to the small distance which is normally the case between the impressed current anodes and the tank. Since such tanks are often buried under asphalt, it is recommended that permanent reference electrodes or fixed measuring points (plastic tubes under valve boxes) be installed. These should be located in areas not easily accessible to the cathodic protection current, for example between two tanks or between the tank wall and foundations. Since storage tanks usually have several anodes located near the tank, equalizing currents can flow between the differently loaded anodes on switching off the protection system and thus falsify the potential measurement. In such cases the anodes should be separated.

3.3.2.3 Potential Measurement under the Influence of Stray Currents

Where stray currents are involved, several measurements have to be taken that are continually changing with time, simultaneously with each other. A double recorder is most suitable for this. Linear recorders with direct indication of the measurements cannot be used for potential measurements because the torque of the mechanism is too small to overcome the friction of the pen on the paper. Amplified recorders or potentiometer recorders are used to record potentials. In amplified recorders, as in amplified voltmeters, the measured signal is converted into a load-independent impressed current and transmitted to the measuring mechanism, which consists of a torque motor with a preamplifier. The amplifier results in an increased torque in order to achieve a response time of 0.5 s with the necessary operating pressure of the recording pen. The energy requirement for amplified recorders is about 3 W. Technical data on recorders are given in Table 3-2.

As shown in Fig. 3-10, in potentiometer recorders with a servomotor, the auxiliary current I_k is supplied by a measuring bridge. The voltage to be measured U_x is compared with the compensation voltage U_k. The voltage difference is converted to an alternating voltage, amplified 10^6-fold and fed to the control winding of a servomotor. This moves the slider s of the measuring bridge to the left until the voltage difference $U_x - U_k$ is zero. The position of the slider indicates the value of the voltage to be measured, which is compensated and indicated by the recorder. Potentiometer recorders have a relatively short response time—up to 0.1 s—and a high accuracy of 2.5% of the maximum scale reading.

The chart movement of the recorder is driven by a synchronous motor with a rating of 2 to 5 W. With rapidly varying readings, a chart speed of 600 mm h^{-1} is needed, but for stray current fluctuations, usually 300 mm h^{-1} is sufficient. For recording over several hours, chart speeds of 120 or 100 mm h^{-1} are advantageous. Optical evaluation of recorder traces is sufficient to obtain average values as well as extreme values important for corrosion protection measurements. In general, recordings at a test point are not made for more than 0.5 to 1 h. Occasional extreme values or potentials at night are not usually recorded. The frequency and length of time that the protection potential is below the required level can be determined with a limiting value counter.

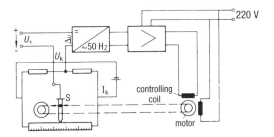

Fig. 3-10 Principle of a compensation circuit for recording potentials (explanation in the text).

Where there are stray currents, the switching method described in Section 3.3.1 cannot be used. Stray current protection stations are usually installed where the pipeline has the most positive pipe/soil potential. When the stray current drainage is cut off, a too-positive stray current exit potential that is not *IR*-free is quickly established. In distant areas a too-negative stray current entry potential that is not *IR*-free will be measured. The determination of the *IR*-free pipe/soil potential is only possible in stray current areas when the origin of the stray current is not operating [17]. In order to avoid more positive potentials than the protection potential, the pipe/soil potential in stray current areas is set much more negative for safety reasons than in installations where there is no stray current. From recordings it can be established at which places the *IR*-free pipe/soil potential should be measured when the stray current source is not active. If the potential measured at these points is more negative than the protection potential, it can be assumed that there is sufficient cathodic protection. To estimate the *IR*-free potential of cathodically protected pipelines during the operation of the stray current generator, the polarization state can be obtained with the help of potential test probes (see Section 3.3.3.2).

3.3.2.4 Potential Measurement under the Influence of Alternating Currents

In cathodically protected pipelines that are affected by high tension transmission lines or electrified railway lines, the pipe/soil potential has superimposed on it an induced alternating voltage. This can considerably falsify the potential measurement if, for example, the induced alternating voltage of some 10 V at 50 or 16⅔ Hz is superimposed on the measured voltage, which is on the order of 1 V [18]. The insensitivity to alternating voltage depends on the mechanism in amplified voltmeters and on their circuitry. If it is insufficient, an RC filter must be included in the circuit. The magnitude of the resistance, R, and capacitor, C, can be calculated to a sufficient approximation from the following equations [19]:

$$R = \frac{F}{100} R_i \qquad (3\text{-}29)$$

and

$$C = \frac{A}{2\pi f R} \qquad (3\text{-}30)$$

where R_i is the resistance of the measuring instrument, F is the allowable measuring error, A is the attenuation factor, R is the resistance of the RC filter, and f is the frequency.

Filters have a time constant $\tau = R_v \times C$ which increases the damping of the measuring instrument. The time constant depends on the required attenuation and the interfering frequency, but not on the internal resistance of the measuring instrument. The time constants of the shielding filter are in the same range as those of the electrochemical polarization, so that errors in the off potential are increased. Since the time constants of attenuation filters connected in tandem are added, but the attenuation factors are multiplied, it is better to have several small filters connected in series rather than one large filter.

Fundamentals and Practice of Electrical Measurements 103

In contrast to direct voltage, alternating voltage can be measured using a grounding rod as reference electrode. The grounding resistance of the rod is considerably lower than that of the reference electrodes in Table 3-1 but can still be too high for measurements with soft iron, moving iron or electrodynamic measuring instruments. It is therefore also recommended that amplified voltmeters or amplified recorders be used which have high internal resistances, great accuracy and a linear scale. Attention to frequency and shape of curve is important in the technique of ac measurements. In general, measuring instruments are gauged for effective values for 50 Hz and sine curves. They can therefore give false readings for diverse frequencies and waveforms (phase control). Measurement errors due to different waveforms can be recognized by the fact that they give different results in different measurement regions.

3.3.3 Application of Protection Criteria

According to the statements in Section 2-3, electrochemical protection exists when the *IR*-free potential corresponds to the criteria in Eqs. (2-39), (2-48) to (2-50). For buried installations of ferrous materials, the potential must be more negative than $U_{Cu/CuSO_4} = -0.85$ V. The details are to be found in Section 2-4. This criterion has been known for a long time [20], with the *IR*-free potential being measured best by the switching-off technique [see Section 3.3.1 and Eq. (3-15)] [21]. In practice, the preconditions for applying the switching-off technique and related methods are sometimes not given: the effect of stray current or galvanic currents from installations that are not isolated (e.g., old rusted pipelines) as well as steel foundations or ground electrodes, or pipelines affected by high voltage (see Chapter 15). In these cases other criteria can be drawn on for information. These are detailed in Table 3-3 for nonalloyed ferrous materials together with their fields of application. Criterion 1 corresponds to Eqs. (2-39), (3-15), and (3-28). This criterion is widely known (see Chapter 1); however, in contrast to Ref. 22, it is not stated in foreign standards how the ohmic voltage drop is to be eliminated (see, for example, Ref. 23). Criterion 7 has the same basis as criterion 1 [24, 25].

Criteria 2 to 6 are pragmatic criteria which are only applicable to qualitative explanations. These criteria only give qualitative information which is dealt with further below (Section 3.3.3.1). Criterion 7 gives better information with potential test probes (see Section 3.3.3.2). In all the criteria it must be remembered that only mixed potentials can be measured for extended objects to be protected [see the explanation to Eqs. (3-19) and (3-28)]. Criteria 5 and 6 are particularly to be observed for objects [22]. A comparison of the different criteria in field experiments has shown that, besides Criterion 1, good results are obtained with Criterion 3 [26].

3.3.3.1 Pragmatic Protection Criteria for Nonalloyed Ferrous Materials

In this section the pragmatic protection criteria of NACE [23] given in Table 3-3 are commented on in light of present knowledge.

Criterion No. 2: $\quad U_{on} - U_R \leq -0.3$ V $\hfill (3\text{-}31)$

This criterion is derived from the fact that the free corrosion potential in soil is generally $U_{Cu\text{-}CuSO_4} = -0.55$ V. Ohmic voltage drop and protective surface films are not taken into consideration. According to the information in Chapter 4, a maximum corrosion rate for uniform corrosion in soil of 0.1 mm a^{-1} can be assumed. This corresponds to a current density of 0.1 A m^{-2}. In Fig. 2-9, the corrosion current density for steel without surface film changes by a factor of 10 with a reduction in potential of about 70 mV. To reduce it to 1 μm a^{-1}, 0.14 V would be necessary. The same level would be available for an ohmic voltage drop. With surfaces covered with films, corrosion at the rest potential and the potential dependence of corrosion in comparison with U_R act contrary to each other so that qualitatively the situation remains the same. More relevant is

Criterion No. 3: $\quad U_{off} - U_R \leq -0.1$ V $\hfill (3\text{-}32)$

Here U_R is measured after switching off of the protection current and after step polarization. The potential difference corresponds to an *IR*-free potential decay. From the slope in Fig. 2-9, a reduction in the corrosion rate of 100 to 4 μm a^{-1} results.

Criterion No. 4: $\quad U <$ potential at the bend in the U (log I) curve $\hfill (3\text{-}33)$

This criterion is understood from the shape of the $I(U)$ curves described in Fig. 2-4 and Eq. (2-35) assuming two cathodic partial reactions according to Eqs. (2-17) and (2-19) [27]. For oxygen corrosion, $J_0 \gg G_C$ so that in the relevant potential range for this reaction there is a limiting current, which also corresponds to the corrosion rate at the rest potential and to the protection currents. For H_2 evolution, $J_0 \ll G_C$. This reaction only occurs at more negative potentials than the protection potential and follows a Tafel slope, which on a logarithmic plot of the $I(U)$ curve shows a marked deviation at the transition from O_2 diffusion current to H_2 evolution [see Fig. 2-21a and the explanation for U_h in Section 24.4 relating to Eq. (24-68b)]. Polarization in this region of the curve shows that the protection current is greater than the O_2 diffusion current and thus according to Eq. (2-40), cathodic protection is occurring.

Criterion No. 5: $\quad U_B^{\perp} < 0$ $\hfill (3\text{-}34)$

Table 3-3 Practical criteria for cathodic protection of plain carbon and low-alloy steels in soil

No.	Basis	Explanation	Application
1	$U \leq U_s$	IR voltage drop "should be considered"	General
2	$\Delta U \geq 300$ mV	300-mV negative change from rest potential on switching on protection current	Pipeline without connections to foreign cathodic structures
3	$\Delta U \geq 100$ mV	Measurement of depolarization on switching off protection current	Uncoated pipelines
4	$U = f(\log I)$	The protection current is determined from the bend in the current density vs. potential curve by a drainage test	Well casings
5	$\Delta U < 0$ at the pipe	All voltage drops perpendicular to the pipeline must be negative to the pipeline	Pipelines with grounds
6	$U_{on} \leq U_s - 0.3$ V (near foreign cathode)	Cancelling the cell voltage	Mixed installation with foreign cathodic structures (steel-reinforced concrete structures)
7	$U_M \leq U_s$	The potential test probe (provided with a near reference electrode) is connnected to the pipeline, and the connection is interrupted for measurement	If the off potential measurement is not possible, contact with foreign cathodic structures (steel-reinforced concrete structures)

This criterion indicates that cathodic current is entering the pipeline and that there is no more cell activity present (see Figs. 3-24 and 3-25 as well as the explanation of Fig. 2-7). The criterion for cathodic protection of pipelines with connected galvanized grounds against high-voltage interference is also comparable with this criterion [22]. In the connection between ground and pipeline, the current must flow to the pipeline.

In DIN 30676 [22] there is a further criterion (No. 6: $U_{on} < U_s - 0.3$ V) for protection of intermeshed objects with foreign cathodic structures. The on potential should be measured in the vicinity of the foreign cathodic structure. Here it is assumed that in spite of a high IR term, the potential of the foreign cathodic structure is so negative that cell formation is not expected to occur and the object also experiences cathodic protection. Verification is only possible with Criterion No. 7, i.e., with potential test probes.

3.3.3.2 Potential Measurement with Potential Test Probes

This measuring technique is applied when there are relatively high IR values due to cell currents with intermeshed objects or contact with steel-concrete structures, as well as the influence of stray currents, which cannot be switched off. The principle corresponds to the information in Fig. 2-3 on the use of measuring probes.

Test coupons of steel of a specified size are buried near the pipeline and connected by cable at the test point with the cathodically protected pipeline. They simulate artificial defects in the coating. The protection current taken from the test coupon can be measured via the cable connection and the true potential determined from a reference electrode in front of the test coupon by momentarily interrupting the cable connection [28]. Ohmic potential drops between the reference electrode and the test coupon are obtained from a measuring test probe that has a built-in reference electrode on the back of it to measure the IR-free potential directly [see Eq. (2-34) with $s \to 0$] without having to switch off the protection current or interrupt the cable connection to the pipeline (see Fig. 3-11). The pipe/soil potential at the plate is measured by means of a permanent reference electrode that is installed in a plastic tube behind the plate filled with electrolyte. Contact is made via a diaphragm in the plate [24]. The steel plate and the diaphragm that is led through the steel plate must be carefully insulated since otherwise the potential of the test coupon unaffected by the protection current inside the plastic tube will be measured instead of that of the external steel probe. A comparison of values from the potential test probe with the external reference electrode, test probes with built-in reference electrodes and the off potentials of the pipeline shows discrepancies of less than 20 mV. Since the potential becomes more negative with defects of decreasing size [see Eq. (3-21)], all coating defects in areas of similar resistivity that are smaller than the test probe must show more negative potentials (Crite-

Fundamentals and Practice of Electrical Measurements

rion 7). Potential test probes are only efficient if they are located in the same soil as the pipeline.

3.4 Current Measurement

3.4.1 General Advice for Measurement of Current

The voltage drops in measurements of pipeline current are on the order of 1 mV. Sensitive amplifier voltmeters are suitable for measuring such small voltages, and they are also used for potential measurements. Recording potentiometers used for corrosion protection measurements have a range of 0 to 1 mV. These have an internal resistance in the balanced state of a few megohms, and in the unbalanced state a few kilohms. In using amplified voltmeters, the measuring leads that pass over a street must be close together or twisted to avoid current impulses induced by the magnetic fields of vehicles.

Voltage drops on the order of 0.1 mV can be caused by thermal voltages that can arise from the sun's radiation on different metals, e.g., between pipeline and

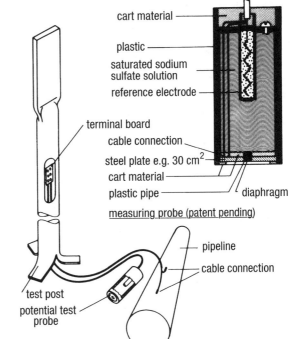

Fig. 3-11 Construction and installation of a potential test probe.

measuring rods. In damp conditions, a cell voltage acting in parallel can falsify the measured value if there is no clean, low-resistance contact. The resistances of measured stretches of the pipeline that are not part of a permanent measurement setup should therefore be checked before every measurement. Today it is usual to install solid measuring connections by cable and pole clamps at test points so that a perfect contact can be quickly achieved (see Section 10.3.2). In excavated pipelines or fittings, good contacts can be made with metal rods, and in the case of small-diameter pipes, with battery or metal clamps. A safer contact can be made above ground with the help of magnets [29].

3.4.2 Pipe Current Measurement

Besides potential measurement, the measurement of currents in the pipeline is of considerable importance, not only for investigating the causes of corrosion, but also for detecting failures in cathodic protection. The current flowing in the pipeline cannot be measured directly because of its low resistance (1 km of pipeline DN* 700, 8-mm wall thickness ≈ 10 mΩ) even if, for example, a measuring instrument can be switched into the electrically interrupted pipeline at insulating sockets or by disconnecting the pipeline. The internal resistance of a low-resistance 60-mV shunt would amount to between 1 and 10 mΩ and therefore would be on the same scale as the resistance of the circuit to be measured. Cable sheathing and pipeline currents are therefore determined indirectly by Ohm's Law from the voltage drop over a standard resistance.

$$R = \frac{\rho_{st} \, l}{S} \tag{3-35}$$

The resistance per unit length can be calculated from the resistance formula of the linear conductor if the outer diameter d_a, the wall thickness s, and wall cross-section $S = \pi s (d_a - s)$ are known:

$$R' = \frac{R}{l} = \frac{\rho_{st}}{S} = \frac{\rho_{st}}{\pi s (d_a - s)} \tag{3-36}$$

A further measurement of current can be obtained from the pipe mass per meter, given in the standards, $m' = m/l = S \rho_s$

$$R' = \frac{\rho_{st}}{S} = \frac{\rho_s \rho_{st}}{m'} \tag{3-37}$$

where ρ_s is the specific gravity of steel.

With $\rho_s = 1.7 \times 10^{-5}$ Ω cm and $\rho_s = 7.85$ g cm^{-3} for steel, it follows from Eqs. (3-37) and (3-38):

$$\frac{R'}{\mu\Omega\ m} = \frac{10^{10}}{\pi} \frac{(\rho_{st}/\Omega\ cm)}{\left(\frac{s}{mm}\right)\left(\frac{d_a - s}{mm}\right)} = \frac{5.4 \times 10^4}{\left(\frac{s}{mm}\right)\left(\frac{d_a - s}{mm}\right)} \qquad (3\text{-}36')$$

$$\frac{R'}{\mu\Omega\ m} = 10^7 \frac{(\rho_{st}/\Omega\ cm)(\rho_s/g\ cm^{-3})}{(m'/kg\ m^{-1})} = \frac{1.33 \times 10^3}{(m'/kg\ m^{-1})} \qquad (3\text{-}37')$$

The specific electrical resistances usually depend on the material and the temperature [31]. For the most important pipe materials these are (in $10^{-5}\ \Omega$ cm):

Steel St 34: 1.7; gray cast iron [32]: 8 to 10,
Steel St 60: 1.8; ductile cast iron: 7.

The values calculated using Eqs. (3-36) and (3-37) are only true for welded pipelines. Extension joints, fittings, and screwed or caulked joints can raise the longitudinal resistance of a pipeline considerably and therefore must be bridged over for cathodic protection.

The usual geometric length for measurements on pipes of 30 m has for DN 700 a resistance of about 0.3 mΩ. This allows, with an easily measurable voltage of 0.1 mV, a current of ≥ 0.3 A to be determined with sufficient accuracy. For a measured distance of pipe current over DN 700, 50 m is sufficient. Since with unwelded steel tubes the wall thickness can vary by 10%, and with welded pipes by 5%, and often the specific conductivity of the steel is not definitely known, it is recommended that long pipelines have built-in calibration stretches. The current measuring sections for the pipe current described in Section 11.2 can be somewhat varied. Separate leads for injecting current and measuring voltage are necessary for accurate calibration of the measuring sections for pipe current. On roads where measurement connections can easily be made above ground, two measuring points with 1 m between measuring connections should be provided for each calibration measurement section on longer distances (about 100 m).

Figure 3-12 shows the current flow and potential distribution on a DN 80 pipe of wall thickness 3.5 mm with a current input of 68 A. Since the pipe was relatively short, no pipe current could flow to the right. At a distance of 15 cm to the left, practically no deviation from the linearly decreasing voltage drop in the pipeline could be noted. Two pipe diameters were therefore a satisfactory distance for separating measuring cables for current and voltage measurements. With short excavations, an improvement can be achieved by alternating the connections at 45° to the right and to the left [33].

3.4.3 Measurement of Current Density and Coating Resistance

The protective cathodic current demand can only be ascertained under stable conditions, that is, on objects that have been cathodically protected over long periods. If two cathodic protection stations are acting on the section to be measured, both stations must be switched off from time to time by using current interrupters. Besides the protection current density, the ohmic voltage drop in the soil at holidays must also be determined. From this the apparent coating resistance of the pipeline can be determined. This corresponds to the sum of the parallel grounding resistances of the holidays (see Section 5.2.1.2). By plotting the pipe current along its length, contacts with foreign lines can often be recognized (see Section 3.6 and Fig. 3-20).

The procedure for determining the protection current density and the average coating resistance is explained in Fig. 3-13. At the current drainage point, the current $2 I_0$ is fed into the anode of the cathodic protection stations or an auxiliary ground electrode and periodically interrupted. With a symmetrical current distribution, the current I_0 flows back from each side of the pipeline. On account of the minimal length resistance of welded pipelines with a good coating, the pipe/soil potential decreases very slowly. Mean values can be linearly approximated according to recommendations by NACE [33,34]. This applies especially if the test point spacings, l_1, l_2 and l_3 are small in comparison with the length of the protected range, L. The current $I_1, I_2, I_3 ..., I_n$ flowing in the pipeline is measured at measuring stations with a spacing $\Delta l = 1$ to 2 km, and the current that enters each section between test points is:

$$\Delta I_n = I_n - I_{n+1} \tag{3-38}$$

This entering current causes ohmic voltage drops in the soil at holidays in the pipe coating, which are specified as $\Delta U_1, \Delta U_2, \Delta U_3..., \Delta U_n$ from which the average value

$$\overline{\Delta U_n} = \frac{1}{2}\left(\Delta U_n + \Delta U_{n+1}\right) \tag{3-39}$$

is given for every measured length l. From this follows the specific coating resistance that corresponds to an average defect resistance according to Section 5.2.1.2:

$$r_u = \frac{\overline{\Delta U_n}}{\Delta I_n} S \tag{3-40}$$

where S is the surface area of the pipeline in the measured section. The sheathing resistance is not only determined by the size and number of holidays in the coating, but also by the specific soil resistivity.

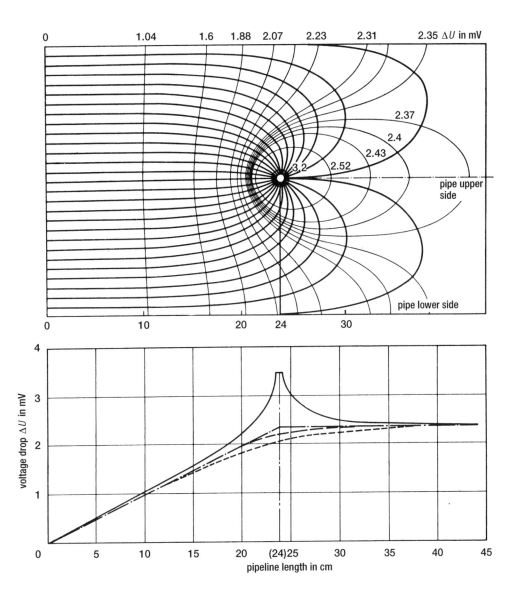

Fig. 3-12 Current and voltage distribution for a DN 80 pipeline with current drainage at point $l = 24$ cm; ohmic voltage drop in the lower figure: upper surface of the pipe (—), middle (— · — · —), underside (- - -).

There is a correlation between the average protection current density of the particular measured section

$$\overline{J}_s = \frac{\Delta I_n}{S} \qquad (3\text{-}41)$$

and the off potentials U_n corresponding to the actual current density–potential curves. Since these are usually not known, calculations are not possible, for example, from the lowest protection current density, J_s, at the protection potential, U_s. The measurement of protection current requirement is based on the fact that the protection criterion Eq. (2-39) $U_n < U_s$ is achieved at every position. In contrast to the coating resistance, the polarization values (U_n and J_s) are strongly influenced by the medium and the duration of polarization.

3.5 Resistivity Measurement

Resistances in and of electrolytes are exclusively measured with low or audio frequency ac so as not to falsify the results with polarization effects. Measurement is mainly by four electrodes, which eliminates voltages due to the grounding field resistances of the measuring electrodes.

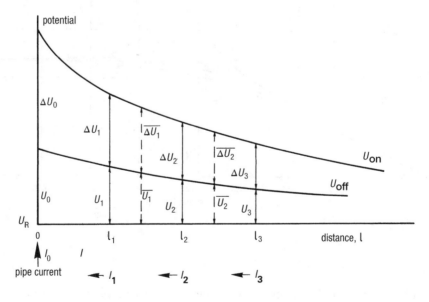

Fig. 3-13 Determination of the protection current density and coating resistance of a pipeline (explanation in the text).

3.5.1 Resistivity Measuring Instruments

Commercial ohm meters with four connection clamps are used for measuring soil resistivity. The measuring ac is produced by means of a chopper or transistor circuit. Figure 3-14 shows the principle of a resistance compensation bridge. The measuring current I from the ac generator flows via a transformer and the two electrodes A and B through the soil. The transformer winding produces a current proportional to the measuring current in the comparison resistance, R. When the number of windings is equal, the current in the resistance, R, is the same as that in the soil. The voltage drop, U, between the two measuring probes, C and D, is compared with the voltage across the resistance, R. A moving-coil galvanometer, N, acts as a null instrument for which the ac is rectified by a contact rectifier that is synchronized with the generator. The sliding adjustment on the comparison resistance, R, is moved until the galvanometer reads zero. The voltage drop across the comparison resistance is then exactly the same as the voltage, U, between the two probes, C and D. The resultant partial resistance corresponds to the measured resistance, $R = U/I^{-1}$ between the probes in the soil and can be read off on the potentiometer. Other measurement ranges are obtained by switching the transformer to other combinations of windings. Microelectronics has made possible the development of digital, automatic instruments. Since these instruments are less sensitive to ac, they can be used with advantage in pipelines affected by high voltages.

With the four-electrode measurement, no errors arise from transmission resistances at rods and subsidiary grounds due to the separate current and voltage leads. With very high transmission resistances, the sensitivity of the galvanometer can at most be insufficient for a null comparison. Since the moving-coil galvanometer responds to external dc voltages possibly present in the soil, a capacitor, C, is connected into the circuit. Other external ac voltages of $16\frac{2}{3}$ Hz or 50 Hz are not able to affect the measured results because the frequency of the ac measuring bridge with a chopper at 108 Hz is about 135 Hz in the transistor circuit. The first harmonic of the cathodic protection rectifier in the bridge circuit (100 Hz) produces

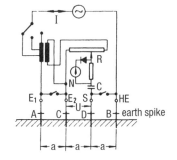

Fig. 3-14 Circuit of a measuring instrument for determining the apparent soil resistance by the Wenner method.

definite beats. If the amplitudes are not too large, it is even possible to make a null comparison by adjusting until the swings either side of the zero are equal. Some data on resistance measuring instruments are given in Table 3-2. All four-electrode resistivity measuring instruments can basically be used to measure grounding resistivities by connecting the two terminals, E_1 and E_2.

The grounding or penetration depth of the electrical resistance in conductors is, according to Eq. (3-42), dependent on the specific resistance and the frequency. The penetration depth, t, is the distance at which the field strength has fallen by $1/e$; μ_r is the relative permeability [35]:

$$t = \sqrt{\frac{\rho}{2\pi f \mu_0 \mu_r}} \tag{3-42}$$

for soil with $\mu_r = 1$:

$$\frac{t}{m} = 500 \sqrt{\frac{\rho/\Omega\,m}{f/s^{-1}}} \tag{3-42'}$$

Equation (3-42) is not valid for conducting systems consisting of several conducting phases (e.g., steel pipeline in soil). Figure 3-15 shows an example for the measured results (3).

The effects of frequency are within the accuracy of the measurement for electrode spacings up to 100 m and at the usual measuring frequency of 110 Hz. Two electrode resistance bridges work mostly with audio frequencies (800 to 2000 Hz) and give strongly deviating values. For measuring grounding resistances of small sections of extended installations, the most suitable is a ground meter with 25 kHz [36]. With plastic-coated casing pipes, the capacitive resistance can be smaller than the ohmic grounding resistance of holidays, which can then be better measured by switching dc on and off.

3.5.2 Measurement of Specific Soil Resistivity

There is a direct and an indirect method of measuring specific soil resistivity. The direct method is carried out in the laboratory on a soil sample using a soil box as shown in Fig. 3-16. The resistivity of a soil specimen of cross-section, S, and length, l, is measured and the specific resistivity determined:

$$\rho = R \frac{S}{l} = \frac{U S}{I l} \tag{3-43}$$

The indirect method is performed in the field with the arrangement shown in Fig. 3-14. In both cases the measurement is made with ac to suppress polarization effects.

When making measurements in the soil box, it has to be remembered that soil samples can change from their original condition and this will have an effect on the resistivity. Soil resistivity measurements in the soil box only give accurate results with cohesive soils. However, the order of magnitude of the specific resis-

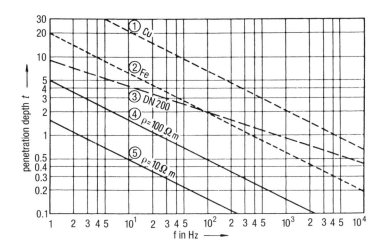

Fig. 3-15 Penetration depth, t, by ac as a function of the frequency, f. (1) Copper, (2) steel; t in millimeters; (3) pipeline of DN 200 steel; (4) and (5) in soil: t in kilometers.

tivity of less cohesive soils can be determined with approximate accuracy if the interfacial resistivity of the end surfaces is eliminated by using the four-electrode process. Current and voltage leads are separated according to the Wenner method [37]. With uniform current distribution, Eq. (3-43) also applies if l stands for the spacing of the inner electrodes [38].

The most commonly applied indirect method of measuring soil resistivity using the four-electrode arrangement of Fig. 3-14 is described further in Section 24.3.1. The measured quantities are the injected current, I, between the electrodes A and B, and the voltage, U, between the electrodes C and D. The specific soil resistivity follows from Eq. (24-41). For the usual measuring arrangement with equally spaced electrodes $a = b$, it follows from Eq. (24-41):

$$\rho = 2\pi a \frac{U}{I} = 2\pi a R \qquad (3\text{-}44)$$

If the specific soil resistivity varies vertically with depth, t, an apparent specific resistivity can be obtained from a combination of the resistivity of the upper and

116   Handbook of Cathodic Corrosion Protection

lower layers. Figure 3-17 shows the ratio of the apparent value, ρ, to the value of the upper layer, ρ_1, in its dependence on the ratio of the soil resistivity ρ_1/ρ_2 for different values of the ratio a/t [39]. It is recognizable that the influence of the lower soil layer is only significant when $a > t$. It is hardly possible to determine the true value of the lower soil layer since even with $a/t = 5$, only about half of the value of the soil resistivity, ρ_2, of the lower layer is obtained. In estimating soil resistivity, different values of a should be used at least up to the lower edge of the buried object.

The Wenner method is chiefly used to determine the grounding resistance along the pipeline track and the installation positions for cathodic structures. Local limited soil resistivity is most clearly determined from the grounding resistance of an inserted Shephard rod (see Fig. 3-18). Soil stratification can be recognized from the apparent specific soil resistivity, ρ, by the Wenner method, if a is varied.

Since the Wenner formula [Eq. (24-41)] was deduced for hemispherical electrodes, measuring errors appear for spike electrodes. To avoid errors in excess of 5%, the depth of penetration must be less than $a/5$. Soil resistivity increases greatly under frost conditions. While electrodes can be driven through thin layers of frost, soil resistivity measurements deeper than 20 cm in frozen ground are virtually impossible.

With four-electrode measurements effected from the surface, an average soil resistivity over a larger area is obtained. The resistivity of a relatively localized layer of earth or pocket of clay can only be accurately measured by using a spike electrode. Figure 3-18 gives dimensions and shape factors, F_0, for various electrodes.

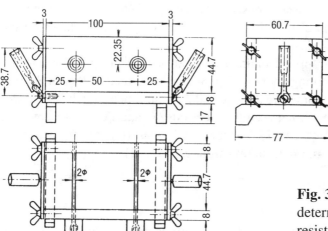

Fig. 3-16 Soil box for determining the specific soil resistivity (dimensions in millimeters).

Fundamentals and Practice of Electrical Measurements

The simplest design of a spike electrode is represented by the Shepard rod in Fig. 3-18a, which uses the right-hand electrode simply as an auxiliary ground, and measures with the left electrode the grounding resistance of the insulated stainless steel tip, which is proportional to the soil resistivity. The Columbia rod (Fig. 3-18b) uses the shaft of the rod as a counter electrode. Since the rod usually moves sideways as it is driven down, this method can easily give values that are too high. Both methods presuppose that the steel point is always in good contact with the soil; if this is not the case, too high values are obtained. To eliminate measuring errors at the electrode point, current and voltage electrodes are separated from each other on the Wenner rod as in Fig. 3-18c. Instead of Eq. (3-44), the following equation is valid for the Wenner rod because the current spreads out in all directions:

$$\rho = 4\pi a R \qquad (3-45)$$

Since the Wenner rod is mechanically somewhat delicate, it is only used in loose soils or in bore holes. For all measuring rods, the specific soil resistivity is equal to the product of the measured ac resistance and the shape factor F_0, which is determined empirically.

Soil resistivity measurements can be affected by uncoated metal objects in the soil. Values that are too low are occasionally obtained in built-up urban areas and in streets. Measurements parallel to a well-coated pipeline or to plastic-coated cables give no noticeable differences. With measurements in towns it is recommended, if

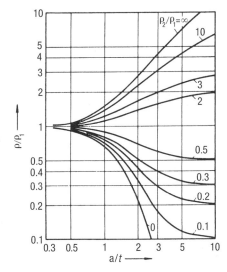

Fig. 3-17 Apparent specific electrical soil resistivity ρ in the case of two different layers of soil with the resistivities ρ_1 down to a depth of t and ρ_2 to a depth below t (explanation in the text).

possible, that two measuring arrangements be selected that are at right angles to each other and when using impressed current anodes, to make the measurements with increased electrode spacings of 1.6, 2.4 and 3.2 m [40].

3.5.3 Measurement of Grounding Resistance

Measurement of grounding resistance (e.g., of sacrificial anodes or impressed current anodes) is effected by the three-electrode method. As shown in Fig. 3-19, the measuring current is fed in the ground electrode to be measured and auxiliary ground and the voltage measured between the electrode and a probe. The auxiliary ground should be about four times (40 m) the distance and the probe twice (20 m) the distance of the spread of the ground. This implies that the ground resistance of pipelines or rails cannot be measured in practice (see Section 10-1). When measuring the resistance of insulating joints, only the sum of the ground resistances of limited line sections that depend on frequency are measured.

If the grounding resistances of the measured object and the auxiliary ground are of the same order of magnitude (Fig. 3-19, curve a), the most accurate value is obtained if the probe is situated at about the midpoint (neutral zone) between the two grounds. However, the resistance of the ground to be measured is often lower than that of the auxiliary ground (curve b), for example, if short ground rods are used. It may then be appropriate to place the probe nearer to the ground electrode than to the auxiliary one. As a rule, grounding resistances are too low with probe spacings that are too small, and too large with spacings that are too large and approach the voltage cone of the auxiliary ground. Other, more limited grounding electrodes (e.g., fence stakes and iron pylons) can be used as auxiliary grounds.

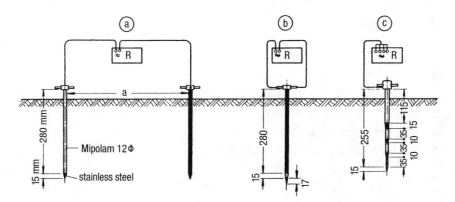

Fig. 3-18 Arrangement and dimensions of rod electrodes (dimensions in millimeters): (a) Shepard rod ($F_0 = 5.2$ cm); (b) Columbia rod ($F_0 = 3.4$ cm); (c) Wenner rod ($F_0 = 38$ cm).

Fundamentals and Practice of Electrical Measurements 119

Grounding resistances for the usual electrode and anode shapes as well as the form of their voltage cones on the soil surface are given in Section 9.1 and in Table 24-1.

3.6 Location of Faults

Older pipelines frequently are in contact at many points with foreign pipelines, cables or other grounded installations, which are only noticed when cathodic protection is switched on. Even with new pipelines there occurs bridging of insulating pieces, contact with unrelated pipelines or cables, contact with casings, connections with grounds of electrical installations, or contact with bridge structures and steel sheathing on pilings. Low-resistance contacts, which often render the cathodic protection of whole sections of pipeline impossible, can be located by dc or ac measurements [41, 42].

The variation in the on and off potentials or the potential difference along the pipeline will usually indicate faults that prevent the attainment of complete cathodic protection. The protection current requirement of the pipeline may be estimated from experience if the age and type of pipeline is known (see Fig. 5-3). Figure 3-20 shows the variation in the on and off potentials of a 9-km pipeline section DN 800 with 10-mm wall thickness. At the end of the pipeline, at 31.84 km, an insulating unit is built in. The cathodic protection station is situated at 22.99 km. Between this and the end of the pipeline there are four pipe current measuring points. The applied protection current densities and coating resistances of individual pipeline sections are calculated from Eqs. (3-40) and (3-41). In the upper diagram the values of

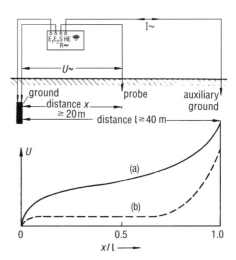

Fig. 3-19 Measurement of the grounding resistance of a soil electrode (explanation in text).

protection current density and coating resistance on individual sections between the protection station and insulating unit are about the same magnitude. So about 25% of the drained current flows back from the insulating unit to the negative point of the protection station.

In the lower part of Fig. 3-20 the types and approximate positions of defects can be recognized from the potential and pipe current curves. Only in the region of the protection station is an off potential more negative than U_s achieved, as a result of the anodic voltage cone. The protection current supply of the protection station has to be raised by 50%. As a result 75% of the current now comes from the direction of the insulating unit. Whereas at 26.48 km almost all the current is indicated as pipe current, at 27.21 km only a much smaller current of 80 mA is flowing. This means that between these two measuring points the total current enters the pipeline via a contact with a low-resistance ground installation. The protection current densities are considerably lower for individual sections than for those without contact, except in the region where there is contact with the foreign installation. In this section the coating resistance is apparently correspondingly low. Coating resistances of the remaining sections are of the same magnitude as those without contact. If there is overbridging of the insulating piece, potential values are similar, but there is a higher pipe current at 27.21 km.

Contacts with other pipelines or grounds can be localized to within a few hundred meters by pipe current measurements. Contacts with foreign pipelines or cables can also be found by measuring potential at the fittings of the other line while the protection current of the cathodically protected pipeline is switched on and off. While the potential of unconnected pipelines will assume more positive values when the protection current is switched on, the cathodic current may also enter any line in contact with the cathodically protected pipeline and thus shift its potential to more negative values. Should the contacting line not be located by this method, fault location can be attempted with direct or alternating current.

3.6.1 Measurement of Foreign Contacts

3.6.1.1 Fault Location using dc

Location of faults by the direct current method is based on the application of Ohm's Law. It is assumed that, because of the good pipe coating, virtually no current passes into the measured span and that the longitudinal resistance R' is known. When the fault-locating current, I, is fed in and takes a direct path via the foreign line to the protected pipeline, the fault distance is determined from the voltage drop ΔU over the measured span:

$$l_x = \frac{\Delta U}{IR'} \tag{3-46}$$

Fundamentals and Practice of Electrical Measurements 121

This simplification is only permissible with very low-resistance contact and if no other current flows in the pipeline. Currents flowing in the pipeline outside the measured span must be measured separately and taken account of in the calculation. This follows also indirectly when locating a contact with an unknown line. Figure 3-21 shows the necessary pipe current measurements on either side of the supposed contact point which, with $U' = Ul^{-1} = IR'$, gives the following calculation of the fault distance [28]:

$$l_x = \frac{\Delta U_2 - I'_1 l_2 R'}{I_F R'} = \frac{(U'_2 - U'_1)}{U'_3 - U'_1} l_2 \qquad (3\text{-}47)$$

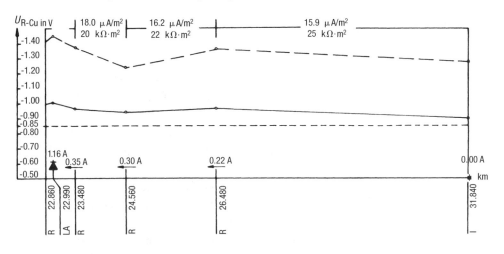

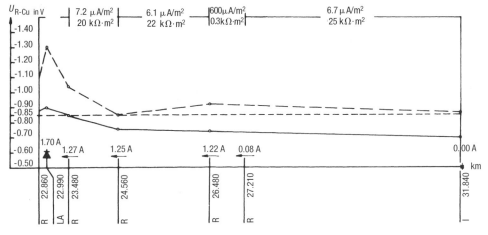

Fig. 3-20 Potential curves and pipe currents for a cathodically protected pipeline without contact with a foreign installation (upper figure) and with contact with a foreign installation (lower figure); o---o U_{on}, o—o U_{off} (explanation in the text).

122 Handbook of Cathodic Corrosion Protection

The direct current method of defect location has also proved useful for contacts between casing pipe and pipeline. Uncoated casing pipes with a low-resistance contact with the pipeline can render cathodic protection impossible [43]. If the position of the fault has been roughly determined using Eq. (3-46), the casing pipe must be excavated at this spot if at all possible. Since the current crosses over from the casing pipe to the pipeline at the contact point, the potential distribution shown in Fig. 3-22 top right results. By taking a measurement of the voltage drop on the pipe surface with two probes, the contact point can be accurately localized [44].

3.6.1.2 Fault Location with ac

Although this method of fault location is more readily affected by parallel-laid pipelines or high-voltage effects, it is as a rule quicker and more convenient to carry out; at least it allows a quick survey. It uses the inductive effect of the electromagnetic field of an audio-frequency current flowing in the pipeline. An audio-frequency generator (1 to 10 kHz) produces a voltage of up to 220 V by means of a chopper over an adjustable resistance between the pipeline and a grounding rod 20 m away, through which a corresponding locating current flows over the ground into the pipeline. A test coil serves as receiver in which a voltage is induced by the electromagnetic field of the ac current flowing in the pipeline. This is rendered audible in an earphone with an amplifier. The receiver contains a selective transmission filter for the locating frequency of 1 to 10 kHz, which reduces interference from frequencies of 50 and 16⅔ Hz by a ratio of 1 : 1000 [45].

Figure 3-23 shows the process of locating the position of a pipeline. The induced voltage in the test coil is lowest when the lines of force are at right angles to the coil axis. The test coil is then directly above the pipeline. A slight sideways displacement is sufficient to produce a component of the lines of force in the direction of the coil axis. This induces a voltage which, after amplification in the earphone or loudspeaker, becomes audible. The volume of sound is represented in Fig. 3-23 by the curve, a. This method is termed the "minimum" method. It allows exact location of the pipeline that is being sought. If the test coil is set at an angle

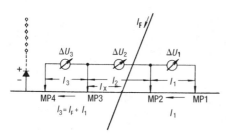

Fig. 3-21 Location of an electrical contact with an unknown pipeline (explanation in the text).

Fundamentals and Practice of Electrical Measurements

of 45°, the minimum lies at a distance from the pipeline axis which corresponds to the depth of the buried pipeline (dashed curve b).

The current from the generator will also flow through the metallic contacts into the foreign pipeline. The electromagnetic field of the pipeline connected to the generator is noticeably lower beyond the contact point, particularly if the contacting pipeline has a much lower ground resistance. A minimum in the sound can be detected with the search coil when it is over the contacting pipeline.

The protection current produced by the usual full-wave rectifier has a 100-Hz alternating component of 48%. There are receivers with selective transmission filters for 100 Hz, which corresponds to the first harmonic of the cathodic protection currents [45]. With such a low-frequency test current, an inductive coupling with neighboring pipelines and cables is avoided, which leads to more exact defect location.

3.6.2 Location of Heterogeneous Surface Areas by Measurements of Field Strength

Heterogeneous surface areas consist of anodic regions at corrosion cells (see Section 2.2.4.2) and objects to be protected which have damaged coating. Local concentrations of the current density develop in the area of a defect and can be determined by measurements of field strength. These occur at the anode in a corrosion cell in the case of free corrosion or at a holiday in a coated object in the case of impressed current polarization (e.g., cathodic protection). Such methods are of general interest in ascertaining the corrosion behavior of metallic construction units

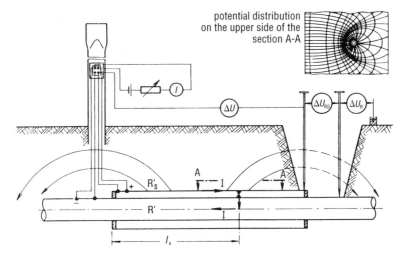

Fig. 3-22 Location of an electrical contact between pipeline and casing (explanation in the text).

124   Handbook of Cathodic Corrosion Protection

inside inaccessible media, e.g., reinforcing steel in concrete (see Chapter 19) or pipelines in masonry. The most usual application is, however, for buried pipelines [24,44,46-49].

3.6.2.1 Location of Local Anodes

The principle of the measurement is described with the help of Fig. 2-7 [50]. Potential measurement is not appropriate in pipelines due to defective connections or too distant connections and low accuracy. Measurements of potential difference are more effective. Figure 3-24 contains information on the details in the neighborhood of a local anode: the positions of the cathodes and reference electrodes (Fig. 3-24a), a schematic representation of the potential variation (Fig. 3-24b), and the derived values (Fig. 3-24c). Figure 2-8 should be referred to in case of possible difficulties in interpreting the potential distribution and sign. The electrical potentials of the pipeline and the reference electrodes are designated by ϕ_{Me}, ϕ_{Bc}, ϕ_{Bx}, $\phi_{Bx'}$ and $\phi_{B\infty}$. The reference electrodes B_M, B_x and $B_{x'}$ are situated over the pipeline as well as B_∞ at a height of B_x in lateral distance. The electrical potential of the soil depends on location and is shown in Fig. 3-24b as a continuous, or a coarse dashed line. At a distance from the pipeline it has the value $\phi_{E\infty}$ (remote ground) which is assumed for the location of electrode B_∞. In the region of the anode, the soil potential is more positive than in the region of the cathode (see RHS Fig. 2-8). With increasing distance from the surface of the pipe, the difference in soil potential becomes smaller due to the *IR* drop. Its variation on the soil surface over the pipeline is shown in Fig. 3-24b as a fine dashed line, and differs from that on the pipe surface by an amount η_Ω which varies with position. The potentials of the reference electrodes are only correct for the locations shown in Fig. 3-24a. The potential differences at the phase boundary of the reference electrode and soil are, according to definition, for reference electrodes equally great and correspond to

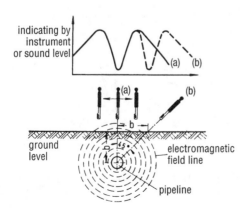

Fig. 3-23 Location of the position of a pipeline with the pipe locator: (a) Locating the position, (b) locating the depth.

the double arrows B_M, B_x, $B_{x'}$ and B_∞. The potential difference at the phase boundary of pipe and soil is represented by the double arrow $\Delta\phi_C$ for cathode and $\Delta\phi_A$ for anode. According to Eq. (2-1,) the *IR*-free pipe/soil potential amounts to $(\Delta\phi_C - B_M)$ and $(\Delta\phi_A - B_x)$ respectively. These values are theoretically only measurable with potential probes. The measurable pipe/soil potentials that are not *IR*-free correspond to the potential differences [see Eq. (2-33)] $(\phi_{Me} - \phi_{Bx}) = U_C$ and $(\phi_{Me} - \phi_{Bx}) = U_A$ for cathode and anode. In Fig. 3-24c these pipe/soil potentials are shown as U_M. The potential curve corresponds to the result shown in Fig. 2-7 for $I = 0$. The potential differences between the reference electrodes $U_B'' = \phi_{Bx} - \phi_{Bx'}$ as well as $U_B^\perp = \phi_{Bx} - \phi_{B\infty}$ are also included in Fig. 3-24c. From the diagram it can be seen that U_M and $U_B^\perp$ have the same shape, only with opposite signs. This is understandable since both values only represent the locational dependence of ϕ_{Bx} and therefore that of the soil potential on the surface because of the locational independence of ϕ_{Me} and $\phi_{B\infty}$. At the local anode U_M has a minimum and $U_B^\perp$ a maximum. For $U_B^\perp$ there is a zero value at the point of inflection.

In Fig. 3-25 the locational dependence of U_M, U_B'' and $U_B^\perp$ is shown together. For practical applications and because of possible disturbance by foreign fields (e.g., stray currents) U_M and $U_B^\perp$ are less amenable to evaluation than U_B'', which can always be determined by a point of inflection between two extreme values [50]. Furthermore, it should be indicated by Fig. 2-7 that there is a possibility of raising the sensitivity by anodic polarization which naturally is only applicable with small objects. In such cases care must be particularly taken that the counter electrode is sufficiently far away so that its voltage cone does not influence the reference electrodes.

3.6.2.2 *Location of Coating Defects*

The same electrical relationships as in Figs. 3-24 and 3-25 can be assumed at the location of defects in the coating. Instead of the cathodic region, a remote impressed current electrode is introduced (e.g., anode). Alternating current or interrupted dc can be used to locate defects; in the latter case, in addition, the defect is cathodically polarized. The ac method has the advantage that the U_B value can be obtained with simple metal electrodes. The sonic frequency generator described in Section 3.6.1.2 is used in the Pearson method [21]. The potential difference is measured by two operators or surveyors with contact shoes or spikes and displayed on instruments or indicated acoustically. Figure 3-26 shows the measuring arrangement (below) and the indication of a defect (above). Curves 1 and 2 correspond to those in Fig. 3-25 for U_B'' and $U_B^\perp$.

For quantitative evaluation of the extent of the damage, it is convenient to use dc because the coating has a lower resistance with ac because of capacity leakage.

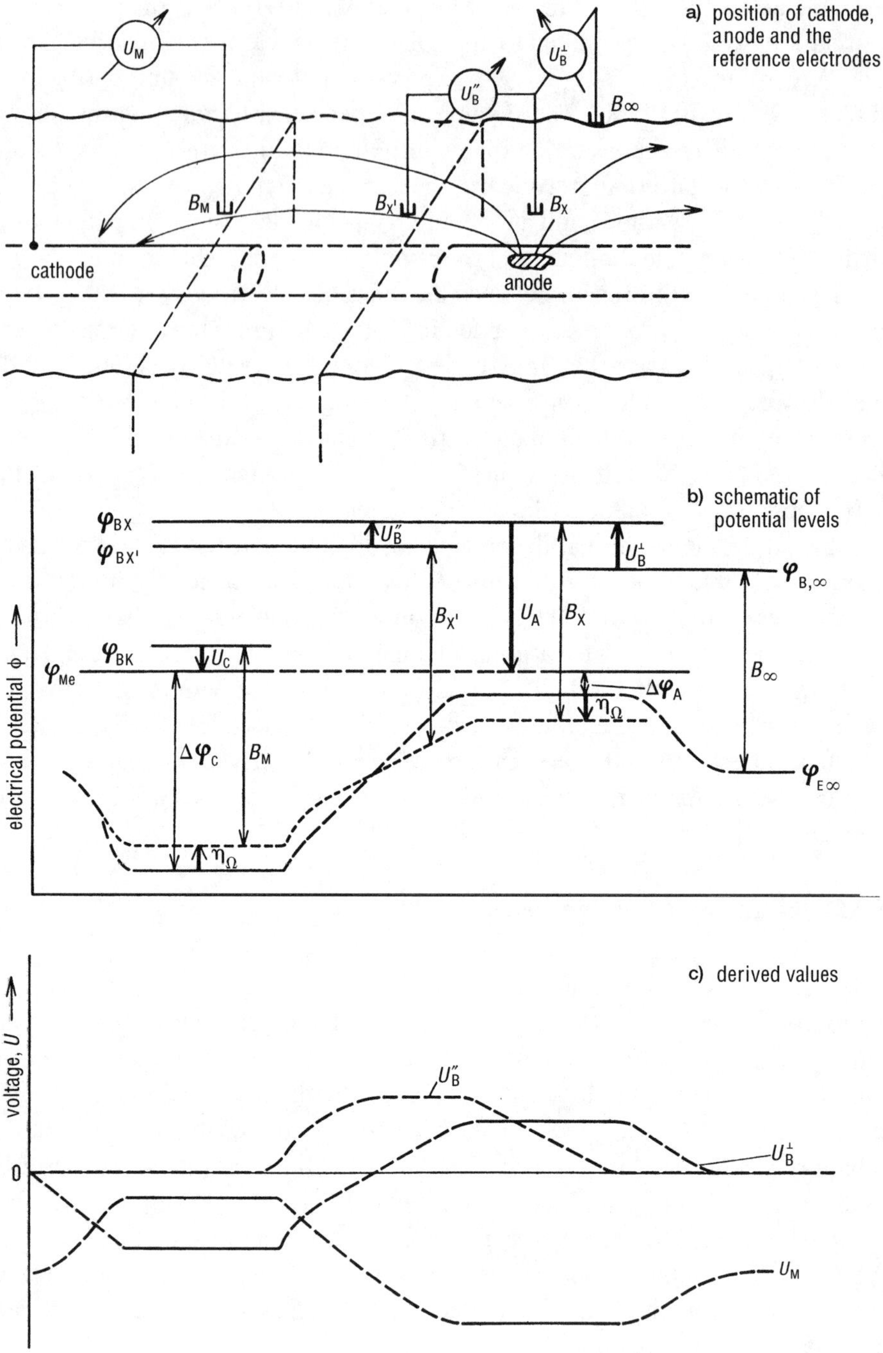

Fig. 3-24 Current and potential distribution in the neighborhood of a local anode on a buried pipeline (explanation in the text).

Fundamentals and Practice of Electrical Measurements

The following considerations are based on the assumption that the part of coating that is free from holidays acts as an insulator. Two limiting cases are considered: a single circular defect and an almost porous coating with numerous small and equally distributed defects.

(a) Circular damage

The impressed current enters the pipe surface at a defect of radius r_0. The defect can be considered as a point source at a sufficiently large distance compared to the diameter of the pipe. The potential on the surface of the soil is from Table 24-1, line 5:

$$\phi(x) = \frac{I_n}{2\pi\varkappa\sqrt{t^2 + x^2}} \qquad (3\text{-}48)$$

where t is the depth of the defect in the soil which approximately corresponds to the average pipe depth and x is the lateral distance on the soil surface of the center of the pipe. From Eq. (3-48) the value of $U_B^\perp$ with a distance x between the reference electrodes is:

$$\Delta U_x = \phi(x=0) - \phi(x) = \frac{I_n}{2\pi\varkappa} f(x,t) \qquad (3\text{-}49a)$$

$$f(x,t) = \frac{1}{t} - \frac{1}{\sqrt{t^2 + x^2}} \qquad (3\text{-}49b)$$

The soil resistance of the defect according to Eq. (24-17) is:

$$R_n = \frac{1}{4r_0\varkappa_0} \qquad (3\text{-}50)$$

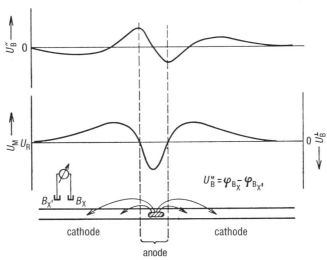

Fig. 3-25 Effect of position on the values of U_M, $U_B^\perp$ and U_B'' in the vicinity of the local anode in Fig. 3-24 (schematic).

Here $\varkappa_0$ is the electrical conductivity of the soil in the vicinity of the defect, which need not necessarily be the same as the average conductivity of the soil, $\varkappa$, in Eq. (3-48). In the area of the defect from Eqs. (3-15) and (3-16)

$$U_{on} - U_{off} = R_n I_n \tag{3-16'}$$

It then follows from Eqs. (3-49a,b), (3-50), and (3-16'):

$$r_0 = F(x,t) \frac{\varkappa}{\varkappa_0} \frac{\Delta U_x}{U_{on} - U_{off}} \tag{3-51a}$$

The factor $F(x,t)$ amounts to, for example, 1.74 m for $t = 1$ m and $x = 10$ m; values for other x's and t's can be derived from Fig. 3-27. For this calculation, $\varkappa$ is set equal to $\varkappa_0$. If the soil has a relatively higher conductivity in the vicinity of the defect, the value obtained for r_0 will be too large. If the defect lies at a significantly lower depth than the assumed average pipe depth, t, the factor according to Fig. 3-27 will be larger and therefore the value determined for r_0 will be too small. If there are several defects with overlapping voltage cones, then $U_{on} - U_{off}$ is greater and r_0 too small.

$$F(x,t) = \frac{\pi}{2f(x,t)} = \frac{\pi}{2} \frac{t\sqrt{t^2 + x^2}}{\sqrt{t^2 + x^2} - t} \tag{3-51b}$$

Information on defects can be obtained with good approximation from Eq. (3-51a). The value of ΔU_x is all that is necessary for an overview. U_{on} should be as high as possible to increase the sensitivity. In addition, to eliminate foreign voltages in the soil, it is necessary to switch the polarization current on and off with the help of a current interrupter; periods of about 2 s off and 18 s on are convenient. Potential differences independent of the polarization current that are the result of foreign currents or electrode faults (see Section 3.2) are totally excluded by this method. On the other hand, the *IR* component of a compensation current can also be

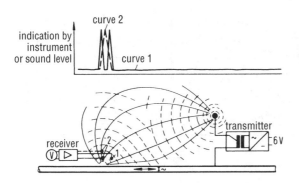

Fig. 3-26 Location of defects in the pipeline coating with ac by the Pearson method: arrangement 1 parallel to the pipeline; arrangement 2 at right angles to the pipeline (see caption to Fig. 3-24).

Fundamentals and Practice of Electrical Measurements

determined from the ΔU_x (off) (see Section 3.3.1). Since these quantities are smaller than ΔU_x (on) [2, 47] the resultant error can be ignored.

Figure 3-28 shows, as an example, results obtained from an intensive measurement of a short section of pipe which could be very strongly polarized to increase the sensitivity of defect location. From Eq. (3-51a) $r_0 = 4$ mm at $\Delta U_x = 0.1$ V. In the results in Fig. 3-29, $r_0 = 9$ cm at $\Delta U_x = 0.1$ V. These results are clear indications of water traps resulting from a poor coating [46]. Further examples are shown in Section 3.7.

(b) Porous coating

For a cylinder diameter, d, the following equations derived from Fig. 24-4 together with Eq. (24-46) to (24-48) are relevant:

$$\phi_3 - \phi_2 = \Delta U_x = \frac{Jd}{2\varkappa} \ln \frac{x^2 + t^2}{t^2} \tag{3-52}$$

$$\phi_2 - \phi_1 = U_{IR} = \frac{Jd}{2\varkappa} \ln \frac{t}{d} \tag{3-53}$$

In contrast to signal spread, according to Eq. (3-48) for a coating with few defects, in this case a locally almost constant conductivity is assumed. For the extreme case of an uncoated pipe and neglecting the ohmic polarization resistances, there is a distance $x = a$ where both voltage drops of Eqs. (3-52) and (3-53) are equal

$$\Delta U = U_{on} - U_{off} = U_{IR} - \Delta U_a \tag{3-54}$$

From Eqs. (3-52) and (3-53) and a soil cover $h = t - d/2$, it follows:

$$a = t\sqrt{\frac{t}{d} - 1} = \left(h + \frac{d}{2}\right)\sqrt{\frac{h}{d} - \frac{1}{2}} \tag{3-55}$$

ΔU_a can be determined from a $U_B^{\perp}$ measurement using this value of a. The true potential then follows from Eq. (3-54) with Eq. (3-15): $U_{IR\text{-free}} = U_{on} - \Delta U_a$. This relation is, however, no longer true when ohmic voltage drops are occurring in surface films or in a porous coating on the pipe surface. Instead of Eq. (3-54), the following relation is applicable:

$$\Delta U = U_{on} - U_{off} = \Delta U_a + (\Delta U_u - DU_{u0}) \tag{3-56}$$

where ΔU_u is the voltage drop in the pores of the coating while the subtracted quantity U_{u0} takes account of the limiting case of no coating. ΔU_{u0} corresponds to the ohmic voltage drop in the medium at the position of the coating. From Ohm's Law it follows that:

$$J = \Delta U_U \frac{\varkappa_0 \theta}{s} = \Delta U_{U0} \frac{\varkappa_0}{s} \quad (3\text{-}57)$$

where s is the thickness of the coating and θ is the population of pores, i.e., the ratio of the sum of the pore cross-section areas to the surface area of the pipeline. It follows from Eqs. (3-56), (3-57) and (3-53) with $U_{IR} = \Delta U_a$:

$$\theta = \frac{\Delta U_a}{\Delta U_a + (\Delta U - \Delta U_a) \dfrac{\varkappa_0 \, d \, \ln t/d}{\varkappa \, 2 \, s}} \quad (3\text{-}58)$$

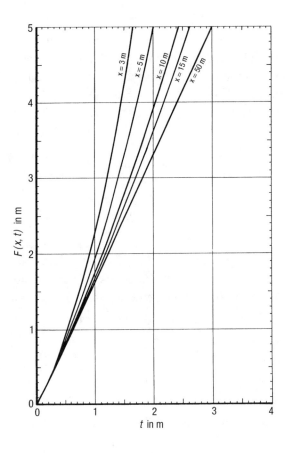

Fig. 3-27 The function $F(x,t)$ from Eq. (3-51b) for evaluating the degree of damage in the pipe coating (spherical field).

It is therefore possible, knowing the coating thickness, s, to determine the pore density θ, i.e., the uncoated surface, by again assuming $\varkappa = \varkappa_0$. Equation (3-58) gives the boundary cases $\Delta U = \Delta U_a$ for $\theta = 1$ or $s = 0$ (no covering) and $\theta = 0$ for $\Delta U_a = 0$ (no pores).

3.7 Intensive Measurement Technique

As explained in Section 10.4, the adjustment and maintenance of cathodic protection of pipelines can only be accomplished in the region of fixed measuring points at spacings of 1 to 2 km. It is assumed that no deviations in the polarization state occur between the test points. According to the explanations of Eqs. (3-19) to (3-21), this is only to be expected with homogeneous soil conditions. This is not always the case. For this reason a special measuring technique has been developed to give a complete assessment of the effectiveness of cathodic protection for pipelines, which provides for the measurement of potentials and potential differences at 5- to 10-m intervals along the pipeline [1,47,48,51,52]. An intensive measurement technique is applied primarily to objects where there is a high necessity for safety (see Chapter 22) and should be carried out every 1 to 2 years after laying the pipeline.

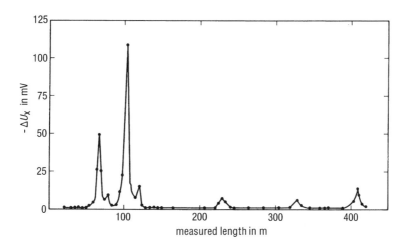

Fig. 3-28 Location and estimation of size of holidays in a polyethylene-coated DN 300 pipeline ($U_{on} - U_{off} \approx 40$ V).

3.7.1 Quantities to be Measured and Objectives of Intensive Measurement Technique

In intensive measurement, the following quantities are measured at short intervals on the pipeline:

pipe/soil potentials: U_{on} and U_{off},
$U_B^\perp$ values (according to Fig. 3-24): ΔU_{on} and ΔU_{off}.

The following information can be obtained from these parameters:

(a) a first overview of the local polarization state of the pipeline;
(b) indications of larger defects in the coating and an assessment of their size;
(c) checking the polarization state in the region of large holidays in the coating.

These three groups of parameters are arranged in order of their priority, so that the necessity for detailed investigation of succeeding parameters depends on the results from the preceding group.

In group (a) pipe/soil potentials are evaluated. Here the explanations in the first half of Section 3.3.1 and criterion No. 1 in Table 3-1 are relevant. If the protection criterion is not fulfilled ($U_{off} > U_s$), the cause must be found and remedied (e.g., removing foreign contacts, repairing major holidays or increasing the protection current).

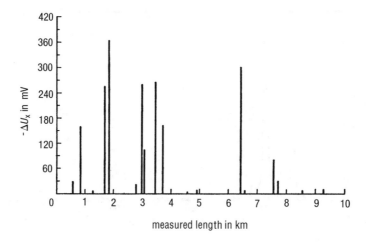

Fig. 3-29 ΔU_x readings on a polyethylene-coated DN 300 pipeline ($U_{on} - U_{off} \approx -2$ V).

Fundamentals and Practice of Electrical Measurements 133

Modern pipelines with plastic coating usually have only a few defects, which can be clearly recognized from differences in the pipe/soil potential (see Fig. 3-30) [47,52]. In support it can be assumed approximately in Fig. 3-24 that:

$$U_{on}(a) + U_B^{\perp}(a) = U_{on}(\infty) \tag{3-59}$$

The terms $U_{on}(a)$ and $U_B^{\perp}(a)$ apply to the region of a defect at position a. $U_{on}(\infty)$ is the measured switching on potential against a remote reference electrode; it corresponds also to the values that can be measured over sections of pipeline with defect-free coating. Since this value can be considered constant, the two measured values have a similar change for variations in a, as in Eq. (3-59) and mentioned in Fig. 3-25. This is also confirmed by the values given in Fig. 3-30 [51,52]. Local changes in on potential indicate voltage cones of defects which can be described in more detail by ΔU values.

In group (b), ΔU values are determined according to the data in Section 3.6.2.2 for circular defects. Examples are shown in Figs. 3-28 and 3-29. Figure 3-31 shows that in fact practically symmetrical voltage cones can occur, as predicted in Eq. (3-48) [47]. Errors in measurement must take into account the information in

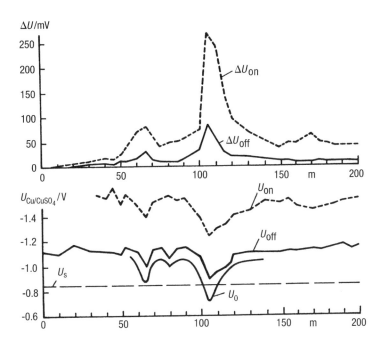

Fig. 3-30 Results of intensive measurement on a long-distance pipeline in the region of a holiday with insufficient polarization (U_0 = calculated true potential).

134   Handbook of Cathodic Corrosion Protection

Eq. (3-51b) and Fig. 3-27 because of the unknown depth, t, of the damage, but not the length, x, in the measurement of $U_B^\perp$. The difference for $t = 1$ m at $x = 10$ m compared with $x = \infty$ only amounts to 10%. Further sources of error which must be heeded are electrode faults and foreign fields. These can be best eliminated by substituting the difference $(\Delta U_{on} - \Delta U_{off})$ for ΔU_x in Eq. (3-51a).

The evaluation of ΔU values together with off potentials are aids in deciding whether coating damage should be repaired. In addition, by comparison with previous results, intensive measurements indicate whether new coating defects have arisen. These could be the result of external foreign forces on the pipeline.

A further objective is the evaluation in group (c) of the local polarization state by taking account of IR errors due to direct currents. Here Eq. (3-28) and the further explanations in the second half of Section 3.3.1 are relevant. In practical application, the error effect of ΔU_{off} must be estimated [2]. When foreign fields are present, it is necessary to substitute for the ΔU value the average of the measurements made on both sides of the pipeline [2,52]. Figure 3-30 gives an example of

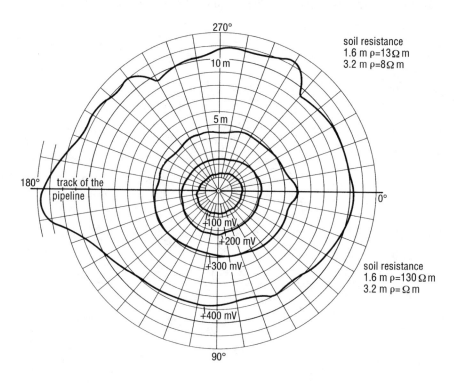

Fig. 3-31 Equipotential lines on the soil surface above a defect in the pipe coating (DN 700, thickness of soil covering 1.5 m).

measured values which show quite clearly that at the position $a = 105$ m the calculated potential U_0 does not fulfill the protection criterion.

3.7.2 Carrying Out an Intensive Measurement

The precondition of an intensive measurement is the exact location of the course of the pipeline. Methods described in Section 3.6.1.2 using the pipe locator are applicable. The harmonics of the protection current can be used. The protection current is <u>interrupted</u> to obtain the pipe/soil potentials and ΔU value for both switching phases. Although the ΔU value can be determined independently of the measuring point, measuring the pipe/soil potentials requires a measuring cable about 1 km in length as the connection to the pipeline. On practical grounds, simplifications can be introduced which certainly in most cases are applicable and low in error [1]. These concern ΔU values for both on and off switching phases as well as the pipe/soil potential.

(a) *Determination of the pipe/soil potential*

The pipe/soil potential is conventionally measured at a test point with a reference electrode B_0:

$$U_0 = \phi_R - \phi_{B0} \tag{3-60}$$

The reference electrode B_1 is set up at a distance, a, over the pipeline. The voltage between the two reference electrodes

$$U_{B1} = \phi_{B1} - \phi_{B0} \tag{3-61}$$

gives the pipe/soil potential at this point:

$$U_1 = \phi_R - \phi_{B1} = U_0 - U_{B1} \tag{3-62}$$

The determination of U_1 only requires the measured value U_0 and no cable connection to the pipeline. This technique can be repeated at equal intervals $\Delta a = a_v - a_{v-1}$:

$$U_n = U_{n-1} - U_{Bn} = U_0 - \sum_{v=1}^{n} U_{Bv} \tag{3-63}$$

where

$$U_{Bv} = \phi_{Bv} - \phi_{B(v-1)} \tag{3-64}$$

A further possibility of error is the addition of electrode errors. For this reason examination of the U_n value according to Eq. (3-62) at the next measuring point is necessary.

(b) Determination of ΔU values

According to Fig. 3-24 the reference electrode B_∞ is remotely positioned, outside the voltage cones of the pipeline holidays. Thus it may also be positioned above a section of the pipeline where there are no defects in the coating, which can be tested by a conventional $U_B^\perp$ measurement. If now the electrode B_0 lies on such a section ($\phi_{B0} = \phi_{B\infty}$), but electrode B_1 already lies in the voltage cone of a defect in the pipe coating, Eq. (3-61) applies for the position, a_1:

$$\Delta U(1) = \phi_{B1} - \phi_{B\infty} = U_{B1} \tag{3-65}$$

and for the general position a_n from Eq. (3-64):

$$\Delta U(n) = \phi_{Bn} - \phi_{B\infty} = \phi_{Bn} - \phi_{B0} = \sum_{v=1}^{n} U_{Bv} \tag{3-66}$$

It is recommended that further evaluation be carried out by direct measurement according to the data in Section 3.7.1 because of possible errors caused by foreign sources in the case of large readings.

The determination of ΔU values as well as the potential is effected by simple measurements of the difference in potential between two reference electrodes over the pipeline. To reduce possible electrode errors, partial sums can also be measured for a larger span, e.g., $(a_k - a_l) \approx 300$ m:

$$U_{k/l} = \phi_{Bk} - \phi_{Bl} = \sum_{v=l}^{k} U_{Bv} \tag{3-67}$$

Here the voltage $U_{k/l}$ is measured directly between reference electrodes at the positions a_k and a_l.

The use of microelectronics for obtaining measurements and evaluating them is rational for intensive measurements with about 1000 values per kilometer [53]. The additions according to Eqs. (3-63) and (3-66) are performed with a pocket calculator using the correct sign, and read off by the engineer.

Equipment has been developed that records measured values in steps of 2.5 mV through a digital converter connected to it, analyzes them and transmits them to a hard disk. The measured values can be fed to a central computer and further analyzed and plotted. This system can be used for intensive measurements and also for monitoring measurements, and offers the following possibilities: listing measured data, preparing potential plots, tracking trends over a timespan of several years, searching for larger coating defects, correcting potential values as well as listing information and data on defects in a protection installation database. At present there are no suggestions for a general need to repeat intensive measurements. It is, however, recommended that in particular local situations (e.g., the construction of parallel lines), repeat intensive measurements be made [54].

3.8 References

[1] W.v. Baeckmann u. W. Prinz, gwf gas/erdgas *126*, 618 (1985).
[2] H. Hildebrand, W. Fischer, W. Prinz u. W. Schwenk, Werkstoffe u. Korrosion *39*, 19 (1988).
[3] W.v. Baeckmann, D. Funk, H. Hildebrand u. W. Schwenk, Werkstoffe und Korrosion *28*, 757 (1977).
[4] W.v. Baeckmann u. J. Meier, Ges. Forschungs-Berichte Ruhrgas *15*, 33 (1967).
[5] W. Schwenk, Werkstoffe und Korrosion *13*, 212 (1962); *14*, 944 (1963).
[6] DIN 50918, Beuth-Verlag, Berlin 1978; W. Schwenk, Electrochemica Acta *5*, 301 (1961)
[7] W.v. Baeckmann, Werkstoffe und Korrosion *20*, 578 (1969).
[8] J.M. Pearson, Trans. Electrochem. Soc. *81*, 485 (1942).
[9] H.D. Holler, J. Electrochem. Soc. *97*, 271 (1950).
[10] W.v. Baeckmann, Forschungsberichtüber IR-freie Potentialmessung vom 10. 5. 1977, Ruhrgas AG, Essen.
[11] W.v. Baeckmann, D. Funk, W. Fischer u. R. Lünenschloß. 3R international *21*, 375 (1982).
[12] T.J. Barlo u. R.R. Fessler, Proceedings AGA Distribution Conf., Anaheim (Calif.) 18. 5. 81.
[13] W.v. Baeckmann, H. Hildebrand, W. Prinz u. W. Schwenk, Werkstoffe u. Korrosion *34*, 230 (1983).
[14] J. Pohl, Bestimmung der Einstellzeit von Meßgeräten, DVGW-Arbeitsbericht vom 25. 2. 1965.
[15] W.v. Baeckmann, Werkstoffe und Korrosion *15*, 201 (1964).
[16] TRbF 408, Richtlinie für den kathodischen Korrosionsschutz von Tanks und Betriebsrohrleitungen aus Stahl, Carl Heymanns Verlag, Köln 1972.
[17] W.v. Baeckmann u. W. Prinz, 3R international *25*, 266 (1986).
[18] J. Pohl, Europ. Symposium Kathod. Korrosionsschutz, s. 325, Deutsche Gesellschaft für Metallkunde Köln, 1960.
[19] W.v. Baeckmann, Taschenbuch für kathod. Korrosionsschutz, S. 134, n. Angaben von Herrn Kampermann, Vulkan Verlag, Essen 1985, und Technische Rundschau 17 v. 22. 4. 1975.
[20] R.J. Kuhn, Bureau of Standards *73* (1928), u. Corr. Prev. Control *5*, 46 (1958).
[21] M. Parker, Rohrleitungskorrosion und kathodischer Schutz, Vulkan-Verlag, Essen 1963.

[22] DIN 30676, Beuth Verlag, Berlin 1985.
[23] NACE Standard RP-01-69, Recommended practices: Control of external corrosion on underground or submerged metallic piping systems.
[24] W.G.v. Baeckmann, Rohre, Rohrleitungsbau, Rohrleitungstranrport *12*, 217 (1973).
[25] J. Polak, 3R intern. *17*, 472 (1978).
[26] T.J. Barlo u. W.E. Berry, Materials performance, *23*, H9, 9 (1984).
[27] H. Kaesche, Korrosion 11, Verlag Chemie, Weinheim 1959.
[28] W.v. Baeckmann u. W. Prinz, gwf *109*, 665 (1968).
[29] A. Baltes, Energie und Technik *20*, 83 (1968).
[30] Stahlrohrhandbuch, 9. Aufg., Vulkan-Verlag, Essen 1982.
[31] F. Richter, Physikalische Eigenschaften von Stählen und ihre Temperaturbhängigkeit, Stahleisen-Sonderberichte, Heft 10, Düsseldorf 1983.
[32] Gießereikalender 1962, Schiele & Schön GmbH, Berlin.
[33] W.v. Baeckmann, gwf *104*, 1237 (1963).
[34] W.v. Baeckmann, Werkstoffe und Korrosion *15*, 201 (1964).
[35] W.v. Baeckmann, gwf, *103*, 489 (1962).
[36] J. Ufermann and K. Jahn, BBC-Nachrichten *49*, 132 (1967).
[37] F. Wenner, U.S. Bulletin of Bureau of Standards *12*, Washington 1919 und Graf, ATM, *65*, 4 (1935).
[38] W.v. Baeckmann, gwf *101*, 1265 (1960).
[39] H. Thiele, Die Wassererschließung, Vulkan-Verlag, Essen 1952.
[40] P. Pickelmann, gwf *112*, 140 (1971).
[41] W.v. Baeckmann und A. Vitt, gwf *102*, 861 (1961).
[42] L. Mense, Rohre, Rohrleitungsbau, Rohrleitungstransport *8*, 11 (1969).
[43] K. Thalhofer, ETZ A *85*, 35 (1964).
[44] W.v. Baeckmann, Technische Überwachung *6*, 78 (1965).
[45] F. Schwarzbauer, 3R international *16*, 301 (1977).
[46] W. Schwenk und H. Ternes, gwf gas/erdgas *108*, 749 (1967).
[47] W. Prinz, 3R international *20*, 498 (1981).
[48] W.v. Baeckmann, gwf gas/erdgas *123*, 530 (1982).
[49] W.v. Baeckmann und G. Heim, Werkstoffe und Korrosion *24*, 477 (1973).
[50] H. Hildebrand und W. Schwenk, Werkstoffe und Korrosion *23*, 364 (1972).
[51] W. Prinz u. N. Schillo, NACE-Conference CORROSION 1987, Paper 313.
[52] W. Schwenk, 3R international *26*, 305 (1987); Werkstoffe und Korrosion *39*, 406 (1988).
[53] H. Lyss, gwf gas/erdgas *126*, 623 (1985).
[54] W. Prinz, gwf/gas/erdgas *129*, 508 (1988).

4

Corrosion in Aqueous Solutions and Soil

G. HEIM AND W. SCHWENK

Corrosion of metals in aqueous solutions and soils is essentially oxygen corrosion with a cathodic partial reaction according to Eq. (2-17). Hydrogen evolution from H_2O according to Eq. (2-19) is severely restricted on very alkaline metals such as Mg, Al and Zn; according to Eq. (2-18) it is possible in acids (e.g., H_2CO_3) or in organic acids in the soil. The aggressive corrosive action of acids is less a participation in the cathodic partial reaction than the result of damage to protective surface films arising from corrosion products. The course of the partial reactions in Eqs. (2-17) and (2-21) is less restricted due to this. An understanding of the properties of surface films is basic to an understanding of corrosion in soils and aqueous media [1-3].

4.1 Action of Corrosion Products and Types of Corrosion

Surface films are formed by corrosion on practically all commercial metals and consist of solid corrosion products (see area II in Fig. 2-2). It is essential for the protective action of these surface films that they be sufficiently thick and homogeneous to sustain the transport of the reaction products between metal and medium. With ferrous materials and many other metals, the surface films have a considerably higher conductivity for electrons than for ions. Thus the cathodic redox reaction according to Eq. (2-9) is considerably less restricted than it is by the transport of metal ions. The location of the cathodic partial reaction is not only the interface between the metal and the medium but also the interface between the film and medium, in which the reaction product OH^- is formed on the surface film and raises the pH. With most metals this reduces the solubility of the surface film (i.e., the passive state is stabilized).

Both partial reactions are stimulated on uncovered areas of the metal surface. Coverage of such a region is determined by whether the corrosion product is formed actually on the metal surface or whether it arises initially as solid oxide at some distance away in the medium. This question can be resolved by the availability of the partners to the reaction. J_A represents the rate of formation of metal ions [see Eq. (2-4)], and J_X is the rate of transport of a partner in the reaction leading to the formation of the surface film. For the case where $J_A > J_X$, precipitation occurs in the medium, and where $J_A < J_X$, it occurs directly on the metal surface. The passive state is favored in alkaline media with a high OH^- ion content because of the low solubility of most metal hydroxides [see Eq. (2-16)]. There are two exceptions to this:

(a) amphoteric metals, e.g., Al:

$$Al^{3+} + 4\,OH^- = Al(OH)_4^- \tag{4-1}$$

(b) easily soluble hydroxides, e.g., $Fe(OH)_2$:

$$Fe^{2+} + 2\,OH^- = Fe(OH)_2 \tag{4-2}$$

Very high pH values are necessary for the reaction in Eq. (4-1), which can only arise as a result of cathodic polarization according to Eq. (2-19) (see Fig. 2-11). No protective films are formed on Fe due to the relatively high solubility of $Fe(OH)_2$, although further oxidation produces solid films:

$$6\,Fe(OH)_2 + O_2 = 2\,Fe_3O_4 + 6\,H_2O \tag{4-3a}$$

$$4\,Fe(OH)_2 + O_2 = 4\,FeOOH + 2\,H_2O \tag{4-3b}$$

The oxidation products are almost insoluble and lead to the formation of protective films. They promote aeration cells if these products do not cover the metal surface uniformly. Ions of soluble salts play an important role in these cells. In the schematic diagram in Fig. 4-1 it is assumed that from the start the two corrosion partial reactions are taking place at two entirely separate locations. This process must quickly come to a complete standstill if soluble salts are absent, because otherwise the ions produced according to Eqs. (2-21) and (2-17) would form a local space charge. Corrosion in salt-free water is only possible if the two partial reactions are not spatially separated, but occur at the same place with equivalent current densities. The reaction products then react according to Eq. (4-2) and in the subsequent reactions (4-3a) and (4-3b) to form protective films. Similar behavior occurs in salt-free sandy soils.

In the presence of dissolved ions, the ion charge at the metal surface can be neutralized by the migration of the counter-ions to the reaction site. The following reactions take place:

Corrosion in Aqueous Solutions and Soil 141

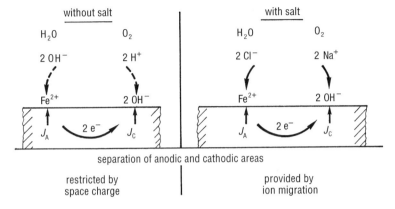

Fig. 4-1 Electrochemical partial and subsequent reactions in corrosion in aqueous media with and without dissolved salt.

$$\{Fe - 2e^-\} + 2\,Cl^- + H_2O = \{Fe(OH)^+ + Cl^-\} + \{H^+ + Cl^-\} \quad (4\text{-}4)$$

$$\{O_2 + 4\,e^-\} + 4\,Na^+ + 2\,H_2O = 4\,\{Na^+ + OH^-\} \quad (4\text{-}5)$$

Both reactions indicate that the pH at the cathode is high and at the anode low as a result of the ion migration. In principle, the aeration cell is a concentration cell of H^+ ions, so that the anode remains free of surface films and the cathode is covered with oxide. The $J_A(U)$ curves in Fig. 2-6 for anode and cathode are kept apart. Further oxidation of the corrosion product formed according to Eq. (4-4) occurs at a distance from the metal surface and results in a rust pustule that covers the anodic area. Figure 4-2 shows the steps in the aeration cell. The current circuit is completed on the metal side by the electron current, and on the medium side by ion migration.

Fig. 4-2 Schematic representation of the electrical and chemical processes in an aeration cell.

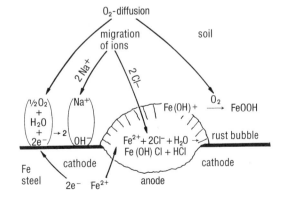

It is a consequence of the action of different pH values in the aeration cell that these cells do not arise in well-buffered media [4] and in fast-flowing waters [5-7]. The enforced uniform corrosion leads to the formation of homogeneous surface films in solutions containing O_2 [7-9]. This process is encouraged by film-forming inhibitors (HCO_3^-, phosphate, silicate, Ca^{2+} and Al^{3+}) and disrupted by peptizing anions (Cl^-, SO_4^{2-}) [10]. In pure salt water, no protective films are formed. In this case the corrosion rate is determined by oxygen diffusion [6,7,10]

$$\left(\frac{w}{\mu m\ a^{-1}}\right) = 16\ \frac{\left[c(O_2)/mg\ L^{-1}\right]}{(K_W/mm)} \tag{4-6}$$

where K_w is the diffusion path in the medium and represents a porous film of corrosion product. K_w is about 0.2 mm in fast-flowing solutions and about 1.5 mm in free convection [10]. Such diffusion rates correspond to protection current densities J_s of 0.1 to 1 A m^{-2}, which have also been measured on moving ships (see Section 17.1.3).

In soils the constituents restrict diffusion so that K_w in general rises to over 5 mm. The removal rate is mostly below 30 μm a^{-1} [11-13]. The danger of corrosion in soil is generally local corrosion through cell formation or by anodic influence (see Fig. 2-5) and can lead to removal rates of from a few tenths of a millimeter to several millimeters/year.

4.2 Determining the Corrosion Likelihood of Uncoated Metals

Corrosion likelihood describes the expected corrosion rates or the expected extent of corrosion effects over a planned useful life [14]. Accurate predictions of corrosion rates are not possible, due to the incomplete knowledge of the parameters of the system and, most of all, to the stochastic nature of local corrosion. Figure 4-3 gives schematic information on the different states of corrosion of extended objects (e.g., buried pipelines) according to the concepts in Ref. 15. The arrows represent the current densities of the anode and cathode partial reactions at a particular instant. It must be assumed that two narrowly separated arrows interchange with each other periodically in such a way that they exist at both fracture locations for the same amount of time. The result is a continuous corrosion attack along the surface.

Figure 4-3a indicates the ideal case of a mixed electrode in free corrosion. Such situations do not arise in soils or aqueous media. Usually the attack is locally nonuniform (see Fig. 4-3b) in which the current balance is not equalized at small regions along the surface. This is a case of free corrosion without extended corro-

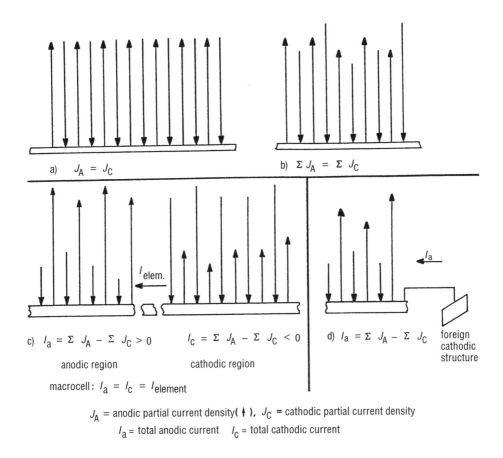

Fig. 4-3 Schematic representation of the partial current densities in corrosion in free corrosion (a-c) and with cell formation with foreign cathodic structures (d).

sion cells [14] with still essentially uniform material loss. Figure 4-3c, on the other hand, describes free corrosion with extended corrosion cells that can be several kilometers apart. Such situations occur usually with soils that vary locally. The free corrosion potential depends on position—there is no single rest potential (see Fig. 2-6).

If one regards the anodic region in Fig. 4-3c as isolated, the condition for free corrosion (i.e., the equalized current balance) is no longer fulfilled for that section. It is a case of cell formation in different regions of a pipeline in different types of soil [12]. This situation is no different in principle from cell formation with foreign cathodic structures in Fig. 4-3d [2,3,14,16], only in the latter case the cell voltages and cell currents can be considerably larger. Foreign cathodic structures include grounding installations and steel in concrete [17,18]. Mostly pitting and nonuniform corrosion are associated with cell formation.

4.2.1 Corrosion in Soils

Estimation of corrosion likelihood results from consideration of the characteristics of the soils and of the installed object, which are tabulated in Table 4-1 for nonalloyed and low-alloy steel products. Rating numbers, Z, are given according to the data on individual characteristics from which a further judgment can be made using the sum of the rating numbers.

$$B_0 = Z_1 + Z_2 + Z_3 + Z_4 + Z_5 + Z_6 + Z_7 + Z_8 + Z_9 \tag{4-7}$$

$$B_1 = B_0 + Z_{10} + Z_{11} \tag{4-8}$$

$$B_E = B_0 - (Z_3 + Z_9) + Z_{12} \tag{4-9}$$

The sum B_0 reflects the corrosion likelihood of objects without extended cells as in Fig. 4-3b. This value also characterizes the class of soil, depending on which type of pipeline coating is selected [16]. The sum B_1 shows the corrosion likelihood of objects with extended cells as in Fig. 4-3c. This indicates that, in the case of extended objects, the class of soil is by itself not sufficiently informative.

Objects with extended concentration cells can be individual lengths of pipeline and storage tanks if the makeup of the soil over the surface changes. The distance between anodic and cathodic areas can lie between a few centimeters and a few kilometers.

The sum B_E can only be obtained with buried objects and provides information on anodic damage through cell formation as in Fig. 4-3d. More detailed considerations can provide information on whether preferential anodic or cathodic regions are formed and how active they are [3,14].

From Table 4-1 it can be seen that a very high corrosion rate can be recognized (1b and 1c) from a few characteristics alone, and also that the action of coke can be seen as a well-aerated foreign cathode. Furthermore, it can clearly be seen that anodic damage can be much reduced (e.g., by laying a pipeline in high-resistance sandy soils, and homogeneous embedding).

A relatively high degree of corrosion arises from microbial reduction of sulfates in anaerobic soils [20]. Here an anodic partial reaction is stimulated and the formation of electrically conductive iron sulfide deposits also favors the cathodic partial reaction.

Table 4-1 Characteristics of soils and their rating number Z (DIN 50929, Pt 3 and DVGW worksheet GW 9).

No.	Characteristic	Z		
1a	Cohesivity (dispersible) (10 to 80%)	+4	to	−4
1b	Peat, marsh and organic carbon			−12
1c	Ash, refuse, coal/coke			−12
2	Soil resistance (1 to 50 kΩ cm)	−6	to	+4
3	Water	−1	to	0
4	pH (4 to 9)	−3	to	+2
5	Buffer capacity $K_{S4.3}$ and $K_{B7.0}$	−10	to	+3
6	Sulfide (5 to 10 mg/kg)	−6	to	0
7	Neutral salts (0.003 to 0.01 mol/kg)	−4	to	0
8	Sulfate extracted in HCl (0.003 to 0.01 mol/kg)	−3	to	0
9	Relative position of object to ground water	−2	to	0
10	Soil inhomogeneity (according to No. 2)	−4	to	0
11a	Heterogeneous impurity	−6	to	0
11b	Soil homogeneity vertically	−2	to	0
12	Potential $U_{Cu\text{-}CuSO_4}$ (−0.5 to −0.3 V)	−10	to	0

Table 4-2 shows as an example of the relationship between the results of field experiments [11] and the class of soil. In general, the time dependence of the corrosion rate can be represented as follows for $t > 4a$:

$$\left[\frac{\Delta s(t)}{\mu m}\right] = \left[\frac{\Delta s(4)}{\mu m}\right] + \left(\frac{w_{lin}}{\mu m\ a^{-1}}\right)\left[\left(\frac{t}{a}\right) - 4\right] \quad (4\text{-}10)$$

The data for the average decrease in metal thickness in 4 years and the linear corrosion rate are given in Table 4-2. In addition, extrapolations of the rate for 50 and 100 years are given, which are of interest for the corrosion likelihood of objects buried in earth. It can be seen from the results that film formation occurs in class I soil. In class II soils, the corrosion rate decreases with time only slightly. In class III soils, the decrease with time is still fairly insignificant.

Table 4-2 Evaluation of field tests [11] by class of soil (average ± standard deviation).

Soil class [14,19]	I	II	III
Number of soils investigated	21	30	27
Decrease in thickness ΔS for 4 years in μm	94±37	22 soils: 137±52 8 soils: 64±36	12 soils: 268±141 15 soils: 220±152
Linear removal rate w_{lin} after years in $\mu m\ a^{-1}$	6±3.3	16±9.0	55±38
ΔS_{50} in μm extrapolated decrease in metal thickness for 50 years	370±189	22 soils: 873±466 8 soils: 800±450	12 soils: 2798±1889 15 soils: 2750±1750
ΔS_{100} in μm extrapolated decrease in metal thickness for 100 years	670±354	22 soils: 1673±916 8 soils: 1536±864	12 soils: 5548±3789 15 soils: 5500±3800

Table 4-3 Data on local corrosion from field experiments (after 12 years [11] and 6 years [12]).

Type of soil/reference	Class of soil	Maximum penetration rate $w_{\ell,max}$ in $\mu m\ a^{-1}$	
		Average	Scatter
Free corrosion [11]	I	30	15 to 120
	II	80	20 to 140
	III	180	80 to 400
Free corrosion [12]	I	133	
	II	250	
	III	300	
Cell formation [12] (sandy soil/clay soil) $S_c : S_a = 10$		400	

The extrapolated values of decrease in thickness for 50 and 100 years in Table 4-2 are relevant in predicting the life of structural components (e.g., buried foundations of roads and steel retaining walls). These structural items lose their functional efficiency if their strength is impaired by too great a loss of thickness.

The size of test specimens in field experiments [11] is only a few square decimeters. In these specimens only micro cells can develop according to Fig. 4-3b. However, the maximum penetration rates in Table 4-3 were greater than expected from the average rate of decrease in thickness. Similar results were obtained from cell experiments with sand and clay soils [12]. Cell action only takes place if the surface ratio S_c/S_a is greater than 10 [13, 18]. In sandy soils, the salt content, and therefore the electrical conductivity, plays a relatively important role [21]. The maximum penetration rates in Table 4-3 are applicable when estimating the life of pipelines and storage tanks. These items lose their functional efficiency if they develop leaks due to shallow pitting or pitting corrosion.

The data in Table 4-1 show the considerable influence of the electrical resistivity of soil. This is particularly so in categories 2, 7, and 10. From a profile of the soil resistance along the course of a pipeline with welded connections or with electrically conducting thrust couplings, one can readily recognize anodic areas, and therefore the locations of areas of increased corrosion. They mostly coincide with minima in the resistance between larger regions of higher resistance [22].

The assessment for nonalloyed ferrous materials (e.g., mild steel, cast iron) can also be applied generally to hot-dipped galvanized steel. Surface films of corrosion products act favorably in limiting corrosion of the zinc. This strongly retards the development of anodic areas. Surface film formation can also be assessed from the sum of rating numbers [3, 14].

Stainless steels in soil can only be attacked by pitting corrosion if the pitting potential is exceeded (see Fig. 2-16). Contact with nonalloyed steel affords considerable cathodic protection at $U_H < 0.2$ V. Copper materials are also very resistant and only suffer corrosion in very acid or polluted soils. Details of the behavior of these materials can be found in Refs. 3 and 14.

4.2.2 Corrosion in Aqueous Media

Corrosion susceptibility in aqueous media is assessed on the basis of the rating numbers [3, 14], which are different from those of soils. An increased likelihood of corrosion is in general found only in the splash zone. Particularly severe local corrosion can occur in tidal regions, due to the intensive cathodic action of rust components [23, 24]. Since cathodic protection cannot be effective in such areas, the only possibility for corrosion protection measures in the splash zone is increased thickness of protective coatings (see Chapter 16). In contrast to their behavior in soils, horizontal cells have practically no significance.

4.3 Enhancement of Anodic Corrosion by Cell Formation or Stray Currents from dc Installations

Anodic enhancement gives rise to high corrosion susceptibility according to Fig. 2-5. It is immaterial whether this is the result of cell formation or stray currents. The danger due to contact with foreign cathodic structures must be treated just as seriously as the danger from emerging stray currents. Stainless steels, brass and bronze, and reinforcing steel in concrete act as foreign cathodic structures for mild and low-alloy steels. The intensity of cell action depends on the aeration of the foreign cathodic structures [25–27] and on the surface ratio (i.e., the ratio of cathodic to anodic surface areas) [18].

Differences in rest potential can be about 0.5 V for cell formation with foreign cathodic structures. The danger increases on coated construction components with coating defects of decreasing size on account of the surface rule [Eq. (2-44)], and is limited, for a given soil resistivity $\rho = 1/\varkappa$, not by the grounding resistance of the defect R_1, but rather by the pore resistance R_2 and the polarization resistance of R_P.

Thus for a circular defect of diameter d in a coating of thickness l and with a cell voltage ΔU in the region of the defect, corresponding to the difference $U_{on} - U_{off}$ from Eq. (3-15), it follows that:

$$J = \frac{4\Delta U}{\pi (R_1 + R_2 + R_P) d^2} \qquad (4\text{-}11)$$

For the individual resistances, the following relations [see Eqs. (24-17)] apply:

$$R_1 = \frac{1}{2\varkappa d}, \quad R_2 = \frac{4l}{\varkappa \pi d^2}; \quad R_P = \frac{4r_P}{\pi d^2}$$

Substituting these values in Eq. (4-11) and applying Eq. (2-45) gives:

$$J = \frac{\varkappa \Delta U}{\dfrac{\pi d}{8} + (l+k)} \qquad (4\text{-}12)$$

As expected, the polarization parameter, $k = \varkappa \times r_P$ is added to the pore length, l (see Section 2.2.5). The polarization resistance is dependent on the current density [Eq. (2-35)]. For pure activation polarization, it follows from Eq. (2-45):

$$k = \frac{U - U_R}{J} \varkappa = \frac{\beta_+ \varkappa}{J} \ln \frac{J}{J_A} \qquad (4\text{-}13)$$

Figure 4-4 is a plot of Eq. (4-12) for $\Delta U = 0.5$ V, $\varkappa = 200$ μS cm^{-1}, $\beta_+ = 26$ mV, $J_A = 10^{-6}$ A cm^{-2} (corrosion rate at the rest potential 0.01 mm a^{-1}). The continuous

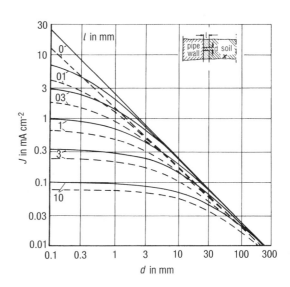

Fig. 4-4 Cell current densities at a defect with an increase in potential $\Delta U = 0.5$ V, $\varkappa = 200$ μS cm^{-1}. —$J(d)$ from Eq. (4-12) without taking into account polarization ($k = 0$); --- $J(d)$ taking into account polarization according to Eq. (4-13) [0.1 mA cm^{-2} ≈ 1 mm a^{-1}; see Eq. (2-5)].

curves are for $k = 0$ and the dashed curves for k from Eq. (4-13). Constant k values where there is film formation according to Eq. (2-44) can be taken care of by adding in the parameter l. In general, the current density increases with cell voltage, conductivity and decreasing defect size. The corrosion rate lies above 1 mm a^{-1}. Cell formation with foreign installations of more positive potential represents a great corrosion danger, which cannot be countered by passive protective measures. Possible protective measures are galvanic separation [14,16,28] and local cathodic protection (see Chapter 12).

Stray currents from foreign sources are to be regarded in the same way as galvanic currents. The explanations for Eq. (4-11) are relevant. Protective measures are described in Chapters 9 and 15.

A particular type of anodic danger arises in the interiors of pipes and storage tanks that are filled with an electrolyte and consist of similar or different metals, which, however, are electrically separated by insulating units. Potential differences are produced from external cathodic protection and are active in the interior [29,30]. These processes are dealt with in more detail in Sections 10.3.5, 20.1.4, and 24.4.6.

A further possibility of anodic danger is attributed to solutions containing O_2 acting on coatings with too low a coating resistance. Bitumen coatings and many thin coatings are, or become in the course of time, electrolytically conducting (see Section 5.2.1.3). A pore-free coated surface of 10^4 m^2 (e.g., a 10-km-long conducting pipeline DN 300) with specific coating resistances $\leq 10^5$ Ω m^2, has a coating resistance $R_u \leq 10$ Ω. The cathodic partial reaction can take place on these coated surfaces, whereas the anodic reaction is inhibited and only occurs in the region of defects. The coated surfaces act as cathodes and can lead to intense pitting corrosion, which has also been observed in saline waters [31-33]. To avoid such danger, the coating resistances must be sufficiently thick, as, for example, with polyethylene coatings (see Section 5.2.1) or the cell voltage must be raised by cathodic protection.

4.4 Corrosion Due to ac Interference

The effects of alternating currents are much less of a corrosion danger than those of direct currents. Experiments on steel have shown that during the positive half wave [34-37] only about 1% contributes to the dissolution of iron according to Eq. (2-21). The remaining 99% is involved in the discharge of capacitances, of redox systems (e.g., Fe^{2+}/Fe^{3+} in surface films) or in the evolution of O_2 by

$$2 H_2O - 4 e^- = O_2 + 4 H^+ \tag{4-14}$$

From the data in Table 2-1 this results in a corrosion rate for Fe of 0.1 mm a^{-1} for an effective ac current density of 20 A m^{-2}. Thus only ac current densities above

50 A m^{-2} are serious. Frequency has an effect which, however, in the region 16⅔ to 60 Hz is small. Generally the danger increases with falling frequency [34].

Even with the superposition of the ac with a cathodic protection current, a large part of the anodic half wave persists for anodic corrosion. This process cannot be detected by the normal method (Section 3.3.2.1) of measuring the pipe/soil potential. The *IR*-free measurable voltage between an external probe and the reference electrode can be used as evidence of more positive potentials than the protection potential during the anodic phase. Investigations have shown, however, that the corrosion danger is considerably reduced, since only about 0.1 to 0.2% contributes to corrosion.

The action of effects in the environment and cathodic current densities on ac corrosion requires even more careful investigation. It is important to recognize that ac current densities above 50 A m^{-2} can lead to damage even when the dc potential is formally fulfilling the protection criterion [40].

Low-frequency potential deviations have to be taken into account with increased corrosion rate if the protection criterion according to Eq. (2-39) occasionally lapses. It is insufficient if only the average value of the potential over time remains more negative than the protection potential [38].

With metals other than Fe, the percent of the ac current leading to corrosion can be considerably different. Cu and Pb behave similarly to Fe [36], whereas Al [36] and Mg [39] corrode much more severely. This has to be watched with sacrificial anodes of these materials if they are subjected to ac.

4.5 References

[1] W. Schwenk, 3R international *18*, 524 (1979).
[2] W. Schwenk, gwf gas/erdgas *123*, 158 (1982).
[3] W. Schwenk, gwf gas/erdgas *127*, 304 (1986).
[4] V. V. Skorchelletti u. N. K. Golubeva, Corrosion Sci. *5*, 203 (1965).
[5] H. Kaesche, Werkstoffe u. Korrosion *26*, 175 (1975).
[6] W. Schwenk, Werkstoffe u. Korrosion *30*, 34 (1979).
[7] A. Rahmel u. W. Schwenk, Korrosion und Korrosionsschutz von Stählen, Verlag Chemie, Weinheim–New York 1977.
[8] K. Bohnenkamp, Arch. Eisenhüttenwes. *47*, 253, 751 (1976).
[9] DIN 50930 Teil 2, Beuth-Verlag, Berlin 1980.
[10] H. Kaesche, Die Korrosion der Metalle, Springer-Verlag, Berlin–Heidelberg–New York 1966, 1979.
[11] M. Romanoff, Underground Corrosion, Nat. Bur. of Stand., Circ. 579, 1957, US Dep. Comm.
[12] H. Hildebrand u. W. Schwenk, Werkstoffe u. Korrosion *29*, 92 (1978).
[13] H. Hildebrand, C.-L. Kruse u. W. Schwenk, Werkstoffe u. Korrosion *38*, 696 (1987).
[14] DIN 50929 Teile 1 bis 3, Beuth-Verlag, Berlin 1985; Kommentar zu DIN 50929 in „Korrosionsschutz durch Information und Normung", Hrsg. W. Fischer, Verlag I. Kuron, Bonn 1988.
[15] DIN 50900 Teil 2, Beuth-Verlag, Berlin 1984.
[16] DIN 30675 Teil 1, Beuth-Verlag, Berlin 1985.
[17] W. Schwenk, gwf gas/erdgas *103*, 546 (1972).
[18] G. Heim, Elektrizitätswirtschaft *81*, 875 (1982).
[19] DVGW Arbeitsblatt GW 9, ZfGW-Verlag, Frankfurt 1986.
[20] H. Klas u. H. Steinrath, Die Korrosion des Eisens und ihre Verhütung, Verlag Stahleisen, 2. Auflage, Düsseldorf 1974, S. 206.
[21] D. Funk, H. Hildebrand, W. Prinz u. W. Schwenk, Werkstoffe u. Korrosion *38*, 719 (1987).
[22] H. Klas u. G. Heim, gwf *98*, 1104, 1149 (1957).
[23] M. Stratmann, K. Bohnenkamp u. H.-J. Engell, Werkstoffe u. Korrosion *34*, 605 (1983).
[24] W. Schwenk, Stahl u. Eisen *104*, 1237 (1984).
[25] H. Arup, Mat. Perform. *18*, H. 4, 41 (1979).
[26] H. J. Abel und C.-L. Kruse, Werkstoffe u. Korrosion *33*, 89 (1982).
[27] H. Hildebrand u. W. Schwenk, Werkstoffe u. Korrosion *37*, 163 (1986).
[28] DVGW Arbeitsblatt G 600 (TRGI), ZfGW-Verlag, Frankfurt 1986.
[29] FKK, 3R international *24*, 82 (1985).
[30] DIN 50927, Beuth-Verlag, Berlin 1985.
[31] H. Hildebrand u. W. Schwenk, Werkstoffe u. Korrosion *30*, 542 (1979).
[32] DIN 50928, Beuth-Verlag, Berlin 1985.
[33] W. Schwenk, farbe+lack *90*, 350 (1985).
[34] W. Fuchs, H. Steinrath u. H. Ternes, gwf gas/erdgas *99*, 78 (1958).
[35] F. A. Waters, Mat. Perform. *1*, H. 3, 26 (1962).
[36] J. F. Williams, Mat. Perform. *5*, H. 2, 52 (1966).
[37] S. R. Pookote u. D. T. Chin, Mat. Perform. *17*, H. 3, 9 (1978).
[38] G. Heim, 3R international *21*, 386 (1982).
[39] G. Kraft u. D. Funk, Werkstoffe u. Korrosion *29*, 265 (1978).
[40] G. Heim u. G. Peez, 3R international *27*, 345 (1988).

5

Coatings for Corrosion Protection

G. HEIM AND W. SCHWENK

5.1 Objectives and Types of Corrosion Protection by Coatings

Corrosion protection measures are divided into active and passive processes. Electrochemical corrosion protection plays an active part in the corrosion process by changing the potential. Coatings on the object to be protected keep the aggressive medium at a distance. Both protection measures are theoretically applicable on their own. However, a combination of both is requisite and beneficial for the following reasons:

- Coatings free of defects cannot be reliably produced, installed and maintained perfectly in operation. Without electrochemical corrosion protection, there is increased danger of corrosion if the coating is damaged (see Section 4.3). To test for damage or defects under operating conditions, a voltage is applied and the current flow measured (drainage test). This is the same electrical circuit that is used in cathodic protection. Thus, a combination of both processes is indicated.

- Without a coating, electrochemical corrosion protection is used if the protection criteria can be fulfilled with sufficiently good current distribution. Usually this can be achieved with extended objects only if the polarization parameter can be considerably increased by a coating (see Eq. 2-45). In addition, the current supply needed can be drastically reduced by a coating, which leads to a considerable reduction in the costs of installation and operation. A decisive advantage of smaller protection currents is a reduction in the interference of foreign installations for which the protection current is to be regarded as a stray current (see Chapters 9 and 15).

Coatings applied for corrosion protection are divided into organic coatings, cement mortar, and enamel and metallic coatings.

5.1.1 Organic Coatings

Organic coatings applied primarily to protect pipelines and storage tanks include paints, plastics and bituminous materials. According to Ref. 1 these can be

154 Handbook of Cathodic Corrosion Protection

further divided into thin and thick coatings. The latter are those with a thickness over 1 mm.

Thin coatings consist of paints and varnishes, which are applied as liquids or powdered resin with a thickness of about 0.5 mm [e.g., epoxy resin (EP) [2]]. Typical thick coatings are bituminous materials [3] and polyolefins [e.g., polyethylene (PE) [4]], thick coating resin combinations [e.g., EP tar and polyurethane (PUR) tar [2]] as well as heat-shrinkable sleeves and tape systems [5].

Thin coatings contain in general many polar groups that promote adhesion. Low polar or nonpolar thick coatings (e.g., PE) are usually combined with polar adhesives to achieve the necessary bond strength against peeling.

All organic coatings show varying degrees of solubility and permeability for components of the corrosive medium, which can be described as permeation and ionic conductivity (see Sections 5.2.1 and 5.2.2). An absolute separation of protected object and medium is not possible because of these properties. Certain requirements have to be met for corrosion protection, which must also take account of electrochemical factors [1] (see Section 5.2).

5.1.2 Cement Mortar Coatings

Cement coatings are usually applied as linings for water pipes and water tanks, but occasionally also for external protection of pipelines [7]. Cement is not impervious to water, so electrochemical reactions can take place on the surface of the object to be protected. Because of the similar processes occurring at the interface of cement and object and reinforcing steel and concrete, data on the system iron/cement mortar are dealt with in this chapter taking into account the action of electrolytes with and without electrochemical polarization. To ensure corrosion protection, certain requirements must be met (see Section 5.3 and Chapter 19).

5.1.3 Enamel Coatings

Enamel coatings are applied for internal protection of storage tanks (see Section 20.4.1). Enamel is impervious to water, i.e., it separates the protected object and corrosive medium. Corrosion protection can fail only at defects in the enamel coating and through corrosion of the enamel (see Section 5.4).

5.1.4 Metallic Coatings

Metal coatings are applied in special cases where the protective action has to be ensured by the coating metal or its corrosion products. Additional electrochemi-

Coatings for Corrosion Protection   155

cal protection can be significant if the coating metal also has the advantage of providing electrochemical protection (see Section 5.5).

5.2 Properties of Organic Coatings

The following requirements must be met for long-term corrosion protection:

(a) high mechanical resistance and adhesion, especially during transport and installation;
(b) chemical stability under service conditions (aging);
(c) sufficiently low permeability for corrosive components under service conditions;
(d) sufficient stability under electrochemical influences, especially with electrochemical protection measures.

The requirements (a) to (c) are obligatory for all types of coating and all objects to be protected; for example, they are well known in the protection of steel structures. The objects discussed in this handbook are continuously in contact with electrolytes. For this reason in addition, requirement (d) is of great importance and is discussed in detail below.

5.2.1 Electrical and Electrochemical Properties

5.2.1.1 Review of the Types of Reactions

Figure 5-1 shows schematically the processes that can take place on coated metals in electrolytes. The following properties are important:

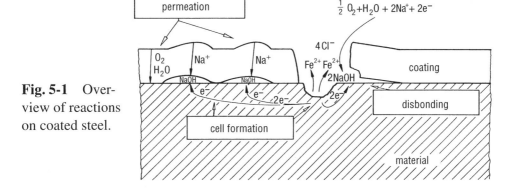

Fig. 5-1 Overview of reactions on coated steel.

- permeability of the coating to corrosive materials
  permeation of molecules (O_2, H_2O, CO_2, etc.),
  migration of ions (anions and cations);
- mechanical damage to the coating that exposes the metal surface, allowing electrochemical corrosion reactions to take place.

Ion migration can be explained by Eq. (2-23). The electrical voltages involved range from a few tenths to several volts and arise from the following causes [8-10]:

Anodic polarization: exiting stray currents, contact with foreign cathodic structures;

Cathodic polarization: entering stray currents, cathodic protection.

A consequence of ion migration is electrolytic blister formation. In the case of anodic blisters the coated surface shows pitting, whereas in the case of cathodic blisters there is no change in the metal surface or there is merely the formation of thin oxide layers with annealing color.

An important consequence of ion migration is the formation of cells where the coated surface acts as a cathode and the exposed metal at the damage acts as an anode (see Section 4.3). The reason for this is that at the metal/coating interface, the cathodic partial reaction of oxygen reduction according to Eq. (2-17) is much less restricted than the anodic partial reaction according to Eq. (2-21). The activity of such cells can be stimulated by cathodic protection.

In the case of free corrosion at the rim of the holidays and in the case of cathodic protection on the entire surface exposed (to the soil), oxygen reduction and production of OH^- ions take place according to Eq. (2-17). The local pH is therefore strongly increased. The OH^- ions are in a position to react with the adhesive groups in the coating and thus migrate under the coating. This process is known as cathodic disbonding, or more appropriately, alkaline disbonding. The parameters influencing the various types of reaction and the requirements for corrosion protection are dealt with in detail in the following sections.

5.2.1.2 Coating Resistance and Protection Current Demand

The following terms apply to the specific coating resistance which is related to the surface, S: r_u^x is the value calculated from the specific resistance ρ_D of the coating material using Eq. (5-1):

$$\left(\frac{r_u^x}{\Omega\,\mathrm{m}^2}\right) = 10^{-5} \left(\frac{\rho_D}{\Omega\,\mathrm{cm}}\right)\left(\frac{s}{\mathrm{mm}}\right) \tag{5-1}$$

Table 5-1 Comparison of specific coating resistances.

Coating material	$\rho_D{}^a$ (Ω cm)	s (mm)	r_u^x (Ω m^2)	r_u^{0b} (Ω m^2)	s (mm)	r_u^c (Ω m^2)
Bitumen [4]	>10^{14}	4	4×10^9	3×10^5	4 to 10	-10^4
PE [4]	10^{18}	2	2×10^{13}	10^{11}	2 to 4	-10^5
EP [4]	10^{15}	0.4	4×10^9	10^8	0.4	-10^4
PUR tar [2]	3×10^{14}	2	6×10^9	10^9	2.5	

[a] From Ref. 11; for PUR tar from Ref. 12.
[b] From Table 5-2.
[c] See Fig. 5-3.

where s is the thickness of the coating, r_u^0 is the value obtained in laboratory or field experiments for a defect-free coating in contact with the medium, and r_u is the value obtained on structures in the medium in service where there are usually pores and holidays present.

Its determination follows from current and potential measurements:

$$r_u = \frac{\Delta U}{\Delta I} S \tag{5-2}$$

With large resistances and at high voltages the replacements $U = \Delta U$ and $I = \Delta I$ can be made (see Refs. 2-5). At lower voltages and for buried objects $(U_{on} - U_{off}) = \Delta U$ can be replaced with $I = \Delta I$ [8]. Such a measurement is described in Section 3.4.3 [see Fig. 3-13 and Eq. (3-40)].

The r_u^x values from Eq. (5-1) for the most important materials for pipe coatings are given in Table 5-1. Table 5-2 contains results of long-term field experiments. For comparison, the values of r_u^0 are included in Table 5-1. It can be seen that r_u^0 values are always smaller than r_u^x values, which is apparently due to the absorption of water when the coating is immersed in the medium. A marked reduction in the coating resistance has been observed with increasing temperature for resins [9,13,14] (see Fig. 5-2 [14]).

Figure 5-3 is a compilation of coating resistances for long-distance pipelines as a function of service life for PE and bitumen [15]. The r_u values for objects in service are considerably lower compared to the r_u^x or r_u^0 values. This is due to pores or holidays in the coating and to poorly coated fittings and defects in the coating of the girth welds, where the metal is exposed to the environment. By neglecting polarization resistances, the resistance R^x of a defect, with a diameter d, in a coating of thickness s, and with a medium of specific resistance ρ, is obtained from the sum of the pore resistances R_F

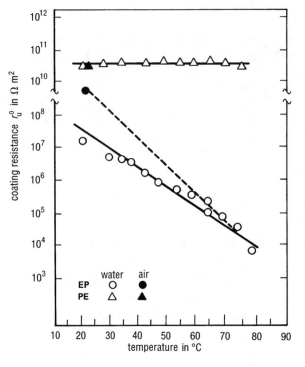

Fig. 5-2 Influence of temperature on the coating resistance r_u^0 when 3-mm-thick PE and 0.4-mm-thick EP coatings are immersed in water.

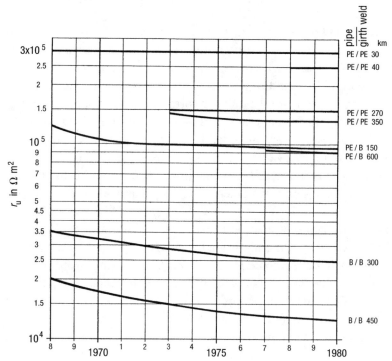

Fig. 5-3 Coating resistances of long-distance pipelines.

Table 5-2 Corrosion rate w_{int} and specific coating resistance r_u^0 after a long exposure of steel pipe sections with damage-free coating.

Coating	Thickness (mm)	Medium[a]	w_{int} ($\mu m\ a^{-1}$)	t (a)	r_u^0 ($\Omega\ m^2$)	t (a)	Remarks
PE (2x), soft adhesive	7	Soil	0.2	10	10^{11}	20	Precursor of DIN 30670
PE (2x), soft adhesive	4	Water	<0.1	19	10^{11}	19	Precursor of DIN 30670
PE molten adhesive	2.4	Soil	0.2	10	3×10^{10}	20	DIN 30670
PE molten adhesive	2.2	Water	<0.1	19	10^{11}	19	DIN 30670
PE melted	4	Soil	0.2	10	3×10^{10}	20	Precursor of DIN 30670
PE melted	2.2	Water	0.4	19	3×10^{10}	19	Precursor of DIN 30670[b]
Bitumen	7	Soil	0.2	10	10^5	20	DIN 30673[c]
Bitumen	4	Water	0.2	19	3×10^5	19	DIN 30673[c]
Bitumen	10	Soil	ND	10	3×10^5	20	DIN 30673[c]
Bitumen	6	Water	0.1	19	3×10^5	19	DIN 30673[c]
EP tar	2	Soil	0.2	10	3×10^6	20	Fiberglass reinforced
EP tar	2	Water	0.1	19	3×10^7	19	Fiberglass reinforced
PUR tar	2.5	Soil	ND	10	10^9	10	DIN 30671
EP	0.4	Soil	ND	10	10^8	10	DIN 30671
PE tape, soft adhesive	0.9	Soil	0.4	10	10^7	20	Precursor of DIN 30672[b]
PE tape, soft adhesive	0.9	Water	0.4	19	10^7	19	Precursor of DIN 30672[b]
PE tape system	3	Soil	ND	10	10^7	20	DIN 30672
PE tape system	3	Water	0.2	19	3×10^9	19	DIN 30672
PE tape system	3.5	Soil	ND	10	10^9	10	DIN 30672
PE tape system	1.5	Soil	ND	10	10^8	10	DIN 30672
PE tape without adhesive	0.7	Soil	ND	10	3×10^4	10	50% overlap
PE heat-shrunk sleeve	3	Soil	ND	10	10^7	10	DIN 30672

ND = Not determined.
[a] Water: pipe length 2 to 4 m, 1.3 m deep; soil: pipe length 12 m, 1 m covering.
[b] The disbonding is clearly time dependent.
[c] The coating resistance decreases slightly with time.

$$R_F = \rho \frac{4s}{\pi d^2} \tag{5-3}$$

and the grounding resistance $R_A = \rho/(2d)$ from Eq. (24-17):

$$R^x = R_F + R_A = \frac{\rho}{2d}\left(1 + \frac{8s}{\pi d}\right) \tag{5-4}$$

If there are several defects, the total resistance R_g of the individual resistances connected in parallel gives from Eq. (5-4):

$$\frac{1}{R_g} = \sum_i \frac{1}{R_i^x} = \frac{2}{\rho}\sum_i \frac{d_i}{1 + \dfrac{8s}{\pi d_i}} \tag{5-5}$$

If the defect diameters are all the same $d = d_i$, it follows from Eq. (5-5):

$$R_g = \frac{\rho}{2nd}\left(1 + \frac{8s}{\pi d}\right) \tag{5-6}$$

where n is the number of defects.

In general, it cannot be assumed that the defect diameters are all the same and larger than the coating thickness ($d_i \gg s$). For an average diameter

$$d = \frac{1}{n}\sum_i d_i \tag{5-7}$$

and from Eq. (5-5):

$$R_g = \frac{\rho}{2nd} \tag{5-8}$$

By introducing the defect density $N = n/S$, the specific total resistance of the defects $r_g = R_g S$ becomes:

(a) equal d_i from Eq. (5-6):

$$r_g = \frac{\rho}{2Nd}\left(1 + \frac{8s}{\pi d}\right) \tag{5-9a}$$

(b) $d_i \gg s$ from Eq. (5-8):

$$r_g = \frac{\rho}{2Nd} \tag{5-9b}$$

From the parallel connection of r_u^0 for the intact coating and r_g for the defects, it follows finally that:

$$r_u = \frac{r_g}{1 + r_g/r_u^0} \tag{5-10}$$

Coatings for Corrosion Protection 161

Because r_u^0 is usually considerably larger than r_g, r_u is practically determined only by the defects (see Table 5-1).

A relation can be derived between r_u and the necessary protection current density J_s from Eq. (5-2) together with the pragmatic protection Criterion 2 in Table 3-3 [see Eq. (3-31)]:

$$r_u = \frac{|U_{on} - U_{off}|}{J} = \frac{0.3 \text{ V}}{J_s} \tag{5-11}$$

$$\left(\frac{J_s}{\mu\text{A m}^{-2}}\right) = \frac{3 \times 10^5}{\left(r_u / \Omega \text{ m}^2\right)} \tag{5-11'}$$

Like Criterion 2, this relation has no theoretical foundation and only serves as a comparison and for the design of protection installations. For this reason J_s in Eq. (5-11') is sometimes less correctly termed the conventional protection current requirement.

If r_g is only the result of defects, there is the question of a connection between r_g and the total area of defects S_0. With $S_0/S = N(\pi d^2/4)$ and Eq. (5-9b), it follows that:

$$\frac{S_0}{S} = \frac{\pi d \rho}{8 r_g} \tag{5-12}$$

With the protection current density J_s^0 for the uncoated surface S_0, the protection current density J_s is given by

$$J_s = J_s^0 \frac{S_0}{S} \tag{5-13}$$

from which, with Eq. (5-12):

$$J_s r_g = \frac{\pi}{8} J_s^0 \rho d \tag{5-14}$$

Equation (5-11) cannot be explained by Eq. (5-14) because J_s^0 and ρ are variables related to the soil and furthermore the average defect diameter, d, is not constant.

If, however, it is assumed from Eq. (2-40) that the protection current density corresponds to the cathodic partial current density for the oxygen reduction reaction, where oxygen diffusion and polarization current have the same spatial distribution, it follows from Eq. (2-47) with $r_g = \Delta\phi/J$:

$$J_s r_g = z\mathcal{F} D \rho c(O_2) \tag{5-15}$$

This equation also includes, as in Eq. (5-14), variables related to the soil, but no data on the defects. A theoretical basis for Eq. (5-11) is also not possible. Equation (5-11) can therefore give at most only approximate information on the expected protection current requirement.

Protection current density and coating resistance are important for the current distribution and for the range of the electrochemical protection. The coating resistance determines, as does the polarization resistance, the polarization parameter (see Sections 2.2.5 and 24.5). For pipelines the protection current density determines the length of the protection range (see Section 24.4.3).

5.2.1.3 Effectiveness of Cathodes and Cell Formation

Cell formation can easily be detected by measuring potential if coated surfaces with no pores have a more positive potential than uncoated material. Usually this is the case with coated steel in solutions containing oxygen. More negative potentials can only arise with galvanized steel surfaces. Figure 5-4 shows examples of measured cell currents [9,10,16].

The cell current is determined by the coating resistance r_u^0 and the size of the coated surface S_c. Neglecting the anode resistance, the cell current from these two quantities is given by [see Eq. (2-43)]:

$$I_e = \frac{\Delta U \, S_c}{r_u^0} \tag{5-16}$$

Dividing by the anode surface S_0 and with $r_g = r_u$, $S = S_c$ and Eq. (5-12) gives:

$$J_a = \frac{I_e}{S_0} = \frac{8}{\pi} \frac{r_u}{r_u^0} \frac{\Delta U}{\rho d} \tag{5-17}$$

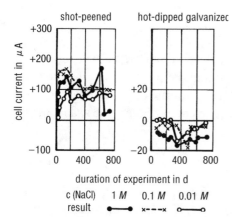

Fig. 5-4 Cell currents between a coated specimen ($S_c = 300$ cm^2) and uncoated steel electrode ($S_a = 1.2$ cm^2) in NaCl solutions at 25°C. Left: shot-peened steel sheet, 150 μm EP-tar. Right: hot-dipped galvanized steel sheet, 150 μm EP-tar.

The importance of large r_u^0 values becomes obvious from Eq. (5-16) but in making an assessment, the size of the object (S_c) must also be considered (see Ref. 8). The r_u^0 values, especially with thin coatings, can lie in a wide range, from 10^2 to 10^7 Ω m². Thus the prime coating and the total coating thickness have a considerable influence [16]. Equation (5-17) can be applied when assessing pipelines [17]. The ratio r_u/r_u^0 has a great effect. The figures in Table 5-1 indicate that in contrast to a bitumen coating, there is certainly no danger of cell formation with PE coating as long as the pipeline is not electrically connected to foreign (cathodic) structures. The assertions in Ref. 18 about sufficient external protection for installations that are to be protected but that are without cathodic protection, are in agreement with these observations.

5.2.1.4 Electrochemical Blistering

Electrochemical blistering is a result of the ion conductivity of the coating material. It is only expected with thin coatings with sufficiently low r_u^0 values which have been immersed in a medium for long periods of service [8-10]. Cathodic blisters are the best known and occur in saline solutions with cathodic protection. What is necessary is the presence of alkali ions and the permeation of H_2O and O_2 so that OH^- ions are formed in the cathodic partial reaction, as in Eq. (2-17) or (2-19) at the metal/coating interface, forming a caustic solution with the migrating alkali ions. H_2O diffuses to the place where the reaction is occurring by osmosis and electro-osmosis, and leads to the formation of relatively large blisters. Anodic blisters can also be observed where there is contact with foreign cathodic objects. The migration of anions is necessary to form soluble corrosion products with the cations of the underlying metal. The cations are formed with anodic polarization according to Eq. (2-21). Osmotic and electro-osmotic processes act against H_2O migration, so that anodic blisters are considerably smaller than cathodic blisters. Pitting corrosion always occurs at anodic blisters.

Blisters are apparently statistically distributed and their formation is connected with paths of increased ion conductivity in the coating material. Whether this is a question of microporosity is a matter of definition. Since the skin of the blister remains impervious to water in such micropores, the term "pore" for conducting regions is avoided in this handbook, especially as the properties of a pronounced pore are the same as those of a damaged area (holiday).

If the metal surface is insufficiently cleaned before coating and contains local salt residues, osmotic blistering is to be reckoned with because it enhances electrochemical blistering and determines the points where it occurs. This is also applicable to the action of ions in the prime coating. Coating systems with an alkali silicate primer coat are particularly likely to form cathodic blisters [16]. Blisters often form in the vicinity of mechanical damage (see Fig. 5-5 [19]). With strong cathodic polarization in limiting cases, complete disbonding can occur.

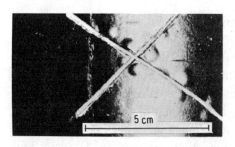

Fig. 5-5 Cathodic blisters in the vicinity of a scratched cross; steel pipe with 70 μm Zn-epoxy resin +300 μm EP-tar, seawater; $U_{Cu\text{-}CuSO_4}$ = –1.1 to –1.2 V, 220 days, 20°C.

There are numerous publications [9,10,16,19-24] and test specifications [8,25] on the formation of cathodic blisters. They are particularly relevant to ships, marine structures and the internal protection of storage tanks. Blister attack increases with rising cathodic polarization. Figures 5-6 and 5-7 show the potential dependence of blister density and the NaOH concentration of blister fluid, where it is assumed that $c(Na^+)$ and $c(NaOH)$ are equal due to the low value of $c(Cl^-)$ [23].

The blister population-potential curves can intersect one another. Thus short-term experiments at very negative potentials, in the region of cathodic overprotection, give no information on the behavior at potentials in the normal protection range. The susceptibility generally increases strongly with overprotection ($U_H < -0.83$ V).

Compared with cathodic blisters, which can be recognized by their alkali content, anodic blisters can be easily overlooked. Intact blisters can be recognized by the slightly lower pH value of the hydrolyzed corrosion product. The pitted surface at a damaged blister cannot be distinguished from that formed at pores.

In general the population of cathodic blisters increases with cathodic polarization and the population of anodic blisters with anodic polarization. Both types of

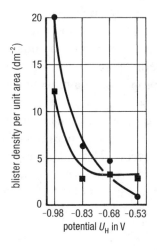

Fig. 5-6 Relation between the density of cathodic blisters and potential; shot-peened steel sheet without primer (■) and with about 40 μm primer coat (Zn ethyl silicate + polyvinyl butyral) (●); top coat: 500 μm EP-tar; 0.5 M NaCl, 770 days at 25°C.

Fig. 5-7 Effect of potential on the composition of the blister liquid, shot-peened pipe with 300 to 500 μm EP-tar, artificial seawater, 1300 days at 25°C.

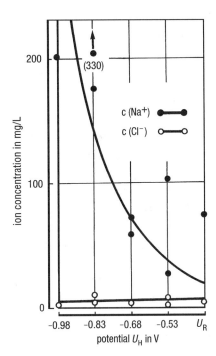

blisters can occur near one another in free corrosion. Since ion conductivity is essential for the formation of these blisters, there is a correlation between r_u^0 and susceptibility to blistering. In a large number of investigations with EP-tar coatings on steel coupons in NaCl solutions, test pieces with an average value of r_u^0 greater than 10^6 Ω m² remained free of blisters. Test pieces with $r_u^0 \leq 10^3$ Ω m² were attacked [16].

Blisters or disbonding can occur with long-term cathodic polarization in solutions without alkali ions and/or in coatings with sufficiently high r_u^0 values where the reaction of the blister liquid is neutral. In the case of adhesion loss on a larger area, as in the right-hand picture in Fig. 5-8, the presence of humidity (water) could only be detected by the slight rusting. In these cases electrolytic processes are not involved but rather electro-osmotic transport of water with cathodic polarization [9,10] based on Cohen's rule [26]. According to this rule, the macrogel coating material is negatively charged compared with water. The contrary process, anodic dehydration, is responsible for the small extent of anodic blisters. It can also be recognized by the tendency of areas in the neighborhood of anodes to dry out (see Section 7.5.1). No blisters are formed in anodic polarization in solutions without such anions, which can form easily soluble corrosion products with cations of the bases metal, or at high r_u^0 values. With thin coatings after long-term anodic treat-

166   Handbook of Cathodic Corrosion Protection

ment of steel, small dark specks can be seen which consist of Fe_3O_4. Pitting corrosion is not observed.

5.2.1.5 Cathodic Disbonding

The production of OH^- ions according to Eq. (2-17) or (2-19) in pores or damaged areas is responsible for cathodic disbonding [9,10], where the necessary high concentration of OH^- ions is only possible if counter-ions are present. These include alkali ions, NH_4^+ and Ba^{2+}. Disbonding due to the presence of Ca^{2+} ions is extremely slight [27].

Disbonding in free corrosion depends on the formation of aeration cells which produce OH^- ions in the cathodic region according to Eq. (2-17). Correspondingly, exclusion of oxygen or inhibiting corrosion can prevent disbonding [19]. No disbonding takes place in a K_2CrO_4 solution because there is no corrosion [27]. Addition of acids also suppresses disbonding by neutralizing OH^- ions, but encourages acid attack at holidays. On the other hand, alkalis strongly promote disbonding [19] (see Fig. 5-9 [10]) but favor corrosion resistance at holidays.

With anodic polarization, the anodic partial reaction predominates at defects so that OH^- ions formed according to Eq. (2-17) are combined in the corrosion

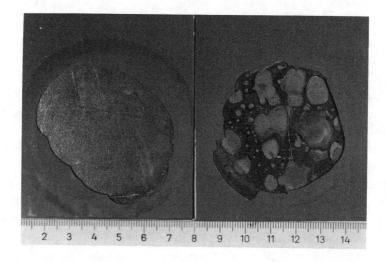

Fig. 5-8   Total adhesion loss of a 500-μm-thick coating of EP (liquid lacquer), 0.2 M NaCl, galvanostatic $J_V = -1.5$ μA m^{-2}, 5 years at 25°C. Left: coating with a pin pore; loss of adhesion due to cathodic disbonding. Right: pore-free coating; loss of adhesion due to electro-osmotic transport of H_2O. In both cases the loose coating was removed at the end of the experiment.

product. Disbonding is thus strongly suppressed [19]. Anodic metal loss leads to pitting rather than uniform corrosion under the coating. Cathodic protection strongly promotes disbonding and cathodically protects the exposed steel surface. Contrary actions with regard to the tendency for disbonding and the danger of corrosion at existing holidays in the coating depend on polarization and the addition of acids or alkalis. From theoretical considerations, laboratory experiments and experience in the field, there is no danger of corrosion in the disbonded region itself [8-10,17,19,28-33]. An example of corrosion resistance in the disbonded region is shown by the steel surface in the left-hand picture in Fig. 5-8. Furthermore, an increase in current requirement is not in question [34].

With cathodic polarization, the depth of disbonding increases with an increase in alkali ion concentration (see Fig. 5-10 [34]) and an increase in cathodic polarization (see Fig. 5-11 [35]). Corrosion inhibitors have no influence. Disbonding even occurs in K_2CrO_4 solutions [27]. Anomalous behavior has only been seen in $AgNO_3$ solutions [36], which can be explained by the cathodic deposition of Ag on the steel surface at the damaged area. Since OH⁻ formation at the steel surface is confined to the Ag dendrites, OH⁻ ions are lacking at the edge of the coating [27].

The disbonding rate decreases with time [35], which can be attributed to the consumption of OH⁻ ions by reaction with adhesive groups. This consumption is obviously partly compensated for by the formation of OH⁻ ions through oxygen reduction; these permeate inward from the outer surface of the coating. If this permeation is hindered by an aluminum foil gas seal, the disbonding rate falls off

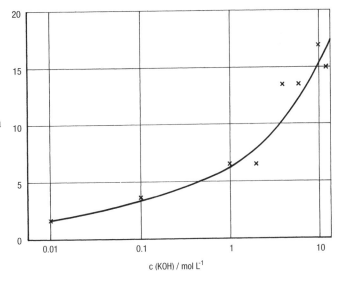

Fig. 5-9 Effect of KOH concentration on disbonding depth; steel pipe with PE coating on fusion adhesive, free corrosion, 10 days, 25°C, defect 1 cm diameter.

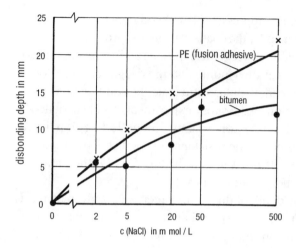

Fig. 5-10 Effect of NaCl concentration on disbonding depth on coated pipes, $U_{Cu\text{-}CuSO_4} = -1.1$ V, 50 days, 25°C, circular artificial holiday 1 cm diameter.

rapidly with time [27]. This indication of the considerable influence of transport processes is confirmed by the observation that disbonding can also be drastically reduced under high mechanical pressure, which keeps the gap between metal and coating very small [17]. The size of holidays has a similar effect. With small areas, the disbonding is considerably less [17,35]. On the other hand, disbonding is encouraged by a rise in temperature.

Many investigations have shown that disbonding is influenced by the choice of coating, where resin coatings with many polar adhesive groups show superiority over thick coatings [19,34,35] (see also Fig. 5-11). This contrasts with the behavior of these coatings in the case of cathodic blistering. By combining thick

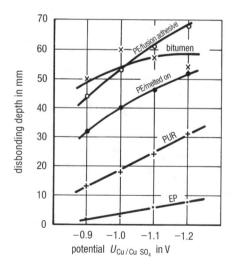

Fig. 5-11 Effect of potential on disbonding depth on coated pipes, 0.1 M Na$_2$SO$_4$, 370 days, 25°C, size of artificial holiday, 1 cm².

coatings with polar adhesives, the difference between thin EP coatings and thick PE coatings in resistance to disbonding can be eliminated [14].

5.2.2 Physicochemical Properties

The properties of coatings must remain constant under service conditions for long-term corrosion protection. The coating materials must be absolutely stable against the corrosive environment. Material standards [2-5] provide the necessary requirements for chemical and aging stability against heat and UV irradiation [37]. With the relatively thin corrosion-protection coatings used for construction steel, the protective action is derived from pigments in the prime coat that inhibit corrosion. In the course of time, this action can be lost due to reaction with corrosive components. Coatings dealt with in this chapter that are constantly exposed to electrolytes have a different protection principle which is based on greater thickness and as low a permeation rate of corrosive agents as possible. Thus, even when the metal under the coating is no longer inhibited or passive, protection is still maintained if the corrosion rate stays sufficiently low in areas of low adhesion.

All organic coatings can dissolve O_2 and H_2O and allow them to penetrate. The following corrosion reaction can take place on an active steel surface under a coating:

$$4\,Fe + 3\,O_2 + 2\,H_2O = 4\,FeOOH \tag{5-18}$$

which on aging of the FeOOH allows the separation of H_2O, so that finally the continuation of the corrosion is solely determined by the permeation of O_2. This neglects corrosion by H_2O with the action of H_2 according to Eq. (2-19). The equation for O_2 permeation is:

$$\frac{j_v}{L\,m^{-2}\,h^{-1}} = 3.6 \times 10^5 \left(\frac{P}{cm^2\,s^{-1}\,bar^{-1}}\right) \frac{(\Delta p / bar)}{(s / mm)} \tag{5-19}$$

where P = permeation coefficient (see Table 5-3); Δp = difference in oxygen pressure, which is equal to the partial pressure in the ambient air of 0.2 bar; and s = thickness of the coating. From Eqs. (2-5) and (2-20) it follows that

$$w = j_v \frac{f_a}{f_v} \tag{5-20}$$

and including the data in Tables 2-1 and 2-3 in Eq. (5-19) gives:

$$\left(\frac{w}{mm\,a^{-1}}\right) = 1.33 \times 10^6 \left(\frac{P}{cm^2\,s^{-1}\,bar^{-1}}\right) \frac{(\Delta p / bar)}{(s / mm)} \tag{5-21}$$

Table 5-3 Permeation coefficients and rate of oxygen corrosion according to Eq. (5-21).

Coating material	Permeation coefficients ($cm^2\ s^{-1}\ bar^{-1}$)			s (mm)	w ($\mu m\ a^{-1}$)
	H_2O	CO_2	O_2		
ND PE	2×10^{-7}	–	10^{-8}	3	0.9
HD PE	6×10^{-7}	7×10^{-8}	2×10^{-8}	3	1.8
PUR tar	4×10^{-6}	–	5×10^{-9}	1	1.3
EP	$(1\ to\ 2) \times 10^{-6}$	3×10^{-8}	$(0.5\ to\ 2) \times 10^{-9}$	0.5	0.3 to 1

The permeation coefficients of the most important coating materials are given in Table 5-3 [17,38]. In addition, the corrosion rates according to Eq. (5-21) for the usual coating thicknesses are calculated. From this it can be seen that in all cases the maximum possible corrosion rates can be completely disregarded. This agrees with the results of long-term experiments in Table 5-2 and with experience in the field [30,39,40]. It should be mentioned here that there is no connection between corrosion rates and coating resistances.

Questions about criteria for judging the effectiveness of protective action can arise with new coating materials where there is no practical experience, and in extreme service conditions, particularly with regard to temperature or electrical voltage. Here the effect of temperature on the r_u^0 value must be considered (see Fig. 5-2) because it can lead to increased blistering. In addition, it must be remembered that alkali is formed in the case of cathodic polarization, which can attack coating materials. Stability with aging and resistance to alkali have to be tested when there are questions of behavior at elevated temperatures. The action of high electrical voltages is not damaging for high r_u^0 values. A PE coating with little damage and subjected to $-2\ A\ cm^{-2}$ ($-100\ V$) was completely unattacked after 100 days, whereas an EP coating was etched after 1 week due to a temperature rise from ohmic heating [17].

5.2.3 Mechanical Properties

Since damage to protective coatings arising in transport, backfilling and installation, as well as in service cannot be excluded, cathodic protection cannot be dispensed with except for short, well-insulated pipelines [18]. The highest possible r_u values and the lowest possible protection current requirements are desirable for the reasons given in Section 5.2.1.2. Therefore, the coatings must be sufficiently resistant to mechanical influences. This demands good adhesion in terms of high peeling and shear strength, high resistance to impact and low penetration under pressure. These requirements have pushed the replacement of bituminous coatings with plastic coatings due to increased pipe dimensions, installation requirements and unfavorable conditions in backfilling.

A series of testing methods relating to products [37] for determining peeling strength, penetration and impact resistance are given in the material standards [2-5]. The results in all these tests serve for comparisons and the establishment of conformity with standards. They do not, however, provide directly applicable information for service use. Since the service loading cannot be completely described, only a statement about the order of magnitude of the defects will be possible. However, as long as even a single defect cannot be safely excluded, electrochemical protection measures should be applied. This applies particularly to objects that are in danger of contact with foreign cathodic structures (e.g., distribution networks in towns with a large number of foreign cathodic structures from concrete foundations) [41]. One must also remember that the evidence from damage-measuring techniques requires the same assumptions as electrochemical protection. Insofar as good mechanical properties are necessary to achieve the highest possible value of r_u, there is a practical limit of about 10^6 Ω m^2 above which there is no advantage to be expected since $r_u = r_u^0$, i.e., freedom from damage cannot be attained with certainty.

In order to choose the type of coating and determine the necessary coating thickness, many practice-oriented tests would have to be carried out in which the evaluation of damage areas and choice of service conditions are not always comparable [42-44]. However, information on the various thickness ranges of the PE coating in Ref. 4 was deduced from such experiments.

Among the different coating materials, bituminous substances have the lowest mechanical strength. In the case of plastics, coating thickness and adhesion are important. Thus, thick coatings with PE are superior to thin coatings of EP [14]. A further increase is possible with an additional cement mortar coating in Ref. 7, but it is used with advantage only in special cases, e.g., stony surroundings or rocks. In such cases rock shields are also usually installed. According to Ref. 43 these are inferior to coatings of increased thickness. Chemical fleeces provide better protection than plastic nets. However, the fleece must have sufficient electrolytic conductivity so that the access of the protection current is not impaired.

5.2.4 Corrosion of the Steel under the Coating

A steel surface can be wetted and corroded by permeation of corrosive agents (see Section 5.2.2) and by cathodic disbonding (see Section 5.2.1.5). In all cases the corrosion rates are negligibly small (see Table 5-2). In the case of cathodic disbonding it could even be shown that cathodic protection is possible in the gap between the pipe surface and the coating provided the steel surface is blast cleaned free of oxide [45-47]. Figure 5-12 shows potentials that were measured under a PE extruded coating without adhesive as a function of the external potential [47, 48].

Corrosion arising from permeation processes through damage-free coatings can lead to a marked reduction of adhesion in service. This is not important in the corrosion protection process since after installation and burying in the ground, the adhesion is no longer necessary for pipes or storage tanks. Shear stress from the soil does not exceed the break strength of the coating. Damage, however, is possible from external causes in the course of other excavation activities. Reduced adhesion can cause more extensive damage than was to be expected. However, in no circumstances is it advisable to remove nonadherent but still tightly attached coatings. Since an attached and sound coating (e.g., of storage tanks or pipes) cannot gap even if the adhesion fails, the corrosion protection is still effective (see Ref. 8 and further explanation in Ref. 4). The condition of coatings on flat surfaces or as linings is to be assessed differently because in the event of a loss of adhesion, lifting of the coating is to be expected.

The region of an adhesion loss caused by cathodic disbonding has led to questions as to a particular susceptibility to intergranular stress corrosion (see Section 2.3.3). Critical components of the media are NaOH, Na_2CO_3 and $NaHCO_3$. The parameters influencing stress corrosion induced by these components were extensively discussed in connection with the properties of the coating, where the relevant critical potential range and the conditions required for the formation of the critical components had to be considered [17, 47-49]. Accordingly, at high temperatures and high cathodic protection, stress corrosion induced by NaOH is

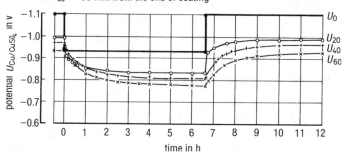

- U_0 = controlled potential above the end of coating
- U_{20} = 20 mm from the end of coating
- U_{40} = 40 mm from the end of coating
- U_{60} = 60 mm from the end of coating

Fig. 5-12 Effect of external polarization (U_0) on potentials in the crevice between pipe surface and nonadherent PE coating, carbonate-bicarbonate solution at 70°C.

not to be excluded where the type of coating has no influence. However, such influence is not to be excluded in the case of stress corrosion caused by $NaHCO_3$ in the disbonded region. Since disbonding is only caused by NaOH, the CO_2 that is necessary for the reaction to $NaHCO_3$ must come from the soil and permeate through the coating or come from the coating itself due to thermal influences. Recent investigations have shown that surface treatment is a very important factor [50]. Accordingly, with shot-peened surfaces, which are necessary with all plastic coatings, one does not have to be concerned with stress corrosion in the disbonded region.

5.3 Properties of Cement Mortar and Concrete

Steel in cement mortar is in the passive state represented by field II in Fig. 2-2. In this state reinforcing steel can act as a foreign cathodic object whose intensity depends on aeration (see Section 4.3). The passivity can be lost by introduction of sufficient chloride ions or by reaction of the mortar with CO_2-forming carbonates, resulting in a considerable lowering of the pH. The coordinates then lie in field I. The concentration of OH^- ions can be raised by strong cathodic polarization and the potential lowered, resulting in possible corrosion in field IV (see Section 2.4).

5.3.1 Corrosion of Mortar

Cement mortar will be attacked by waters that have an excess of free carbon dioxide compared with that of waters that are in a lime-carbonic acid equilibrium. There is a two-step mechanism with a carbonization process according to

$$CaO + CO_2 = CaCO_3 \tag{5-22}$$

and finally a deliming process (loss of lime) according to

$$CaCO_3 + CO_2 + H_2O = Ca^{2+} + 2\,HCO_3^- \tag{5-23}$$

Of the two reactions, only the deliming results in softening and removal of a mortar lining. The information given in the comment in Ref. 6 on the lining of water pipes with mortar is relevant here. The carbonization hardens the mortar somewhat and is essentially only detrimental to the coated steel. Lime-dissolving conditions are frequently found in soils so that both reactions occur. As long as there is no strong flow of ground water, the mechanical forces are insufficient to cause abrasion of a softened surface.

5.3.2 Corrosion of Steel in Mortar

Due to both carbonization and penetration of chloride ions, steel will pass from a passive to an active condition and (consequently) may corrode. If the mortar is completely surrounded by water, oxygen diffusion in wet mortar is extremely low so that the situation is corrosion resistant because the cathodic partial reaction according to Eq. (2-17) scarcely occurs. For this reason the mortar lining of waste pipes remains protective against corrosion even if it is completely carbonated or if it is penetrated by chloride ions.

A danger exists with anodic polarization (e.g., by neighboring aerated passive regions or by the presence of oxygen), which is possible in the gas phase through unwetted pores. It should be understood that just above the water/air zone, severe corrosion occurs in which the growing corrosion products can break off the mortar coating. Galvanizing has a beneficial influence on this process because internal cell formation is increased by the galvanic anode [52]. This protective action is limited in time by the thickness of the zinc coating. Permanent corrosion protection of steel in mortar can therefore only be given by an insulating organic coating.

The cathodic effectiveness of the passive steel in cement mortar can be seen in Fig. 5-13. The cell current is measured between a mortar-coated DN 100 pipe section and an uncoated steel ring 16 mm broad as anode. It can be clearly seen that the cell current immediately falls and after 100 days goes toward zero. The same result is obtained by removing the specimens and aerating the mortar coating and repeating the experiment with the same components [51].

Cathodic protection can be used to protect steel in concrete (see Chapter 19). There is no fear of damage by H_2 evolution due to porosity of the mortar. Local corrosion attack can be observed under extreme conditions due to porosity (water/cement ratio = 1) and polarization ($U_H = -0.98$ V) with portland cement but not with blast furnace cement, corresponding to field IV in Fig. 2-2 [53]. However, such conditions do not occur in practice.

Anodic polarization can occur in the presence of stray currents. Oxygen is evolved on the passive steel according to:

$$2\,H_2O - 4\,e^- = O_2 + 4\,H^+ \tag{5-24}$$

The anodically produced acid is neutralized by the alkaline mortar (CaO). Corrosion is then possible only if the supply of alkali at the steel surface is consumed and the steel becomes active. This process is possible only under certain circumstances after a very long incubation period. Apparently in steel-concrete foundations the possible current densities are so small that this case never arises. The possibility of danger has to be verified with thin outer coatings where deliming has been noticed on the steel surface.

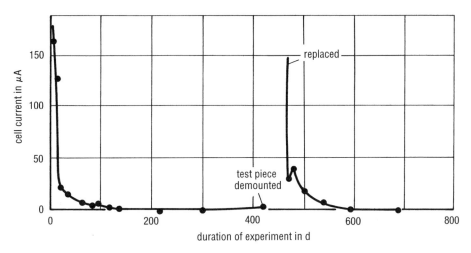

Fig. 5-13 Cell current in a pipe with cement-mortar lining; anode is an uncoated ring, tapwater at 15°C.

5.4 Properties of Enamel Coatings

Enamel coatings are completely electrochemically inert, but in general are not free of pores. Pores are mostly difficult to recognize. Electrical pore testing reveals places covered with conducting oxide covering the metal. In free corrosion, pores are mostly blocked up with corrosion products so that corrosion comes to a standstill. This process is not so definite with larger areas of damage in the enamel.

Enamel coatings are used for the internal protection of storage tanks that in most cases have built-in components (e.g., fittings with exits, probes, temperature detectors) that usually exhibit cathodic effectivity. These constitute a considerable danger of pitting corrosion at small pores in the enamel. Corrosion protection is achieved by additional cathodic protection which neutralizes the effectiveness of the cathodic objects.

Enamel coatings usually consist of several layers in which the prime coating is applied for adhesion but does not have the chemical stability of the outer layers. With cathodic polarization at holidays, attack on the exposed prime coating is possible as the cathodically produced alkali causes the defects to increase in size. This particularly cannot be excluded in salt-rich media.

Enamels have very varied properties where their chemical stability is concerned. Relevant stability testing must be carried out for the different areas of application. Enamel coatings for hot water heaters, their requirements and combination with cathodic protection are described in Section 20.4.1.

It must be remembered that the interface of steel and enamel reacts very sensitively to hydrogen recombination, which causes cracks (fish scales) and spalling.

The hydrogen can originate from the enamel itself, or can be produced during service. The rear surfaces should not be treated with acid as far as possible. Cathodic protection of these surfaces can lead to weak H absorption (see Fig. 2-20), which leads to long-term fish-scaling and spalling. With reference to this effect, there is an electrolytic test method for determining the susceptibility to fish-scaling [54].

5.5 Properties of Metallic Coatings

Coatings of more noble metals than the substrate metal (e.g., Cu on Fe) are only protective when there are no pores. In other cases severe local corrosion occurs due to cell formation (bimetallic corrosion). Cathodic protection is theoretically possible. This protection combination is not very efficient since the coating usually consumes more protection current than the uncoated steel.

Coatings of less noble metals than the substrate metal (e.g., Zn on Fe) are only protective if the corrosion product of the metal coating restricts the corrosion process. At the same time, the formation of aeration cells is hindered by the metal coating. No corrosion occurs at defects. Additional cathodic protection to reduce the corrosion of the metal coating can be advantageous. Favorable polarization properties and low protection current requirements are possible but need to be tested in individual cases. The possibility of damage due to blistering and cathodic corrosion must be heeded.

Blistering of the zinc coating has been observed with cathodic protection of hot-dipped galvanized steel in warm water. The cause was absorption and recombination of hydrogen. The blisters arise in the Fe-Zn alloy phases [55]. With Zn and especially with Al, the protection potential region must be limited at the negative end because corrosion-forming zincates and aluminates are possible (see Figs. 2-10 and 2-11).

5.6 References

[1] DIN 50927, Beuth-Verlag, Berlin 1985.
[2] DIN 30671, Beuth-Verlag, Berlin 1987.
[3] DIN 30673, Beuth-Verlag, Berlin 1987.
[4] DIN 30670, Beuth-Verlag, Berlin 1980.
[5] DIN 30672, Beuth-Verlag, Berlin 1979.
[6] DIN 2614, Beuth-Verlag, Berlin 1989.
[7] DIN 30674 (Vornorm) Teil 2, Beuth-Verlag, Berlin 1984.
[8] DIN 50928, Beuth-Verlag, Berlin 1985.
[9] W. Schwenk, Metalloberfläche *34*, 153 (1980); NACE Conf. "Corrosion Control by Organic Coatings," Houston 1981, s. 103B110.
[10] W. Schwenk, farbe+lack *90*, 350 (1984).
[11] J. D'Ans u. E. Lax, Taschenbuch f. Chemiker u. Physiker, Springer, Berlin 1949, S. 772, 773 und 1477.
[12] W. Klahr, Fa. Th. Goldschmidt, persönl. Mitteilung 1978.
[13] H. Landgraf, 3R international *20*, 283 (1981).
[14] R.E. Sharpe u. R.J. Dick, J. Coating Techn. *57*, 25 (1985).
[15] W.v. Baeckmann, 3R international *20*, 470 (1981).
[16] H. Hildebrand u. W. Schwenk, Werkstoffe u. Korrosion *30*, 542 (1979).
[17] W. Schwenk, 3R international *19*, 586 (1980).
[18] TRbF 131, C. Heymanns-Verlag KG, Köln 1981.
[19] W. Schwenk, 3R international *15*, 389 (1976).
[20] H. Determann, E. Hargarter, H. Sass u. H. Haagen, Schiff & Hafen *28*, 729 (1974).
[21] H. Determan, E. Hargarter u. H. Sass, Schiff & Hafen *28*, 729 (1976); *32*, H. 2, 89 (1980).
[22] W. Bahlmann, E. Hargarter, H. Sass u. D. Schwarz, Schiff & Hafen *33*, H. 7, 50 (1981); W. Bahlmann u. E. Hargarter, Schiff & Hafen, Sonderheft Meerwasser-Korrosion, Febr. 1983. S. 98.
[23] H. Hildebrand u. w. Schwenk, Werkstoffe u. Korrosion *33*, 653 (1982).
[24] W. Fischer, EUROCORR '87, DECHEMA, Frankfurt 1987, S. 719B724; W. Fischer, U. Hermann u. M. Schröder, Werkstoffe u. Korrosion *42,* 620 1991.
[25] Schiffbautechnische Gesellschaft, STG-Richtilinien Nr. 2220, Hamburg 1988.
[26] J. Eggert, Lehrbuch der physikalischen Chemie, Leipzig 1948, S. 559B570.
[27] W. Schwenk, gwf gas/erdgas *118*, 7 (1977).
[28] W. Schwenk, Rohre – Rohrleitungsbau – Rohrleitungstransport *12*, 15 (1973).
[29] W. Takens, J. van helden u. A. Schrik, Gas *93*, 174 (1973).
[30] P. Pickelmann u. H. Hildebrand, gwf gas/erdgas *122*, 54 (1981).
[31] W. Schwenk, 3R international *23*, 188 (1984).
[32] W.v. Baeckmann, D. Funk u. G. Heim, 3R international *24*, 421 (1985).
[33] J.M. Holbrook, Unveröffentl. Untersuchung, Battelle, Columbus, 1982/84.
[34] W. Schwenk, 3R international *18*, 565 (1979).
[35] G. Heim, W.v. Baeckmann u. D. Funk, 3R international *14*, 111 (1975).
[36] H.C. Woodruff, J. Paint Techn. *47*, 57 (1975).
[37] G. Heim u. W.v. Baeckmann, 3R international *26*, 302 (1987).

[38] N. Schmitz-Pranghe u. W.v. Baeckmann, Mat. Perf. *17*, H. 8, 22 (1978).
[39] H. Fisher, gwf gas/erdgas *113*, 283 (1972).
[40] P. Pickelmann, gwf gas/erdgas *116*, 119 (1975).
[41] F. Schwarzbauer, 3R international *25*, 272 (1986); gwf gas/erdgas *128*, 98 (1987).
[42] N. Schmitz-Pranghe, W. Meyer u. W. Schwenk, DVGW Schriftenreihe Gas/Wasser 1, ZfGW Verlag, Frankfurt 1976, S. 7B19.
[43] H. Landgraf u. W. Quitmann, 3R international *22*, 524 u. 595 (1983).
[44] A. Kottmann u. H. Zimmermann, gwf gas/erdgas *126*, 219 (1985).
[45] R.R. Fessler, A.J. Markoworth u. R.N. Parkins, Corrosion *39*, 20 (1983).
[46] A.C. Toncre, Mat. Perform. *23*, H. 8, 22 (1984).
[47] W. Schwenk, 3R international *26*, 305 (1987).
[48] W. Schwenk, gwf gas/erdgas *123*, 158 (1982).
[49] G. Herbsleb, R. Pöpperling u. W. Schwenk, 3R international *20*, 193 (1981).
[50] W. Delbeck, A. Engel, D. Mhller, R. Spörl u. W. Schwenk, Werkstoffe u. Korrosion *37*, 176 (1986).
[51] H. Hildebrand u. M. Schulze, 3R international *25*, 242 (1986).
[52] H. Hildebrand u. W. Schwenk, Werkstoffe u. Korrosion *37*, 163 (1986).
[53] H. Hildebrand, M. Schulze u. W. Schwenk, Werkstoffe u. Korrosion *34*, 281 (1983).
[54] H. Hildebrand u. W. Schwenk, Mitt. VDEfa *19*, 13 (1971).
[55] S. Schwenk, Werkstoffe u. Korrosion *17*, 1033 (1966).

6

Galvanic (Sacrificial) Anodes

H. BOHNES AND G. FRANKE

6.1 General Information

In contrast to impressed current anodes, the possibility of using galvanic anodes is limited by their electrochemical properties. The rest potential of the anode material must be sufficiently more negative than the protection potential of the object to be protected so that an adequate driving voltage can be maintained. Rest potentials (see Table 2-4) and protection potentials (see Section 2-4) can often be approximately related to the standard potentials (see Table 2-1). This relation is frequently given by Eq. (2-53′) for the protection potential as long as there is no complex formation [see Eq. (2-56)]. On the other hand, there is no clear relationship for rest potentials according to the explanation for Fig. 2-5, where there is a definite influence of the environment according to Table 2-4. Furthermore, temperature can have an influence. The potential of zinc in aqueous media becomes increasingly more positive with a rise in temperature, due to film formation, which is attributed to a retardation of the anodic partial reaction (passivation) according to Fig. 2-5. The passivatability of galvanic anodes should be as low as possible. The anode metals form numerous less easily soluble compounds from which, under normal circumstances, hydroxides, hydrated oxides and oxides, carbonates, phosphates and several basic salts can arise. With the use of galvanic anodes for the internal protection of a chemical plant (see Chapter 21), still further compounds that are less easily soluble can be formed. In general the insoluble compounds do not precipitate on the working anode because the pH is lowered there by hydrolysis [1]. If the anode is not heavily loaded or the concentration of the film-forming ions is too high, the less easily soluble compounds can cover the anode surface (see also Section 4.1). Many surface films are soft, porous and permeable and do not interfere with the functioning of the anode. However, some films can become as hard as enamel, completely blocking the anode, but on drying out alter their structure and become brittle and porous. They can then be removed by brushing.

Galvanic anodes should exhibit as low a polarizability as possible. The extent of their polarization is important in practice for their current output. A further anode

180   Handbook of Cathodic Corrosion Protection

property is the factor Q for the equivalence between charge and mass according to Eq. (2-3). This factor is termed current content. It is larger the smaller the atomic weight and the higher the valence of the anode metal [see Eq. (2-6)]. The theoretical current content is not sufficient to determine the practical application because from the data in Fig. 2-5 the total anode current density J_a from Eq. (2-38) is smaller by an amount J_C than the anodic partial current J_A. Here J_A corresponds to the theoretical and J_a to the usable current content. J_C corresponds to the self-corrosion of the anode, which is derived from the cathodic partial reaction in which intermediate cations of anomalous valence can be involved (see Section 6.1.1).

For application in flowing media (e.g., for ships) it is also necessary that the usable current content be as large as possible not only per unit of mass but also per unit of volume, so that the volume of the installed anodes becomes as small as possible.

The cathodic protection of plain carbon and low-alloy steels can be achieved with galvanic anodes of zinc, aluminum or magnesium. For materials with relatively more positive protection potentials (e.g., stainless steels, copper, nickel or tin alloys), galvanic anodes of iron or of activated lead can be used.

The anodes are generally not of pure metals but of alloys. Certain alloying elements serve to give a fine-grained structure, leading to a relatively uniform metal loss from the surface. Others serve to reduce the self-corrosion and raise the current yield. Finally, alloying elements can prevent or reduce the tendency to surface film formation or passivation. Such activating additions are necessary with aluminum.

6.1.1   Current Capacity of Galvanic Anodes

The theoretical current content Q' per unit mass, from Eqs. (2-3) and (2-6) is given by:

$$\frac{Q}{\Delta m} = Q' = \frac{z\mathcal{F}}{M} = \frac{1}{f_b} \tag{6-1}$$

f_b values are given in Table 2-1. In practical units:

$$\frac{Q'}{\text{kg A}^{-1}\,\text{h}^{-1}} = \frac{2.68 \times 10^4\,z}{\left(M/\text{g mol}^{-1}\right)} \tag{6-1'}$$

Correspondingly, the theoretical current content Q'' related to volume is given by Eqs. (2-3) to (2-7):

$$\frac{Q}{\Delta V} = Q'' = \frac{z\mathcal{F}\rho_s}{M} = \frac{1}{f_a} = \frac{\rho_s}{f_b} = Q'\rho_s \tag{6-2}$$

f_a values are given in Table 2-1. In practical units:

$$\frac{Q''}{\text{dm}^3\,\text{A}^{-1}\,\text{h}^{-1}} = \frac{2.68 \times 10^4 z(\rho_s/\text{g cm}^{-3})}{(M/\text{g mol}^{-1})} \tag{6-2'}$$

Current content values Q' and Q'' for the most important anode metals are given in Table 6-1. These data apply only to pure metals and not to alloys. For these the Q values corresponding to the alloy composition can be calculated from:

$$Q' = \frac{\sum_i x_i Q'_i}{100} \tag{6-3}$$

where x_i is the mass fraction in percent of the alloying element, i, with current content, Q'_i. With the usual anode materials, the influence due to alloying is negligibly small.

On the other hand, the difference between the theoretical and the usable current content due to self-corrosion is not negligible. This effect is taken care of

Table 6-1 Properties of pure metals as anode materials

Metal	z	Current content of pure metal		α
		Q'/A h kg	Q''/A h dm^{-3}	
Al	3	2981	8049	(~0.8)
Cd	2	477	4121	–
Fe	2	960	7555	>0.9
Mg	2	2204	3835	(~0.5)
Mn	2	976	7320	–
Zn	2	820	5847	>0.95

Rest potential U_H/V:	Soil	Seawater
Fe	−0.1 to −0.3	~−0.35
Zn	−0.6 to −0.8	~−0.80

by a correction factor α which, according to data in Fig. 2-5 and Eq. (2-10), is defined as follows:

$$\alpha = \frac{I_a}{I_A} = \frac{I_a}{I_a + I_C} \tag{6-4}$$

The sum of all the cathodic partial reactions is included in I_C, e.g., oxygen reduction according to Eq. (2-17) and hydrogen evolution according to Eq. (2-19). The intermediate formation of anode metal ions of anomalous valence is also possible:

$$\text{Me} \rightarrow \text{Me}^{(z-n)^+} + (z-n)\text{e}^- \tag{6-5a}$$

with subsequent reaction in the medium,

$$\text{Me}^{(z-n)^+} + n\text{H}^+ \rightarrow \text{Me}^{z+} + \frac{n}{2}\text{H}_2 \tag{6-5b}$$

can be treated in the overall reaction as if one of the reactions according to Eq. (6-5b) had occurred as an equivalent cathodic partial reaction at the anode.

The factor α varies considerably with individual alloys and is also dependent on the application conditions. It varies between wide limits from about 0.98 for zinc to below 0.5 for magnesium anodes. The α values given in Tables 6-1 to 6-4 apply to cold seawater. Deviations in application (i.e., cooling water) temperature and loading can lead to considerable changes.

The influence of loading on α is due to the dependence of I_C on potential or current or is due to self-corrosion. Whereas oxygen corrosion should be independent of material and potential, hydrogen evolution decreases with increased loading. Furthermore, it is strongly dependent on the material, where the more noble alloying elements promote self-corrosion. Since in both cases I_C is not proportional to the current output, I, there can be no value for α independent of I according to Eq. (6-4) or for the self-corrosion. In contrast, the anodic reaction according to Eq. (6-5a) and the equally fast reaction according to Eq. (6-5b) increase with potential and loading. Then I and I_C are proportional to each other and α becomes independent of loading. This is roughly the case with magnesium anodes where $\alpha = 0.5$ given by $z = 2$ and $n = 1$ [2]. Another explanation of this α value refers to an active region on the anode surface being proportional to the current I [3,4]. As to the consequence of hydrolysis analogous to Eq. (4-4), in this region acid corrosion with hydrogen evolution occurs. In this case values deviating from or smaller than $\alpha = 0.5$ are understandable. The two

Galvanic (Sacrificial) Anodes

mechanisms can hardly be differentiated if the reaction sites for the partial reactions in Eqs. (6-5a) and (6-5b) are very close together.

The practical usable current content values are given from Eqs. (6-1) and (6-2) as

$$Q'_{pr} = Q' \alpha = \frac{\alpha}{f_b} \tag{6-6a}$$

$$Q''_{pr} = Q'' \alpha = \frac{\alpha}{f_a} \tag{6-6b}$$

With the help of Q'_{pr} and Q''_{pr} the anode masses or volumes can be calculated from the necessary charge Q:

$$\frac{Q}{A\,h} = \frac{Q'_{pr}}{A\,h\,kg^{-1}} \frac{m}{kg} = \frac{Q''_{pr}}{A\,h\,dm^{-3}} \frac{V}{dm^3} \tag{6-7}$$

6.1.2 Current Discharge from Galvanic Anodes

The current-density-potential graph for a working galvanic anode is given by Eq. (6-8) in which the polarization resistance r_p is dependent on loading:

$$U = U_R + r_p J \tag{6-8}$$

U_R is the rest potential. The difference between the potential of the working anode and the protection potential U_s of the object to be protected is termed the driving voltage U_T:

$$U_T = U_s - U \tag{6-9}$$

It follows from Eqs. (6-8) and (6-9) that:

$$U_T = (U_s - U_R) - r_p J \tag{6-10}$$

This function together with the linear behavior of the resistance of the protection system is shown in Fig. 6-1.

$$U_T = (J \times S_a)(R_a + R_c) \tag{6-11}$$

Here $J \times S_a$ is the protection current, S_a is the surface area of the anode, and R_a and R_c are the grounding resistances of the anode and the object. Information on the

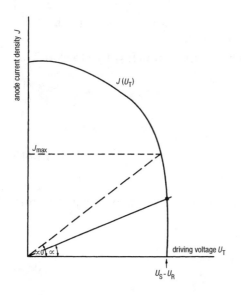

Fig. 6-1 Characteristic $J(U_T)$ and resistance straight lines of a galvanic anode. Slope of the resistance lines: $\tan \alpha = [1/S_a(R_a + R_c)]$. Slope of the resistance line at maximum current output: $\tan \alpha_0 = (1/S_a R_a)$.

measurement and calculation of grounding resistances is given in Sections 3.5.3 and 24.1 (see also Table 24-1). The grounding resistance of the cathode R_c can be neglected compared with R_a on account of the large cathode area. With well-coated and not too large objects, R_c can, however, be significantly large. According to Eq. (6-11), an increase in the current discharge is to be expected if R_c decreases in the course of time. The anode has therefore a current reserve.

The intersection of the graphs of Eq. (6-10) and (6-11) gives the working point for the protection system. The driving voltage decreases with increasing J. With anodes that are less polarizable, it remains more or less constant over a wide range of current density. The anode graph of Eq. (6-10) indicates the efficiency of the anode. It depends on the composition of the anode and the electrolyte. The working point of the system should lie in the unpolarizable part of the $J(U_T)$ curve, so that it can be assumed that $r_p \approx 0$. The maximum current discharge at $R_c = 0$ is given from Eqs. (6-10) and (6-11) as:

$$J_{max} = \frac{U_s - U_R}{S_a R_a} \qquad (6\text{-}12)$$

If the working point does not lie in the unpolarizable region (e.g., on account of the action of surface films) then instead of Eq. (6-12),

$$J_{max} = \frac{U_s - U_R}{S_a R_a + r_p(J)} \qquad (6\text{-}12')$$

This value can be considerably smaller. It corresponds in Fig. 6-1 to the ordinate of the intersection of the resistance graph of slope α_0 with a $J(U_T)$ curve that deviates markedly to the left of that plotted. The maximum current density is an important quantity for the setting up of cathodic protection with galvanic anodes and is dependent on the anode geometry and conductivity of the medium.

6.2 Anode Materials

6.2.1 Iron

Galvanic anodes of cast iron were already in use in 1824 for protecting the copper cladding on wooden ships (see Section 1.3). Even today iron anodes are still used for objects with a relatively positive protection potential, especially if only a small reduction in potential is desired, e.g., by the presence of limiting values U'' (see Section 2.4). In such cases, anodes of pure iron (Armco iron) are mostly used. The most important data are shown in Table 6-1.

6.2.2 Zinc

Zinc was also already in use for protection in seawater in 1824 (see Section 1.3). In the beginning zinc material that was available from the hot-dip galvanizing industry was used but was less suitable because it became passive. Passivation does not occur with high-purity zinc. Super high grade zinc is the anode material with the least problems [5] and consists of 99.995% Zn and less than 0.0014% Fe without further additions. It is specified in Ref. 6 and permitted by the German Navy [7]. The most important properties of pure zinc are listed in Table 6-1.

Super high grade zinc is mostly coarse grained, has a columnar crystal structure and tends to nonuniform removal. Alloy additions of up to 0.15% Cd and 0.5% Al are made for grain refinement [6,7] (see Fig. 6-2). These also compensate for the detrimental influence of higher Fe content up to 0.005% [8]. Zinc anodes for use in salt-rich media do not need additional activating elements. Additions of Hg to repel oily and waxy coatings on anodes in crude oil tanks have no advantage because these coatings do not seriously hinder current transfer (see Section 17.4). Today zinc alloys containing Hg or In for this purpose are no longer used [9]. The properties of zinc alloys for anodes are listed in Table 6-2.

The rate of self-corrosion of zinc anodes is relatively low. In fresh cold water, it amounts to about 0.02 g m^{-2} h^{-1}, corresponding to a corrosion rate of 25 μm a^{-1}. In cold seawater, the value is about 50% higher [10]. These figures refer to stagnant water. In flowing water the corrosion rates are significantly greater. Zinc is not practically suited for use in warm waters because of its tendency to passivate.

186 Handbook of Cathodic Corrosion Protection

Fig. 6-2 Microstructure of zinc anodes. Top: columnar crystals in pure zinc. Bottom: fine-grained structure of MIL zinc alloy.

Figure 6-3 shows current-density-potential curves for zinc anodes in stagnant, nonaerated 3.5% NaCl solution. The Tafel slope of the current-density-potential curve for activation polarization according to Eq. (2-35) is $b_+ = 50$ mV. Untreated castings with a skin behave similarly in stirred and aerated 3.5% NaCl at the start of the experiment, but the polarization increases markedly with time (see Fig. 6-5).

Table 6-2 Composition (wt.%) and properties of zinc alloys for anodes

Type	Material No. 2.2302/6/	Material No. 2.2301/6/	Dow (USA)
Aluminum	<0.10	0.10 to 0.50	0.1 to 0.5
Cadmium	<0.004	0.025 to 0.15	0.023 to 0.15
Iron	<0.0014	<0.005	<0.0014
Copper	<0.005	<0.005	<0.001
Lead	<0.006	<0.006	<0.003
Silicon	–	<0.125	–
Mercury	–	–	0.1 to 0.15
Rest potential U_H/V (seawater):		−0.8 to −0.85	
α		0.95 to 0.99	
Q'_{pr}/A h kg^{-1}		780 to 810	
Q''_{pr}/A h dm^{-3}		5540 to 5750	

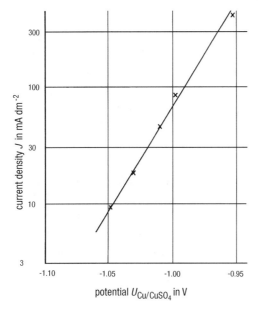
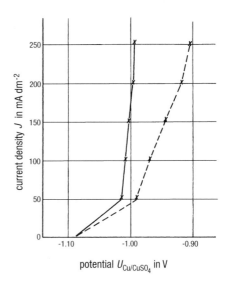

Fig. 6-3 $J(U)$ curves for pure zinc (machined surface) in 3.5 wt.% NaCl solution, free convection, not aerated.

Fig. 6-4 $J(U)$ curves for zinc anodes (material No. 2.2301) in 3.5 wt. % NaCl solution, aerated and stirred;— at the start of the experiment; ----- after 90 h.

In seawater, HCO_3^- ions lead to surface films and increased polarization. In aqueous solutions low in salt and with low loading of the anodes, less easily soluble basic zinc chloride [10] and other basic salts of low solubility are formed. In impure waters, phosphates can also be present and can form $ZnNH_4PO_4$, which is very insoluble [11]. These compounds are only precipitated in a relatively narrow range around pH 7. In weakly acid media due to hydrolysis at the working anode, the solubility increases considerably and the anode remains active, particularly in flowing and salt-rich media.

A change in structure of the surface films [e.g., by a transition from $Zn(OH)_2$ to electron-conducting ZnO] is the cause of the relatively positive potentials observed in oxygen-bearing fresh water at elevated temperatures around 60°C. In such cases the rest potential of the zinc can become more positive than the protection potential of the iron [12,13]. This process, also termed potential reversal, can be encouraged by Fe as an alloying element and has been observed on hot-dipped galvanized steel in cold water [14]. As a result of the potential reversal, an engine block in a ship with a closed cooling water circuit can suffer local corrosion in the neighborhood of zinc anodes due to cell formation with the zinc as cathode.

In spite of a low driving voltage of about 0.2 V, about 90% of all galvanic anodes for the external protection of seagoing ships are zinc anodes (see Section 17.3.2). Zinc alloys are the only anode materials permitted without restrictions for the internal protection of exchange tanks on tankers [16] (see Section 17.4).

For the external protection of pipelines in seawater (see Section 16.6), zinc anodes in the form of bracelets are installed; in the length direction these are welded on to clips attached to the pipe or in the form of half-shells. In brackish or strongly saline water, such as that found in oil drilling or in mining, zinc anodes can be installed for the internal protection of storage tanks. The use of zinc anodes in fresh water is very limited because of their tendency to passivate. This also applies to installing them in soil. Apart from occasional installation of rod or band anodes as ground electrodes, zinc anodes are only used for soil resistivities below 10 m. In order to reduce passivation and lower the grounding resistance, the anodes have to be surrounded with special bedding materials as backfill (see Section 6.2.5).

6.2.3 Aluminum

Pure aluminum cannot be used as an anode material on account of its easy passivatability. For galvanic anodes, aluminum alloys are employed that contain activating alloying elements that hinder or prevent the formation of surface films. These are usually up to 8% Zn and/or 5% Mg. In addition, metals such as Cd, Ga, In, Hg and Tl are added; as so-called lattice expanders, these maintain the long-term activity of the anode. Activation naturally also encourages self-corrosion of the anode. In order to optimize the current yield, so-called lattice contractors are added that include Mn, Si and Ti.

The various aluminum alloys behave very differently as anodes. The potentials lie between about $U_H = -0.75$ V and -1.3 V; the α value is 0.95 for alloys containing Hg and lies between 0.7 and 0.8 for alloys with Cd, In and Sn. Three types of alloy are particularly important for aluminum anodes. All of them contain a few percent of Zn. In, Hg, Sn and Cd are present as activators. Anodes containing Hg can give a high current yield but they are seldom used because of the toxicity of Hg salts. For this reason, aluminum anodes with Zn and In as activators are acquiring increased importance. In spite of an α value of about 0.8 and a rest potential of only $U_H = -0.8$ V, their low polarizability is a special advantage. They are therefore employed preferentially in offshore applications. Aluminum anodes with Zn and Sn as activators occupy a middle position with regard to α value. Their rest potentials are similar to those of alloys containing In or are somewhat more positive. Their polarizability is, however, considerably greater. Heat treatment is necessary according to the alloy composition. Table 6-3 gives the

Table 6-3 Composition (wt.%) and properties of aluminum alloys for anodes

Type	Hg-Zn (X-Meral)	In-Zn (Galvalum III)	Sn-Zn
Zinc	2.0 to 2.2	3.0	5.5
Mercury	0.045 to 0.055	–	–
Indium	–	0.015	–
Tin	–	–	0.1
Iron	<0.1	–	<0.1
Copper	<0.02	–	<0.005
Silicon	<0.05	0.1	<0.1
Manganese	0.25 to 0.3	–	<0.005
Titanium	0.02 to 0.03	–	<0.04
Magnesium	0.04 to 0.05	–	<0.005
Rest potential in seawater			
U_H/V	–0.8/–1.0	~ –0.85	~ –0.86
α	0.92	0.88	0.80
Q'_{pr}/A h kg^{-1}	2700	2550	2280
Q''_{pr}/A h dm^{-3}	7560	7140	6385

properties of three different aluminum alloys, one each of the Hg, In and Sn-containing types.

The rate of self-corrosion of aluminum alloys as well as their dependence on loading and medium fluctuates between wide limits according to the alloy type and is always greater than that of zinc anodes. In addition, the anode material can behave differently in the region of the casting skin than in the interior of the anode. This particularly applies to anodes containing Sn if temperature control was not optimal in their production. With some aluminum alloys, the potential becomes more negative after a short period in service, reaching a stationary value after a few hours or days. On the other hand, alloys containing Hg with a higher Mg content can have a very negative potential to begin with which, however, increases after a short period in service. The polarizability of aluminum alloys is also very variable. Figure 6-5 shows the relation between potential and loading for an Al-Zn-Sn alloy. The deviation of individual points to more negative potential values is due apparently to increasing activation with time. In Fig. 6-6 the relation between potential and loading is shown for aluminum alloys with Zn as well as In,

190   Handbook of Cathodic Corrosion Protection

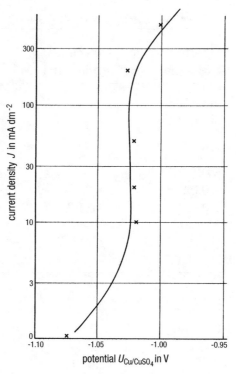

Fig. 6-5 $J(U)$ curves for an aluminum alloy with 2% Zn and 0.1% Sn in 3.5 wt. % NaCl solution, free convection, not aerated.

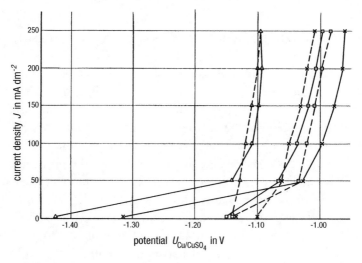

Fig. 6-6 $J(U)$ curves for aluminum alloys in 3.5 wt. % NaCl solution, aerated and stirred; — at the start of the experiment, ----- after 1 week; △ Al-In-Zn, × Al-In-Zn, □ Al-In-Zn.

Hg and Sn at the beginning of the experiment and after 1 week. Alloys containing In are the least polarizable, and those containing Sn the most polarizable.

Oxide, hydroxide and basic salts of aluminum are less soluble at pH values of about 7 than those of zinc [17], which explains the easy passivatability. Galvanic anodes of aluminum alloys are primarily employed in the area of offshore technology. The anodes work in relatively pure seawater flowing with a high velocity so that by using suitable alloys, passivation phenomena are rare. Their low weight is particularly favorable in view of a service time of 20 to 30 years.

Their polarizability increases in saline muds and their current efficiency markedly decreases e.g., for the alloy galvalum III in Table 6-3, from 2550 A h kg^{-1} to 1650 A h kg^{-1}.

6.2.4 Magnesium

Magnesium is considerably less passivatable than zinc and aluminum and has the highest driving voltage. On account of these properties and its high current content, magnesium is particularly suitable for galvanic anodes. Magnesium, however, is prone to self-corrosion of considerable extent, which increases with increasing salt content of the medium [18]. The available current content of pure magnesium is therefore much less than the theoretical current content. It is influenced by the impurity content of the anode metal, by the type of material removal (uniform or pitting), and by the current density as well as the electrolyte.

Even in good alloys and under favorable conditions, the α value does not lie above about 0.6. In enamelled storage tanks where the current requirement is low, the α value can fall to as low as about 0.1. The cause of the high proportion of self-corrosion is hydrogen evolution, which occurs as a parallel cathodic reaction according to Eq. (6-5b) or by free corrosion of material separated from the anode on the severely craggy surface [2-4, 19-21].

Magnesium anodes usually consist of alloys with additions of Al, Zn and Mn. The content of Ni, Fe and Cu must be kept very low because they favor self-corrosion. Ni contents of >0.001% impair properties and should not be exceeded. The influence of Cu is not clear. Cu certainly increases self-corrosion but amounts up to 0.05% are not detrimental if the Mn content is over 0.3%. Amounts of Fe up to about 0.01% do not influence self-corrosion if the Mn content is above 0.3%. With additions of Mn, Fe is precipitated from the melt which on solidification is rendered harmless by the formation of Fe crystals with a coating of manganese. The addition of zinc renders the corrosive attack uniform. In addition, the sensitivity to other impurities is depressed. The most important magnesium alloy for galvanic anodes is AZ63, which corresponds to the claims in Ref. 22. Alloys AZ31 and M2 are still used. The most important properties of these alloys are

Table 6-4 Composition (wt.%) and properties of magnesium alloys for anodes

Type	AZ 63	AZ 31	M 2
Aluminum	5.3 to 6.7	2.5 to 3.5	<0.05
Zinc	2.5 to 3.5	0.7 to 1.3	<0.03
Manganese	>0.3	>0.2	1.2 to 2.5
Silicon	~0.1	~0.1	–
Copper	<0.05	<0.05	<0.05
Iron	<0.01	<0.005	<0.01
Nickel	<0.001	<0.001	<0.001
Lead	–	<0.01	–
Tin	–	<0.01	–
Properties in 10^{-3} M NaCl at 60°C			
α	0.1 – 0.5	0.1 – 0.5	0.1 – 0.5
Q'_{pr}/A h kg^{-1}	220 – 1100	220 – 1100	220 – 1100
Q''_{pr}/A h dm^{-3}	420 – 2100	420 – 2100	420 – 2100
Working potential at $J_C = 50$ μA cm^{-2}			
U_H/V	–0.9/–1.1	–0.9/–1.1	–1.0/–1.2
Rest potential			
U_H/V	–1.0/–1.2	–1.0/–1.2	–1.1/–1.3
Properties in cold seawater			
α	0.50	0.52	0.53
Q'_{pr}/A h kg^{-1}	1100	1150	1150
Q''_{pr}/A h dm^{-3}	2100	2200	2200
Rest potential			
U_H/V	–1.2/–1.3	–1.1/–1.3	–1.0/–1.3

given in Table 6-4. While AZ63 is used predominantly for cast anodes, AZ31 and M2 are chiefly produced for extruded rod anodes.

Figure 6-7 shows the effect of water conductivity on the rest potential of type AZ63, and M2 as well as of zinc and aluminum [23]. In cold waters with chlorides or sulfates, the polarization of magnesium anodes is low — even in the case of high current densities. This is demonstrated for 70°C with current-density vs.

Galvanic (Sacrificial) Anodes 193

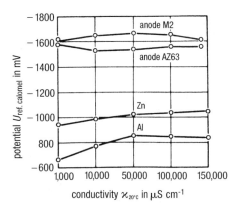

Fig. 6-7 Rest potentials of various galvanic anodes as a function of the salt content of the medium at 20°C.

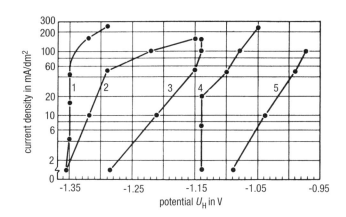

Fig. 6-8 $J(U)$ curves for magnesium in aqueous solutions at 70°C.

Curve	Anode	Medium	$\varkappa$ (20°)
1	M2	0.1 M MgSO$_4$	–
2	M2	200 mg/L Na$_2$SO$_4$	300 μS cm^{-1}
3	M2	150 mg/L NaCl	300 μS cm^{-1}
4	AZ31	150 mg/L NaCl	300 μS cm^{-1}
5	AZ63	150 mg/L NaCl	300 μS cm^{-1}

potential curves in Fig. 6-8 [24]. A marked polarization takes place in the presence of borates, carbonates, bromates, fluorides, oxalates, and phosphates [25].

Low pH values favor self-corrosion, displace the rest potential to more negative values, reduce polarization, and lead to uniform material consumption; pH values above 10.5 act opposite to this. Below pH 5.5 to 5.0, the current yield is so low that their use is impracticable.

In the application of magnesium anodes for enamelled boilers, the consumption rate of the anodes is determined less by current supply than by self-corrosion. The calculation of life from data on protection current requirement, I_s, and anode mass, m, is difficult because the α value is so low.

According to Eq. (6-4):

$$\alpha = \frac{I_s}{I_s + J_C \times S_a} \qquad (6\text{-}4')$$

where S_a is the anode surface. The cathodic partial current density, J_C, and the self-corrosion rate, w, are proportional to each other according to Eq. (2-5). From Eq. (6-1) and taking account of (6-6a) it follows that:

$$I_s\, dt = dQ = dm Q'_{pr} = dm Q' \frac{I_s}{I_s + S_a \dfrac{w}{f_a}}, \quad dt = \frac{Q'\, dm}{I_s + S_a \dfrac{w}{f_a}} \qquad (6\text{-}13)$$

For cylindrical anodes, S_a equals $2\pi rL$, where the reduction in radius with time is assumed to be $-dr/dt = k$. Substitution in Eq. (6-13) and integration with boundary conditions $r = r_0$ and $m = 0$ for $t = 0$ as well as $r = 0$ and $m = M$ for $t = T$ gives:

$$T = \frac{Qm'}{I_s + \dfrac{S_a w}{2 f_a}} \qquad (6\text{-}13')$$

and substitution for $Q' = 2204$ A h kg^{-1} and $f_a = 22.8$ (mm a^{-1})/(mA cm^{-2}) gives

$$\frac{T}{a} = \frac{(m/g)}{3.97(I_s/\text{mA}) + 0.087 \dfrac{S_a}{\text{cm}^2} \times \dfrac{w}{\text{mm/a}}} \qquad (6\text{-}13'')$$

From this equation it follows that for a given mass, the life of an anode is that much greater the smaller the anode surface S_a. This optimization is quite possible

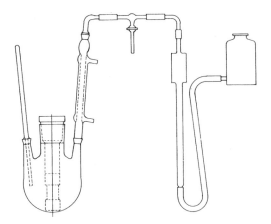
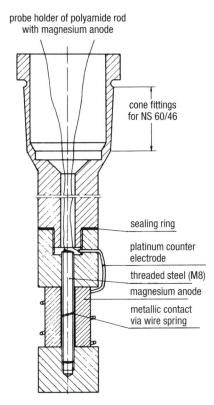

Fig. 6-9 (above) Apparatus for determining the weight loss of galvanic anodes by measuring hydrogen evolution.

Fig. 6-10 (right) Specimen holder for measuring current yield.

for anodes in enamelled boilers, because position and size of the anodes is of little importance for current distribution because of the small extent of defects in the enamel. This only holds, however, for the enamel coatings given in Ref. 26 with a limit on the surface area of defects and their distribution (see Section 20.4.1).

On the basis of available practical experience, the consumption rate of magnesium anodes in enamelled boilers is lower than 3 mm a^{-1}. For a rod anode with a diameter of 33 mm, this corresponds to a life of over 5 years. As a guideline for the required anode mass, 200 to 250 g per m^2 of internal surface is recommended [27].

In oxygen-free water, the self-corrosion is practically solely due to hydrogen evolution

$$Mg + 2\,H_2O = Mg^{2+} + 2\,OH^- + H_2 \qquad (6\text{-}14)$$

or to the consecutive reactions of Eq. (6-5 a,b). The amount of hydrogen evolved can be determined in the apparatus shown in Fig. 6-9. The specimen holder shown in Fig. 6-10 is suitable for measurements at given anode currents, I_a. Such measurements serve to determine the current yield α and test the quality of the anode (see Section 6.6). In addition, determining the potential vs. current supply allows the tendency to passivation of the magnesium to be recognized [27], which, for example, exists for pure magnesium in low-salt water and which should be tested generally for the formation of less easily soluble compounds.

The low-solubility compounds that form on magnesium anodes under normal loading include hydroxides, carbonates and phosphates. However, the solubility of the hydroxides and carbonates is relatively high. Only magnesium phosphate has a low solubility. The driving voltage with magnesium anodes for steel with not too low conductivity $\varkappa > 500\ \mu S\ cm^{-1}$ with 0.65 V is almost three times as great as that for zinc and aluminum. Magnesium alloys are used if the driving voltage of zinc and aluminum anodes is not sufficient or where there is a danger of passivation.

Magnesium anodes are used in the case of higher specific resistivity of the electrolyte and higher protection current densities. Objects to be protected include steel-water structures in fresh water, ballast tanks for fresh water, boilers and drinking water tanks. The physiological assurance concerning corrosion products is important in containers for drinking water (see Section 20.5). The total weight content of Sb, As, Pb, Cr and Ni must not exceed 0.1%, and that of Cd, Hg and Se must not exceed <0.01% [27]. In the soil, smaller objects can be protected with magnesium electrodes for resistivities up to 250 Ω m and large tanks and pipelines for resistivities up to 100 Ω m if the current requirement is not too high.

Hydrogen is involved in cathodic protection with magnesium anodes on account of the high contribution of self-corrosion. This must be considered in its use in closed containers, e.g., boilers. In enamelled boilers there is no danger from deflagration of the oxy-hydrogen gas under normal service conditions [2]; however safety requirements must be observed [28,29], particularly with routine maintenance work.

6.3 Backfill Materials

According to Table 24-1 there is a relationship between the grounding resistance of anodes and the specific soil resistivity, which is subjected to annual variations. To avoid these variations and to reduce the grounding resistance, anodes in soil are surrounded by bedding materials, so-called backfill. Such materials, besides restricting the formation of surface films and preventing electro-osmotic dehydration, act to provide uniform current delivery and uniform material consumption. The latter is primarily due to the presence of gypsum in the backfill, while bentonite and kieselguhr retain moisture. The addition of sodium sulfate reduces the specific resistivity of the backfill. By changing the amounts of the individual constituents, particularly sodium sulfate, leaching can be controlled and

Table 6-5 Composition of backfill materials

Specific soil resistivity in Ω m	Backfill						
	for Mg anodes				for Zn anodes		
	Gypsum	Bentonite	Kieselguhr	Na_2SO_4	Gypsum	Bentonite	Na_2SO_4
Up to 20	65	15	15	5	25	75	–
	25	75	–	–	50	45	5
20 to 100	70	10	15	5	75	20	5
	75	20	–	5	–	–	–
	50	40	–	10	–	–	–
Above 100	65	10	10	15	–	–	–
	25	50	–	25	–	–	–

thus the specific soil resistivity in the neighborhood of the anodes reduced. The composition of various backfill materials is given in Table 6-5.

Backfill containing a large proportion of bentonite has a tendency to change its volume with variations in water content of the surrounding soil. This can lead to formation of hollow cavities in the backfill with a considerable decrease in the current delivery. A standard backfill consists of a mixture of 75% gypsum, 20% bentonite and 5% sodium sulfate. The specific resistivity of this backfill is initially 0.5 to 0.6 Ω m and can rise with increased leaching to 1.5 Ω m.

The backfill is either poured into the borehole, or the anodes are enclosed in sacks of permeable material filled with backfill. Such anodes are sunk into the borehole and backfilled with water and fine soil. Anodes installed in this way deliver their maximum current after only a few days.

Galvanic anodes must not be backfilled with coke as with impressed current anodes. A strong corrosion cell would arise from the potential difference between the anode and the coke, which would lead to rapid destruction of the anode. In addition, the driving voltage would immediately collapse and finally the protected object would be seriously damaged by corrosion through the formation of a cell between it and the coke.

6.4 Supports

Usually all cast galvanic anodes have specially shaped feeder appendages as anode supports by which the anodes are fixed by screws, brazing or welding. This guarantees a very low resistance for current flowing from the anode to the object to be protected. Anode supports usually consist of mild steel. Supports of nonmagnetic steels or bronze are used on warships. Wire anodes of zinc can have an aluminum core. Sheet iron supports 20 to 40 mm in breadth and 3 to 6 mm thick are used for plate anodes, and cast iron rods 8 to 15 mm in diameter for rod anodes. For larger anodes such as those used, for example, in the offshore field (see Chapter 16), heavier supports are necessary. Here pipes of suitable diameter are used as appendages and section steel as core material.

With correctly constructed and prepared anodes, nearly all the available anode material is consumed. In other circumstances a more or less greater part of the anode material can fall off during service and thereby be lost for cathodic protection. For this reason a strong bond between anode alloy and the core is necessary, which should extend over at least 30% of the contacting surfaces [6]. With good anodes this percentage is widely exceeded because an intermediate layer between anode and support is formed by alloying. In order to encourage this, the support must be well cleaned by degreasing and pickling. After washing and drying, the support must have the anode metal cast onto it. The support can also be shot blasted according to standard purity grade Sa $2\frac{1}{2}$ [30] and then must immediately have the metal cast onto it.

If the supports are provided in a state ready for casting, they must be given temporary corrosion protection. For this purpose coatings are provided by, for example, aluminizing, cadmium plating, galvanizing and phosphating. The two last-named processes are mostly used. A minimum thickness of 13 μm is prescribed for zinc coatings [31]. Phosphate coatings must be very thin because they can impair the bond between anode alloy and support. Correctly prepared anodes have a contact resistance between anode alloy and support of less than 1 mΩ. Larger contact resistances must be avoided as they reduce the current delivery. This also applies to the mounting to the object to be protected.

In most cases galvanic anodes are fixed to the object to be protected by welding or brazing, less often by bolting to the supports. If for technical reasons or on the grounds of safety (e.g., internal protection) they have to be bolted, care must be taken to provide a sufficiently low resistance and reliable joint. Particularly in the internal protection of tankers, where vibration and shaking is to be expected, the bolts must be secured against gradual loosening.

In soil, anodes are connected by cables to the object to be protected. The cable must be low resistance in order not to reduce the current delivery. Therefore with long lines, the cable cross-section must be proportionately large. A cable with NYM sheathing with 2.5 mm^2 Cu is mostly sufficient. Occasionally stronger cables and special insulation are required, e.g., NYY* 4 mm^2 Cu.[1] Power supply cable buried in soil should have a noticeably light color. For use in seawater, occasionally temperature, oil and seawater-resistant cable is demanded, e.g., HO7RN.[2]

6.5 Forms of Anodes

The number of different anode shapes has decreased considerably with the passage of time. Magnesium anodes for rapid prepolarization, so-called booster anodes, are no longer used today. Guidelines are laid down for the form of galvanic anodes in various standards [6,22,31,32]. These guarantee good exchangeability of anodes, particularly for the external protection of ships.

---

[1] In the United States high molecular weight polyethylene is used, and in the United Kingdom, cross-linked polyethylene with polyvinylchloride is used.

[2] In the United States and the United Kingdom, KYNAR or CATHORAD cable is used.

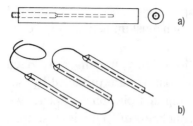
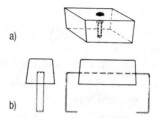

Fig. 6-11 (a) Rod-shaped anode and (b) chain of rod anodes.

Fig. 6-12 (a) Block-shaped anode with screw fixing and (b) block-shaped anode for internal protection.

6.5.1 Rod Anodes

Rod anodes are supplied as lengths with a core wire or as parts with the holder through them or through one side (see Fig. 6-11a). They are either cast or extruded. Their main uses are for protecting objects buried in the soil or the internal protection of storage tanks. The diameter varies between that of a thin wire anode and about 70 mm. Individual anodes can be up to 1.5 m long. The weight of the anode depends on the alloy. For zinc they can be up to 100 kg. Separate lengths of anode with outer core on both sides can be combined to form anode chains if they are not already supplied as several anodes connected by cable (see Fig. 6-11b). Rod anodes with support on one side can be inserted in storage tanks with a tube bung by bolting from the outside.

6.5.2 Plates and Compact Anodes

Plate and compact anodes with cast-on supports are predominantly used for the external protection of ships, for steel-water structures and for the internal protection of large storage tanks. This form of anode is not suitable in soils due to the large grounding resistance. Compact anodes are supplied with either square, rectangular or round cross-sections, often with iron tubes cast into them for fixing them with bolts (see Fig. 6-12a). Such anodes are usually magnesium alloys. They also comprise compact anodes for the internal protection of storage tanks (see Fig. 6-12b). Plate anodes are primarily used where the flow resistance should be as low as possible, e.g., the external protection of ships. They are more or less long anodes and have a teardrop-shaped contour with flat iron holders sticking out at the ends or are lateral

Galvanic (Sacrificial) Anodes 201

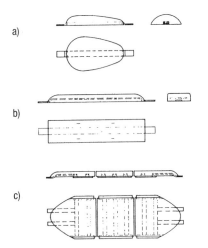

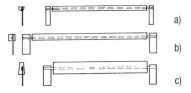

Fig. 6-13 (left) Anode shapes for the external protection of ships. (a,b) Individual anodes and (c) groups of anodes.

Fig. 6-14 (above) Anode shapes for the internal protection of tanks. (a) Semi-circular cross-section, (b) rectangular cross-section, (c) trapezoid cross-section.

shackles welded on. The edges are bevelled as long as the individual anodes are not combined in larger groups (see Fig. 6-13). In addition, there are plate anodes that can be bolted on [6,22].

The individual weights range from less than 1 kg to several 100 kg; the latter is particularly used in steel-water structures. Anodes for the external protection of ships are heavier than 40 kg only in exceptional cases. According to need, anodes are combined into groups whose total weight amounts to several hundred kilograms.

Finally there are large plate anodes up to 1 m square with cable connections as hangers. Such anodes serve to drain stray currents in ships in fitting out or repairs (see Section 15.6).

6.5.3 Anodes for Tanks

Tank anodes are elongated, with a round iron core running through them. For production reasons the cross-sections are half round, almost rectangular or trapezoidal and also occasionally triangular (see Fig. 6-14). The iron cores are welded onto fasteners or crimped onto them. The shackles or the crimped core serve for welding to the object to be protected. The weight extends from a few kilograms for magnesium alloys to about 80 kg for zinc. The lengths, excluding supports, are mostly between 1.0 and 1.2 m.

The anodes for internal protection of containers and tanks are frequently fixed by screws; because of the danger of explosion, welding or brazing is not allowed.

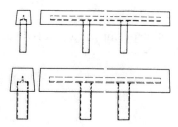

Fig. 6-15 Anode shapes for offshore applications.

Anodes that fall can produce sparks, depending on the material. For this reason the choice of anodes for tankers is governed by certain regulations (see Section 17.4).

6.5.4 Offshore Anodes

Offshore anodes are similar in shape to tank anodes. They are, however, much larger and weigh about 0.5 t. They are predominantly manufactured from aluminum alloys. On the basis of strength, most of them are cast onto pipes or profile iron as supports on which lateral protruding shackles are welded. The cross-section is usually trapezoidal (see Fig. 6-15).

For pipelines, so-called bracelets are used (see Section 16.6). They consist of groups of plate anodes of more or less large breadth, partly curved, which are secured near to one another on an iron band support like links of a bracelet and laid around pipelines. With the same objective, half-shells are used which are fixed in pairs under stress around the pipe and then welded.

6.5.5 Special Forms

The special forms consist of the many types of anode which are used for protecting smaller containers. Boilers, heat exchangers and condensers belong to this group. Besides the rod anodes already mentioned with tube screw joints which can be screwed into the container from outside, there are also short and round anode supports as well as more or less flat ball segments which are bolted onto the protected surface with cast-on supports. These shapes are mostly manufactured from magnesium alloys. In addition, there are star-shaped or circular anodes for installation in condensers and pipes. The weight of these anodes lies between 0.1 and 1 kg.

In enamelled boilers, magnesium anodes for the most part are insulated and connected with the container by a ground lead (see Fig. 6-16). This method of

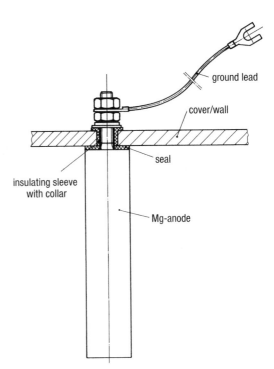

Fig. 6-16 Insulated Mg anode mounting for simple operational control.

insertion has the advantage that the anode function can be checked by simply measuring the protection current.

6.6 Quality Control and Performance Testing

The quality control of galvanic anodes is reduced mainly to the analytical control of the chemical composition of the alloy, to the quality and coating of the support, to an adequate joint between support and anode material, as well as to restricting the weight and size of the anode. The standards in Refs. 6, 7, 22, 27, 31 refer to magnesium and zinc anodes. Corresponding specifications for aluminum anodes do not exist. In addition, the lowest values of the rest potentials are also given [16]. The analytical data represent the minimum requirements, which are usually exceeded.

With objects requiring long-term (up to 30 years) protection and/or large amounts of anode material, a particular alloy is occasionally specified whose properties are strictly controlled in production. A test certificate indicating the type of anode material is usually required (e.g., Ref. 33).

Specifications for testing the electrochemical properties are given in an appendix to Ref. 33. The test consists of a long-term free-running test with flowing natural seawater. For production control, an additional short-term galvanostatic test [34] is prescribed on account of the long experimental duration of the free-running test. In addition, there are further specifications (e.g., for large-scale applications) which prescribe special experimental conditions relating to practice. The results obtained in such tests, however, only represent comparative values; they cannot be applied to the available data in a practical application in an individual case.

To estimate the current yield of magnesium alloys, the weight loss is determined indirectly over the volumetric measurement of the evolved hydrogen in the apparatus in Fig. 6-9 in Section 6.2.4. The mass balance for oxygen-free media follows from Eqs. (6-1) and (6-5a,b) or (6-14):

$$\Delta m = \frac{Q}{Q'(\mathrm{Mg})} + V(\mathrm{H}_2)\frac{\mathbf{M}(\mathrm{Mg})}{\mathbf{V}(\mathrm{H}_2)} \tag{6-15}$$

Here Q is the charge delivered by the anode, $V(\mathrm{H}_2)$ is the measured hydrogen volume and $\mathbf{V}(\mathrm{H}_2)$ its molecular volume under the same test conditions.

From Eq. (6-15) it follows analogously with Eq. (6-14) and Eq. (6-1):

$$\alpha = \frac{Q}{Q'(\mathrm{Mg})\Delta m} = \frac{Q}{Q + Q'(\mathrm{Mg})V(\mathrm{H}_2)\dfrac{\mathbf{M}(\mathrm{Mg})}{\mathbf{V}(\mathrm{H}_2)}}$$

$$= \frac{Q}{Q + 2\mathcal{F}\dfrac{V(\mathrm{H}_2)}{\mathbf{V}(\mathrm{H}_2)}} \tag{6-16}$$

Anodes for boilers can be tested by such methods. Good-quality magnesium anodes have a mass loss rate per unit area < 30 g m^{-2} d^{-1}, corresponding to a current yield of >18% under galvanostatic anode loading of 50 μA cm^{-2} in 10^{-2} M NaCl at 60°C. In 10^{-3} M NaCl at 60°C, the potential should not be more positive than $U_\mathrm{H} = -0.9$ V for the same polarization conditions [27].

6.7 Advantages and Disadvantages of Galvanic Anodes

Cathodic protection with galvanic anodes requires generally no great technical expenditure. Once the protection system has been installed, it operates almost without maintenance and only needs occasional checking of potential or protection current. Galvanic protection systems need no ac electric supply and cause no damaging interference on neighboring objects because of the low driving voltage. There are also no safety problems on account of the low voltages. Galvanic systems can therefore be installed in areas where there is a danger of explosions.

For corrosion protection in soils, the anodes can be brought close to the object to be protected in the same construction pit so that practically no further excavations are needed. By connecting anodes to locally endangered objects (e.g., in the case of interference by foreign cathodic voltage cones) the interference can be overcome (see Section 9.2.3).

In poorly conducting media or soils, however, the low driving voltage can limit the use of galvanic anodes. Raising the current delivery, which becomes necessary in service, is only practically possible with the help of an additional external voltage. In special cases this is used if an installed galvanic system is overstretched or if the reaction products take over additional functions (see Section 7.1).

6.8 References

[1] E. Eberius u. H. Bohnes, Metall *14*, 785 (1960).
[2] W. Schwenk, Techn. Überwachung *11*, 54 (1970).
[3] L. Robinson und P. F. King, J. electrochem. Soc. *108*, 36 (1961).
[4] P. F. King, J. electrochem. Soc. *113*, 536 (1966).
[5] N. N., Cathodic protection of pipelines with zinc anodes, Pipe & Pipelines International, S. 23, August 1975.
[6] MIL-A-18001, USA Deptmt. of Defense, Washington DC 1968.
[7] Normenstelle Marine, VG-Norm 81 255, Zink für Anoden, Beuth Verlag, Köln 1983.
[8] K. I. Vassilew, M. D. Petrov u. V. Z. Krasteva. Werkstoff und Korrosion *25*, 587 (1974).
[9] E. de Bie, DOS 2411608, 1974.
[10] W. Wiederholt, Das Korrosionsverhalten von Zink, Band 2: Verhalten von Zink in Wässern, S. 33, Metallverlag, Berlin 1965.
[11] Gmelins Handbuch der anorganischen Chemie, Band Zink, System-Nr. 32 (1924).
[12] G. Schikorr, Metallwirtschaft *18*, 1036 (1939).
[13] C. L. Kruse, Werkstoffe und Korrosion *27*, 841 (1976).
[14] W. Friehe u. W. Schwenk, Werkstoffe und Korrosion *26*, 342 (1975).
[15] F. Jensen, Cathodic protection of ships, Proc. 7th Scand. Corr. Congr., Trondheim, 1975.
[16] Lloyds Register of Shipping, Guidance Notes on Application of Cathodic Protection, London 1966.
[17] Gmelins Handbuch der anorganischen Chemie, 8. Aufl., Band Aluminium, System-Nr. 35, Teil B, 1934.
[18] H. H. Uhlig, Corrosion Handbook, J. Wiley & Sons Inc., New York 1948, S. 224.
[19] R. L. Petty, A. W. Davidson u. J. Kleinberg, J. Amer. Chem. Soc. *76*, 363 (1954).
[20] D. V. Kokulina u. B. N. Kabanov, Doklady Akad. Nauk. SSSR *112*, 692 (1957).
[21] R. Tunold et al., Corr. Science *17*, 353 (1977).
[22] MIL-A-21412 A, USA Deptmt. of Defense, Washington DC 1958.
[23] E. Herre, aus Kap. XV, dieses Handbuch 1. Aufl.
[24] W. Schwenk, aus Kap. XV, dieses Handbuch 1. Aufl.
[25] W. Rausch, Kathodischer Schutz – Grundlagen und Anwendung, Metallgesellschaft, Frankfurt 1954.
[26] DIN 4753 Teil 3, Beuth-Verlag, Berlin 1987.
[27] DIN 4753 Teil 6, Beuth-Verlag, Berlin 1987.
[28] DIN 50 927, Beuth-Verlag, Berlin 1985.
[29] Unfallverhütungsvorschriften der Berufsgenossenschaft der chemischen Industrie vom 01.04.1965, Abschn. 16, Druckbehälter § 2.4 in Verbindung mit AD-Merkblatt A3.
[30] DIN 55 928 Teil 4, Beuth-Verlag, Berlin 1978.
[31] British Standards Institution, Code of Practice for Cathodic Protection, London 1973.
[32] VG 81 257 Teil 1, Beuth-Verlag, Köln 1982.
[33] Det Norske Veritas, No. IOD-90-TA 1, File No. 291.20, Oslo 1982.
[34] Det Norske Veritas, Technical Notes TNA 702, Abschn. 2.7, Oslo 1981.

7

Impressed Current Anodes

H. BOHNES AND D. FUNK

7.1 General Comments

Galvanic anodes should have, besides a sufficiently negative potential, as high a current efficiency as possible and be as little polarizable as possible. Impressed current anodes, on the other hand, can deliver a much higher current supply, as is to be expected from Eq. (2-4) and Table 2-1, when anodic redox reactions run in parallel. Impressed current anodes, which are connected to the protected object from a dc source, in operation usually have a more positive potential than the protected object. Materials for impressed current anodes should not only be of as low solubility as possible, but should also not be damaged by impact, abrasion or vibration. Finally, they should have high conductivity and be capable of heavy loads. Anodic loading with only little material loss or none according to Eq. (2-4) is possible if an anodic redox reaction occurs according to Eq. (2-9) and Table 2-3 [e.g., oxygen evolution according to Eq. (2-17)]. This usually results in a reduction of the pH in the vicinity of the anode:

$$2 H_2O - 4 e^- = O_2 + 4 H^+ \tag{7-1}$$

$$2 Cl^- - 2 e^- = Cl_2 \tag{7-2a}$$

$$Cl_2 + H_2O = HCl + HOCl \tag{7-2b}$$

There are two types of impressed current anodes: either they consist of anodically stable noble metals (e.g., platinum) or anodically passivatable materials that form conducting oxide films on their surfaces. In both cases, the anodic redox reaction occurs at much lower potentials than those of theoretically possible anodic corrosion.

This general statement does not of course mean that materials with stoichiometric weight are completely unsuited as impressed current anodes.

Sometimes scrap iron is used as a material for anode installations; moreover, aluminum anodes are used in electrolytic water treatment (see Section 20.4.2). Zinc alloys are used in electrolytic derusting in order to avoid the formation of explosive chlorine gas. Magnesium anodes are sometimes supplied with impressed current in the internal protection of storage tanks containing poorly conducting water to raise the current delivery. In the so-called Cathelco process, copper anodes are deliberately used in addition to aluminum anodes to prevent algal growths, as well as to provide corrosion protection by incorporation of toxic copper compounds into the surface films. All these cases, however, involve special applications. In practice, the number of materials suitable for impressed current anodes is relatively limited. The main materials are graphite, magnetite, high-silicon iron with various additions, and lead-silver alloys as well as coated anodes of so-called valve metals. These are metals that at very positive potentials have stable passive films, which are not electron conducting, e.g., titanium, niobium, tantalum and tungsten. The best known in this group are the platinized titanium anodes.

The anodes developed mainly for electrochemical processes, with electron-conducting surfaces of platinum metal oxides on valve metals, have become important in cathodic protection. In contrast, anodes of valve metals with coatings of electron-conducting ceramic material provide considerably better anode properties. There are a large number of such ceramic oxides (spinel ferrite) which are suitable as coating materials; one of them—magnetite—has been used for a long time as an anode material. Magnetite anodes produced by sintering are of course brittle, restricted in shape and size and cannot be machined. Applying ferrites to valve metals, leads on the other hand, to anodes that are durable and can be produced in any shape. They exhibit low consumption rates, they are impermeable to oxygen, have a relatively low electrical resistance, do not passivate in service and are mechanically strong. Titanium anodes coated with lithium ferrite are used in corrosion protection.

Anodes similar to cables are used which consist of a copper conductor covered with conducting plastic. This creates an electrolytically active anode surface and at the same time protects the copper conductor from anodic dissolution.

7.2 Anode Materials

Data on impressed current anodes are assembled in Tables 7-1 to 7-3.

7.2.1 Solid Anodes

Scrap iron is seldom used as anode material today. When applied, it includes old steel girders, pipes, tram lines or railway lines (30 to 50 kg m^{-1}) which are welded

Table 7-1 Data on impressed current anodes for use in soil

Anode material	Iron		High-silicon iron			Graphite			Magnetite	Lithium ferrite on titanium
Length (m)	1 m NP 30 double T girder	1 m rail	0.5	1.2	1.5	1	1.2	1.5	0.9	0.5
Diameter (m)	ht. 0.3 br. 0.13	0.14 0.13	0.04	0.06	0.075	0.06	0.06	0.08	0.04	0.016
Weight (kg)	56	43	16	26	43	5	6	8	6	0.2
Density (g cm^{-3})	7.8	7.8	7	7	7	2.1	2.1	2.1	5.18	6 to 12
Practical loss without coke backfill (kg A^{-1} a^{-1})	10	10	0.2 to 0.3			1	1	1	0.002	0.001
Practical loss with coke backfill (kg A^{-1} a^{-1})	5	5	ca. 0.1			ca. 0.2 to 0.5			–	< 0.001
Life at 1 A per anode without coke backfill (yrs)	5	4	50	80	140	5	6	8	200	120
Life at 1 A per anode with coke backfill (yrs)	10	8	160	260	430	10	12	16	–	> 120
Danger of fracture	None		Moderate			High			Moderate	None
Recommended installation site	Extended anode installation with coke backfill in poorly conducting soils, very economical		Mostly used for impressed current anodes with long life, also without coke backfill			Aggressive soils and aqueous solutions also without coke backfill; relatively economical			Soils Seawater	Deep anodes Seawater

together. A suitable cable is brazed onto them (e.g., NYY-O 2 H 2.5 mm² Cu for use in soil or NSHöu of different diameters in water). The anode in the vicinity of the cable connection must be well insulated with bitumen (in soils) or cast resin (in water) so that the exposed copper cable cannot be anodically attacked at defects in the insulation.

Since the anodic dissolution of iron takes place with almost 100% current efficiency through the formation of Fe(II) compounds, 1 kg of iron gives about 960 A h. This high weight loss requires the use of a large amount of scrap iron [1]. For a protection system that requires 10 A, at least 2 tons of scrap iron are necessary for 20 years of service life. Iron anodes in the ground are always embedded in coke. Low cost and good stability during transport make up for the high material consumption.

Magnetite, Fe_3O_4, conducts electrons because of its defect structure. The resistivity of pure magnetite lies between 5.2×10^{-3} Ω cm [2] and as high as 10×10^{-3} Ω cm. Magnetite occurs as the mineral (kirunavara and gellivare) in northern Sweden and is mined in large amounts as iron ore. The melting point can be lowered by adding small amounts of other minerals. Cast magnetite is glass hard and free of pores. Formerly only cylindrical compact anodes were produced because of the difficulty of casting this material. According to production data [3], the anode consumption is 1.5 g A^{-1} a^{-1} calculated from an anode current density of 90 to 100 A m^{-2}. Consumption increases with increasing anode current density but it is still very low even at 160 A m^{-2}, giving 2.5 g A^{-1} a^{-1}. Magnetite anodes can be used in soils and aqueous environments, including seawater. They endure high voltages and are insensitive to residual ripple. Their disadvantage is their brittleness, the difficulty in casting, and the relatively high electrical resistance of the material.

Graphite has an electron conductivity of about 200 to 700 Ω^{-1} cm^{-1}, is relatively cheap, and forms gaseous anodic reaction products. The material is, however, mechanically weak and can only be loaded by low current densities for economical material consumption. Material consumption for graphite anodes initially decreases with increased loading [4, 5] and in soil amounts to about 1 to 1.5 kg A^{-1} a^{-1} at current densities of 20 A m^{-2} (see Fig. 7-1). The consumption of graphite is less in seawater than in fresh water or brackish water because in this case the graphite carbon does not react with oxygen as in Eq. (7-1),

$$C + O_2 \rightarrow CO_2 \qquad (7\text{-}3)$$

but with HOCl corresponding to Eq. (7-2b) to give

$$C + 2\ HOCl \rightarrow CO_2 + 2\ HCl \qquad (7\text{-}4)$$

Fig. 7-1 Material consumption from impressed current anodes. ● graphite anode without coke backfill, ○ graphite anode with coke backfill, ▲ FeSi anode without coke backfill, △ FeSi anode with coke backfill.

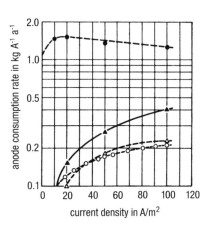

Because a considerable part of the anodically evolved chlorine is carried away by the flowing water, the amount of hypochlorous acid is small and therefore the removal of graphite is less than in fresh water or soil, where direct oxidation according to Eq. (7-3) predominates.

Under medium loading of from 10 to 50 A m^{-2} in seawater, the consumption is between 300 and 1000 g A^{-1} a^{-1}. The density of the material is between 1.6 and 2.1 g cm^{-3}. The material loss of impregnated graphite anodes is much less at comparable current densities.

High-silicon iron is an iron alloy with 14% Si and 1% C. Its density is between 7.0 and 7.2 g cm^{-3}. Surface films containing silicic acid are formed on the passage of anodic current and restrict the anodic dissolution of iron in favor of oxygen evolution according to Eq. (7-1). The formation of surface films is insufficient in seawater and brackish water. Alloys with about 5% Cr, 1% Mn and/or 1 to 3% Mo have greater stability for use in salt water. High-silicon iron anodes behave worse than graphite in chloride-rich water because chloride ions destroy the passive film on the alloy. The main areas of their use are therefore soil and brackish and fresh water. Average loading is between 10 and 50 A m^{-2} so that the consumption rate, depending on service conditions, is below 0.25 kg A^{-1} a^{-1}. High-silicon iron anodes are frequently buried directly in the soil because of their low loss rate [6]; however, an outlet for the evolved gases must be provided to prevent an increase in anode resistance [7].

Lead-silver is primarily used in seawater and strong chloride-containing electrolytes. PbAg anodes are particularly suitable for use on ships and in steel-water constructions, especially as they are relatively insensitive to mechanical stresses. The original alloy developed by Morgan [8,9] consists of 1% Ag and 6% Sb, with the remainder Pb. It is represented as alloy 1 in Table 7-2. A similar alloy developed by Applegate [10] has 2% Ag and the remainder Pb. Another alloy

Table 7-2 Composition and properties of solid impressed current anodes (without coke backfill)

Type	Composition (wt.%)	Density (g cm^{-3})	Anode current density (A m^{-2}) max.	Anode current density (A m^{-2}) avg.	Anode consumption (g A^{-1} a^{-1})
Graphite	100°C	1.6 to 2.1	50 to 150	10 to 50	300 to 1000
Magnetite	Fe$_3$O$_4$ + additions	5.2	–	90 to 100	1.5 to 2.5
High-silicon iron	14 Si, 1 C remainder Fe (5 Cr or 1 Mn or 1 to 3 Mo)	7.0 to 7.2	300	10 to 50	90 to 250
Lead-silver alloy 1	1 Ag, 6 Sb, remainder Pb	11.0 to 11.2	300	50 to 200	45 to 90
Alloy 2	1 Ag, 5 Sb, 1 Sn, remainder Pb	11.0 to 11.2	300	100 to 250	30 to 80
Lead–platinum	Lead + Pt pins	11.0 to 11.2	300	100 to 250	2 to 60

(alloy 2 in Table 7-2) consists of 1% Ag, 5% Sb and 1% Sn, with the remainder Pb. It is not only capable of bearing more load but has a somewhat lower consumption rate than alloy 1. The average loading capacity lies between 50 and 200, or 100 and 250 A m^{-2}, with a maximum value of 300 to 350 A m^{-2}. The anode consumption rate is between 45 and 90, and 30 and 80 g A^{-1} respectively. Finally, there are also lead anodes containing platinum pins [11]. The information in the literature about this system is contradictory. The consumption rate at a loading of up to 500 A m^{-2} should be between 2 and 60 g A^{-1} a^{-1}. According to the producer's information, an anode with this low consumption rate should have a life of 20 years on full load. This, of course, is only if the platinum pins remain in the material, which is not a foregone conclusion.

While antimony improves the mechanical properties of the soft lead, additions of silver and tin as well as platinum pins, act to form a thick and good conducting layer of PbO_2 that actually constitutes the current leakage surface in operation. If the alloying elements or the Pt pins are absent, the PbO_2 remains full of cracks, is porous and adheres badly, so that in electrolytes containing chloride, a reaction occurs on the metallic lead under the film, forming $PbCl_4^{2-}$ ions so that the anode is consumed too quickly. Even with the presence of the additions or pins, the formation of the black-brown, strongly adherent, uniformly growing film occurs over the $PbCl_2$ and is also bonded in the presence of chloride ions. If a tolerable anode consumption is to be ensured, then healing of damage to the PbO_2 film which cannot be avoided in operation must be sufficiently certain. Anodes based on lead, such as silver alloy or platinum pins, can only be used in chloride-rich electrolytes [12]. The advantage of lead anodes is their easy formability. Disadvantages—besides their limitation in strong chloride electrolytes—are the high density of 11 to 11.2 g cm^{-3} and, for the exterior of ships, their relatively low anode current densities.

Finally it must be remembered with these anodes that PbO_2 film, which acts to provide the current leakage, can be detached even when no current is flowing. With renewed anodic loading, the film has to be reformed, which leads to a corresponding consumption of anode material. The anodes should therefore be operated as continuously as possible with a basic load. An exhaustive treatment of the composition and behavior of lead alloy anodes can be found in Ref. 13.

7.2.2 Noble Metals and Valve Metals Coated with Noble Metals

Impressed current anodes of the previously described substrate materials always have a much higher consumption rate, even at moderately low anode current densities. If long life at high anode current densities is to be achieved, one must resort to anodes whose surfaces consist of anodically stable noble metals, mostly platinum, more seldom iridium or metal oxide films (see Table 7-3).

Cotton proposed solid platinum [14] as a material for impressed current anodes. Such anodes are capable of delivering current densities of 10^4 A m^{-2} under suitable conditions. The driving voltage is practically unlimited, and the consumption rate, assuming optimum conditions, is very low—in the region of less than a few milligrams A^{-1} a^{-1}. This applies chiefly at relatively low current densities in seawater where there is good dispersion of the hypochlorous acid that is formed. When noble materials have to be used to achieve high anode current densities in poorly conducting electrolytes, the anodic consumption of platinum increases due to the formation of chlorocomplexes and is then directly dependent on the current density [15-17]. In addition, in solutions containing little chlorine, the more soluble PtO_2 instead of PtO is preferentially formed due to the dominance of oxygen evolution at the anode surface, so that the consumption of platinum also

Table 7-3 Composition and properties of noble metals with metal oxide coatings

Substrate metal	Density (g cm^{-3})	Coating	Density (g cm^{-3})	Coating thickness (μm)	Anode current density (A m^{-2}) max.	avg.	Allowable maximum driving voltage (max/V)	Loss (mg A^{-1} a^{-1})
Platinum	21.45	Platinum	21.45	Solid	>10^4			<2
Titanium	4.5	–		–			12 to 14	
Niobium	8.4	–		–			about 50 (<100)	
Tantalum	16.6	–		–			>100	
Ti, Nb, Ta		Platinum	21.45	2.5 to 10	>10^3	600 to 800		4 to 10
Ti, Nb, Ta		Lithium-ferrite	6 to 12	<25	>10^3	100 to 600		<1 to 6

rises. However, the loss is small so that solid platinum is actually an ideal anode material. Such anodes are very heavy because of the high density of platinum (21.45 g cm^{-3}) and are uneconomical because of the very high purchase price of platinum. For this reason anodes of another metal are used whose effective surfaces are coated with platinum.

Platinum on titanium is the best-known anode of this type. The use of platinum on the so-called valve metals was first mentioned in 1913 [18]. Titanium is a light metal (density 4.5 g cm^{-3}) that is capable of anodic passivation. The passive film is practically insulating at driving voltages of less than 12 V. According to Ref. 16, the current in an NaCl solution is very small up to a potential of $U_H = +6$ V, but rises steeply above this value. The oxide film on titanium, which initially consists of the orthorhombic brookite, is replaced by tetragonal anatase as the thickness of the film increases [19]. If the oxide film, however, is damaged with increasing potential, then it is only regenerated at potentials of $U_H < 1.7$ V. This critical self-healing potential is thus less positive than the above-mentioned critical breakdown potential of +6 V.

If titanium is coated with platinum, the oxide film must first of all be removed by careful etching of the surface. Then the platinum must be applied by electrochemical or thermal means, or mechanically by cladding. The film thickness in electrochemical or thermal platinizing is from 2.5 to 10 μm. Clad platinum coatings are usually thicker. They are also, in contrast to the former, free of pores and therefore more efficient from the point of view of resistance, but of course they are considerably more expensive. The platinum surface behaves practically like solid platinum and can be subjected to higher driving voltages as long as the unclad titanium surface is isolated from the medium. The thinner porous platinum coatings, on the other hand, possess the limitations of titanium if with lower conductivity and higher driving voltages the TiO_2 film in the pores can be destroyed and the platinum film can be disbonded and removed [20-22].

Whereas with impressed current anodes of passivatable materials no special demands are imposed on the quality of the direct current that is fed in, this is quite different with platinized anodes. Investigations [23-25] showed that the loss of platinum with residual ripple of the direct current increased over 5%. Basically the demand on the residual ripple of the direct current increases with increasing driving voltage and anode current density and is dependent on the removal of the electrolysis products and therefore on the electrolytic flow at the anode. It can be assumed as certain that increased consumption occurs for low-frequency residual ripple <50 Hz. Above 100 Hz, the influence of the residual ripple is small. The residual ripple lies in this frequency range for bridge rectifiers operated with 50-Hz alternating voltage; with three-phase rectifiers the frequency is much higher (300 Hz), and the residual ripple is controlled by the circuit at 4%. Experience has shown that the influence of residual ripple is small under optimum conditions at the anode.

Platinum on other valve metals is used where the low critical breakdown potential of titanium presents intolerable limitations. There are several reasons for this in cathodic protection. With good conducting electrolytes, the high anode current densities of on average between 600 and 800 A m^{-2}, and even up to 10^3 A m^{-2} and above, can be made good use of without further problems. With poorly conducting electrolytes (e.g., fresh water) the allowable driving voltage for an economic layout with platinized titanium anodes is not sufficient. In addition, at higher temperatures above about 50°C, the critical breakdown potential decreases still further; at 90°C it is about $U_H = 2.4$ V. Finally, the chemical composition of the electrolyte can influence the breakdown potential. This is for instance the case for acid electrolyte containing hydrogen halites. In such cases materials are chosen as the substrate metal that have properties similar to titanium but are considerably more stable, particularly niobium and tantalum. The breakdown potentials under the limiting conditions described are considerably higher. Driving voltages up to 100 V in chloride solutions are permissible with niobium and tantalum. Platinized niobium and tantalum anodes can be used almost without any limitations. They only break down in electrolytes containing fluoride or boron fluoride, as does titanium, because all three substrate materials cannot form passive films in such media. Niobium has a density of 8.4 g cm^{-3} and tantalum 16.6 g cm^{-3}. A further superiority of these valve metals is their high electrical conductivity, which is almost three times that of titanium at 2.4×10^{-4} Ω^{-1} cm^{-1}.

Iridium on valve metals is suitable if the consumption rate of platinum is too high at elevated temperatures or critical composition of the medium. Mostly platinum-iridium alloys are used with about 30% Ir, because coating valve metals with pure iridium is somewhat complicated. For the same reason, other noble metals such as rhodium cannot be used [21]. At present there is little price difference between platinum and iridium.

7.2.3 Metal Oxide-Coated Valve Metals

Good results are obtained with oxide-coated valve metals as anode materials. These electrically conducting ceramic coatings of p-conducting spinel-ferrite (e.g., cobalt, nickel and lithium ferrites) have very low consumption rates. Lithium ferrite has proved particularly effective because it possesses excellent adhesion on titanium and niobium [26]. In addition, doping the perovskite structure with monovalent lithium ions provides good electrical conductivity for anodic reactions. Anodes produced in this way are distributed under the trade name Lida [27]. The consumption rate in seawater is given as 10^{-3} g A^{-1} a^{-1} and in fresh water is about 6×10^{-3} g A^{-1} a^{-1}. With bedding in calcium petroleum coke, the consumption rate is less than 10^{-3} g A^{-1} a^{-1}. The anodes are very abrasion resistant and have a hardness of 6 on the Mohs scale; they are thus much more abrasion resistant than platinized titanium anodes. There are no limitations in forming these anodes as is also the case with platinized titanium anodes. The polarization resistance of these anodes is very low. Therefore, voltage variations at the anode surface resulting from alternating voltage components in the supplying rectifier are very small. The current loading in soil with coke backfill is 100 A m^{-2} of active surface, in seawater 600 A m^{-2} and in fresh water 100 A m^{-2}. They offer great advantages over high-silicon iron anodes as deep well anodes because of their low weight.

7.2.4 Polymer Cable Anodes

Polymer cable anodes are made of a conducting, stabilized and modified plastic in which graphite is incorporated as the conducting material. A copper cable core serves as the means of current lead. The anode formed by the cable is flexible, mechanically resistant and chemically stable. The cable anodes have an external diameter of 12.7 mm. The cross-section of the internal copper cable is 11.4 mm^2 and its resistance per unit length R' is consequently 2 mΩ m^{-1}. The maximum current delivery per meter of cable is about 20 mA for a service life of 10 years. This corresponds to a current density of about 0.7 A m^{-2}. Using petroleum coke as a backfill material allows a higher current density of up to a factor of four.

Without coke backfill, the anode reactions proceed according to Eqs. (7-1) and (7-2) with the subsequent reactions (7-3) and (7-4) exclusively at the cable anode. As a result, the graphite is consumed in the course of time and the cable anode resistance becomes high at these points. The process is dependent on the local current density and therefore on the soil resistivity. The life of the cable anode is determined, not by its mechanical stability, but by its electrical effectiveness.

7.3 Insulating Materials

Impressed current anodes must be insulated from the surface that is being protected. Also, the current connections must be well insulated to prevent the free ends of the cable from being attacked and destroyed.

The demands on insulating materials in soil and fresh water are relatively low. Anodically evolved oxygen makes the use of aging-resistant insulating materials necessary. These consist of special types of rubber (neoprene) and stabilized plastics of polyethylene, and polyvinylchloride, as well as cast resins such as acrylate, epoxy, polyester resin and many others.

218   Handbook of Cathodic Corrosion Protection

Much greater demands are made on insulating materials for seawater and other electrolytes containing halides because anodic chlorine is evolved, depending on the chloride content and the current density, which is particularly aggressive and destructive to many insulating materials. HCl and HOCl formed according to Eq. (7-2b) attack the materials forming the anode supports. Chloride-resistant insulating materials are polypropylene, neoprene, chloroprene, and special types of polyvinylchloride (e.g., Trovidur HT) as well as special variations of epoxy and unsaturated polyesters. A particular requirement is a reliable bond between anode material and insulating material. Even with scrupulous cleanliness, it is often difficult to produce a strong and lasting bond that will be exposed (e.g., in cathodic protection in and on ships) to considerable mechanical stresses from streaming, vibration, impact, etc. Where gluing or casting with cast resin is not possible, elastic sealing materials such as silicone cement are used. Disbonding of the insulating material, especially in the area of the current leads, can result in breakdown of the anode in a short time.

Especially resistant insulating materials have to be provided for the following severe service conditions: aggressive environments, high temperatures and high pressure. Organic insulating materials that can withstand severe chemical conditions are fluorinated plastics, e.g., polytetrafluoroethylene (Teflon). At higher temperatures and pressures, ceramic insulating materials are used (e.g., porcelain insulators, or glass inserts for high-pressure-resistant, screwed-in anodes). The brittleness and different coefficients of thermal expansion have to be considered with ceramic materials.

7.4   Cables

A single or multicored plastic-coated cable of the type NYY* or NYY-O is used as the connecting cable between a protected object and an anode in soils and fresh water, and particularly in seawater, medium heavy or heavy rubber-sheathed connections of an NSHöu or NSSHöu* type are used. Heavy welded connections of type NSLFSöu* are used for severe mechanical loading. In addition to these, for ships, marine cable of type MGCG* or watertight cables must be considered.

Ohmic voltage drops resulting in losses cannot be ignored in the connecting cables with long anode cables and high protection currents [28]. Cable costs and losses must be optimized for economic reasons. The most economic calculated cable dimension depends primarily on the lowest cross-section from the thermal point of view. For various reasons the permitted voltage drop usually lies between 1 and 2 V, from which the cross-section of the cable to be installed can be calculated from Eq. (3-36).

Impressed Current Anodes

7.5 Forms of Anode

7.5.1 Anodes Suitable for Soil

The anodes most suitable for burying in soil are cylindrical anodes of high-silicon iron of 1 to 80 kg and with diameters from 30 to 110 mm and lengths from 250 to 1500 mm. The anodes are slightly conical and have at the thicker end for the current lead an iron connector cast into the anode material, to which the cable connection is joined by brazing or wedging. This anode connection is usually sealed with cast resin and forms the anode head (see Fig. 7-2). Ninety percent of premature anode failures occur at the anode head, i.e., at the cable connection to the anode [29]. Since installation and assembly costs are the main components of the total cost of an

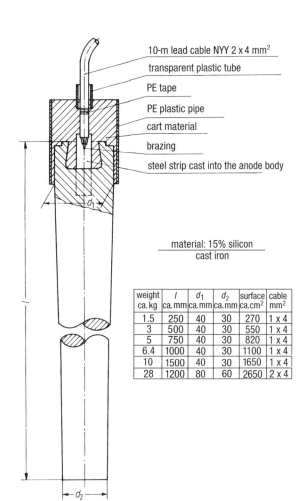

Fig. 7-2 Anode head and dimensions of high-silicon anode.

material: 15% silicon cast iron

weight ca. kg	l ca. mm	d_1 ca. mm	d_2 ca. mm	surface ca. cm^2	cable mm^2
1.5	250	40	30	270	1 x 4
3	500	40	30	550	1 x 4
5	750	40	30	820	1 x 4
6.4	1000	40	30	1100	1 x 4
10	1500	40	30	1650	1 x 4
28	1200	80	60	2650	2 x 4

anode system, very careful attention should be given to the construction of a reliable and durable anode head. With heavier anodes, support for the cable is necessary (e.g., attachment to a carrying rope) and a device is needed at the exit point from the anode head to prevent kinking and damage during installation.

Impressed current anodes in soil are usually bedded in coke. Blast furnace coke No. 4 that is 80 to 90% C, has a resistivity $\rho = 0.2$ to $0.5 \, \Omega$ m and a particle size between 2 and 15 mm is particularly suitable as backfill. Anodically evolved gases (O_2, CO_2 and Cl_2, e.g., in soils with high chloride content) can escape so that no increase in anode resistance arises from a gas lock. In addition, by bedding in coke, the actual dimensions of the anode can be increased and the grounding resistance considerably reduced (see Chapter 9). The grounding resistance of anodes bedded in coke remains constant over many years, whereas for anodes not bedded in coke, it can double in a few years due to electro-osmotic dehydration and surface film formation. Anodes with a low current loading can also be bedded in clay mud (see Section 6.3). Coke as backfill must be lightly compacted to give good electrical connection with the anode. As a result, loss from the anode is mostly transferred to the outer surface of the coke backfill. To avoid the formation of larger amounts of acid at the anode according to Eqs. (7-1) and (7-2b), slaked lime can be added to the coke backfill. The consumption rate of the coke backfill is about 2 kg A^{-1} a^{-1}. Practical consumption rates for scrap iron, graphite, and high-silicon iron anodes with and without coke backfill are given in Table 7-1. The reduction in consumption rate by coke backfill is particularly important with scrap iron anodes because material consumption can be reduced by ca. 50%.

In addition to anodes with a simple connecting head, there are cylindrical double anodes that have cable connectors cast on at both ends and that can be used in the construction of horizontal or vertical anode chains. Anodes of graphite or magnetite are more compact than anodes of high-silicon iron because of the danger of fracture.

Cylindrical anodes are preferred for the oxide-coated titanium anodes, with diameters of 16 and 25 mm and lengths between 250 and 1000 mm [31]. The anodes are supplied as single anodes or as finished anode chains with a connecting cable. The carrier material consists of a titanium tube through which a special cable (EPR/CSPE[1]) is threaded. This cable is insulated in the middle of the titanium tube and is equipped with a copper outer covering. Electrical contact with the anode is achieved by crimping the tube at the ends and sealing it to the cable in the middle. To protect the metal oxide coating, copper shoes are fixed at the contact points. These anodes are particularly suitable for the construction of deep well anode installations because of their low weight as well as the simple production of anode chains. The anodes

---

[1]A trade name of CEAT CAVI Ind., Turin, Italy. It is ethylene propylene rubber/chlorosulfonated polyethylene.

have to be inserted into a backfill of calcined petroleum coke to give a low ground resistance and increased life. This coke has a bulk weight of 0.8 tons m^{-3} and a particle size of 1 to 5 mm. The resistivity is between 0.1 and 5 Ω cm, depending on the compressive loading. Detergents are added to the coke to give a rapid compaction time.

With deep anodes, a tube that allows gases to escape must be included to avoid a build-up of gas (see Section 9.1.3). The anode chains are supplied with cable, a centering device and a gas escape tube.

Cable anodes of conducting polymers have an advantage when there are site problems with the installation of other anodes. They are extensively used for the cathodic protection of reinforcing steel in concrete (see Section 19.5.4).

7.5.2 Anodes Suitable for Water

Cylindrical anodes with a construction similar to those in Section 7.5.1 are also suitable for use in water to protect steel-water constructions and offshore installations, and for the inner protection of tanks. In addition to graphite, magnetite and high-silicon iron, anodes of lead-silver alloys are used as well as titanium, niobium or tantalum coated with platinum or lithium ferrite. These anodes are not usually solid, but are produced in tube form. In the case of lead-silver anodes, the reason is their heavy weight and relatively low anode current density; with coated valve metals, only the coating suffers any loss. Finally, the tubular shape gives larger surfaces and therefore higher anode currents. The same types of connection apply to lead-silver anodes as those given in Section 7.5.1. The cable can be directly soft soldered onto the anode if a reduction in the tensile load is required. This is not possible with titanium. Such anodes are therefore provided with a screw connection welded on where appropriate, which is also of titanium. The complete connection is finally coated with cast resin or the whole tube is filled with a suitable sealing compound. Because of the poor electrical conductivity of titanium, with long and highly loaded anodes it is advisable to provide current connections at both ends.

Disc and ingot-shaped anodes are also used in water besides the cylindrical or conical shapes. Several parallel-connected rod anodes as well as hurdle-shaped racks are sometimes used for the protection of larger objects such as sheet steel linings and loading bridges if sufficient space is available and there is no likelihood of the anodes being damaged, e.g., by anchors. These are situated on the ground and contain several anodes, mostly rod anodes, next to one another in insulated fixtures. In calculating the ground resistance of such groups of anodes, the influence of neighboring anodes must be taken into account (see Section 24.2). Floating anodes are used for offshore installations in which the current outflow surface is attached to the outside of a cylindrical or spherical float which is attached to the seabed by the anchor rope and grounded cable, so that the anode body floats at a predetermined depth in the water. The advantage of this is the ability to carry out repairs without interrupting the operation of the offshore installation (see Section 16.6). Furthermore, a desired uniform current distribution can be achieved by distancing the anode from the protected object. Special shapes of anodes are used for the external protection of ships and these are described in Section 17.3.3.2.

7.5.3 Anodes for Internal Application

In addition to the types of anodes already described, which can also be used for the internal protection of installations, there are some special types for this purpose. These include thin, rod-shaped anodes. These are not sufficiently stable if they are long and in fast-flowing liquid and so they must be attached to special holders, mostly of plastics. Rod anodes are particularly suitable for the inner protection of containers where the design is complicated, or components interfere with the uniform propagation of the protection current. Most commonly used are platinized titanium anodes consisting of a more or less long titanium rod, completely or partly covered with a platinum film with a current connection on one side. An optimum current distribution can be achieved by having a suitable rod length and partial platinization corresponding to the geometrical proportions of the container (see Fig. 7-3). Installation is particularly simple if the current connections can be made from outside the container. The titanium rod passes through an insulated bushing externally via a screw thread on which the cable is fixed (see Fig. 7-4). At the required position, all that is needed is a screwed socket welded into the container wall, the rod anode screwed in and sealed with a suitable sealant.

Although rod anodes screwed into the side walls of containers are preferred, plate anodes and basket anodes are mostly necessary on the base. They are particularly suitable for large containers without being built in. Plate anodes are supplied with a flat plastic holder to which the cable conductor is attached and

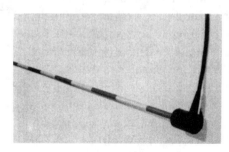

Fig. 7-3 Partially platinized rod anode.

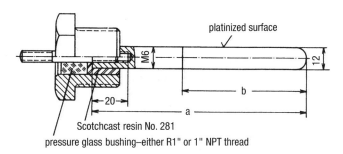

Fig. 7-4 Standard rod anodes for internal protection of containers and pipes (dimensions in millimeters).

type	a	b	capacity
A	80	30	0.5 A
B	100	50	0.75 A
C	120	70	1.0 A
D	160	100	1.5 A
E	200	100	1.5 A
F	250	100	1.5 A

insulated. Large plate anodes are seldom used because the current distribution at anode current densities from 600 to 800 A m^{-2} becomes too inhomogeneous and can lead to premature breakdown of the anode due to preferential removal of platinum at the edges.

In such cases basket anodes are frequently used. These have a relatively large surface and work at a low driving voltage due to their special construction. A cylinder of platinized titanium-expanded metal serves as the basket to which a titanium rod is welded. This serves as the current lead and carrier, and ends in a plastic foot that contains the cable lead and at the same time serves as the mounting plate. The expanded metal anode exhibits a very uniform anode current density distribution, even at large dimensions, in contrast to the plate anode. The reason is the many corners and edges of the metal that make the point effect only evident at the outer edges of the anode.

In internal protection, attention must be paid to the fact that mixed gases that contain hydrogen and oxygen are evolved by the protection current. Noble metals and noble metal coatings can catalyze an explosion. The safety measures required in this case are presented in DIN 50927 [32] (see also Section 20.1.5).

7.6 References

[1] M. Parker, Rohrkorrosion und kathodischer Schutz, Vulkan-Verlag, Essen 1963.
[2] H. Bäckström, Öjvers. Akad. Stockholm *45*, 544 (1888).
[3] Bergsøe, Anticorrosion AS, Firmenschrift „BERA Magnetite Anodes".
[4] W. v. Baeckmann, gwf *107*, 633 (1966).
[5] S. Tudor, Corrosion *14*, 957 (1958).
[6] NACE, Publ. 66-3, Corrosion *16*, 65 t (1960).
[7] W. T. Bryan, Mat. Perf. *9*, H. 9, 25 (1970).
[8] J. H. Morgan, Corrosion *13*, 128 (1957).
[9] J. H. Morgan, Cathodic Protection, Leonhard Hill Ltd, London 1959.
[10] L. M. Applegate, Cathodic Protection, 90, Mc Graw-Hill Book Co, New York 1960.
[11] R. L. Benedict, Mat. Protection *4*, H. 12, 36 (1965).
[12] G. W. Moore u. J. H. Morgan, Roy. Inst. Nav. Arch. Trans. *110*, 101 (1968).
[13] J. A. v. Frauenhofer, Anti-Corrosion *15*, 4 (1968), *16*, 17 (1969).
[14] J. B. Cotton, Platinum Metals Rev. *2*, 45 (1958).
[15] R. Baboian, International Congress of Marine Corrosion and Fouling, Juni 1976.
[16] R. Baboian, Mat. Perf. *16*, H. 3, 20 (1977).
[17] C. Marshall u. J. P. Millington, J. Appl. Chem. *19*, 298 (1969).
[18] R. H. Stevens, USP 665 427 (1913).
[19] J. B. Cotton, Werkstoffe und Korrosion *11*, 152 (1960).
[20] E. W. Dreyman, Mat. Perf. *11*, H. 9, 17 (1972).
[21] v. Stutternheim, Metachem, Oberursel, pers. Mitteilung März 1978.
[22] M. A. Warne u. P. C. S. Hayfield, Mat. Perf. *15*, H. 3, 39 (1976).
[23] R. Juchniewicz, Platinum Metals Rev. *6*, 100 (1962).
[24] R. Juchniewicz, Corrosion Science *6*, 69 (1966).
[25] R. Juchniewicz und Bogdanowicz, Zash. Metal *5*, 259 (1969).
[26] A. Kumar, E. G. Segan u. J. Bukowski, Mat. Perf. *23*, H. 6, 24 (1984).
[27] D. H. Kroon u. C. F. Schrieber, NACE-Confer. Corrosion 1984, Paper Nr. 44.
[28] H. G. Fischer u. W. A. Riordam, Corrosion *13*, 19 (1956).
[29] F. Paulekat, Kath. Innenschutz, HDT-Veröffentlichung 402, Vulkan-Verlag, Essen 1978.
[30] G. Heim u. H. Klas, ETZ A *77*, 153 (1958).
[31] C. G. Schrieber u. G. L. Mussinelli, Mat. Perf. *26*, H. 7, 45 (1987).
[32] DIN 50927, Beuth Verlag, Berlin 1985.

8

Impressed Current Equipment and Transformer-Rectifiers

W. V. BAECKMANN AND H. KAMPERMAN

The current needed for cathodic protection by impressed current is supplied from rectifier units. In Germany, the public electricity supply grid is so extensive that the CP transformer-rectifier (T-R) can be connected to it in most cases. Solar cells, thermogenerators or, for low protection currents, batteries, are only used as a source of current in exceptional cases (e.g., in sparsely populated areas) where there is no public electricity supply. Figure 8-1 shows the construction of a cathodic impressed current protection station for a pipeline. Housing, design and circuitry of the rectifier are described in this chapter. Chapter 7 gives information on impressed current anodes.

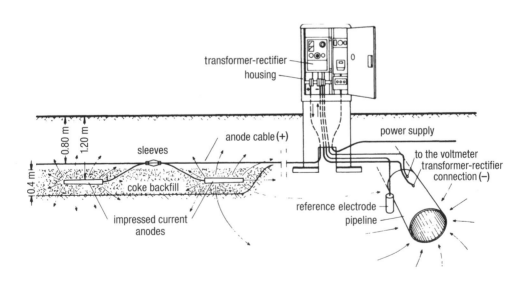

Fig. 8-1 Construction of an impressed current cathodic protection station.

8.1 Site and Electrical Protection Measures

Impressed current stations for buried tanks are connected to the electricity supply of the site on which the tank stands. The location of a protection installation for local distribution and for transmission lines, on the other hand, is determined principally by the availability of connection to a public electricity supply grid because considerable costs are involved in connecting a very long voltage supply to the grid. With a very large current requirement, it is of only secondary importance that the anodes be installed in a region of low soil resistivity. In choosing the location, the following considerations are relevant and are given in their order of importance:

- availability of a grid connection,
- as low an electrical soil resistivity as possible in the area of the anode bed,
- sufficient distance from buried foreign installations to reduce their interference (see Chapter 9),
- as little disturbance of the site owner's property as possible,
- good access for vehicles,
- with stray current installations, a location that has as large a leakage area as possible for the stray current of the pipe network.

With impressed current stations on privately owned sites, it is advisable to conclude contracts with the property owner in order to guarantee access to the installation at all times, to prevent damage to the cables and to exclude construction or modifications in the area of the anode beds.

In choosing a site for protection installations for steel-water structures, decisive factors are the location in the harbor area and the need to keep the lengths of cable to the protected object and anode as short as possible where very high protection currents are involved.

The units and ancillary equipment must be protected from mechanical damage and the effects of weather to ensure the reliable operation of an impressed current station. This is achieved by installing it in a weatherproof plastic housing (see Fig. 8-2). Sufficient ventilation must be provided to disperse heat. The ventilation holes should be protected with brass gauze to keep out animals. Transformer-rectifier units must be connected to a circuit that is continuously energized, especially if they are in a building where the current is turned off at night, e.g., gas stations that are not open for 24 h.

In work places, T-R units should be installed where there is no danger of explosion, since explosion-proof measures are very costly and units in such installations are difficult to service and maintain. It is more convenient to bear the

Fig. 8-2 Impressed current unit in a plastic housing with a meter panel.

cost of longer cables. If this is especially impossible, then the pertinent rules and regulations for installing electrical systems in situations where there is a danger of explosion must be adhered to (see Section 11.4). However, it must be remembered that TRs with an output over 200 W cannot be installed in standard explosion-resistant housing because of their size and heat output [2].

Protection against electrocution (through accidental contact with live components or indirect contact) must be provided according to regulations [3-5]. Protection with dc is provided by "functionally low voltage with safe disconnection" [3]. This requires a grid transformer with 3.5-kV test voltage for T-Rs, to prevent an overload from the grid side being transmitted to the dc side. Fan current (FI) and ground fault protective switches are used in leakage current circuit breakers, which are also tripped by a pulsating dc current [6].

8.2 Design and Circuitry of Impressed Current

Transformer-rectifiers are produced for use in corrosion protection stations with a rated dc output of about 10 W for storage tanks and short pipelines as well as up to a few kilowatts for large steel-water constructions. In general, T-Rs for pipelines have an output of 100 to 600 W. It is recommended that the rated current of the T-R be twice that of the required current so that sufficient reserve is available to cope with potential future enlargement of the installation, a possible decrease in the coating resistance, increase in stray currents or other changes. The necessary rated output voltage results from the required protection current and the resistance of the anode/soil/protected object circuit, which can be estimated or measured after installation of the final anode bed. Sufficient reserve should also be provided for the output voltages.

In rail installations, a drainage test of sufficient duration during the most severe service conditions (such as commuter traffic) must always be carried out where there is stray current interference. This should be carried out as far as possible from the previously installed anode bed and only in exceptional circumstances with provisional anodes, or in the case of planned stray current forced drainage, by a connection to the rails (see Section 15.5.1). A portable self-regulating T-R serves for the drainage test. During the test, protection current and output voltage as well as the potential at the test site and at other significant points in the pipe network are recorded. It is particularly recommended that attention be paid to the output data of the T-R and that there is sufficient reserve.

In stray current forced drainage against rails, current peaks up to 100 A can occur, particularly in the protection of old pipe networks with poor coating, although the necessary output voltage remains quite low. This current depends on the voltage between rail and pipe as well as on the resistance of the circuit. With rails laid in the street and thus earthbound tracks, this is mostly between 5 and 10 V. The dc supply for such installations is therefore considerably less than for normal impressed current anode stations.

The behavior is quite different with modern tramways, which run on their own track bed with wooden ties and clean ballast. Here the grounding resistance of the track would have to be very high to produce even a small stray current interference. Due to this and the high operating current of the heavy locomotives, the voltage between rail and soil is very high and can reach 40 to 50 V in short pulses. Therefore, the above-mentioned advantage of the necessary low output voltage of the T-R is lost with forced stray current drainage. It is then more beneficial to use impressed anode beds, especially when protecting well-coated pipelines with low protection current requirements. It has also been shown that connecting several protection installations to a stretch of track with a high ground resistance has a

8.3 Rectifier Circuit

In single-phase bridge circuits for ac connections and for very low ac output voltages below 5 V, single-phase center tap circuits are used as rectifier circuits for CP transformer-rectifiers. They have an efficiency of 60 to 75% and a residual ripple of 48% with a frequency of 100 Hz. A three-phase bridge circuit for three-phase alternating current is more economical for outputs of about 2 kW. It has an efficiency of about 80 to 90% and a residual ripple of 4% with a frequency of 300 Hz. The residual ripple is not significant in the electrochemical effect of the protection current so that both circuits are equally valid.

The following factors can impose limits on the residual ripple:

- A current with ripples can lead to a high consumption rate for platinized titanium anodes and their premature failure. The residual ripple should be limited to 5% (see Section 7.2.2) [7].

- In the cathodic protection of asymmetrically connected communication cables, distortions are coupled into the transmission lines coming from the ripple of the sheath current. In this case also, limiting the residual ripple to 5% is usually sufficient.

- In cathodically protected water pipelines, inductive flow meters are affected by the ripple on the pipe current. The producers of these instruments mostly say that a ripple of 5% is permissible.

Reducing the residual ripple from single-phase rectifiers for currents up to about 20 A and voltages of up to about 20 V can be achieved by filter circuits of choke coils and condensers. For greater output and constant residual ripple independent of load, the only possibility is the three-phase bridge circuit. It is always more satisfactory than a filter circuit.

The rating of a transformer-rectifier unit refers to an external temperature of 35°C. Higher temperatures require the special design of some components and this should be discussed with the manufacturer. Usually T-Rs are outfitted with self-cooling by natural ventilation. Forced ventilation with a fan leads to considerable contamination and is not used for this reason. In particularly difficult climatic conditions (e.g., for steel-water constructions in the tropics), oil cooling is necessary for larger units. This provides good protection for the rectifier cells and transformers, removes heat, and provides protection against atmospheric effects for variable ratio transformers with current collectors.

Selenium columns or silicon stacks are an important component in the rectifier unit. Selenium rectifiers are insensitive to overcurrent, short circuit and overvoltage, and can be inertly fused. They have an outstanding operational safety and are preferred for all normal applications of CP transformer-rectifiers. Several plates of selenium are connected in series for output voltages over 20 V due to their low barrier voltage of 25 to 30 V. This results in rapid increases in losses and volume, which are much greater than with silicon, where one cell per section in the bridge is sufficient even for high-output voltages. This is the basis of the widely known advantage of silicon in rectifier design, but it is not effective in CP transformer-rectifiers with low output voltages. Silicon cells are very sensitive to overcurrent and overvoltage, and require protection by quick-response safety devices and voltage limiters. They have indeed a very high barrier voltage, but are not able to suppress short, low-energy voltage peaks as do selenium rectifiers, owing to their extremely small barrier current, and they blow out when this occurs. On the other hand, blowouts with selenium rectifiers are self-healing. Silicon is preferred under severe climatic conditions since the cells can be hermetically sealed and made insensitive to atmospheric effects. Furthermore, the permitted operating temperature is somewhat higher than for selenium. Silicon cells are used exclusively for high-voltage T-Rs.

Rectifiers cannot be completely protected against overvoltage by gas discharge arresters even if the barrier voltage is considerably above the response voltage of the arrester. The reason for this is the relatively long response time of the arrester (a few microseconds), whereas semiconductors blow out much quicker, in the region of a few nanoseconds. Voltage limiters based on selenium and condensers have proved themselves in protecting silicon cells against overvoltage.

8.4 Adjustable Transformer-Rectifiers

Transformer-rectifiers with fixed output voltages are used in all cathodic protection stations in which the current circuit resistance and the current requirement remain constant. For low output and low currents, this is achieved by taps and terminals on the secondary of the transformer. For greater output and for a simple setting, it is convenient to provide an isolating transformer with a fixed secondary voltage for the maximum output dc voltage of the device and to include a control transformer in the circuit of the primary that serves as an autotransformer. It can be installed as a toroidal core or pile control transformer, variable or with taps, combined with a multiple contact switch. It is recommended that the contacts of control transformers and switches be cleaned thoroughly during maintenance and occasionally activated to keep them clean.

Usually the annual variations in the specific resistance of the soil lead only to small changes in the current circuit resistance so that adjustment of the protection

Impressed Current Equipment and Transformer-Rectifiers 231

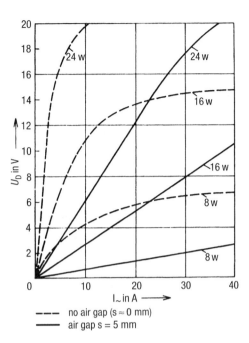

Fig. 8-3 ac voltage, U_D, in a current-limiting choke as a function of the ac flowing through it, $I_\sim$, for different numbers of windings, w, with and without an air gap.

current in protecting storage tanks and pipelines is not necessary, or necessary only in very few circumstances. A control T-R can be used for large changes in the current circuit resistance (see Section 8.6). Transformer-rectifiers for high-voltage cables in steel pipes, which are connected to the low voltage of the dc decoupling device, have a very small output voltage of 1.5 V maximum and must be very finely adjustable. This can be achieved by a combination of a multiple contact switch and transformer with a variable control transformer to distribute the output voltage in two ranges. Since transient overvoltages occur at the resistance of the dc decoupling device during switching operations, the device must be protected against them by special switching measures, such as overvoltage limiters, reverse chokes for high frequencies and insulated assembly.

If an adjustable T-R is connected as forced stray current drainage between pipeline and rails and its output voltage is fixed at a definite level, the protection current and the pipe/soil potential can undergo considerable fluctuation.

If an additional resistor is included in the current circuit and equalizes the previously mentioned voltage drop by raising the output voltage, changes in the pipe/rail voltage have less effect on the current. This arrangement can be regarded as a rectifier with increased internal resistance, which is similar to a galvanostat [8]. In order to avoid the heat losses due to ohmic resistance, it is advisable to connect a choke as an inductive resistance in the T-R between the transformer and the rectifier bridge. Figure 8-3 shows the relation between the flowing alternating current I and the declining alternating voltage U_D resulting from a choke with different numbers of windings with and without an air gap. It can be seen, particularly with a choke with an air gap, that the alternating voltage increases almost linearly with increasing current. With this circuit, current fluctuations are also reduced because the alternating current flowing from the transformer via the choke to the rectifier bridge corresponds quantitatively to the direct current flowing in the forced stray current drainage. With such units the previously mentioned current and potential variations from the rail/pipe voltage are not completely compensated. Their adjustment takes time and depends on experience.

8.5 Rectifiers Resistant to High Voltage

A very high alternating voltage between pipeline and anode, and therefore at the output terminals of the T-R, can arise rapidly in the protection station in the case of failures in the high-tension line. A T-R that is resistant to high voltage must meet the following requirements:

- it must not be destroyed by high voltage,
- it must prevent an overshoot in the current supply grid,
- it should have a limiting action on the pipe/anode voltage.

A cascade of chokes, overvoltage arresters and a condenser between bridge rectifier and output act so that the response to the overvoltage taking place is damped by the "long" response time of the arrester (about 1 μs), so that in this period the permissible inverse voltage of the rectifier cells is not overstepped. Since the inverse voltage must be greater than the response voltage of the arrester, silicon diodes with a surge peak inverse voltage of 2000 V are used. The isolating transformer is provided with increased insulation and shows a test voltage of 10 kV. A cathode drop arrester is situated immediately at the output terminals and limits the pipe/anode voltage for arrester currents of up to 5 kA to 1.5 kV (see Fig. 8-4). This circuit also protects the apparatus against thunderstorm overvoltages [9]. Chokes with an inductance of 1 mH and overvoltage arresters with response voltages of ≤1 kV are sufficient for protection against lightning in the absence of high-voltage interference [10].

Under normal working conditions, voltages are induced in the pipeline where there are long stretches of pipeline closely parallel to high-voltage overhead transmission lines or railway lines operating on ac, which can reach a few tens of volts at the end of the parallel stretch if the coating resistance is high (see Chapter 23). This induced ac voltage experiences a one-way rectification in the rectifier bridge of the cathodic protection station, reinforces the protection current and further depresses the pipe/soil potential. Since the operating current in the overhead

Impressed Current Equipment and Transformer-Rectifiers

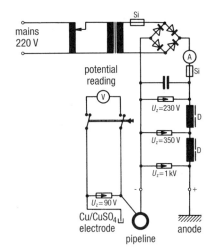

Fig. 8-4 Circuit diagram of a transformer-rectifier with overvoltage protection (protected against high voltage).

transmission line or the railway changes with time, the induced voltage and therefore also the rectified alternating current varies synchronously so that the pipe/soil potential is continuously fluctuating. Optimal adjustment of the cathodic protection station becomes extremely difficult. High-voltage-resistant rectifiers here have the advantage that their chokes severely reduce the rectified alternating current.

In the case of an accident, steel pipes containing high-voltage cables can take on a high voltage against the ground (see Chapter 14). Therefore high-voltage T-Rs are used in the cathodic protection of these pipes whenever they are used outside of the switchplants [11].

8.6 Control Rectifiers

Cathodic protection stations frequently operate under conditions that are continually changing. These include:

- stray current interference,
- frequent and large changes in the grounding resistance of an impressed current station due to the varying specific electrical resistance of the medium, e.g., in steel-water constructions,
- fluctuation in the current requirement with changes in the aqueous condition and oxygen access (flowing velocity), e.g., in the internal protection of storage tanks,
- protection potential ranges with two limiting potentials (see Section 2.4).

In such conditions, it is recommended that the T-R be equipped with an electrical control circuit, which primarily keeps the potential constant, and, in exceptional circumstances, also the protection current. These pieces of equipment are potentiostats (for controlling potential) and galvanostats (for controlling current) [8].

The inclusion of control units in a protection station has the considerable advantage that the station is always working under optimal conditions. As an example, with potential control, sufficient protection current is always available for forced stray current drainage and for stray current peaks resulting from large negative rail/soil potential during periods when the railway is not operating; with more positive rail/soil potentials, only the necessary current flows to attain the protection potential. This ensures that the interference of foreign installations is, on average, small. Furthermore, a reference point in the course of the potential along the pipeline is held constant, namely the potential at the protection stations. Also, limiting values of other fluctuating potentials at the remaining measuring points can be covered.

The adjustment of a protection station or of a complete protection system where there is stray current interference is made much easier by potential control. Potential control can be indispensable for electrochemical protection if the protection potential range is very small (see Sections 2.4 and 21.4). This saves anode material and reduces running costs.

Current control can be more advantageous where rail/soil potentials are predominantly positive. Current control is also preferred in the cathodic protection of steel-water construction if the anode resistance fluctuates due to changes in electrical conductivity.

Figure 8-5 shows the main circuit diagram of a potential control rectifier provided with magnetic amplifiers (transducers). The chosen potential is set at the nominal value with a potentiometer. The actual potential is compared with this value, which corresponds to the voltage between a reference electrode and the protected object.

The difference between the nominal and the actual voltage controls a prestepped magnetic amplifier that sets the primary alternating voltage for the rectifier transformer via a final-step magnetic amplifier. The output voltage and, with it, the protection current of the rectifier, is raised or lowered by this means whenever the potential of the protected object deviates from the nominal value in either direction. The response time is between 0.1 and 0.3 s. The control current amounts to about 50 μA. At this loading, the reference electrode must have sufficiently low resistance and low polarizability.

Potential control rectifiers can also be constructed using thyristors. However, these produce strong high-frequency harmonic waves that can be transmitted to

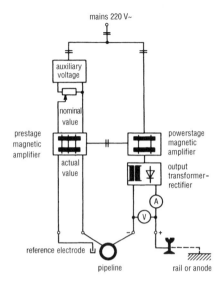

Fig. 8-5 Circuit diagram illustrating the principle of a potential-controlled transformer-rectifier.

neighboring telephone cables and cause considerable interference in them as well as in radio and television reception. Transistors are only relevant as control devices for low protection currents (e.g., internal protection of storage tank). In addition, they can interrupt the protection current briefly, measure the off potential at the anode as a measuring electrode, and store and afterward adjust the protection current. Rectifiers with transistors as control elements are not suitable for controlling pipelines influenced by stray currents. If a long reaction time can be tolerated (e.g., in the cathodic protection of steel-water constructions), high-power rectifiers with simpler two-point controls can be used. The actual potential is connected to a voltmeter that has limiting value contacts and controls a motor-driven control transformer via relays, which adjusts the necessary output voltage. This circuit is simple and cost effective. It has the advantage that there is no phase cutting and residual ripple is constant. Special reference electrodes are used as control electrodes for incorporation in the soil or hanging in water. They consist of metals with constant rest potentials and must not be polarized by the control current. They should be installed where the ohmic voltage drop is as small as possible. Depending on the electrolyte, copper, iron or zinc are possibilities.

Rectifiers working according to the control diagram in Fig. 8-6 are used for anodic corrosion protection in passivatable systems that go spontaneously from the passive to the active state when the protection current is switched off [12]. The predetermined nominal voltage U_S between reference electrode and protected object is compared with the actual voltage U_I in a differential display unit D. The difference $\Delta U = U_S - U_I$ is amplified in a voltage amplifier SV to $V_0 \Delta U$. This

236 Handbook of Cathodic Corrosion Protection

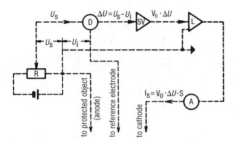

Fig. 8-6 Principle of a potentiostatic transformer-rectifier (V_0 = amplification factor, S = power factor).

amplified differential voltage controls a power amplifier L, which delivers the necessary protection current I_S to the impressed current cathode. Occasional control oscillations occur with transducer- or transistor-controlled rectifiers. Slower acting potentiostats with mechanical control devices can be used to avoid this. These are particularly suitable for systems with only slow activation on failure of the protection current.

Figure 8-7 shows a control device that switches the protection current on or off if there is an under- or overstepping of a previously set limiting value and also gives a visible or acoustic signal [13]. The dc input signal U_m is converted to an ac signal in a transistor modulator whose amplitude is proportional to U_m. After amplification, two demodulation circuits are connected to the output. In the first, the signal is conducted via a measuring instrument to the modulator where it acts as negative feedback that raises the input resistance of the modulator. In the second circuit, the signal is led to the controller after demodulation, where it is compared with the set critical limiting potential, U_S. When $U_m < U_S$, an activating signal is riggered if, for example, a limiting value at too-negative a potential has been chosen.

Current-controlling rectifiers are constructed in general on the same circuit principles as potential-controlling rectifiers; only with them, the protection current is converted to a voltage via a constant shunt in the control circuit and fed in as the actual value. With devices with two-point control, the ammeter has limiting value contacts that control the motor-driven controlled transformer.

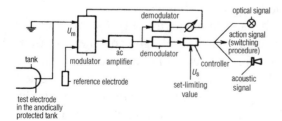

Fig. 8-7 Principle of a transformer-rectifier control device with set-limiting potentials.

8.7 Transformer-Rectifiers without Mains Connections

In inaccessible regions where an impressed current installation is not sufficiently close to a low voltage supply, the protection current can be supplied from batteries, thermogenerators, and if there is sufficient radiation from the sun, solar cells. Wind generators and diesel units, on the other hand, are less suitable because of the maintenance necessary for continuous operation.

The limit for the application of an accumulator battery is a protection current of about 20 mA. This is calculated to be supplied by a 100-A h capacity battery over a period of about 7 months. It is sufficient to change the discharged battery for a fully charged one every 6 months.

In a thermogenerator, heat energy is transformed into electrical energy by thermocouples. A dc voltage is produced by the temperature difference between the junctions of two different metals (thermocouples), its magnitude depending on the temperature difference. Essential parts of a thermogenerator are the combustion chamber, the thermopile with the thermocouples, which consist of semiconductors, as well as the cold junction. The efficiency is about 3%. Propane, butane or natural gas can be used as the fuel. The last can be obtained, for example, from the long-distance gas pipeline that is being cathodically protected. Its operating pressure must first be reduced and controlled at about 2 bars. The controller belonging to the thermogenerator reduces this pressure to 0.5 bar. One generator that was tried out in practice had a maximum electrical capacity of 130 W. The protection installation driven by it supplied a protection current of 9 A at an output voltage of 12 V and has been working for about 15 years without any problems.

Solar cells can also be used to produce current for protection installations of low capacity [14]. Such an installation consists of a panel of solar cells mounted on a mast, an accumulator battery to bridge the periods when the sun is not shining, a charge regulator connected between the solar cells and the battery, as well as an output controller between the battery and the protection current circuit. It can either be current controlled or potential controlled. These construction elements are designed according to the local sun radiation and the electrical supply necessary for the protection installation. Computer programs are available for this purpose. Solar generators are mainly used for the cathodic protection of pipelines in desert regions. An experimental unit is also operating in Germany [15].

8.8 Equipment and Control of Transformer-Rectifiers

Transformer-rectifiers should have an ammeter to indicate the current and a high-resistance voltmeter to indicate the potential at the protection station.

Table 8-1 Measurement protocol for a cathodic protection installation (example)

1. General Data

Installation No. 526	Location: Lohr	Pipeline No. 71 Length 13.5 km
Type of installation:	Impressed current anodes ☒	
	Stray current forced drainage ☐	
	Stray current drainage ☐	Rectifier type MEK 20/10
	Potential bond ☐	Factory No. 1477/326
	Magnesium anodes ☐	
Date: 2.8.76		Meter reading: 8345.3 kWh

2. Buried Reference Electrodes

ΔU buried-in against set electrode (protective current "off"): 50 mV
Polarity of the buried electrode: (+)

3. Pipe/soil Potentials $U_{Cu\text{-}CuSO_4}$ in Volts

External instrument against buried electrode		External instrument against set electrode		Built-in instrument against buried electrode	
U_{on}	U_{off}	U_{on}	U_{off}	U_{on}	U_{off}
−1.53	−1.08	−1.6	−1.13	−1.5	−1.0

4. Protection Current I in Amperes

Total current: External instrument	5.1	Total current: Built-in instrument	5
Partial currents pipeline:	4.1	cable:	0.9

5. Rectifier Output Voltage in Volts: 12.7

Resistances in ohms	Pipe/anode	2.2
	Pipe/rails	–
	Pipe/buried electrode	253
	Pipe/auxiliary earth	52
	Pipe/PEN conductor	–

6. Contact Voltage Protection

Failure voltage (FU)	☒	Overvoltage protection against:	
Failure current (FI)	☐	High-voltage protection device	☒
Zero voltage	☐	Lightning protection device	☐
Terminating at 24 V		Current supply (cathode drain at pylon)	☒
Overvoltage drain:		checked ☒	replaced ☐

Table 8-2 Troubleshooting at cathodic protection stations

Trouble	Cause	Remedy
Installation out of service	Cutout switch or fuse, excess current	Switch on cutout, replace fuse, investigate cause
No grid voltage at rectifier	Contact voltage protection, FU, FI switch off, insulation failure, action of lightning or high voltage	Insulation measurement, install cathode arrester against lightning or high-voltage interference, check resistance at auxiliary ground
Too low protection current or none	Cable or contact break	Measure pipe/anode resistance, locate cable failure, test connecting terminals
	Rise in anode resistance, anode exhausted	Raise rectifier voltage or install extra anodes, test anode connection
	Output fuse failure	Test current limitation, remove overloading or short circuit, set control device
Protection current too high	Anode resistance reduced by ground water or soil moisture	Do not alter rectifier setting as anode resistance will increase in summer
	Contact with unprotected pipelines, bridged isolating joints	Locate contact area, remove cause
Forced drainage	Rail fracture, change in tram current supply	Locate failure, speak to traffic organization
Protection potential not reached	Larger protection current requirement through foreign contact or bridged isolating joints	Remove foreign contact, look for bridged isolating joints, change ground of electrically operated equipment (isolating transformer)
	Increased anode resistance	Increase rectifier voltage, check anode bed
	Measuring cable to pipeline or reference electrode interrupted	Measure resistance of cable and electrode
No potential control	Failure of rectifier control	Test the instrument installation, ac interference
	Control electrode resistance too high	Check connections, measure resistance and potential of electrode, possibly renew

Measuring terminals parallel to the instruments facilitate their control and the connection to voltage and current recorders.

With single-phase transformers, large surges can occur when they are switched on if this is done when the ac voltage supply is in an unfavorable phase. This can trip fuses or automatic cutouts in switching on fixed-value rectifiers of about 600 W capacity. This leads to, for example, unnoticed breakdown of the protection station when the current is switched on again after an interruption. It is recommended that a starting attenuator be installed so the rectifier is connected to the mains through a resistor. This is then bridged after a short time by a contactor. Rectifiers with a control current generally do not experience current surges when switched on.

In rectifiers with fuses as protection against overloading and short circuits, only those of the type recommended by the supplier must be used. This is particularly the case with silicon rectifiers. Potential-controlling rectifiers can be protected against overloading and short circuits by limiting current in the control circuit.

Stray currents can flow from pipe to rails where a rectifier is used for forced stray current drainage when there is a small secondary voltage in the transformer and the negative voltage between rails and pipe is greater than the open circuit voltage of the transformer. This state of affairs can be recognized when the voltmeter of the rectifier reads in the opposite direction, so that a very high current can flow. Overloading of the rectifier can then be prevented by an automatic circuit. An overcurrent relay provides protection that separates the output current circuit pipe/rectifier/rails and in this case produces a direct connection between pipe and rails. The protection is switched on again by an adjustable time switch. By this means the installation is maintained in operation. A counter mechanism records the number of shutdowns. Frequent shutdowns give information about interferences in the operation of the railway.

Switching of protection installations to measure the off potential should always be on the dc side, particularly if there are ac voltage effects between anode and pipeline that are rectified in the rectifier and provide a protection current, or when it is a matter of forced drainage where the drainage current flows from the pipeline via the rectifier to the rails. If the rectifiers control potential, the control circuit must also be switched off; otherwise current surges will occur when it is switched on.

Impressed current stations must be inspected regularly [16,17]. This should take place generally every month because:

- considerable anodic corrosion can occur on failure of the stray current protection system,
- installation can be overloaded by possible disturbances in railway operation,
- changes in the operation of tramways or the section supply can create other stray current behavior.

On inspection, the built-in measurements and the counter readings should be read off and recorded. Any breakdown should be reported to the authorized headquarters of the cathodic protection organization. Table 8-1 illustrates a form for use in the supervision of cathodic protection stations. Causes of trouble and remedial measures are given in Table 8-2.

With impressed current installations supervised by technical telecommunications, it is sufficient to carry out operational control annually [17]. With gas pipeline grids, it is advisable to build in inspection instruments in the control installations or connecting stations since connection with the telecommunication system is possible.

8.9 References

[1] DVGW-Arbeitsblatt GW 12, ZfGW-Verlag, Frankfurt 1984.
[2] AfK-Empfehlung Nr. 5, ZfGW-Verlag, Frankfurt 1986.
[3] DIN VDE 0100 Teil 410, Beuth-Verlag, Berlin 1983.
[4] AfK-Empfehlung Nr. 6, ZfGW-Verlag, Frankfurt 1985.
[5] VBG 4, Elektrische Anlagen und Betriebsmittel, Heymanns-Verlag, Köln 1979.
[6] DIN VDE 0664 Teil 1, Beuth-Verlag, Berlin 1983.
[7] V. Ashworth u. C. J. L. Booker, Cathodic Protection, Theory and Practice, E. Horwood Ltd., Chichester/GB 1986, S. 116, 265.
[8] DIN 50918, Beuth-Verlag, Berlin 1978.
[9] H. Kampermann, ETZ-A *96*, 340 (1975).
[10] W. v. Baeckmann, N. Wilhelm u. W. Prinz, gwf gas/erdgas *107*, 1213 (1966).
[11] AfK-Empfehlung Nr. 8, ZfGW-Verlag, Frankfurt 1983.
[12] H. Gräfen u. a., Die Praxis des Korrosionsschutzes, expert-Verlag, Grafenau 1981, S. 304.
[13] J. Prušek, K. Mojžiš u. M. Macháček, Werkstoffe u. Korrosion *20*, 27 (1969).
[14] G. Gouriou, 3R internat. *26*, 315 (1987).
[15] R. Wielpütz. Neue Deliwa-Z. *38*, 16 (1987).
[16] DIN 30676, Beuth Verlag, Berlin 1985.
[17] DVGW-Arbeitsblatt GW 10, ZfGW-Verlag, Frankfurt 1984.

9

Impressed Current Ground Beds and Interference Problems

W. VON BAECKMANN AND W. PRINZ

Cathodic protection of long pipelines, distribution networks, pipelines in industrial plants and other buried installations with a high protection current requirement is achieved mostly by means of impressed current anodes. With a high protection current requirement, the grounding resistance of the anodes, as the highest resistance in the protection current circuit, determines the necessary rectifier voltage and also the power of the protection station. The lowest grounding resistance practically possible should be designed for in order to keep down the electric power and therefore the operating costs (see Section 9.1.4). According to Table 24-1, the grounding resistance, R, is directly proportional to the specific soil resistivity, ρ. Anode beds with protection current output >1 A are installed wherever possible in areas of low specific soil resistivity [1]. Today the anode bed is generally arranged in continuous coke backfill in horizontal or vertical positions [2].

The protection current is transmitted to the soil via the impressed current anode bed. Therefore the current density and the field strength, the voltage drop per meter, are greatest in the vicinity of the anode bed; they decrease with increasing distance from it. At a distance from the anode bed where no appreciable field strength due to the protection current is detectable, the soil potential $\phi_{E\infty} = 0$ (see Fig. 3-24). This potential is termed that of the remote ground. The voltage between the remote ground and the anodes is the anode voltage. Because of the cone-shaped $\Delta U(x)$ curve of the voltage distribution on the soil surface (see Fig. 9-1), this is called the voltage cone of the anode bed. The height of the voltage cone depends on the anode voltage and its shape depends on the arrangement of the anodes. This is important for possible effects on buried foreign installations.

Voltage cones also occur where the protection current enters through defects in the pipe coating (see Sections 3.6.2 and 24.3.4). Figure 9-1 shows schematically the variation of the voltage cone of an anode bed and a cathodically protected pipeline that results from the raising and lowering of potential.

244 Handbook of Cathodic Corrosion Protection

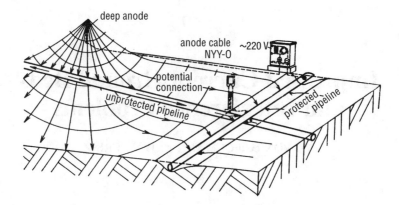

Fig. 9-1 Anodic and cathodic voltage cones. (See the list of codes in the front of the book for NYY-O cable.)

9.1 Impressed Current Ground Beds

9.1.1 Continuous Horizontal Anode Beds

Where there is available ground and the specific resistivity of soil in the upper layers is low, the anodes are laid horizontally [3]. A trench 0.3 to 0.5 m wide and 1.5 to 1.8 m deep is dug with, for example, an excavator or trench digger (see Fig. 9-2). A layer of coke 0.2-m thick is laid on the bottom of the trench. The impressed current anodes are placed on this and covered with a 0.2-m layer of coke. Finally the trench is filled with the excavated soil. No. IV coke with a particle size of 5 to 15 mm and specific gravity of 0.6 t m^{-3} is backfilled at a rate of 50 kg per meter of trench. The anodes are connected in parallel and every three to four anode cables are connected to the anode header cable by a mechanical cable crimp encapsulated in an epoxy splice kit to give an economical service life at high current output.

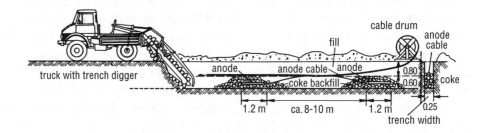

Fig. 9-2 Burying anodes with a mechanical digger.

Impressed Current Ground Beds and Interference Problems

The header cable between anode bed and rectifier must be particularly well insulated. For this reason cables with double plastic sheathing of type NYY-O are used. The cable sheath must not be damaged during installation because the copper core at the defects will be anodically attacked in a very short time and the connection to the rectifier broken. Damage to the cable sheath is not so serious if a multicored cable is used. Usually not all the core insulation is damaged so that the operation of the anode bed is not interrupted. In addition, measurement of resistance and detection of defects is easier.

The grounding resistances of continuous anode beds with a diameter of 0.3 m and a covering of 1 m are given in Fig. 9-3. All the values are calculated for a specific soil resistivity of $\rho_0 = 10\ \Omega$ m. To determine the grounding resistance for a particular ρ, the value derived from the curve should be multiplied by ρ/ρ_0. With horizontal individual anodes in a continuous coke bed, almost the same favorable grounding resistance can be achieved as with long continuous anodes. The effective grounding resistance R_W according to Eq. (24-63) and the data in Section 24.4.3 is given by

$$R_w = Z \coth(L/l_k) \qquad (9\text{-}1)$$

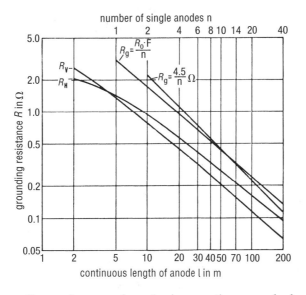

Fig. 9-3 Grounding resistance of anodes in a continuous coke bed with a covering of earth $t = 1$ m and a diameter $d = 0.3$ m for a specific soil resistivity of $\rho = 10\ \Omega$ m. Horizontal anodes: R_H from Eq. (24-23), see line 9 in Table 24-1; vertical anodes: R_v from Eq. (24-29), see line 7 in Table 24-1; anode bed as in Fig. 9-7 with n vertical anodes with $l = 1.2$ m ($R_0 = 3.0\ \Omega$ m) at a spacing of $s = 5$ m: R_g from Eqs. (24-31) and (24-35) with an average value of $F = 1.5$.

where Z is from Eq. (24-60) and the characteristic length l_k from Eq. (24-58)

$$l_k = \frac{1}{\alpha} = \sqrt{\frac{1}{R'\,G'}} \qquad (9\text{-}2)$$

The effective grounding resistance from Eq. (9-1) does not decrease further if the length, L, exceeds the characteristic length, l_k. R' and G' are given by Eqs. (24-56) and (24-57). The relevant data for cylindrical horizontal electrodes of radius r and length L are: $\varkappa = 1/\rho_C$ (ρ_C = specific resistivity of the coke backfill); $S = \pi r^2$; $l = 2\pi r$; $r_p = (2\pi r L)R$. R is the grounding resistance from Table 24-1 and is proportional to the specific soil resistivity ρ_E. Putting these values in Eq. (9-2) and using Eqs. (24-56) and (24-57) gives:

$$l_k = C\sqrt{\frac{\rho_E}{\rho_K}} \quad \text{where} \quad C = r\sqrt{\pi L \left(\frac{R}{\rho_E}\right)} \qquad (9\text{-}3)$$

The constant, C, depends on the dimensions of the anode bed. It follows from line 4 of Table 24-1 for a horizontal ground in the half space

$$C = r\sqrt{\ln \frac{L}{r}} \qquad (9\text{-}4a)$$

and for a horizontal ground buried at a depth, t, from line 9 of Table 24-1:

$$C = r\sqrt{\frac{1}{2}\ln\left(\frac{L^2}{2rt}\right)} \qquad (9\text{-}4b)$$

According to Eq. (9-3), the characteristic length l_k increases with rising specific soil resistivity. Figure 9-4 represents this relationship. Therefore, the characteristic length is designated as an effective increase in length of an anode in the coke bed. In the case of an infinitely long coke bed, the current in the anode, from Eq. (24-65), and therefore the current density at the point $x = l_k$, is only the e'th part of the initial value at the point $x = 0$.

For installations with continuous coke backfill, the anodes can be installed at double the spacing of the anode bed extension. The lower the ratio ρ_C/ρ_E (i.e., the higher the specific soil resistivity), the further apart the anodes can be placed.

Impressed Current Ground Beds and Interference Problems 247

The equipotential lines of the voltage cone of continuous anode beds are represented first as ellipses on the soil surface, and with increasing distance, as circles. Therefore, the voltage cone decreases with r^{-1} in the direction of the anode axis, and with $\ln(r^{-1})$ in a direction at right angles to the anode axis (see Table 24-1). Figure 9-5 shows the voltage ratio U_Z/U_A for different length horizontal anodes. U_Z is the potential of a point on the soil surface, taken as the remote ground; this is also the voltage of the pipeline reduced by its polarization voltage η; z is the perpendicular distance to the middle of the anode bed. U_A is the applied anode voltage, which corresponds to the output voltage of the rectifier U_{rect}, reduced by the on potential of the protected object. Figure 9-6 shows the corresponding voltage ratio U_x/U_A in the direction of the anode axis; x is the perpendicular distance to the end of the anodes. Horizontal impressed current anode beds of 100 m in length and more usually behave as extended voltage cones. Figures 9-5 and 9-6 show that at a distance of ca. 100 m from the anode beds, 7 to 10% of the anode voltage still exists as measured against the remote ground. This voltage only falls

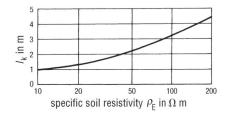

Fig. 9-4 Effective anode lengthening l_k by coke backfill with Eq. (9-3) with $\rho_C = 1\ \Omega$ m and $C = 0.31$.

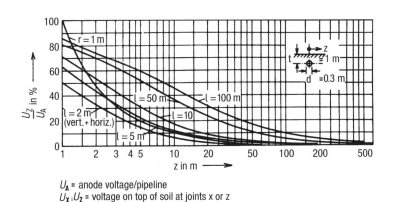

U_A = anode voltage/pipeline
$U_x ; U_z$ = voltage on top of soil at joints x or z

Fig. 9-5 Voltage cone of horizontal anodes with the z axis at right angle to the anode; U_A = voltage anode/pipeline; U_Z = voltage of ground at Z against remote ground.

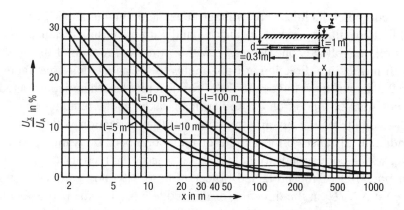

Fig. 9-6 Voltage cone of a horizontal anode in the direction of the anode axis, x; U_A = voltage anode/pipeline; U_X = voltage of ground at X against remote ground.

to 1% at a distance of 1 km. The size of the voltage cone is of great importance in its interference with foreign lines.

9.1.2 Single Anode Installations

Installations of single anodes today are generally used only for small protection currents or where a continuous anode bed is not possible, e.g., in woods or swampy ground. Vertical anodes generally 1.2 m in length are centrally mounted in boreholes 2 m deep and 0.3 m in diameter. The holes for vertical anodes are either dug by hand or by motor-driven drilling machines. The bottom of the hole is covered with a 0.2-m thick layer of coke on which the impressed current anode is placed. The surrounding space is filled with ca. 75 kg of No. IV coke to a depth of 0.2 m above the anode (see Fig. 9-7). Single anodes can also be installed in horizontally excavated pits 0.5 × 1.5 × 2 m in dimension. About 200 kg of coke are required as backfill for horizontal single anodes.

The grounding resistance, R_g, of an anode bed of n single anodes with a spacing, s, is only a little greater than that of a continuous anode bed of length, $l = sn$. Since the single anodes with finite spacing, s, have mutual interference up to about 10 m, the total grounding resistance, R_g, of an anode installation of n single anodes is, however, noticeably greater than the resistance of anodes connected in parallel with a spacing $s \to \infty$. The influence factor, F, which the grounding resistance magnifies, is shown in Fig. 9-8 as a function of the spacing, s, of single vertical anodes [see Eq. (24-35)]. The dimensions of the single anode beds are $l = 1.5$ m and $d = 0.3$ m. The interference factor also applies with good accuracy to short horizontal anodes with a covering of 1 m. For an anode installation of n single

Impressed Current Ground Beds and Interference Problems

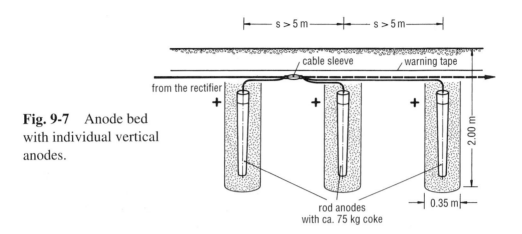

Fig. 9-7 Anode bed with individual vertical anodes.

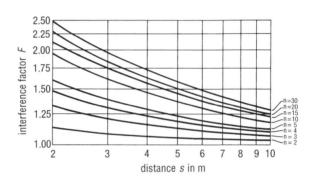

Fig. 9-8 Interference factor, F, for an anode bed with n individual anodes (for $\rho/R = \pi \text{ m} \approx 10\ \Omega\ \text{m}/3.2\ \Omega$).

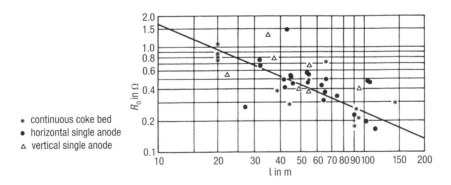

* continuous coke bed
• horizontal single anode
△ vertical single anode

Fig. 9-9 Measured grounding resistances of anode installations [the straight line corresponds to Eq. (9-5) with R_g from Fig. (9-3)].

anodes with grounding resistance, R_0, for a specific soil resistivity $\rho_0 = 10\ \Omega$ m, the total grounding resistance is given by

$$R_g = F \frac{R_0}{n} \frac{\rho}{\rho_0} \tag{9-5}$$

Figure 9-9 represents measured and calculated grounding resistances for $\rho_0 = 10\ \Omega$ m of anode beds. The relatively large deviations of the values from the calculated straight line occur because the specific soil resistivity in the region of the anode installation is locally not constant and cannot be measured with sufficient accuracy by the Wenner method at that depth (see Section 3.5.2). The variation of the voltage cone of an anode bed with single anodes arranged in a straight line corresponds sufficiently accurately to Figs. 9-5 and 9-6.

9.1.3 Deep Anode Beds

Deep anodes are installed where the resistivity is high in the upper layers of soil and decreases with increasing depth. This type of installation is recommended for densely populated areas and for local cathodic protection (see Chapter 12) on account of the small space needed and the smaller voltage cone, which avoids interference with foreign structures.

Deep anodes consist of parallel-connected single anodes which are set in boreholes 50 to 100 m deep with a diameter of 0.3 m. The boreholes can be produced by a variety of boring methods, but the "air lift" method has proved particularly suitable (see Fig. 9-10). The borehole is filled with water to ground level according to the principle of a mammoth pump. Compressed air is fed via a pipe to

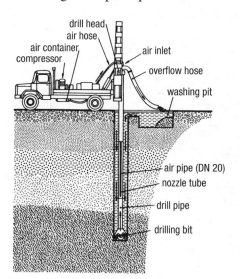

Fig. 9-10 Drilling for a deep anode bed with the air lift method.

Impressed Current Ground Beds and Interference Problems 251

the nozzle at the lower end of the standpipe so that it fills the whole circumference of the standpipe. As the water-material mixture in the bore is displaced by the air, it rises upward due to the lower specific gravity and is led to the washing pit via the washing head and hose. The bore material is deposited here. The washing water flows back to the borehole. In the air lift process, enlarging of the borehole is avoided, which is possible with other processes. In stable soils, the borehole is piped to a depth of 10 m to avoid damage by washing water flowing in. If the borehole is piped due to, for example, quicksand, and if the piping has to remain in the ground, this is then connected in parallel to the impressed current anodes. As the borehole gets deeper, it is recommended that the grounding resistance of the bore linkage be measured in the borehole (see Section 3.5.3). It can be determined by this whether the calculated low grounding resistance has been reached and whether it is worthwhile boring to a greater depth.

Centering equipment is used to ensure that the impressed current anode is centrally situated in the borehole. The anode with the centering device can be inserted in the borehole by use of, for example, plastic-insulated wire ropes (see Fig. 9-11). After each of the anodes is inserted, the free space is filled with No. IV coke up to the level of the next anode; about 50 kg of coke are necessary per meter of anode bed. The wire rope is fixed to a support above the borehole and provides off-loading to the anode cable. The anode cables are laid to a junction box so that the

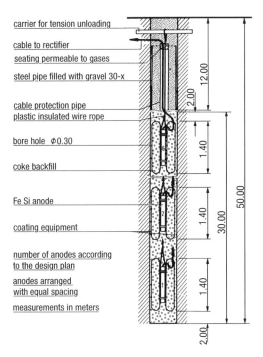

Fig. 9-11 Construction of a deep anode bed.

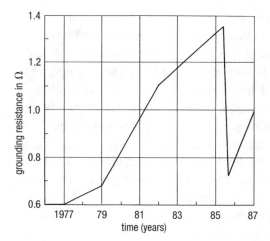

Fig. 9-12 Increase in grounding resistance of a deep anode due to formation of a gas pocket.

individual anode currents can be measured. An overall anode cable leads from here to the rectifier. With deep embedded anodes in calcined petroleum coke, it is strongly recommended that a perforated plastic pipe be included and the borehole filled above the anode bed with gravel so that the oxygen anodically evolved according to Eq. (7-1) (1.83 m^3 A^{-1} a^{-1} in Table 2-3) can be evacuated and does not lead to an increase in the grounding resistance [5]. Figure 9-12 shows the increase in grounding resistance that results from this process in a 100-m deep anode bed. When the voltage is switched off, the resistance regains its original value after a few days, but then starts to climb again with renewed current loading.

With impressed current anodes, an increase in resistance cannot only be caused by a gas cushion or electro-osmotic dehydration, but also, in unfavorable circumstances (no ground water, high soil resistivity and high anode voltages), by a rise in temperature of the anode bed caused by resistance heating. It has been observed in some cases that the coke bed can be set on fire by this, or in anodes of high-voltage dc transmission installations, decomposition of water to oxyhydrogen gas results.

A significant rise in temperature ΔT is calculated in a volume of soil at a distance up to 3 r_0 from the anode, where r_0 is the anode radius. Factors are the thermal conductivity of the soil, k, length of the anode, L, the grounding resistance, R_0, and the current, I. The rise in temperature can be calculated from these parameters [10]. For deep anodes it amounts to:

$$\Delta T = \frac{I^2 R_0}{4\pi \, k(L + r_0)} \tag{9-6}$$

and for anodes near the surface:

Impressed Current Ground Beds and Interference Problems

$$\Delta T = \frac{I^2 R_0}{2\pi k(L + r_0)} \tag{9-7}$$

It follows approximately from Eqs. (9-6) and (24-24):

$$U^2 = \Delta T \, 2 \, k\rho \ln(L/r_0) \tag{9-8}$$

The permissible anode voltage can be calculated from Eq. (9-8). For unfavorable conditions, with $\Delta T = 20$ K, for dry soil with $k = 0.2$ W m^{-1}, $\rho = 50$ Ω m and $L/r_0 = 500$, Eq. (9-8) gives $U = 50$ V.

The grounding resistance of deep anodes can be read off from Fig. 9-3 in which the length of the coke bed is to be inserted. The covering has practically no effect on the grounding resistance. From Eq. (24-29) the grounding resistance of a 30-m-long anode in homogeneous soil covered with 10 m of soil is reduced only 10% compared with a covering of 1 m. The interference factor for deep anodes can be calculated from Eq. (24-37). The value is smaller than the value in Fig. 9-8. Therefore it is safer to use the values in Fig. 9-8 in practice.

In contrast to the grounding resistance, a reduction of the voltage cone on the soil surface can be achieved in an area up to 20 m from the anode axis by increasing the covering, which is important where foreign installations are concerned. Figure 9-13 represents the voltage cone for 30-m-long deep anodes for different levels of covering.

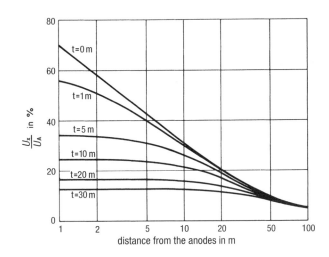

Fig. 9-13 Voltage cone of 30-m-long deep anodes with various depths of earth coverings t.

9.1.4 Design of Anodes

The major part of the output power of a cathodic protection rectifier is required to conduct the protection current from the anode bed into the soil. If only a few single anodes or a short continuous horizontal anode is installed, the installation costs are low but the annual cost of current is very high. If, on the other hand, very many anodes are installed either singly or in a long continuous coke bed, the annual cost of current is low but the installation costs correspondingly high. The total annual costs must be determined to give an optimal design of an anode installation for a given protection current and soil resistivity [1].

The installation costs for a single impressed current anode of high-silicon iron can be taken as K_A = DM 975 (\$550).[1] This involves about 5 m of cable trench between anodes so that the costs for horizontal or vertical anodes or for anodes in a common continuous coke bed are almost the same. To calculate the total costs, the annuity factor for a trouble-free service life of 20 years (a = 0.11, given in Fig. 22-2) should be used. For the cost of current, an industrial power tariff of 0.188 DM/kWh should be assumed for t = 8750 hours of use per year, and for the rectifier an efficiency of w = 0.5. The annual basic charge of about DM 152 for 0.5 kW gives about 0.0174 DM/kWh for the calculated hours of use, so that the total current cost comes to

$$k = 0.21 \text{ DM/kWh} = 2.1 \times 10^{-4} \text{ DM V}^{-1} \text{ A}^{-1} \text{ h}^{-1}$$

The wattage $R_g I_s^2$ is directly proportional to the grounding resistance of the whole anode bed R_g and therefore to the specific soil resistivity. Equation (9-5) gives the grounding resistance of the anode installation which either consists of n horizontal or vertical single anodes or of anodes with a horizontal continuous coke bed of total length $l \approx ns$. The total cost function is given by [1]:

$$K = aK_A n + F \frac{R_0}{n} \frac{\rho}{\rho_0} I_s^2 \frac{kt}{w} \qquad (9\text{-}9)$$

Differentiating with respect to n gives the desired minimum cost for the economical number of anodes, n_w:

$$\frac{n_w}{I_s} = \sqrt{\frac{FR_0 kt}{aK_A \rho_0 w}} \qquad (9\text{-}10)$$

With average values of F = 1.45 and R_0 = 3.1 Ω as well as the known data for k, t, a, K_A, ρ_0 and w, it follows from Eq. (9-10)

---

[1] These amounts are in deutsche mark values at the time the original text was written (the late 1980s) and therefore should be used only as relative numbers.

$$n_w = 0.12 \left(\frac{I_s}{A}\right) \sqrt{\frac{\rho}{\Omega\,m}} \qquad (9\text{-}10')$$

The economical number of anodes n_w is plotted in Fig. 9-14 according to Eq. (9-10′). It must be emphasized that areas with low soil resistivity should be chosen for siting the anode beds.

Losses in the connecting cables and in the rectifier are no longer negligible with long anode cables and particularly with high protection currents [7]. The rectifier losses can usually be kept low by a slight overdesign that gives the most favorable rectifier efficiency of between 70 and 80%. The lowest losses in the

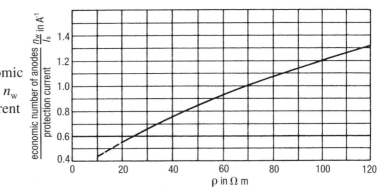

Fig. 9-14 Economic number of anodes n_w for impressed current anode beds.

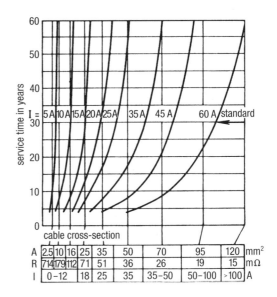

Fig. 9-15 Economic cable cross-sections for anode cable.

anode cable can be calculated from considerations of economy similar to those for impressed current anodes [8]. The design of the cables from such considerations depends essentially on the lowest thermally permissible cross-section. It is shown in Fig. 9-15 for various operating lives of a cathodic protection station.

The current requirement of the protected object basically determines the design of the anode bed. For example, for a pipeline requiring 10 A with horizontal anodes laid in soil with $\rho = 45\ \Omega$ m, according to Fig. 9-14, eight anodes are necessary. The grounding resistance of one anode amounts to $R_0 = 14\ \Omega$. From Fig. 9-8, the grounding resistance of the anode bed with an interference factor $F = 1.34$ for 8 anodes spaced at 5 m comes to $R_g = 2.34\ \Omega$.

9.2 Interference with Foreign Pipelines and Cables

Under normal operating conditions, cathodic corrosion protection stations lead to the passage of a direct current through soil, which is introduced at the anodes and conducted to the protected object. According to DIN VDE 0150, there are direct current installations from which stray currents emerge which can bring about corrosion damage on foreign buried metallic installations, e.g., pipelines and cables [9]. The protection current causes an anodic voltage cone in the area of the anodes. This results in the potential of the soil being raised above the potential of the remote ground. The protection current produces cathodic voltage cones at defects in the pipe coating. Here the soil potential is lowered with respect to the potential of the remote ground. Buried foreign metal installations take up current in the area of the anodes and give it up at a distance from the anodes, i.e., they are cathodically polarized in the neighborhood of the anodes and anodically polarized at a distance from the anodes. Anodic corrosion occurs where the stray current emerges.

Such interference can be measured by switching the protection stations on and off and observing the changes in potential of the foreign installation and the voltage cone in the soil (see Section 3.6.2.1). Anodic damage to neighboring installations is to be reckoned with in such areas if the voltage between the influenced installation and a reference electrode placed directly above it on the soil surface changes with the flow of protection current by an average of more than 0.1 V in a positive direction [9], particularly

- with installations that are not cathodically protected against the potential when the protection current is switched off.
- with installations cathodically protected against the protection potential, U_s.

The measured potential contains, besides the true object/soil potential, an ohmic voltage drop that is proportional to the specific soil resistivity and the current density.

Impressed Current Ground Beds and Interference Problems

The *IR*-free anodic potential change actually caused by the emergence of the stray current is in practice not usually measurable if external measuring probes are not used (see Section 3.3.3.2). Furthermore, there is also no direct information on the danger of corrosion, because the shape of the anodic partial current-potential curve is unknown. According to Fig. 2-9, an increase in the *IR*-free potential of 100 mV or 20 mV leads to an increase in corrosion rate by a factor of 30 or 2, respectively, if there are no surface films. In the presence of surface films, the increase is considerably lower. This effect and the ever-present *IR* portion explain why the 0.1-V criterion derived from experience is sufficient for soils that do not have particularly low resistance. In England and Switzerland, on the other hand, only an almost *IR*-free potential change at the phase boundary of 20 mV is allowed [10,11]; in this case, however, the accuracy of the measurement is unclear. Some other European countries allow a potential change with an *IR* contribution of 50 mV [12].

9.2.1 Interference from the Voltage Cone of Anodes

The greatest effect on the potential of foreign pipelines and cables is usually the result of the voltage cone of impressed current anodes, where a high current density and a large potential drop are present in the soil. Since there is only a negative shift in potential near the anodes, there is no danger of anodic corrosion. With corrosion systems of group II in Section 2.4 (e.g., aluminum and lead in soil), of course cathodic corrosion can occur. The magnitude of the current uptake depends on the

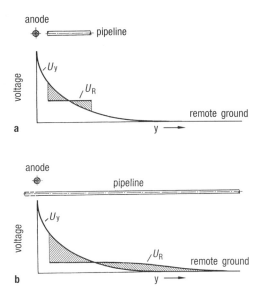

Fig. 9-16 Interference of a short (a) and a long (b) pipeline by the anode voltage cone; ///// region of cathodic polarization; \\\\\ region of anodic polarization.

interfering voltage (i.e., on the potential of the voltage cone at the particular place versus the potential of the remote ground) and on the coating resistance of the influenced installation. Basically a distinction is made between short and long objects in interference from the anodic voltage cone, as is apparent in Fig. 9-16 [13].

The rest potential of a particular pipeline is U_R and only exists with the cathodic protection station switched off in the area of the remote ground. When the anode installation is in operation in soil, different potentials exist according to the voltage cones, whose values U_y are referred to the remote ground. The object/soil potential, ignoring the ohmic potential drop in the object, is directly dependent on U_y (see Fig. 3-24). The values of U_y and U_R are shown in Fig. 9-16 as functions of the position coordinate, y. To give a clearer picture, it is assumed that the pipe is not polarized. The difference between U_y and U_R indicates the voltage that is available for polarization of the pipe, including the ohmic voltage drop. So in the region $U_y > U_R$ the pipeline is cathodically polarized and in the region $U_y < U_R$ anodically polarized. The resultant change in the pipe/soil potential is not represented.

Figure 9-16a shows the situation for a short object (pipeline or storage tank). The anodic interference is relatively large in this case. Such interference occurs at insulating joints in pipelines, where usually no shorting is necessary at low voltages. With voltages about 1 V and with a positive potential shift of >0.1 V, the insulating unit has to be shorted with a balance resistance. With water pipelines, the inclusion of an additional internal cathodic protection installation or the installation of a length of pipe with an insulating internal coating may be necessary (see Section 10.3.5).

Figure 9-16b shows the situation for a very long pipeline that reaches to the area of the remote ground. Here the voltage drops in the protected object have to be taken into account, i.e., U_R decreases with increasing values of y. It should be recognized that the interference in the anodic region is considerably less than in the case of the short pipeline in Fig. 9-16a. Figure 9-17 shows the measured pipe/soil potentials of a long and a short pipeline subject to interference.

Experience shows that in interference by the anodic voltage cone, no unacceptable anodic interference is caused by an interference voltage $U_y < 0.5$ V versus the remote ground. The distance from the impressed current anode at which $U_y = 0.5$ V depends on the anode voltage, U_A, and the anode length, L [13]. It amounts to:

$$\left(\frac{r_A}{m}\right) = \left(\frac{L}{m}\right)^{0.65} \left(\frac{U_A}{V}\right) \tag{9-11}$$

No damaging interference to foreign objects has to be considered in cathodic protection with galvanic anodes because of the small current densities in soil and the lower anode voltages.

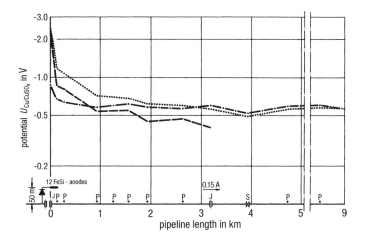

Fig. 9-17 Interference of a long (by shorting insulating joints) and a short pipeline by an anode voltage cone 1 km in length; P = potential measuring point; I = insulating joint; ······ "on" protection current, long pipeline; ------ "on" protection current, short pipeline; –·–·–. "off" protection current, both pipelines.

9.2.2 Interference from the Cathodic Voltage Cone of the Protected Object

The current entering at defects in the coating of cathodically protected pipelines causes a cathodic voltage cone in the soil (see Section 3.6.2). With pipelines that have coatings of high mechanical strength (e.g., PE coating), usually only a few widely spaced defects occur. In the vicinity of these defects the shape of the potential is similar to a grounded circular disc; at a greater distance it can be approximated from the potential of a buried spherical ground [see case (a) in Section 3.6.2.2]. The size of small defects can be assessed by measuring the voltage cone ΔU_x and the difference between the on and off potential with the help of Eq. (3-51a). If the pipe coating has very many defects close together, the individual voltage cones coalesce into a cylindrical voltage field around the pipeline [14] [see case (b) in Section 3.6.2.2]. A potential distribution according to Eq. (3-52) is likely, especially with old pipelines with coatings of jute- or wool felt-bitumen for average protection current densities of a few mA m^{-2}. The high protection current requirement of older pipelines is often caused by uncoated fittings, badly insulated welds and conducting contacts with foreign pipelines or uncoated casing tubes. Since protection current densities of up to 500 mA m^{-2} are necessary to cathodically protect bare iron surfaces in soil, particularly with good aeration, voltage cones of a few hundred millivolts occur.

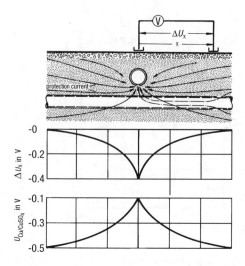

Fig 9-18 Current distribution and voltage cone ΔU_x at a defect in the pipe coating of a cathodically protected pipeline and the variation in the pipe/soil potential of a pipeline subjected to interference.

Buried pipelines that cross cathodically protected pipelines in the voltage cone region pick up protection current outside of the voltage cone; this leaves the pipeline in the area of the cathodic voltage and causes anodic corrosion at the unprotected pipeline. The potential of the affected pipeline measured with a reference electrode above the crossing point is essentially the ohmic voltage drop caused by the protection current flowing through the soil to the defect in the cathodically protected pipeline. Figure 9-18 shows schematically the potential distribution in the soil, the shape of the voltage cone and the potential distribution on the affected pipeline.

Since interference of foreign pipelines results from the voltage cone of the cathodically protected pipeline, this can be assessed by measuring the voltage drop at the soil surface (see also Section 24.3.4). It is therefore not necessary to set up potential measuring points at every crossing point to determine the interference of foreign pipelines by cathodically protected installations. Thus ΔU_x is measured as $U_B^\perp$ according to Fig. 3-24 or according to Fig. 9-18 with $x = 10$ m. The voltage drop ΔU_x of a cylindrical field at the soil surface is plotted in Fig. 9-19 according to Eq. (3-52) as a function of the nominal width for a covering of 1 m at a protection current density of 100 μA m^{-2} for a high specific soil resistivity of $\rho = 100$ Ω m. From this it can be seen that for pipelines with protection current densities below 100 μA m^{-2}, the voltage drop in the soil remains below 10 mV so that no deleterious effects are produced on foreign pipelines. If the average protection current densities are caused by poorly coated or uncoated fittings that require protection current densities of several hundred μA m^{-2}, then very large voltage cones will arise.

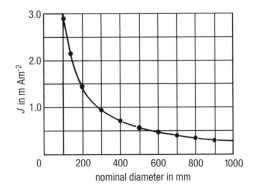

Fig 9-19 Protection current density as a function of the nominal diameter causing a voltage drop of $\Delta U_x = 100$ mV ($x = 10$ m) for $\rho = 100$ Ω m.

9.2.3 Avoidance of Interference

Interference with foreign structures due to anodic and cathodic voltage cones can in every case be prevented by a continuity bond with the cathodically protected installation. Figure 9-1 shows a potential connection at the crossing point of a cathodically protected and an unprotected pipeline. The current entering the unprotected pipeline from the voltage cone of the anode bed no longer flows through the soil as a corrosion current at the crossing point, but via the continuity bond to the cathodically protected pipeline. By this means an unallowable positive potential shift at the crossing point is converted to a reduction in potential. Usually a balancing resistance of about 0.2 to 2 Ω is connected into the potential connection in order to limit the balance current and the potential reduction in the affected pipeline. The balancing resistance is connected so that on switching on a small potential shift in a negative direction, only a few millivolts are produced. The cathodic protection can be impaired by a continuity bond if too much protection current has to be taken up by the affected installation. In addition, checking the cathodic protection by measurement of *IR*-free potentials by the switching technique can be faulty due to the considerable equalizing currents. So a connection with the affected pipeline should only be resorted to if there is no other possible economical means available to reduce the interference to a tolerable level. Figures 9-5, 9-6, and 9-13 show that interference from the anodic voltage cone can be avoided by remaining a sufficiently large distance from foreign installations or low anode voltages. The site of the anode bed is therefore not only chosen for a convenient current supply and low specific soil resistivity, but also with consideration of the distance from foreign pipelines. Low anode voltages can be achieved by having several protection stations with lower current output, by lengthening the anode bed to reduce the grounding resistance and depressing the required anode voltage, or by deep anodes. Deep anodes with a covering of 20 m as in Fig. 9-13 are therefore particularly suitable for the cathodic protection of pipelines in urban areas, thus essentially reducing the distance from foreign installations.

The uptake of current by foreign installations should be avoided as far as possible in the region of anodic voltage cones. Foreign pipelines in an anodic voltage cone should have a coating of high resistance, no uncovered fittings and no electrical contacts to steel-reinforced concrete pits, foundations, or electrically grounded structures. Foreign pipelines that are laid near existing anode beds must have a coating of the best possible insulation, e.g., polyethylene, to keep the uptake of current as small as possible. Figure 9-20 shows the pipe/soil potentials of a pipeline that is situated 5 m parallel to an anode installation and that is provided with a particularly good coating of polyethylene in the region of the anodic

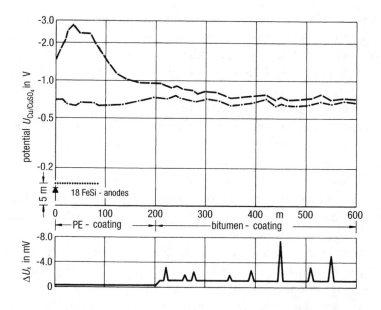

Fig. 9-20 Pipe/soil potential of a pipeline with defect-free PE coating (up to 200 m) and with bitumen coating (200 to 600 m) in the region of an anodic voltage cone on switching the protection current on (------) and off (–·–·–); the ΔU_x values show that no current enters in the region of the voltage cone.

potential cone. By this means interference can be avoided. With cables with plastic insulation laid near anode beds, there also is no danger of interference.

The exit current density, J_a, at coating defects of the affected pipeline can be calculated from Eqs. (24-51) or (24-51'), where r_1 is the radius of a defect causing the cathodic voltage cone. Since the quantities in these equations are generally unknown, no quantitative assessment of possible corrosion danger can be made, but only a qualitative assessment of tendencies, e.g., measures to reduce interference. The most important measure is to remove the cause of the interference by improving the coating of the affected pipeline in the crossing region. This measure is, however, economical only if the voltage cone is the result of a single defect and not of very many defects (cylindrical field).

In addition to a continuity bond, the following further measures can be taken to reduce the interference:

- apparent increase of the defect area by connection of galvanic anodes; in addition, the anodes will reduce the effect of interference with the aid of their own protection current output,

- increasing the distance between both defects, e.g., by using a plastic casing,

- electrical separation at the crossing point or of the parallel course by insulating joints.

With increasing improvement in pipe coatings, the interference problem is becoming less severe. Interference by foreign objects is excluded in pipelines with polyethylene coating. Interference by foreign objects can only occur with badly coated fittings and uncoated welds. However, with old pipelines with average protection current densities of 0.5 to 1 mA m^{-2}, interference can occur only as the result of the cathodic potential cone of large defects, or electrical contacts with uncoated casing pipes or insufficiently coated fittings and uncoated girth welds. In these cases it is necessary for the safety of the cathodic protection to remove electrical contacts or to repair defects in the coating.

9.3 References

[1] W. v. Baeckmann, gwf/gas *99*, 153 (1958).
[2] F. Wolf, gwf/gas *103*, 2 (1962).
[3] J. Backes u. A. Baltes, gwf/gas *117*, 153 (1976).
[4] W. Prinz, 3R internat. *17*, 466 (1978).
[5] W. T. Brian, Mat. Perf. *9*, H. 4, 25 (1970).
[6] W. H. Burkhardt, Corrosion *36*, 161 (1980).
[7] R. G. Fischer u. M. A. Riordam, Corrosion *13*, 519 (1956).
[8] R. Reuter u. G. Schürmann, gwf/gas *97*, 637 (1956).
[9] DIN VDE 0150, Beuth-Verlag, Berlin 1983.
[10] J. H. Gosden, 19. Tagung der CIGRE in Paris, Archiv für Energiewirtschaft 1962, Bericht 207.
[11] Schweizerische Korrosionskommission, Richtlinien für Projekte, Ausführung und Betrieb des kathodischen Korrosionsschutzes von Rohrleitungen 1987.
[12] Technisches Komitee für Fragen der Streustrombeeinflussung (TKS): Technische Empfehlung Nr. 3, VEÖ-Verlag, Wien 1975.
[13] AfK-Empfehlung Nr. 2, ZfGW-Verlag, Frankfurt 1985.
[14] W. v. Baeckmann, Techn. Überwachung *6*, 170 (1965).

10

Pipelines

W. PRINZ

Buried steel pipelines for the transport of gases (at pressures >4 bars) and of crude oil, brine and chemical products must be cathodically protected against corrosion according to technical regulations [1-4]. The cathodic protection process is also used to improve the operational safety and economics of gas distribution networks and in long-distance steel pipelines for water and heat distribution. Special measures are necessary in the region of insulated connections in pipelines that transport electrolytically conducting media.

10.1 Electrical Properties of Steel Pipelines

The electrical characteristic of a buried pipeline corresponds to that of an extended ground with a longitudinal resistance (see Section 24.4.2). The longitudinal resistance, R_l, related to the length l is described by Eq. (24-70) and is termed the resistance per unit length (resistance load), R'.

$$R' = \frac{R_l}{l} = \frac{\rho_{st}}{\pi d s} = \frac{1}{\pi d s \varkappa_{st}} \tag{10-1}$$

Here $\rho_{st} = 1/\varkappa_{st}$, the electrical resistance of the pipeline material; d is pipe diameter and s is the pipe wall thickness. In contrast to grounds, anode installations, storage tanks and other spatially limited objects, pipelines have no definable grounding resistance. For a limited length of pipe, l, the grounding resistance, R_E, follows from a drainage test from Eq. (3-10) with $r_M = r_u$ [see Eq. (5-10)]:

$$R_E = \frac{U_{on} - U_R}{I} = \left(r_p + r_u\right)\frac{J}{I} = \frac{r_p + r_u}{\pi d\, l} \tag{10-2}$$

266 Handbook of Cathodic Corrosion Protection

The reciprocal grounding resistance is the leakage, G. The leakage related to the length, l, is the leakage per unit length (leakage load), G' [see Eq. (24-71)]:

$$G' = \frac{1}{R_E l} = \frac{\pi d}{r_p + r_u} \tag{10-3}$$

For coated pipelines, $r_p \ll r_u$, which modifies Eq. (10-3):

$$G' = \frac{\pi d}{r_u} \tag{10-3'}$$

Equation (10-2) is true only for pipelines of relatively short lengths since otherwise the longitudinal resistance will be noticeable at longer distances. Thus in Eq. (10-2) R_E will approach Z, not zero, for a long length. Here Z denotes the characteristic resistance defined in Eq. (24-66). Z is therefore the minimum grounding resistance for the case where current is fed at one end of the pipeline. This quantity is reduced by half for current fed in at a central point on the pipe. It follows from Eq. (24-66)

$$Z = \sqrt{\frac{R'}{G'}} = \frac{1}{\pi d} \sqrt{\frac{r_p + r_u}{s \, \rho_{st}}} \tag{10-4}$$

In Section 24.4.3, the cathodic protection length ($2L$) of a pipeline from a current drainage point is derived for locally constant values of R' and G' [see Eq. (24-75)]:

$$(2L)^2 = \frac{8 \Delta U}{R' G' \eta_L} = \frac{8 \Delta U_S}{\rho_{st} J_s} \tag{10-5}$$

The voltages ΔU and η_L are defined by Eqs. (24-69) and (24-68a) and have a constant value of about 0.3 V. It is shown in Section 24.4.4 that with overprotection (i.e., by polarization into the range of hydrogen evolution) the cathodically protected range cannot be markedly lengthened. Therefore Eq. (10-5) is basic for the cathodic protection of pipelines.

Protection ranges from Eq. (24-75) are represented in Fig. 10-1 as a function of current density J_s. Here ΔU is taken as 0.3 V, which is the difference between the pipe/soil potentials at the end of the protected range ($U_{Cu-CuSO_4} = -0.85$ V) and the current drainage point ($U_{Cu-CuSO_4} = -1.15$ V) (see Eq. 24-69). Figure 10-2 shows the protection current requirement of pipelines of protected length, $2L$, as a function of the protection current density according to Eq. (24-76).

Pipelines   267

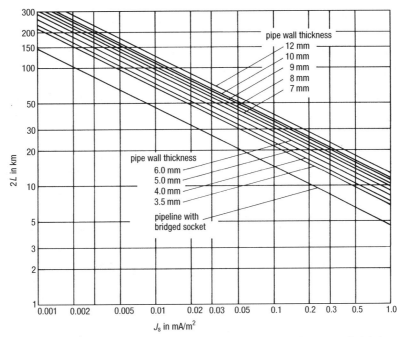

Fig. 10-1 Length of the protected range $2L$ as a function of pipe wall thickness and protection current density J_s from Eq. (10-5); the pipeline ($\rho_{st} = 0.18 \times 10^{-6}\ \Omega$ m, $s = 8$ mm; DN 600) with insulated sockets is bridged by 0.5-m NYY cable of 16 mm² copper.

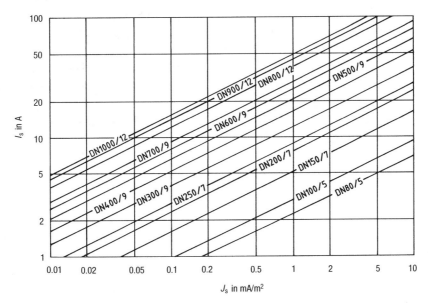

Fig. 10-2 Protection current requirement I_s for protected length $2L$ and various pipe dimensions (DN/s) as a function of protection current density J_s ($\rho_{st} = 0.18 \times 10^{-6}\ \Omega$ m).

10.2 Preconditions for Pipeline Protection

Conclusions can be drawn from Eq. (10-5) to achieve the greatest possible protected length. The resistance and leakage loads, R' and G', must be as low as possible:

(a) The pipeline must not have any additions that would raise the longitudinal resistance (e.g., couplings, expansion joints, flanged valves) or they must be bridged with metal conductors.

(b) The pipeline must have a good insulating coating with as few defects as possible. It should have no contact with grounded installations that would raise the leakage load, e.g., grounding installations, pipelines, casings, valve supports, or interconnections.

(c) Insulating connections should be fitted at the grounded end points of short pipelines and at the end of the protected range (where $I = 0$) of long pipelines, or by parallel-connected additional cathodic protection stations to limit the protected range electrically [see the details of Eqs. (24-77) to 24-80)].

10.2.1 Measures for Achieving a Low Resistance Load

Pipelines with welded connections always have a low resistance load, R'. This is, however, considerably higher with higher resistance pipe connections such as flanges, couplings with insulating spacers (e.g., rubber screw sockets) or expansion joints. They must be bypassed using insulated copper cable as short as possible with a cross-section of at least 16 mm² and with a bridging resistance <1 mΩ. The cable should be connected to the pipeline by a suitable method such as Cadweld, stud welding or pin brazing [5-7] in order to guarantee a low-resistance bridge. The connection points must be covered with perfect coating [8]. Earlier, common pipeline connections with lead as the sealing material were usually sufficiently conducting, but reduced the protection range.

10.2.2 Measures for Achieving a Low Leakage Load

10.2.2.1 Pipeline Coating

In order to achieve a low leakage load, the direct ground contact (i.e., the effect of the ground connection of the pipeline) must be reduced. This is done by coating the pipeline and avoiding or preventing conducting connections with installations with low grounding resistances.

To achieve a low leakage load for a pipeline, it is necessary first of all to have a coating with a sufficiently high resistance (see Section 5.2.1.2). The leakage load is increased by damage to the pipe coating during construction or by thermal effects in operation. Information on the application of coating for steel pipes is given in Ref. 9.

Where greater stress is placed on the pipe coating (e.g., in placement in rocky areas) a coating with greater impact and penetration resistance is necessary without additional protection measures. The important factors (see Section 5.2.3) are:

- increased coating thickness;
- additional coating with PE tape for bitumen-coated pipelines;
- a rock shield;
- additional coating of fiber cement mortar;
- ingrowth of roots; this can be prevented by coating of polyethylene, epoxy resin, polyurethane tar or tapes with plastic foils.

The low leakage load of a good pipe coating can be nullified by contact with low-resistance grounded installations. Such contacts occur at custody transfer stations with electrically actuated valves, tubular annuli at casings and concrete constructions, at interconnection points and valve supports, and at foreign lines and cables. Such contacts are to be avoided by constructive measures.

10.2.2.2 Insulating Joints

Electrical contact through station piping to installations with low grounding resistance is present at all stations along the pipeline. Insulating joints are built into these station pipes which prevent low grounding resistance. The insulating joints can be ready-made insulating pieces (insulating couplings) (see Fig. 10-3) or insulating flange joints (see Fig. 10-4). They must conform to certain mechanical, electrical and chemical requirements which are laid down in technical regulations [10,11].

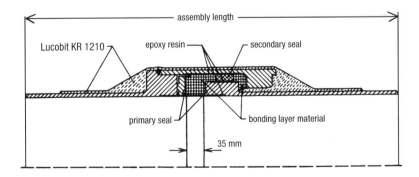

Fig. 10-3 Section through an insulated coupling for direct burial.

With underground installations in the soil, it must be ensured that no water can penetrate in gaps between cathodically protected and unprotected parts since the cathodically unprotected side of the coupling can be destroyed by anodic corrosion. Sections of pipe behind the insulator must be particularly well coated.

In areas where there is danger of explosion, insulating joints in above-floor installations in buildings must be bridged by explosion-proof spark gaps in order to avoid open arcing which can lead to the detonation of an explosive mixture [12]. The response flash impact voltage (1.2/50 μs of the spark gap) should not be greater than 50% of the 50-Hz arcing ac voltage (effective value) of the insulating connection to be protected. This requirement must be ensured by construction if an existing pipeline is provided with flanges with insulating gaskets and by insulation of the screws to insulating flanges. The connecting cables must be short so that the response of the spark gap is not impaired by loop induction (see Fig. 10-4).

In installing protected spark gaps in lines influenced by high voltage, care must be taken that the long-term interference voltage of the pipeline lies below the burning voltage of about 40 V of the spark gap, since otherwise the arc in the spark gap will not be extinguished. The spark gap welds together so that the insulating action is continuously bridged and the cathodic protection is damaged. Insulating

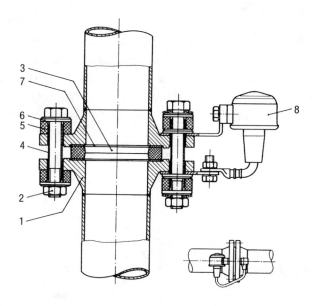

Fig. 10-4 Insulating joint with explosion-proof spark gap ($U_{50\,Hz} = 1$ kV, $U_{1/50\,\mu s} = 2.2$ kV; surge current = 100 kA). 1, insulating flange; 2, hexagonal nut; 3, insulator ring; 4, insulator sleeve; 5, insulator disc; 6, steel disc; 7, blue asbestos gasket; 8, explosion-proof spark gap.

pieces for service pipes of the gas and water authorities should not be installed in areas where there is danger of explosion [12].

10.2.2.3 *Electrically Operated Valves and Regulators*

The grounding resistance of protection or PEN conductors (protective conductor with neutral conductor function) with connected grounds of electrically operated valves and regulators is very low. Due to this, the leakage load of the pipeline is raised at these points and the cathodic protection severely compromised or impaired. The following possibilities exist for protection against currents dangerous to the touch while maintaining cathodic protection:

- for TN-, TT- and IT grids (i.e., protective system with PEN conductor, protective grounding system, and protective system with isolated starpoint, respectively) use of isolation according to Fig. 10-5 where the remaining electrically operating equipment is connected to its own ground [13];

- for TN- and TT-grids, use of the fault current circuit (FI) according to Ref. 14 (Fig. 10-6), where grounding resistances $R_E < 3\ \Omega$ lead to damage to the cathodic protection;

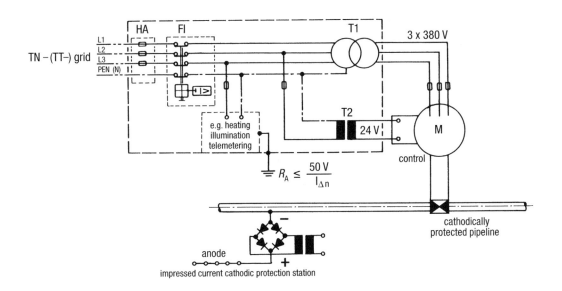

Fig. 10-5 Protection measure by separation of electrical operational equipment that is connected to the cathodically protected object via the housing, with an FI protection circuit leakage current circuit breaker (see Ref. 14); T_1 and T_2: isolating transformers (see Ref. 15).

- separation of the electrically operated equipment from the pipeline by insulated connections and bridging the devices with cable connections; thus, the valve is not cathodically protected and an excellent holiday-free coating is required;
- separation of the electrical grid by including an insulated drive mechanism between the driving motor and the devices.

10.2.2.4 Casings

The cathodic protection of pipelines can be compromised by contact with casings [16]. Transport pipes must therefore be provided with sufficiently high and mechanically stable spacers to exclude electrical contacts. In addition, only casing seals that prevent ingress of ground water into the annulus between the casing and the pipeline should be used. With uncoated casings, the cathodic protection can be lost over wide stretches by electrical contact with the carrier pipe arising from the increased leakage load. Also, no cathodic protection is possible inside the casing because it acts as a Faraday cage. The presence of water in the annular space is not detectable by external electrical measurement where there is electrical contact

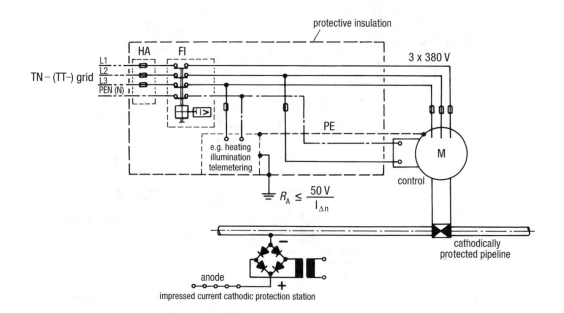

Fig. 10-6 FI-protected circuit for electrical operational equipment that is connected to the cathodically protected object via the housing (see Ref. 14).

between carrier pipe and casing. Corrosive water in the annular space can cause corrosion damage. In this case one must consider the possibility of cell formation where the rusted inner surface of the casing acts as a cathode.

If well-coated casings are used, local impairment of cathodic protection through electrical contact is small due to the low drainage load. Cathodic protection is not possible for the carrier pipe inside the annular space. Even without electrical contacts with casings whose coating has fewer defects than the carrier pipe, cathodic protection cannot be achieved in the annular space. Measures for dealing with contacts between casings and the pipeline are given in Table 10-1. Casings should

Table 10-1 Measures for casings

Type of failure	Measures	
	Uncoated casing	Coated casing
Electrical contact between pipe and casing[a]		
No electrolyte in annulus	Locate and eliminate contact or provide local cathodic protection. Take note of interfering foreign installations	Locate and eliminate contact. If not possible for structural reasons—depending on the individual case—take other protection measures, e.g. insert hydraulic sealing medium or organic materials into the annulus.
Electrolyte in annulus	Locate and eliminate contact or insert hydraulic sealing material or organic materials into the annulus. If necessary, install local cathodic protection	
Nonelectrical contact between pipe and casing		
No electrolyte in annulus	No measures necessary, since cathodic protection of pipe acting	Usually no measures necessary
Electrolyte in annulus	Usually no measures necessary	

[a] With electrical contact of pipe and casing, the absence of electrolyte within the annulus is not detectable from the outside by electrical measurements.

only be provided on cathodically protected pipelines where the latter cannot be laid at crossing points without risk of damage, and where the use of reinforced concrete casings is not possible.

In installing carrier pipes by thrust boring, a drainage test should be conducted before tie-in of the pipe to see whether, on imposition of cathodic protection, the pipe can be sufficiently negatively polarized according to Section 10.4. Table 10-2 shows a data sheet for such a drainage test. If the coating resistance of the pipeline is so high that no polarization can be achieved, it is recommended that an uncoated steel rod be driven into the soil and directly connected electrically to the pipe. If the drainage current increases and the off potential is more negative than the protection potential, then there are no defects present in the coating and the carrier pipe can be included in the cathodic protection of the pipeline.

10.2.2.5 *Special Installations on the Pipeline*

To ensure water tightness in concrete foundations, pipelines are usually embedded in a concrete ring. The danger here is of an electrical contact with the reinforcement of the concrete, which not only damages the cathodic protection but also endangers the pipeline by cell formation with the large steel-concrete interface (see Section 10.3.6). The supporting ring has to be insulated with a pressure-resistant plastic (e.g., epoxy resin) so that contacts can be prevented. With restraint points, valve supports, river crossings, and bridge constructions, connection with the ground must be prevented by an intermediate layer of a mechanically strong insulating material. Gas and water crossing regulations have to be taken into account in the area of overhead conductors of railway systems [17]. Pipelines may only be connected over spark gaps with flashover protection equipment [18]. Grounds of pipelines influenced by high voltage must be connected to the pipeline via dc coupling devices (see Section 23.5.4).

10.2.2.6 *Prevention of Electrical Contact with Foreign Objects*

Metallic contact with pipelines and cables that are not cathodically protected must be avoided on account of the high leakage load. Where pipelines and cables are running parallel, a spacing of at least 0.4 m must be observed for safety reasons. A distance of not less than 0.2 m must be adhered to in narrow passages and crossings. If such a distance is not possible, electrical contact must be prevented by interposing insulating shells or plates. Such materials can be PVC or PE. The dimensions of these plates should not be less than the diameter of the larger pipeline [19].

Table 10-2 Form for recording measurements on an extruded pipe

Measurement Data Date _____
Measurements on coated pipes installed by thrust boring

Pipeline: casing/pipe
Designation of pipeline:_____
Location of thrust boring:_____
Type of coating:_____ Type of girth weld coating:_____
Diameter; DN_____
Wall thickness:_____ mm   Total length of pipe:_____ m
Length in earth: _____ m   Surface area of pipe:_____ m²
Appearance of pipe coating in the receiving pit after thrust boring:_____

Type of soil: _____

Readings
Rest potential, launching pit $U_{\text{Cu-CuSO}_4}$: _____ V
Rest potential, receiving pit $U_{\text{Cu-CuSO}_4}$: _____ V
Grounding resistance R:_____ Ω; Specific soil resistivity ρ:_____ Ω m
f: 135; 105 Hz

Time min	$U_{\text{Cu-CuSO}_4}$ "on" V	$U_{\text{Cu-CuSO}_4}$ "off" V	ΔU V	I_s µA	R_A Ω	J_s µA/m²	r_u Ω m²
3	−1.5						
6	−1.5						
9	−1.5						
12	−1.5						
15	−1.5						
30	−2.0						
60	−2.0						

Switch-off time: <5 s

$$\Delta U = U_{\text{on}} - U_{\text{off}} \qquad R_A = \frac{\Delta U}{I_S}(\Omega)$$

$$J_S = \frac{I_S}{A}(\mu A / m^2) \qquad r_u = R_A \times A (\Omega\, m^2)$$

A = surface area of pipe (m²); Cathodic protection possible: yes/no

Measurements carried out by: _____

10.3 Design of Cathodic Protection

10.3.1 Design Documents

The following documents are necessary for planning cathodic protection [20]:

- a general map showing the position of lines and details of valves, regulating stations, casings, river crossings, insulating joints, expansion joints, and lines in bridges; flowing media, operating conditions (temperature and pressure);
- length, diameter, wall thickness and material of the pipe, type of pipe connections, with buried pipelines, year of burial;
- type of mill-applied as well as field coating, coating of fittings and valves;
- routing of dc railway lines, location of transformer substations and negative feeder points, encroachments, parallel trajectories and crossings with overhead power lines of >100 kV as well as ac railway lines.

If the projected pipeline is situated in an area with dc railway lines, rail/soil potential measurements should be carried out at crossing points and where the lines run parallel a short distance apart, particularly in the neighborhood of substations, in order to ascertain the influence of stray currents. Potential differences at the soil surface can give information on the magnitude of stray current effects in the vicinity of dc railway lines. It is recommended that with existing pipelines the measurements be recorded synchronously (see Section 15.5) and taken into account during design.

10.3.2 Test Points

Test points are necessary for measuring pipe/soil potentials, and pipe currents as well as resistances of insulating connections and casings. Potential test points should have a maximum spacing of 1 to 2 km, and every fifth test point should be designed as a pipe-current test point. In built-up areas, the spacing of test points should be reduced to about 0.5 km. At the initial point of long, branched pipelines, it is recommended that pipe-current test points be installed so that the current consumption of this pipeline can be controlled. The installation of pipe-current test points before insulating is advisable in order to determine internal electrolytic bridging, which can lead to the destruction of insulating connections.

In general NYY-O cable is used with a minimum copper cross-section of 2×2.5 mm. The cable is connected to the pipeline by a suitable process [5-7], and the connections carefully coated. The cable is usually connected to aboveground test points and covered with hoods, tiles or a cable ribbon.

In order to be able to recognize the type and function of the test point even without an inscription, it is recommended that the design of the test point conform to factory standards. Aboveground test points are usually installed in marker posts with a closable flap. The measuring cable is attached to a plastic plate with terminals (see Fig. 10-7).

The plastic plates and terminals are fitted into cast housings inside concrete columns or in built-up areas, in cable junction boxes installed at walls. Belowground test points should be installed in built-up areas only in exceptional circumstances. In this case watertight, flush-mounted test stations are installed under a street-level covering and can be kept dry only by the most careful construction.

10.3.3 Determination of Current Demand

In choosing a protection method, the magnitude of the required protection current, which depends on the necessary protection current density, is of considerable importance. From Section 5.2.1.2 a rough estimate of the current demand can be made using Eq. (5-11′).

Table 5-1 gives values of electrical resistance r_u for different coating materials. It is advisable to increase these by 100% at the planning stage because of the uncertainty in estimating the protection current density.

The maximum protected length $2\,L$ is given in Fig. 10-1 and the required protection current in Fig. 10-2. For pipelines with carefully mill-applied PE and excellent field-applied coating of the girth weld area, the protection current densities lie between 1 and 3 $\mu A\ m^{-2}$. With carefully buried pipelines with bitumen (or coal tar) coating, the protection current densities lie between 10 and 30 $\mu A\ m^{-2}$.

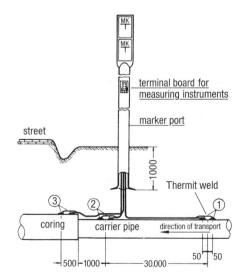

Fig. 10-7 Aboveground test points, potential measuring points (2) and (3), pipe current measuring point (1/2).

In the case of older pipelines and offshore pipelines, protection current densities can amount to several mA m^{-2}. For older onshore pipelines, the protection current densities are determined by a drainage test according to Section 3.4.3.

The drainage test should be carried out if possible in the area in which the planned protection installation is to be provided. The connection for the drainage test can be provided via already installed test points or at fittings that have low-resistance connections with the pipeline. Provisional grounding rods can be driven in to act as anodes. Since the resistance of these grounds is different from that of the final anode installation, the output voltages of the drainage test and of the final dc current supply units are also different. However, this is not important in designing an anode installation. The drainage current is periodically switched on and off and the on and off potentials at all possible measuring points on the pipeline (e.g., standpipes, valve spindles and hydrants) are measured. Because of the short polarization period, pipelines with high drainage load will mainly have $U_{on} > U_s$ at most of the measuring points. From experience, it can be assumed that at all measuring points where $U_{on} \leq U_s$, after a sufficient polarization period of a few weeks $U_{off} \leq U_s$ will be attained. The drainage test, in which all the measuring points $U_{on} < U_s$, gives the conventional current demand of the pipeline. In order to be able to increase the protection current output of the anode installation and to provide, for example, additional connecting pipelines with cathodic protection, it is recommended that the projected protection current demand be increased by a factor of 1.5. Pipe/soil potentials and pipe currents in a drainage test are shown in Fig. 10-8, together with those after 1 year of operation.

10.3.4 Choice of Protection Method

The current output of galvanic anodes depends on the specific soil resistivity in the installation area and can only be used in low-resistivity soils for pipelines with a low protection current requirement because of the low driving voltage. Impressed current anode installations can be used in soils with higher specific soil resistivities and where large protection currents are needed because of their variable output voltage.

10.3.4.1 Galvanic Anodes

Cathodic protection with magnesium anodes can be just as economical as impressed current anode assemblies for pipelines only a few kilometers in length and with protection current densities below 10 μA m^{-2}, e.g., in isolated stretches of new pipeline in old networks and steel distribution or service pipes. In this case, several anodes would be connected to the pipeline in a group at test points. The distance from the pipeline is about 1 to 3 m. The measurement of the off potential is difficult because of the necessary synchronous switching of several groups of anodes. In the immediate surroundings of pipelines with high operating temperatures, soils become electrolytically conducting even in conditions of severe frost, while the frozen soil acts as an insulator. Since for this reason impressed current installations cannot be used, zinc anode wires are connected to such pipelines as protection anodes [21]. Where there are stray current effects, galvanic anodes cannot be used since they do not allow sufficient stray current drainage and in unfavorable circumstances even increased stray current entry is possible. With ac effects, high alternating-current densities can lead to an increase in the magnesium potential, thereby resulting in the pipeline becoming anodic.

10.3.4.2 *Impressed Current Anodes*

Where there is a high protection current requirement, and for long pipelines, the impressed current method is almost always recommended, since it can provide for the increased protection current requirements resulting from branched pipelines by raising the output voltage. The following factors should be taken into

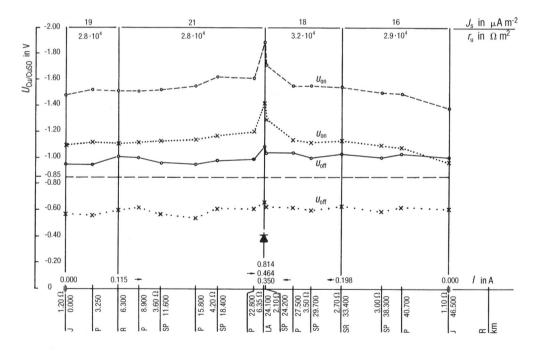

Fig. 10-8 Pipe/soil potentials and protection currents for a pipeline. Drainage test: x-x; after 1 year o-o. P = potential test point; R = pipe current test point; LA = cathodic protection station; J = insulating joint; SP = pipe casing potential test point.

account when choosing the location of an impressed current cathodic protection station:

- presence of a low voltage supply;
- as low as possible specific electrical soil resistivity in the area of the anode ground bed;
- the maximum protection length as shown in Fig. 10-1 and in Eq. (10-5);
- sufficient distance between the anode ground bed and foreign lines as in Eq. (9-11) to keep interference low;
- good access to the protection installation.

The housing of the transformer-rectifier unit should be erected in an area with a right-of-way for the pipeline.

The information in Section 9-1 covers the anode installation. Housing, layout and circuitry of the transformer-rectifier unit are described in Chapter 8, and type and possible choices of anode materials in Chapter 7.

The results of the design should be assembled in a report which should include the following documents:

- a general map of the location of the proposed impressed current cathodic protection station and test points as well as the position of insulating connectors and casings;
- site plan of the proposed protection station;
- circuit diagram of the protection station and test points;
- calculations of the protected range and protection current densities;
- rating of the anode ground bed and transformer-rectifier unit.

It is appropriate to assemble the results in a data sheet as shown in Table 10-3 for planning and documentation.

10.3.5 Pipelines for Electrolytically Conducting Liquids

Practically no electric currents occur inside pipes for electrolytically conducting media such as potable water, district heating or brine [22]. On the other hand, the situation at insulating connections or joints is completely different (see Section 24.4.6). In cathodic protection there is a difference in the pipe/soil potentials of the protected and unprotected parts of the pipeline, leading to a voltage of between 0.5 and 1 V. This leads to the danger of anodic corrosion on the inner surface of the part of the pipeline that is not cathodically protected, which with good conducting media (e.g., salt solution) rapidly leads to damage. The danger of internal corrosion increases with the size of the pipe diameter and the conductivity of the transported medium. It decreases with the length of the

Table 10-3 Form for design of cathodic protection installation

Estimated parameters	Unit	Plan	Check
Length of pipeline	km	200	
Diameter/wall thickness	mm	600/8	
Type of coating/thickness	mm	PE/2.5	
Surface area of pipe per meter	m^2	2	
Total surface area	m^2	4×10^5	
Current density (from drainage test or estimate)	µA m^{-2}	10	8.7
Maximum protected length (2 L)	km	103	
Number of cathodic protection stations required	n	4	
Actual protected length per station	km	50	
Protection current required per station	A	1.0	0.9
Protection current including safety per station	A	2-4	
Specific soil resistivity at the anodes at depth $a = 1.6$ m	Ω m	50	
Type of anode	Type	FeSi	
Dimensions of anode (l/d)	mm	1200/40	
Weight of anode	kg	10	
Anode spacing	m	5	
Anode voltage	V	8	9.0
Number of anodes	n	3	
Length of anode bed	m	10	
Anode grounding resistance	Ω	6	7.5
Backfill, 3 or 4 coke nuggets (per anode)	kg	100	
Total anode weight	kg	30	
Life of anodes	a	150	
Rectifier rating, current	A	4	
Rectifier rating, voltage	V	20	
Distance of anodes from foreign lines/type of line	m	80/telephone	
Interference measurement/object		telephone cable	
Interference measurement +ΔU values	mV	—	60.0

insulating section [see Eqs. (24-102) and (24-105)]. For pipelines with a nominal diameter of DN 200 for transporting brine with a specific resistivity of $\rho \approx 1\,\Omega\,cm$, a 10-m length of tube must be insulated at the insulating joint on the cathodically protected side. With pipelines for brine, the length of the insulated section at the insulating joint becomes very large with increasing pipe diameter. In this case additional local internal cathodic protection is recommended as in Fig. 10-9. Platinized titanium is used as the impressed current anode and pure zinc as the reference electrode.

Softened water in district heating pipelines has a much lower conductivity than saline water. Therefore the insulation for avoiding internal damage on the cathodically unprotected side is much smaller. There is no danger with completely desalinated hot water. No danger of corrosion is to be expected in pipelines for drinking water with cement-mortar lining on both sides of the insulating joint, provided the lining is sufficiently thick and free from cracks [22].

As in the case of corrosion at the insulating connection due to different potentials caused by cathodic protection of the pipeline, there is a danger if the insulating connection is fitted between two sections of a pipeline with different materials, e.g., mild and stainless steel. The difference between the external pipe/soil potential is changed by cell currents so that the difference between the internal pipe/medium potential has the same value, i.e., both potential differences become equal. If the latter is lower than the former for the case of free corrosion, the part of the pipe with the material that has the more positive rest potential in the soil is polarized anodically on the inner surface. The danger increases with external cathodic protection in the part of the pipeline made of mild steel.

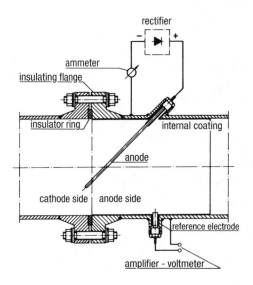

Fig. 10-9 Internal cathodic protection to avoid the danger of anodic corrosion behind an insulating joint in a brine pipeline.

In the previous case, it was a question of a type of bimetallic corrosion without electrical contact between the metals involved. The process initially is difficult to understand; it is, however, easy to understand if one considers it as a battery with an external short circuit caused by an electrolyte behaving in the same way. The results of measurements on a model and the action of cathodic protection are described in Ref. 23.

10.3.6 Distribution Networks

Distribution networks in towns have been built over periods of decades with various phases of pipeline technology. Networks can consist of welded steel pipes, steel pipes with lead sockets, different types of cast iron and plastic pipes.

Steel pipelines in urban networks are at greater risk of corrosion than those outside built-up areas. This is chiefly due to the different types of cell formation caused by many electrical connections to low-resistance grounded installations and the numerous possibilities of damage to the pipe coating caused by relatively frequent excavation in the vicinity of pipelines. In addition, the presence of strongly aggressive soils due to pollution by slag, domestic builders' rubble, and sewage, or by infill and mixtures of soils [24] is important in urban districts. There is a great danger of corrosion damage to supply lines in the area of influence of stray currents, mostly caused by dc railways (see Chapter 15).

A similar danger of corrosion lies in cell formation in steel-concrete foundations (see Section 4.3). Such steel-concrete cells are today the most frequent cause of the increasing amount of premature damage at defects in the coating of new steel pipelines. The incidence of this type of cell formation is increased by the connection of potential-equalizing conductors in internal gas pipelines and domestic water pipelines [25], as well as by the increased use of reinforcing steel in concrete foundations for grounding electrical installations [26].

Figure 10-10 shows the voltage cones for four different steel-concrete foundations [27]. Pipelines in the vicinity of such foundations are affected by these voltage cones (see Section 9.2), which can quickly lead to corrosion damage, particularly in pipelines that have some defects in their coatings.

The requirements derived in Eq. (10-5) are relevant in the cathodic protection of distribution networks for low and as uniform as possible values of resistance and leakage loading. The second requirement is often not fulfilled with old pipeline networks on account of their different ages and the type of pipe coating. When setting up cathodic protection, a distinction must be made between old and new steel distribution networks.

Insulating connections or joints are used in new distribution networks for gas service pipes. Their installation in these pipes has been compulsory in Germany since 1972 [28]. Therefore, use of cathodic protection here is no problem and is

284   Handbook of Cathodic Corrosion Protection

economically possible. The cost of installing cathodic protection can of course be considerable if there are numerous pipeline connections between new and old supply lines. Networks installed after 1972 must be separated from older networks by additional insulating connections. Furthermore, test points must be installed for supervision of the cathodic protection (see Section 10.5). With new networks with pipeline lengths greater than 5 km, it is recommended that additional insulating connections be built in to provide electrical separation of certain measuring regions for supervision and location of defects. Since the protection current requirement of such new networks is generally very small, the networks can usually be connected to the cathodic protection of the high-pressure gas supply line. In this potential connection, a fuse must be included so that if the low-pressure grid fails, the fuse blows and breaks the connection to the high-pressure gas line so that the latter's cathodic protection is not damaged [29].

In distribution networks installed before 1972, all service pipes must be provided with insulating connections. The following documents are useful in addition to those given in Section 10.3.1:

- lists of insulating connections in the service pipes;
- plans of the cathodically protected superimposed high-pressure gas lines by impressed current cathodic protection stations and stray current protection systems;
- renovation measures planned for the pipe network.

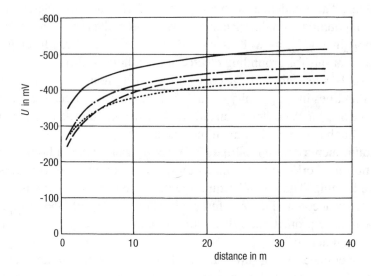

Fig. 10-10   Voltage cone of reinforced concrete structures.

The protection current density is usually considerably higher than that for a new network and can amount to several mA m^{-2}. The protection current requirement of the pipe network is determined from drainage tests. Contacts with foreign pipelines lead to a high protection current demand. After foreign contacts are located and eliminated, the current, which is now too high, can be used to extend the protection range. Interference measurements are of special importance because of the high current densities (see Section 9.2).

Factors that are important for the limitation of protected areas are the pipe network structure, degree of mesh, number of service pipes, type of pipe connections, quality of the pipe coating and availability of protection current as well as stray current effects. A protected area in a distribution network is shown in Fig. 10-11 with separate parts of the network (NT I to NT IV). Previous experience has shown that protected areas of 1 to 2 km^2 with lengths of pipeline from 10 to 20 km are advantageous [30].

If the distribution network consists of pipes with screwed rubber sockets and only the service pipes are of steel, these can be protected with zinc or magnesium anodes. This requires an insulating service clip near the insulating connection. It is usually sufficient just to insulate the stirrup of the service clip since the rubber seal of the service pipe acts as an insulator. In the United Kingdom, the protection of numerous service pipes has been proved with magnesium anodes [31].

In urban districts, installation of impressed current anodes near the surface is usually very difficult because of the interference of nearby installations. Here the installation of deep anodes in suitable soils is recommended; these also have the advantage of being able to be installed in the track of the supply line (see Section 9.1.3).

10.4 Commissioning the Cathodic Protection Station

After verifying the effectiveness of the protective measures against contact voltage of an impressed current cathodic protection station, before the current is switched on, the ground resistance of the anode ground bed, the resistance between the pipeline and the anode ground bed, the ac voltage between pipeline and anode ground bed in the presence of high-voltage interference, as well as the pipe/soil potential should be measured. Where there are stray current effects, measurement of pipe/soil potentials at all test points is recommended. The protection current should be adjusted so that U_{on} at the cathodic protection station falls to $U_{Cu\text{-}CuSO_4} = -1.5$ V. If the necessary protection current is somewhat higher than the estimated value, the cause should be clarified by further measurements.

If the avoidance of anodic interference with other pipelines (see Section 9.2) is involved in the cathodic protection of the main pipeline, all lines must be con-

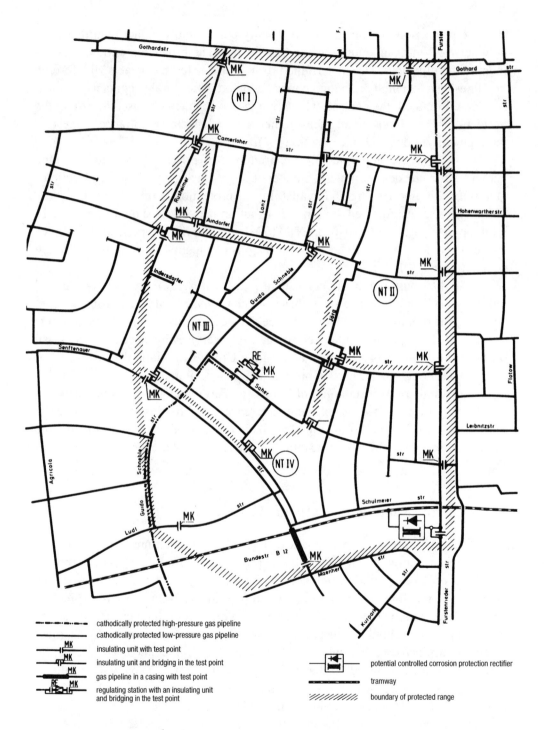

Fig. 10-11 Protected regions in a distribution network.

nected via diodes and variable resistors to the cathodic protection station to avoid equalizing currents. After the final adjustment of the protection current where the protection criterion is achieved at the ends of the protection range, the output voltage, the meter reading and the on potential in the vicinity of the cathodic protection station outside the voltage cone of the anode ground bed as well as at the end points of the protection range must be measured and recorded. A few weeks after commissioning and setting up the cathodic protection, if the final polarization is attained, a final check is made. In this all the necessary and important data of the cathodic protection station, the protection current output, pipe/soil potential, rectifier output voltage, grounding resistance of the anode ground bed, meter reading, and the state of the current supply must be measured and recorded in a register (see Table 8-1). The pipe/soil potentials are checked at all test points to ensure that the protection criterion $U_{off} \leq U_s$ [Eq. (2-39)] is fulfilled at all measuring points. In addition, the resistance of the insulating joints and the resistances between casing and the pipeline as well as pipe currents are measured and recorded [32]. It is recommended that for further evaluation and clear representation, the values be presented in potential plots (see Fig. 3-30). These potential plots can today be conveniently computerized to show the protection current densities and coating resistances of individual sections of the pipeline (see Fig. 10-8).

This check provides a good survey of the polarization state of the pipeline, assuming that the protection current distribution is approximately independent of location. Uniform protection current distribution does not, however, exist in areas of serious damage to the coating. For evidence that the protection criterion is fulfilled at such locations, intensive measurements are carried out on new pipelines after reaching complete contact with the surrounding soil, about 1 year after being laid (see Section 3.7). With pipelines where cathodic protection has subsequently been applied, intensive measurements can be carried out after attaining the necessary polarization.

Deviations from the protection criterion determined by the detailed measurements must be rectified either by digging up the pipe and repairing the coating or by resetting the cathodic protection station or constructing an additional station. The detailed measurements should then be repeated and recorded.

10.5 Monitoring and Supervision

According to Ref. 32, the functioning of impressed current cathodic protection stations should be monitored every 2 months, and the stray current protection station every 1 month. If protection installations are provided with measuring instruments for current and potential, this supervision can be carried out by operating staff, so that the readings are recorded and sent to the technical department for examination. Deviations of the actual values from the nominal values can reveal faults that must be located and dealt with by the operating staff. By noting the readings, the type of damage or the deficiency in the protection installation and its causes can be identified and measures taken to achieve the safe operation of the protection system. Causes of damage and remedial measures are described in Table 8-2.

A check on the cathodic protection of the pipeline should be carried out annually according to Section 10.4, where, of course, only the on potential should be measured. This value should also be compared with the values of the measurements in Section 10.4. If there are no changes in the on potentials and the protection current densities for the individual sections of the pipeline, it can be concluded that the off potential has not changed. The values can easily be compared using computers and represented in plots. If the protection current and potential distribution have changed, or in any case every 3 years, the off potentials as well as the on potentials should be measured.

Remote monitoring of the pipe/soil potential is worthwhile for extended networks of cathodically protected pipelines where test points show noticeable changes in a positive direction of the pipe/soil potential due to deficiencies in neighboring cathodic protection installations, the shorting of insulating connectors or contact with foreign pipelines. The limit should be determined by investigation at those locations from which a fault in the cathodic protection system can be detected. This limit is to be supervised and announced, if exceeded, at a control room. Monitoring of the limit every 24 h is sufficient [33]. Daily inspection is particularly suitable for the hours of the night between 1 and 4 A.M. since fluctuations in the pipe/soil potentials caused by dc railways do not occur because they are not operating. Failures recorded in the monitoring unit due to telluric currents can be recognized by long-term recording of the pipe/soil potential at a location not affected by stray currents and transmitting the readings to the control room. Also, periodic large changes due to switching off the protection station to determine off potentials should not be recorded as failures.

The cathodic protection of pipelines is best monitored by an intensive measurement technique according to Section 3.7, by an off potential survey every 3 years and by remote monitoring of pipe/soil potentials. After installation of parallel pipelines, it can be ascertained by intensive measurements whether new damage of the pipe coating has occurred. These measurements provide evidence of possible external actions that can cause mechanical damage.

10.6 References

[1] TRbF 301, Carl Heymanns Verlag, Köln 1981.
[2] TRGL 141, Carl Heymanns Verlag, Köln 1977.
[3] DVGW Arbeitsblatt G 463, ZfGW-Verlag, Frankfurt 1983.
[4] DVGW Arbeitsblatt G 462/2, ZfGW-Verlag, Frankfurt 1985.
[5] F. Giesen, J. Heseding und D. Müller, 3R intern. *9*, 317 (1970).
[6] H. J. Arnholt, Schweißen + Schneiden *32*, 496 (1980).
[7] Firmenschrift, Bergsoe Anti Corrosion, Landskrona (Schweden) 1987.
[8] DVGW Arbeitsblatt GW 14, ZfGW-Verlag, Frankfurt 1989.
[9] DIN 30675 Teil 1, Beuth Verlag, Berlin 1985.
[10] DIN 2470 Teil 1, Beuth Verlag, Berlin 1987.
[11] DIN 3389, Beuth Verlag, Berlin 1984.
[12] AfK-Empfehlung Nr. 5, ZfGW-Verlag, Frankfurt 1986.
[13] AfK-Empfehlung Nr. 6, ZfGW-Verlag, Frankfurt 1985.
[14] DIN VDE 0664, Teil 1, Beuth Verlag, Berlin 1985.
[15] DIN VDE 0551, Beuth Verlag, Berlin 1980.
[16] AfK-Empfehlung Nr. 1, ZfGW-Verlag, Frankfurt 1985.
[17] Gas- und Wasserkreuzungsrichtlinien der Deutschen Bundesbahn (DS 180).
[18] DIN VDE 0185 Teil 2, Beuth Verlag, Berlin 1982.
[19] DIN VDE 0101, Beuth Verlag, Berlin 1988.
[20] DVGW Arbeitsblatt GW 12, ZfGW-Verlag, Frankfurt 1984.
[21] A. W. Peabody, Materials Perform. *18*, H. 5, 27 (1979).
[22] H. Hildebrand u. W. Schwenk, 3R intern. *21*, 387 (1982).
[23] Fachverband Kathodischer Korrosionsschutz, 3R intern. *24*, 82 (1985).
[24] A. Winkler, gwf/gas *120*, 335 (1979).
[25] DIN VDE 0190, Beuth Verlag, Berlin 1986.
[26] DIN VDE 0100 Teil 540, Abschn. 4, Beuth Verlag, Berlin 1986.
[27] F. Schwarzbauer, pers. Mitteilung 1988.
[28] DVGW Arbeitsblatt G 600 (TRGI), ZfGW-Verlag, Frankfurt 1972 u. 1987.
[29] DVGW Arbeitsblatt GW 412, ZfGW-Verlag, Frankfurt 1988.
[30] F. Schwarzbauer, gwf/gas *121*, 419 (1980).
[31] W. v. Baeckmann, ndz. *22*, 42 (1971).
[32] DVGW Arbeitsblatt GW 10, ZfGW-Verlag, Frankfurt 1984.
[33] H. Lyss u. W. Prinz, 3R intern. *24*, 472 (1985).

11

Storage Tanks and Tank Farms

K. HORRAS AND G. RIEGER

11.1 Special Problems Relating to the Protection of Tanks

The external cathodic protection of underground storage tanks [1], particularly older tank installations, presents difficulties compared with buried pipelines for the following reasons: the tanks are often close to buildings or grouped close together. In many cases, buried storage tanks are mounted on large concrete foundations as a protection against buoyancy. With older tank installations, the containers are mostly situated in secondary containments that formerly were provided as collecting facilities for single-wall containers and therefore may show signs of leaking. Depending on the method of construction, such arrangements can affect the distribution of the protection current when there are larger areas of damage in the coating in which the protection current is restrained, but to which corrosive constituents of the soil have uninhibited access. In this case the conditions for Eq. (2-47) do not apply. By careful planning and construction of new installations, factors interfering with the cathodic protection of the tank can be safely avoided.

11.2 Preparatory Measures

The preparatory work for tank installations with single-wall containers begins with checking whether cathodic protection is prescribed or is appropriate on the grounds of economy [2]. The information in Chapter 4 is relevant in judging the risk of corrosion. Corrosion risk in storage tanks consists of the formation of cells with foreign cathodic structures via connecting lines, e.g., pipelines of copper or stainless steel or rusted steel pipes embedded in concrete as well as steel-reinforced concrete structures.

For economical and complete cathodic protection against external corrosion without harmful effects on nearby installations, the storage tank to be protected must have good coating and therefore require a low protection current density. In addition, it must have no electrical contacts with other buried installations, such as pipelines and cables. In this case the interconnected installations take up much more current than the object to be protected because of their generally lower grounding resistance so that the higher protection current required has damaging effects on neighboring installations (see Section 9.2). If the interconnected installations are very close to the tank to be protected, they can have a shielding effect so that the latter does not receive sufficient protection.

In new installations the coating of the tank should be tested and repaired if necessary before backfilling. Also, all metallic components that are electrically connected to the storage tank which have to be included within the cathodic protection must be well coated. These components are filling and emptying pipework and aeration pipes, as well as possible steel domes and the lifting points on the tank. Tanks and their connected pipework have to be embedded in soft soil. Good coating must comply with the data in Section 5.1 for the safe application of cathodic protection. Guidelines and regulations for the installation of storage tanks must be adhered to [3,4].

The tank and auxiliary pipework that is to be cathodically protected must be electrically isolated from all other metallic installations. This is achieved by arranging insulating couplings so that all the steel and coated copper connection pipes of the storage tank which are underground can be covered by the cathodic protection. The insulating couplings should be inside the building where the pipework enters and at the base of the pumps in service stations. To protect against corrosion by cell formation with the above-named parts of the installation, electrical separation of the service area and the building through use of insulating joints that can be examined is emphatically recommended [2,5].

Recently in some types of refuelling stations, corrosion-resistant materials have been used between the filter-water separator and the outlet pipeline and valves, usually stainless steel, rarely aluminum. If these are buried, they must have good insulating coating and be electrically separated from other tank installations by insulating couplings.

At the point where pipes enter buildings or pits, accidental electrical contact between the pipes and the ducting must be strenuously avoided. Accidental aboveground contact of aeration pipes and grounded metal components, which frequently occurs, can be prevented simply by ensuring that all the fixings and supports are mounted on the aeration pipes with a mechanically strong insulating insert. If underground crossing between the protected pipeline and other installations (e.g., cables, lightning conductors) is unavoidable, care should be taken that no contact can occur if compaction or later subsidence of the soil takes place. All additional equipment in contact with the tank installation (e.g., leak prevention equipment and filling gauges) must be installed so that no electrical connections that would damage the cathodic protection arise in the protection leads of the current supply, grounds, metal structures, etc. For the same reasons, in the case of underground storage tanks that have to be secured against buoyancy, the concrete bases and foundations must have no contact with the tank; strap restrainers should be provided with sufficiently large area and mechanically stable insulating seating. A layer of sand at least 5 cm thick should be provided between the tank and the concrete base.

Also for consideration in the preparation measures is determining the soil resistivity where the anodes are to be located on the site. Electrical safety measures must be considered for impressed current installations [6]. Complete cathodic protection of tanks with small protection current densities can be easily achieved with new installations, without damaging interference with neighboring installations. Relatively higher protection current densities are to be expected with existing older tanks, depending on the state of the tank and pipe coating. From experience it is possible in most cases, even with older tank installations, to achieve sufficient cathodic protection, but in comparison with new tank installations, expenditures on the work and the cost of the protection installation are higher.

11.3 Storage Tanks

11.3.1 Determination of Current Demand, Evaluation, and Connections of the Protection Equipment

Based on past experience, it has been found that the protection current density for buried storage tanks coated with bitumen is over $100\ \mu A\ m^{-2}$. With coatings in very good condition, it can amount to a few tens of $\mu A\ m^{-2}$ but for coatings in a very poor state, it can rise to a level of $mA\ m^{-2}$. The protection current demand can be very different for storage tanks of the same size so that in the design of the protection system it cannot be estimated with sufficient accuracy as with pipelines. For this reason it is necessary to determine the protection current requirement by a drainage test.

In this test the protection current is increased in a stepwise fashion. At a given time after a polarization period of about 1 h, which from experience is sufficient for cathodic polarization of a new tank installation with good coating, on and off potentials are measured. The latter should be determined within 1 s of switching off the current and compared with the protection criterion (see Section 3.3). The measurements should be made at at least three locations on every storage tank, at the connecting pipework by positioning the reference electrode on the soil, and also, as shown in Fig. 11-1, at a location beneath the soil that is remote from the potential position of the protection current entry. The grounding resistance of the object to be protected is also determined in the drainage test.

Storage Tanks and Tank Farms 293

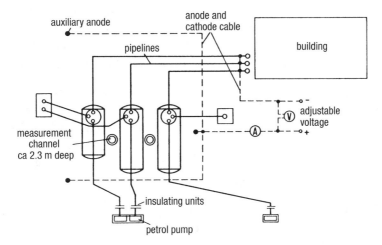

Fig. 11-1 Determination of protection current requirement at a gas station using a drainage test.

If the protection current densities of bitumen-coated pipelines [see Fig. 5-3 and Eq. (5-11′)] and storage tanks are compared, it can be seen that the values for pipelines are usually below 100 μA m^{-2} but are considerably higher for storage tanks. This is because the installation of a storage tank in the ground, depending on its weight and dimensions, is considerably more difficult and leads to greater damage to the coating than is generally the case with pipelines. It should also be understood that the protection current demand increases rapidly with increasing service life. From experience, the protection current densities can be several mA m^{-2} after 25 years. With new storage tanks, a guideline value of about 200 μA m^{-2} can be assumed if the measures given in Section 11.2 are adhered to.

Horizontal, cylindrical steel tanks with volumes of 300 m^3 are used in special refuelling stations. These single-wall containers are coated externally with a fiber glass reinforced resin. Internally they are provided with a fuel-resistant coating. These types of tanks are mostly provided with welded or flanged steel domes and produced in standard sizes. Protection current densities of a few μA m^{-2} through the plastic coating are usual, assuming that the dome is coated in the same way. The protection current demand for a 300-m^2 tank with two domes and a total surface of about 400 m^2, assuming a protection current density of 10 μA m^{-2}, should only amount to 4 mA. If, however, the domes are only coated with bitumen, the current demand can be considerably higher.

For normally buried tanks, tanks with measures against buoyancy and tanks with a secondary containment, no difficulties are to be expected for protection current densities below 200 μA m^{-2}. On the other hand, the protective action can be impaired by much higher protection current demand and serious obstacles to access to the protection current (e.g., plastic-coated secondary containments). The following advice for judging protective action and the possible interference with neighboring installations is derived from a number of case studies:

1. If the protection current density for underground tank installations is not much over 200 μA m^{-2} and the protection current is no more than a few 10 mA, complete cathodic protection can usually be achieved even in unfavorable site conditions, e.g., if the anodes can only be arranged on one side of the object to be protected. Influence on foreign installations is not to be expected as long as these are not situated in the voltage cone of the anodes.

2. For installations with several storage tanks and a protection current of several tens of an mA, uniform protection current distribution should be the goal, so that the current injection occurs via a number of anodes distributed over the site or via a more distant anode bed. Dividing up the protection current over several anodes avoids large local anodic voltage cones and therefore effects on neighboring installations.

3. Protection currents of a few amperes are needed for the cathodic protection of assemblies of storage tanks or refuelling stations. In this case, electrical contact with grounded installations is the main problem. For cathodic protection, these contacts must be located and electrically separated. If this is not possible, then local cathodic protection should be installed (see Chapter 12).

In the case of higher protection current densities and protection currents, interference can occur on nearby installations not covered by the protection. The danger of anodic interference must be investigated by making measurements and prevented by taking appropriate measures [7] (see Section 9.2). For the same reasons, anode systems should not be installed near steel-reinforced concrete foundations.

A cathode cable is adequate for the return path of the protection current for individual storage tanks. In pumping stations with several storage tanks, each tank must be provided with a cable connection. If the tanks are connected to each other by electrical connections, then two cathode cable connections must be provided [2].

Suitable cable for cathode and anode cables to be laid in earth is NYY-0 (in the United States, HMWPE on Kynar). The cable must be protected in the earth and connected (by straps to the dome supports of tanks; see Ref. 3) to structural parts of the object to be protected, which must not become detached in operation.

The lowest cross-section of 4 mm^2 Cu (equivalent to No. 12 AWG) is for cable to the object to be protected, and 2.5 mm^2 Cu (equivalent to No. 14 AWG) to the anodes [2]. It is recommended that two-core cathode connecting cable, 2 $\times$ 4 mm^2, be used for each object to be protected. If two or more anodes are required, the anodes must be connected via separate cables or cable cores so that the protection current output of each anode can be measured. The individual cable cores are to be connected to separate terminals which can be contained in a junction box or, with impressed current installations, in the housing of the transformer-rectifier. The transformer-rectifier and the external junction boxes must be sited outside the area where there is danger of explosion. NYKY* type cables (in the United States, Halar cable) must be used if petrol or solvents can come in contact with the cable sheathing.

11.3.2 Choice of Protection Method

Magnesium anodes are generally used for cathodically protected buried storage tanks with galvanic anodes. Protection with zinc anodes has been tried in a few cases [8], but in general these have too low a driving voltage (see Section 6.2.2). The attainable protection current, I_s, depends on the driving voltage U_T, the voltage between the object to be protected and the anodes, as well as the grounding resistances of the object to be protected R_c and the anodes R_a, from Eq. (6-13):

$$I_s = \frac{U_T}{R_c + R_a} \tag{11-1}$$

Correction for anode distance and conductor resistances can be neglected.

The grounding resistance of different types of anodes can be calculated from the equations in Section 24.1 (see Table 24-1). The use of magnesium anodes is convenient and economical for relatively small protection currents. In the case of an increase in the protection current demand, because the voltage is fixed at about 0.6 V, the current can only be raised by lowering the grounding resistance of the anodes, i.e., by installing more anodes. Alternatively, the voltage can also be increased by an impressed current system.

In contrast, impressed current installations have the advantage of a choice of voltage, so that the protection current can be imposed stepwise or without steps. Formerly, impressed current systems were only used for protection currents above 0.1 A. Today impressed current installations are employed almost exclusively.

Tank/soil potential measurements cannot be made on objects to be protected with very high coating resistances which are found in rare cases of defect-free coating, and particularly with resin coatings. Off potentials change relatively quickly with time, similar to the discharge of a capacitor, and show erroneous values that are too positive [8]. This is the case with coating resistances of $10^5\ \Omega\ m^2$. If there are defects, the resistance is clearly much lower. The advice in Section 3.3.2.2 is then applicable for potential measurement.

11.3.3 Examples of the Design of Protective Installations

11.3.3.1 Equipment Using Galvanic Anodes

As can be seen in Fig. 11-2, the fuel oil tank to be protected is close to a building and is buried. On the side of the tank facing away from the building, the boundary of the premises is a few meters from the tank. The steel pipelines connected to the tank and included in the cathodic protection are coated. The necessary insulating couplings for electrical separation of the fuel oil tank are situated in the building. The overall data necessary for the design of the cathodic protection installation, determined by a drainage test, are included in the following list:

- storage tank 20 m^3, tank and pipework surface area: 50 m^2;
- grounding resistance of the fuel oil tank installation: 30 Ω;
- resistance at the insulating coupling: 28 Ω;
- soil resistivity at the anode sites measured with probe at distances of 1.6 and 3.2 m, average value of 8 measurements: 35 Ω m;
- protection current at an off potential $U_{Cu\text{-}CuSO_4} = -0.88$ V: 10 mA;
- current density: 200 $\mu A\ m^{-2}$.

Magnesium anodes were chosen as the source of the protection current in this old example, because on one hand a sufficient current, including current reserve, could be achieved due to the relatively low soil resistivity, and on the other hand, use of an impressed current protection system would have required much greater expenditure.

Since to achieve the necessary protection current, including a current reserve (in total about 15 mA), the total resistance of the protection current circuit at a voltage of 0.6 V between steel and magnesium anodes [Eq. (11-1)] must not exceed 40 Ω, two 5-kg magnesium block anodes were provided whose grounding resistance R_A for each anode amounted to about 20 Ω. The grounding resistance of the two anodes connected in parallel, taking into account an interference effect factor, $F = 1.1$ with $n = 2$ in Eq. (24-35), comes to about 11 Ω.

Basically the maximum anode currents given by Eq. (6-14) indicate that the grounding resistance of the cathode is considerably smaller than that of the anode. Since this is not necessarily the case, particularly with storage tanks with good coating, the current output of the anodes is considerably lower [see Eqs. (6-13) and (11-1)].

Although an anode arrangement on both sides of this tank had been planned to ensure uniform protection current distribution, the two anodes had to be installed in the strips of land between the storage tank and the boundary of the site, taking into account the grounding resistance and the required life of 25 years [2], because there were several foreign installations between the building and the storage tank (water pipes, electric cables and telephone cables). The anodes were installed at a distance

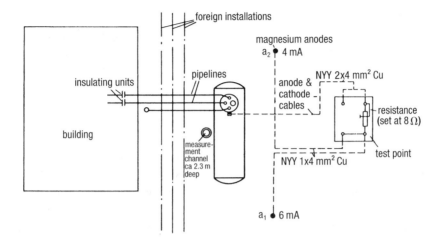

Fig. 11-2 Cathodic protection of a fuel tank with magnesium anodes.

of 11 m and about 2.8 m deep in the ground. Burying the anodes at a shallower depth, 1.5 to 2.0 m, is only practicable if the soil in the area remains damp in long dry periods. If the anodes are situated at an even shallower depth, the grounding resistance and also the resistance of the protection current circuit rises so sharply due to drying out of the surface of the soil that the current necessary to protect the tank surfaces at a greater depth can no longer be supplied by galvanic anodes.

At the relatively low protection current density of 200 μA m^{-2} and with the anode positioned on one side, it is to be expected that with this storage tank sufficient reduction in potential would be achieved on the other side of the tank from the anode. The off potential was measured using a measurement point at a depth of about 2 m as $U_{Cu-CuSO_4} = -0.88$ V at the tank. At the other side of the tank as well as above it, off potentials of -0.90 to -0.94 V were found. These potentials were measured with a protection current of 10 mA (anode 1: 6 mA, anode 2: 4 mA) with an additional resistance of 8 Ω in the protection current circuit (see Fig. 11-2). With a direct connection between the tank and the group of magnesium anodes, the initial current was about 16 mA, which after 1 h of polarization decreased to about 14 mA. The reserve current, based on a long-term current of 10 mA, amounted to ca. 40% in the operation of the cathodic protection installation.

The life of the magnesium anodes with a current content of about 1.2 A a for 10 mA according to Table 6-4 was calculated from Eq. (6-9) as 120 years. This assumes that the protection current is equally distributed over both anodes. The calculated life would certainly not be reached because uniform anode current distribution cannot be achieved over a long period of time. It would, however, be substantially longer than the minimum required life of 25 years. For this length

298 Handbook of Cathodic Corrosion Protection

of life, one anode would have been sufficient. Taking into account the anode grounding resistance, which to achieve sufficient current reserve has to be about 10 Ω, the installation of a second anode was necessary. Measurements showed, as expected, that with the small currents no damaging effects on foreign installations were experienced.

11.3.3.2 Impressed Current Station

A cathodically protected fuel tank installation with three storage tanks is represented schematically in Fig. 11-3. The tank installation is connected at the pumps with the metal sheathing and the neutral of the electricity supply cable. In addition, there is an electrical contact between the aerating and venting pipes attached at the building and its reinforcement. These electrical connections must first of all be done away with. The fuel pipes leading to the pumps must be electrically isolated from them by installing insulating couplings. Accidental contact with the deaerating pipes is avoided by giving the securing clamps an insulating coating. In addition, the tubes in the two filling chambers are isolated from the grounded filling support by insulating couplings and these joints are bridged by explosion-proof spark gaps (see Section 11.5). The efficiency of the insulating couplings is checked by subsequent resistance measurements. It was established by this that all the tanks and pipelines were electrically connected to each other. After these preliminary

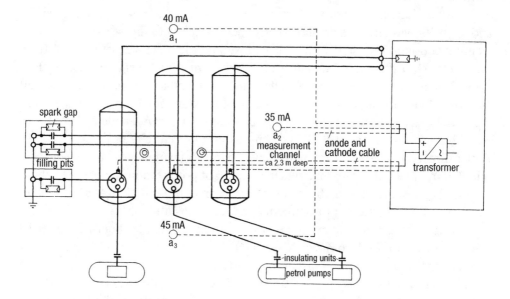

Fig. 11-3 Cathodic protection of a gas station with a mains-fed transformer-rectifier.

measures, a drainage test was carried out. The following data are relevant in the design of the protection installation:

- total surface area of the tanks and pipelines: about 190 m^2;
- resistance of the built-in insulators: 11 Ω;
- resistance between the aeration pipes and the connecting clamps after insulating: 13 Ω;
- average soil resistivity of the site of the tank installation: 70 to 80 Ω m;
- current requirement of the installation: 120 mA;
- protection current density: about 630 μA m^{-2}.

In this case, impressed current protection with several anodes was chosen on the one hand to achieve uniform current distribution with the relatively high protection current density, and on the other hand to avoid large anode voltage cones. A transformer-rectifier with a capacity of 10 V/1 A was chosen.

In total, three high-silicon iron anodes of 3 kg each were installed at points a_1, a_2 and a_3 as shown in Fig. 11-3. The anodes were bedded vertically in fine-grained coke in boreholes about 2.3 m deep and $d = 0.2$ m so that the length of the coke backfill was about 1 m. Each anode was connected by a separate cable to the anode bus bar of the transformer-rectifier to allow the current of individual anodes to be monitored. Three cathode cables 2 × 4 mm^2 were installed for the return path of the protection current and attached on the tank end to the connecting clamps of the dome support.

The grounding resistance of the three anodes with the stated dimensions of the coke backfill, a soil resistivity of 75 Ω m and an interference factor, $F = 1.2$, was calculated from Eq. (24-35) as about 14 Ω. After the anode installation was in operation, measurements of the grounding resistance gave a value of about 12 Ω.

During commissioning, the cathodic protection system gave a protection current of 120 mA with a voltage of about 4 V. The adequate off potential of $U_{Cu-CuSO_4} = -0.88$ to -0.95 V was found at all the potential test points and also between the tanks where the potential was measured by means of a measurement point at a depth of 2.3 m at places where the spacing between the tanks was closest. The anode currents are given in Fig. 11-3. The voltage cones of the anodes were kept small by the anode arrangement chosen and the current distribution so that foreign installations at the pumping station were not affected.

11.4 Tank Farms and Filling Stations

Efforts are made in the construction of new, large tank farms to achieve electrical isolation of buried and cathodically protected fuel installations from all grounded installations, and to provide at the planning stage a corresponding number of insulating couplings. If there is an extended pipe network, it is recommended that the pipe system be electrically separated from the tanks by insulating couplings. The eventual trouble-shooting of faults is made much easier. Also, the current uptake of the tanks and pipework can be determined separately. In such cases, cathodic protection usually presents no difficulties.

When installing insulating joints, a measuring lead should be provided to ensure that the joints can be monitored. Insulating units in areas of explosion danger must be bridged with explosion-proof spark gaps (see Section 11.5). Bridging with spark gaps of low response voltage ($U_{aw} \leq 1$ kV) is particularly important with pipelines that normally carry explosive mixtures of air and vapor (ventilation pipes, empty-running waste pipes, etc.). It is important to ensure that the breakdown voltage of the insulating joints as well as the wall entrances of the pipes into the building and similar arrangements are at least twice the response voltage of the spark gaps. Since an electrical voltage arises at insulating joints in cathodic protection, it is particularly important to provide sufficient isolation in explosion-hazardous regions so that the insulating joints in the pipe cannot be electrically bridged.

Only local cathodic protection can be used for large installations and old installations with electrical contact to components with low grounding resistances that cannot be isolated (see Section 12.6). The measures necessary for tank installations are described in Ref. 10.

As an example, a tank farm that is to be cathodically protected by this method is shown schematically in Fig. 11-4. As can be seen in the figure, injection of the protection current occurs with two current circuits of a total of about 9 A, via 16 vertically installed high-silicon iron anodes embedded in coke. These are distributed over several locations in the tank farm to achieve an approximately uniform potential drop. The details of the transformer-rectifier as well as the individual anode currents are included in Fig. 11-4. Anodes 4, 5 and 6 have been placed at areas where corrosion damage previously occurred. Since off potentials for *IR*-free potential measurements cannot be used, external measuring probes should be installed for accurate assessment (see Section 3.3.3.2 and Chapter 12).

11.5 Special Problems in Cathodic Protection Near Railways

11.5.1 General Comments

Tank installations below ground in the neighborhood of railway lines are often in aggressive soil that is contaminated with slag. Cathodic protection is therefore particularly important. Because of the particularities of railway operations, company standards must be followed in addition to the generally required directions and definitions. The instruction leaflets of the International Railway Association [11] give general advice on tank installations. The regulations for electric railways must be observed for electrified lines [12]. Instructions and service regulations for installations containing flammable liquids are issued by the German Railway as well as by national agencies in other countries [13].

11.5.2 Equipotential Bonding and Insulating Joints

In tracks to tank installations of danger categories AI, AII and B, the rail joints must be connected by good conducting cross connections [12]. There must be a bond between the tracks and the filling nozzles of the tank installation.

If the protection current becomes too high due to this connection in cathodically protected tank installations, then insulating joints are usually installed in the pipeline from the filling nozzle. Care must be taken that the continuity bond is not broken. If there is a danger of stray currents with dc railways due to a permanent connection between track and filling equipment, the continuity bond should be applied only during the filling process.

On stretches of electric line, according to Ref. 12, filling tracks without overhead conductors are usually isolated from the rest of the rail network with insulating joints in order to keep the rail currents flowing to the tank installation as

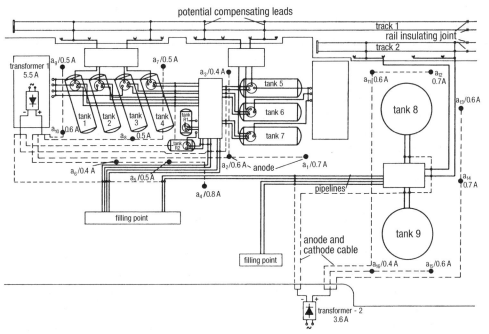

Fig. 11-4 Local cathodic protection of a tank farm with impressed current.

low as possible. The insulating joints are installed outside the danger zone, in particular at the beginning of a dead-end line, and on both sides of the danger zone for lines that are connected to other lines on both sides. Installing insulating units in the pipeline to the filling nozzle is then not necessary if the protection current for the tank installations is not noticeably raised by connecting the filling equipment to these lines. Insulating joints in the filling tracks can be dispensed with if there are no voltages above 50 V between the rail network and the tank installation, so long as they are not connected by a conductor and the return current of the railway will not endanger the tank installation (see Section 11.5.3).

Insulating joints must not be built into filling tracks with overhead conductors. The overhead conductors must be switched off during filling and can be connected to the railway ground. In these cases, insulating couplings must always be built into the pipeline to the filling nozzle in order to enable cathodic protection of the tank installation.

11.5.3 Protective Grounding with Electrified Railways

The railway lines (tracks) of electrified railways serve as the railway ground. Metallic parts of filling equipment that are aboveground should be grounded to rails directly or via breakdown fuses (voltage fuses) if these parts are situated in the overhead conductor area of electric railways with nominal voltages of more than 1 kV for ac or 1.5 kV for dc. As can be seen in Fig. 11-5, directly grounded parts must be electrically separated from the cathodically protected installation. Where grounding is via breakdown fuses, the parts and also the cathodically protected installation are not connected to the railway ground under normal operating conditions.

Monitoring of the breakdown fuses is necessary. Tracks of electrified railways conduct return currents and will have a voltage against the remote ground; this voltage is also termed rail potential. The transformer-rectifier must be provided according to Ref. 14.

11.5.4 Lightning Protection

Grounded installations, including buried metallic pipelines, that are within 2 m of a lightning conductor should be connected directly to it or via a spark gap [15]. If pipelines connected to a lightning conductor are provided with insulating couplings, the latter must be bridged by spark gaps. Tracks of electrified railways and tracks of all railways up to 20 m distant from the connection of a lightning conductor to the track must be dealt with in the same way as lightning conductors.

With AI, AII and B-classified installations on electrified railways, spark gaps in the danger zone must be explosion proof [12]. Insulating couplings and spark

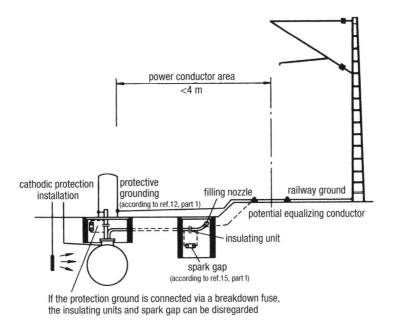

Fig. 11-5 Protection measures on electrified railways close to cathodically protected tank farms.

gaps must, in addition, be protected by insulating coating against accidental bridging, e.g., by tools. The response surge voltage of the spark gaps, according to Ref. 16, for a pulse voltage of 1.2/50, should not be more than 50% of the arcing ac voltage (effective value) of the insulating couplings.

With AIII installations, enclosed fireproof spark gaps are sufficient. These must respond before arcing at the insulating coupling. If a tank installation is near the ground of a mast, it must be carefully investigated to see whether there is a proximity problem in the sense of Ref. 15.

11.5.5 Interference and Working in the Area of Railways

If the protection current requirement for below-ground storage tanks and pipelines is large, then the protection current must be fed in through several anodes, in order to reduce the interference of the numerous underground installations near railways. Where space is restricted and with low current output per anode, impact anodes, e.g., round bar steel, should be used.

Drainage tests in the neighborhood of tracks should only be carried out with the agreement of the railway personnel because of the possible interference with the railway signalling system. Since pipelines near to electrified railway systems can carry railway return currents, connections with the railway ground at both sides or bridging of the pipeline must avoid spark formation in the case of pipe cutting or removal of metallic components [12].

11.6 Measures in the Case of Dissimilar Metal Installations

In some service stations, stainless steel or aluminum materials are used for all the filters, pipes and fittings to maintain the purity of the fuel. The rest potentials of these materials are different from that of plain carbon steel (see Table 2-4).

Cathodic protection of different materials in installations of dissimilar metals is only possible if the protection potential ranges of the individual materials overlap. Section 2.4 gives information on the protection potential ranges of various systems. If there is no overlapping, then insulating couplings must be installed. This is also appropriate and even necessary if the protection current densities are very different.

If the individual materials are separated by insulating couplings but connected to a protection system, the connections must be made through diodes to avoid bimetallic corrosion when the protection system is shut down (see Fig. 11-6). Furthermore, the different protection currents should be adjusted via variable resistors.

11.7 Internal Protection of Fuel Tanks

Cathodic protection of the interior of flat-bottomed tanks is possible if the medium is conducting [17]. Segregation of water can occur in the bottom of fuel tanks with constituents (salts, e.g., chloride) that lead to corrosion attack. This corrosion at the bottom of tanks can be prevented by cathodic protection in the design according to DIN 4119 if the electrolyte covers the bottom of the tank to a sufficient depth. Fuel producers and suppliers usually prevent access of electrolyte solutions because they see it as endangering the purity of the oil according to DIN 51603 and they fear damage to the heating circuit. However, damage caused by introducing electrolyte solutions or by the reaction products of cathodic protection has not been observed to date.

A proposal for a draft standard on the requirements for the internal cathodic protection of fuel tanks has been put forward by a working party entitled "Internal cathodic protection of fuel tanks" [18]. This contains the following information: an electrolyte is produced by dissolving sodium bicarbonate in drinking water with a resistivity not greater than 2000 Ω cm. The solution should completely cover the anodes in the tank.

The suction pipe of the tank must be arranged so that the electrolyte cannot be sucked out. Also, the filling pipe must be designed so that in filling, the electrolyte

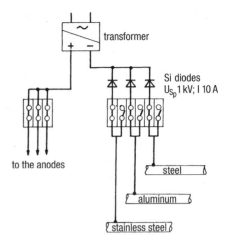

Fig. 11-6 Diode circuit for the cathodic protection of various materials.

is disturbed as little as possible. Undesirable turbulence can be prevented if the velocity of the fuel as it enters the tank does not exceed 0.3 m s^{-1}. The end of the suction pipe must not be lower than 5 cm above the level of the electrolyte. Magnesium anodes such as type AZ63 (see Section 6.2.4) are used as galvanic anodes. They have a current capacity of about 1100 A h kg^{-1} and a weight of 0.8 kg for a length of 35 cm per m^2 of tank surface to be protected. They are electrically separated at a distance of 3 to 5 mm from the bottom of the tank and arranged in uniform distribution. The anodes are connected electrically with each other and to the dome of the tank with an insulated cable. Galvanic anodes produce relatively large amounts of corrosion product, which leads to the formation of sludge and impurities. These disadvantages do not arise with inert impressed current anodes. These are usually metal oxide-coated titanium anodes (see Section 7.2.3). Platinum and anodes coated with platinum may not be used because of the danger of explosion [17]. Impressed current anodes are installed in the same way as galvanic anodes. The length of anode per 2 m^2 of surface to be protected is 10 cm.

Anodes are connected to the object to be protected or to the transformer-rectifier by insulated conductors that are resistant to mineral oil (e.g., Teflon-coated cable) with a cross-section of 2.5 mm^2 of Cu. The transformer-rectifier must meet the demands according to Ref. 6 and have the capability for monitoring and controlling its operation. The life of the anodes is in every case designed to be at least 15 years.

Cathodic protection installations must be tested when commissioned and at least annually. The potentials should be measured at several points on the bottom of the tank with special probes under the oil, and the height of the electrolyte solution should be checked. The off potential and the protection potential as in Section 2.4 are the means for checking the protection criterion according to Eq. (2-39).

11.8 Consideration of Other Protection Measures

Other protection measures (e.g., lightning protection, protection against electrical contact voltage) must not be affected by the use of cathodic protection. Therefore, all the necessary protection measures must be coordinated before setting up cathodic protection installations. With a lightning protection installation, if the cathodically protected tank installation or its pipework is situated near lightning conductor grounds [15], the relevant tank installation should be connected to the lightning conductor grounding system with spark gaps and, in areas where there is danger of explosion, with explosion-proof spark gaps. These should be installed so that the connections are as short as possible [19]. The information in Section 8.1 should be noted in the installation of ground fault circuit breakers for protection against contact.

Tanks that use the connected pipelines as grounding systems are not permitted [20]. It has been shown that magnesium anodes connected to the tanks as grounds are effective in reducing the cathode grounding resistance, avoiding the need for higher protection currents. The grounding resistance of the anodes should be ≤ 50 $V/I_{failure}$. The protection current should be so arranged that a small entry current (a few mA) into the magnesium anodes is detectable to avoid their corrosion. The circuit can be tripped by failure of a circuit breaker if the auxiliary ground is in the voltage cone of the anodes, although there is no failure voltage. In such cases, the corresponding measures could be avoided in the construction of the cathodic protection station, but a remedy can be obtained by connecting a properly sized capacitor to the auxiliary ground's conductor.

The relevant instructions and regulations, among others, are to be observed in regions where there is danger of explosion [21-23]. When carrying out work on the dome of cathodically protected tanks, the cathodic protection station should be switched off and the tank itself grounded to achieve potential equalization. In such cases, fitting so-called cover grounding switches is worthwhile. These break the cathode connection to the tank when the cover dome is opened and connect the tank to the grounding system of the installation. The method can be used where all the tanks have separate cathode connections and when the pipelines from the tanks are electrically separated with insulating couplings.

Figure 11-7 shows the basic circuit diagram for a tank with two domes. The protection current flows via the two interconnected openers of the cover grounding switch to the cathode connection. If one of the covers is opened, the protection current circuit is broken and the tank grounded via the closing contact. The unconnected cable connection of the tank is without current and can be used for measuring potential. By this method, only one tank at a time is separated from the protection system while the other parts of the installations are still supplied with protection current.

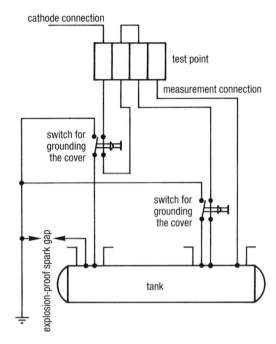

Fig. 11-7 Grounding a cathodically protected tank with a switch for grounding the cover.

11.9 Operation and Maintenance of Cathodic Protection Stations

For commissioning and monitoring of cathodic protection stations, the advice in Refs. 1 and 2 is relevant. For potential measurement, the explanations in Section 3.3 are valid.

After adjustment, the tank and pipe/soil potentials have to be checked annually and recorded. The results of the measurements are the basis of the maintenance process. If the check measurements differ from the values established by the control, the cause must be determined and the defect remedied.

A direct current flows in the installations during operation of cathodic protection stations; therefore, the transformer-rectifier must be switched off when pipes are out or other work on the fuel installation is carried out, and the separated areas must be bridged with large cross-section cables before the work is started in order to avoid sparking that could come from the current network.

11.10 References

[1] DIN 30676, Beuth-Verlag, Berlin 1986.
[2] TRbF 521, Carl Heymanns Verlag, Köln 1984; C.-H. Degener u. G. Krause, Kommentar zu VbF/TRbF „Lagerung und Abfüllung brennbarer Flüssigkeiten", Abschn. 6.2 „Kommentar zu TRbF 521", Carl Heymanns Verlag, Köln 1986.
[3] DIN 6608, Beuth-Verlag, Berlin 1981.
[4] Verordnung über Anlagen zur Lagerung, Abfüllung und Beförderung brennbarer Flüssigkeiten zu Lande (VbF), Carl Heymanns Verlag, Köln 1980.
[5] DIN 50928 Teil 3, Beuth-Verlag, Berlin 1985.
[6] DIN VDE 0100 Teil 410, Beuth-Verlag, Berlin 1983.
[7] DIN VDE 0150, Beuth-Verlag, Berlin 1983.
[8] G. Burgmann u. H. Hildebrand, Werkst. Korros. 22, 1012 (1971).
[9] AfK-Empflg. Nr. 9, ZfGW-Verlag, Frankfurt 1979.
[10] TRbF 522, Carl Heymanns Verlag, Köln 1988.
[11] Merkblätter des internationalen Eisenbahnverbandes UIC, Internationaler Eisenbahnverband, 14 rue Jean Rey, F 75015 Paris, Juli 1987.
[12] DIN VDE 0115 Teile 1 bis 3, Beuth-Verlag, Berlin 1982.
[13] Dienstvorschriften der Deutschen Bundesbahn, 901c, 954/2 und 954/6, Drucksachenzentrale DB, Karlsruhe, 1970/72/73.
[14] DIN VDE 0551 Teil 1, Beuth-Verlag, Berlin 1975.
[15] DIN VDE 0185 Teile 1 und 2, Beuth-Verlag, Berlin 1982.
[16] DIN VDE 0433 Teil 3, Beuth-Verlag, Berlin 1966.
[17] DIN 50927, Beuth-Verlag, Berlin 1985.
[18] Normvorlage Nr. 941-978024.2 TÜV Rheinland e.V., Köln 12/1987.
[19] AfK-Empfehlung Nr. 5, ZfGW-Verlag, Frankfurt 1986.
[20] TRbF 100, Carl Heymanns Verlag, Köln 1984.
[21] Verordnung über elektrische Anlagen in explosionsgefährdeten Räumen (Elex V), Bundesgesetzblatt I, S. 214, v. Juni 1980.
[22] DIN VDE 0165, Beuth-Verlag, Berlin 1983.
[23] Richtlinie für die Vermeidung der Gefahren durch explosionsfähige Atmosphäre mit Beispielsammlung (Ex-RL), ZH1/10, Carl Heymanns Verlag, Köln 1986.

12

Local Cathodic Protection

W. v. Baeckmann and W. Prinz

12.1 Range of Applications

The basis of conventional cathodic protection (see Section 10.2) is the electrical separation of the protected object from all installations which have a low grounding resistance. This separation, however, creates technical difficulties in industrial installations because of the large number of pipes with very large nominal diameters. These measures are not only very expensive but also susceptible to trouble due to possible foreign contacts or bridging of insulating joints during their operational life. This is particularly the case during alteration to and extension of the piping system. Technical difficulties arise in installations where there is danger of explosion and with pipelines that transport electrolytic solutions (e.g., cooling water, heating water, waste water, brine). With electrolytes of low resistivity and large pipe diameters, there is a danger of internal corrosion by the cathodic protection current at the unprotected side of insulating joints (see Sections 10.3.5 and 24.4.6). Altogether the monitoring and maintenance of the cathodic protection for such pipelines is very expensive.

The danger of corrosion is in general greater for pipelines in industrial installations than in long-distance pipelines because in most cases cell formation occurs with steel-reinforced concrete foundations (see Section 4.3). This danger of corrosion can be overcome by local cathodic protection in areas of distinct industrial installations. The method resembles that of local cathodic protection [1]. The protected area is not limited, i.e., the pipelines are not electrically isolated from continuing and branching pipelines.

Very high protective currents are usually required because of the very low grounding resistance of the total installation. The high cost of setting up the anode ground beds is offset by the saving in insulating joints and by the increased operational safety. Typical cases where it is used are pipelines, grounds, cables and storage tanks in power stations, refineries and tank farms. It is also used for piping in pumping and compressor stations, and measuring and control stations as well as for pipelines that cannot be electrically separated from concrete [2,3].

12.2 Special Features of the Local Cathodic Protection

The danger of corrosion on buried installations in industrial plants is increased by various soils and by cell formation with cathodes of steel in concrete. The rest potentials of these foreign cathodes are between $U_{Cu\text{-}CuSO_4} = -0.2$ and -0.5 V [4-6]. Factors that affect cell formation are the type of cement, the water/cement ratio and the aeration of the concrete [6]. Figure 12-1 shows schematically the cell action and the variation of the pipe/soil potential where there is contact with a steel-concrete structure. The cell current density is determined by the large area of the cathode [see Fig. 2-6 and Eq. (2-44)]. In industrial installations the area of steel surface in concrete is usually greater than 10^4 m^2.

Local cathodic protection has the objective, not only of compensating for the cell current of the foreign cathodic structures, but of sufficiently cathodically polarizing the protected object so that the protection criterion of Eq. (2-39) is fulfilled. A disproportionately large part of the protection current flows to the foreign cathodes due to the low contact resistance between the protected object and the foreign cathodic structures, and the very low grounding resistance of the latter. In setting up the impressed current anode beds, the objective is to increase the part of the current for the protected object. In addition to the geometrical dimensions of the protected object and foreign cathodic structures, the specific soil resistivity has a large influence. In contrast to conventional cathodic protection, the protected object lies mainly within the voltage cone of the impressed current anodes. For this reason and the very different protection current requirement of the individual components (object to be protected and foreign cathodic structures), the soil cannot be regarded as an equipotential space. The pipe/soil potential in local cathodic protec-

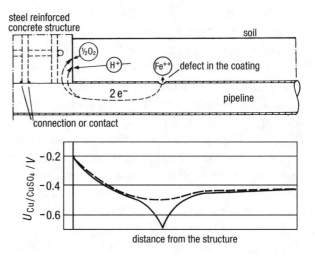

Fig. 12-1 Danger of corrosion caused by cell formation with steel-reinforced structures and of the pipe/soil potential.
——— at the pipe surface,
----- at ground level.

tion changes only in relation to a nearby reference electrode and less in relation to the potential of a remote ground. This involves considerable measurement problems since the off potential for interpretation cannot be directly measured. The pragmatic criterion No. 6 in Table 3-3 and monitoring with a measuring probe (No. 7 in Table 3-3) must be considered.

In the case of very strong polarization of steel in concrete, it was feared that corrosion field IV in Fig. 2-2 could be reached [7]. Tests have shown, however, that there is no danger of corrosion of the steel in concrete and any evolved hydrogen would be dispersed through the porous concrete (see Section 5.3.2).

In order to achieve complete cathodic protection for all pipes, the foreign cathodic structures (foundations, grounds) must be polarized to the protection potential, i.e., U_{on} in the neighborhood of the foreign cathodic structures must be definitely more negative than U_s (see No. 6 in Table 3-3). In order to polarize the reinforcement to the protection potential, an average protection current density of 5 to 10 mA m^{-2} is necessary, which decreases with time to 3 mA m^{-2}. To evaluate the protection current requirement with a reinforcement factor of 1, the total area of concrete exposed to the soil must be used. In comparison with this, the protection current requirement of the object to be protected is negligibly small. The protection current requirement in industrial installations is generally over 100 A.

Deep anodes are mainly used for injecting such high protection currents (see Section 9.1). The advice given in Section 9.1 on resistances and potential distribution relates to anodes in homogeneous soils. Large deviations are to be expected in soil used as backfill and in the neighborhood of structures [2]. This is generally the case with local cathodic ("hot-spot") protection.

Additional individual anodes must be installed at points on the protected object where a sufficiently negative pipe/soil potential cannot be achieved. Since usually only the voltage cone is of interest, the place of installation does not depend on the specific soil resistivity. Coke backfill is not necessary, and the place of installation is determined by the local circumstances. Individual horizontal anodes are conveniently installed parallel to the pipeline at the depth of the pipe axis. The voltage, length and distance of the anodes from the protected object are chosen according to Section 9.1 so that criterion No. 6 or No. 7 in Table 3-3 is fulfilled.

With local cathodic protection, the off potential measurement cannot be used directly to check the protective action because, due to the mixed type of installation of the protected object and foreign cathodic structures in the soil, there is a considerable flow of cell currents and equalizing currents. The notes to Eq. (3-28) in Section 3.3 are relevant here, where the *IR*-free potentials must be substantially more negative than the off potential of the protected object. If U_{off} is found to be more positive than U_s, this does not confirm or conclusively indicate insufficient polarization since in every case the *IR*-free potential, depending on the intensity of the equalizing currents, must be more negative. For this reason, the U_{off} value can only be used as a comparison and not for direct estimate. The results of measurements given below are to be used in the same way (see Figs. 12-2, 12-5, and 12-7).

Therefore, for controlling local cathodic protection, U_{on} values are generally measured with the reference electrode arranged as close as possible to the protected object. The measured values should be substantially more negative than $U_{Cu\text{-}CuSO_4} = -1.2$ V. It can, however, be that with a value of $U_{on} = -0.85$ V, the damaging cell formation with steel in concrete is avoided [2]. The least negative potentials occur where the protected object is in the wall entrance or near the steel-reinforced concrete foundation. It is recommended therefore that potential test points be installed at these locations as well as external measuring probes.

For efficient current distribution, steel-reinforced concrete walls should be provided at the wall entrance of pipes and at least 1 m around them and up to the soil surface with at least 2 mm thick electrically insulating layers of plastic or bitumen. This is also recommended if the pipelines are laid in soil parallel to steel-reinforced concrete foundations and the closest spacing is smaller than twice the pipe diameter or smaller than 0.5 m [2].

12.3 Power Stations

Cooling water pipes are essential for the operation of power stations and must not cease to function. Pipelines for fire fighting are also important for safety reasons. Such steel pipelines are usually well coated. At areas of unavoidable damage to the pipe coating, there is an increased danger due to cell formation between steel and concrete where local corrosion rates of ≥ 1 mm a^{-1} are to be expected [4]. Damage to pipelines for fire fighting has frequently been observed after only a few years in service.

Figure 12-2 shows as an example the arrangement of the anode installation for the local cathodic protection of pipelines in a power station. The cooling water pipelines have a nominal diameter of DN 2000 and 2500 and a covering of earth up to 6 m. The fire-fighting pipelines have a nominal diameter of DN 100 and a covering of 1 m. All the pipelines have a bitumen coating.

Since it relates to an existing power station, the points at which the pipeline enters the steel-reinforced concrete foundation have no electrically insulating coating. The soil has a high specific resistivity of 150 to 350 Ω m. To polarize the steel-reinforced concrete foundation, the required protection current of about 120 A is injected via eight deep anode installations according to Fig. 9-11 with six high-silicon iron anodes, each in well-conducting soil layers 20 to 50 m deep. Table 12-1 contains the information on the installations taking up the current and the loading

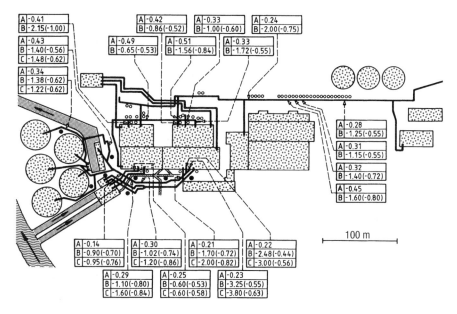

Fig. 12-2 Local cathodic protection in a power station. ● deep anodes; ○ horizontal anodes; Potential readings $U_{\text{Cu-CuSO}_4}$ in volts: (A) free corrosion potential before commissioning the cathodic protection; (B) U_{on} (U_{off}) 4 months after switching on the protection current; (C) 1 year later.

on the anodes. To achieve the desired current distribution in the area of the wall entrance of the cooling water pipelines to polarize the steel-concrete foundation, the deep anodes were arranged on the ingress side. The protection current for four deep anodes at a time was provided by one rectifier. For the necessary polarization of the cooling water pipelines in the area of the pipeline wall entrance, the additional installation of one vertical anode at a distance of 1 m before the wall entrance was necessary.

Only the cooling water pipelines were protected by the deep anodes and not the more distant fire-fighting pipelines. To protect the latter, a total of 45 horizontal anodes were installed along the pipeline which provided a protection current of 9 A. The arrangement and number of these anodes were determined from drainage cone tests. Since a large voltage cone was necessary to lower the pipe/soil potential, the otherwise usual coke backfill was dispensed with. The individual anodes were combined in four groups which were from time to time connected via variable resistances to a rectifier. This enabled the current and potential distribution to be adequately regulated.

External measuring probes were installed at the pipeline wall entrance points. The potentials of the measuring probes and the currents flowing between the probes

Table 12-1 Data for examples of local cathodic protection

Example	Designation	Area of surface concerned (m^2)	Protection current requirement (mA m^{-2})	Protection current requirement (A)	Example	Designation	Area of surface concerned (m^2)	Protection current requirement (mA m^{-2})	Protection current required (A)
Power station (Section 12.3)	Structural steel-reinforced concrete in soil	19,000	5	95	Refinery (Section 12.4)	Concrete foundation	95,000	2.5	237.5
	Cooling water pipeline	5500	0.1	0.55		Pipelines	10,000	0.05	0.5
	Buried cable	250	2	0.5					
	Copper grounding grid	2000	10	20					

| | Total current for local cathodic protection | | | | | Total current for local cathodic protection | | | |
Deep anode	Grounding resistance (Ω)	Protection current output (A)	Anode voltage (V)		Deep anode	Grounding resistance (Ω)	Current output (A)	Anode voltage (V)	
No. 1	0.8	17.1	15		No. 1	0.12	32.5	8.5	
No. 2	1.0	12.0	15		No. 2	0.10	30.7	5.0	
No. 3	0.8	15.3	15		No. 3	0.10	38.0	6.0	
No. 4	1.0	11.8	14		No. 4	0.16	33.0	7.5	
No. 5	0.9	13.5	14		No. 5	0.11	26.5	5.0	
No. 6	0.6	20.0	14		No. 6	0.12	25.0	5.0	
No. 7	1.0	14.4	15		No. 7	0.12	25.0	5.0	
No. 8	0.8	15.5	14						

and the pipeline give good information on the danger of corrosion and the protective action after setting up the local cathodic protection [9]. As an example, Fig. 12-3 shows the time dependence of current and potential for two measuring probes after cathodic protection was set up. One probe indicated that the current uptake and sufficient lowering of the potential could only be achieved by installation of an additional anode.

To ensure long durability, electrical grounding equipment in power stations is constructed of corrosion-resistant materials that have a very positive rest potential (e.g., copper with $U_{Cu-CuSO_4} = -0.1$ to -0.2 V). These grounds, like steel in concrete, lead to cell formation. Since copper can be polarized less easily than steel in concrete, local cathodic protection is difficult under these circumstances. Copper behaves more favorably with lead sheathing or hot-dipped galvanized steel, where according to Fig. 2-10 the zinc is cathodically protected at $U_{Cu-CuSO_4} = -1.2$ V. Thus considerably lower current densities are necessary than for the polarization of other grounding materials.

12.4 Oil Refineries

In contrast to power stations with small and large pipes, in refineries there are many small pipelines for fire-fighting water, waste water, products, etc. Furthermore, refineries have a large area. Conventional cathodic protection can possibly be used for the small pipelines, but it is actually susceptible to trouble because of possible electrical bridging of the numerous insulating joints required.

The protection current required for polarization of the steel-reinforced concrete foundation can be determined by a drainage experiment. In new installations this is relatively simple with individual electrically separated foundations [11]. In the foregoing case the results showed various current densities between 1.5 and

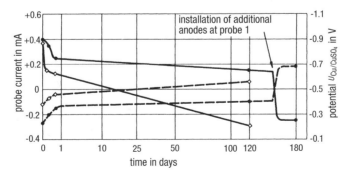

Fig. 12-3 Change in potential (― ― ―) and current (―――) with time at external measuring probes for local cathodic protection. ● probe 1; ◇ probe 2.

3.5 mA m^{-2}. With an average current density of 2.5 mA m^{-2}, the total area of the steel-reinforced concrete foundation of 95,000 m^2 was polarized with 200 A via seven deep anode installations and six transformer-rectifier units. The details are given in Table 12-1 and Fig. 12-4.

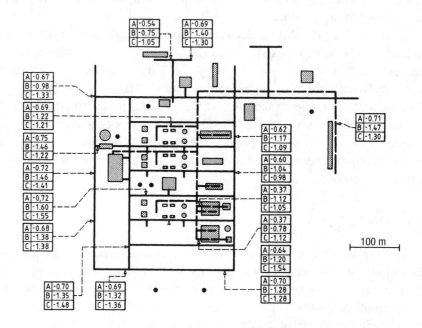

Fig. 12-4 Local cathodic protection in a refinery. ● deep anodes; potential readings $U_{Cu-CuSO_4}$ in volts: (A) free corrosion potential before commissioning the cathodic protection, (B) U_{on} 4 months after switching on the protection current, (C) U_{on} 1 year later.

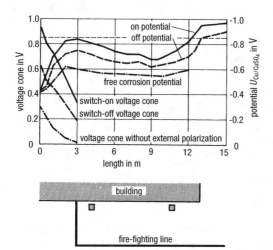

Fig. 12-5 Voltage cone ΔU and pipe/soil potentials at a wall entrance in a steel-reinforced concrete foundation.

In soil with a low specific resistivity, sufficient protection could be obtained with deep anodes alone. Only in the neighborhood of the foundation was the protection potential not completely reached. Figure 12-5 shows the voltage drops in the soil of a steel-reinforced concrete foundation, together with the pipe/soil potentials of a nearby pipeline. From this, it can be seen that the steel-concrete foundation-pipeline cell produces a voltage drop of 0.3 V in the soil (voltage cone without external polarization). This voltage drop is raised by the protection current and remains even after an interruption noticeably greater than before the injection of the protection current (switched-off voltage cone). This is due to the equalizing to the pipeline and clearly indicates the error in the U_{off} values. This case shows the technical difficulties in measuring local cathodic corrosion protection. Only a built-in measuring probe can give a clear indication of the pipe/soil potential.

Within 2 years the pipe/soil potential U_{on} immediately at the wall entrance of the pipeline changed from $U_{\text{Cu-CuSO}_4} = -0.45$ V to -0.7 V. Favorable potential values at these points are to be expected if the foundations have insulating coatings.

12.5 Installations with Small Steel-Reinforced Concrete Foundations

Pumping or compressor stations are necessary for the transport of material in pipelines. These stations are usually electrically separated from the cathodically protected long-distance pipeline. The concrete foundations are much smaller than in power stations and refineries. Since the station piping is endangered by cell formation with the steel-reinforced concrete foundations, local cathodic protection is recommended.

In soils with high resistivity, it is advisable to locate the impressed current anodes immediately next to the pipeline [12]. The pipelines then lie within the voltage cone of the anodes. Figure 12-6 shows the arrangement of the anodes for local cathodic protection of a pumping station. The distance of the anodes from the protected objects should be chosen according to Figs. 9-5 and 9-6 so that the pipe/soil potential is reduced by the protection current to $U_{\text{on}} = -1.2$ V. The voltage cones of the individual anodes will thus overlap.

Structures or pits for water lines are mostly of steel-reinforced concrete. At the wall entrance, contact can easily arise between the pipeline and the reinforcement. In the immediate vicinity of the pit, insufficient lowering of the potential occurs despite the cathodic protection of the pipeline. Figure 12-7 shows that voltage cones caused by equalizing currents are present up to a few meters from the shaft. With protection current densities of 5 mA m^{-2} for the concrete surfaces, even for a small pit of 150 m^2 surface area, 0.75 A is necessary. A larger distribution pit of 500 m^2 requires 2.5 A. Such large protection currents can only be obtained with additional impressed current anodes which are installed in the immediate vicinity of the pipe entry into the concrete. The local cathodic protection is a necessary completion of the conventional protection of the pipeline, which would otherwise be lacking in the pit.

318 Handbook of Cathodic Corrosion Protection

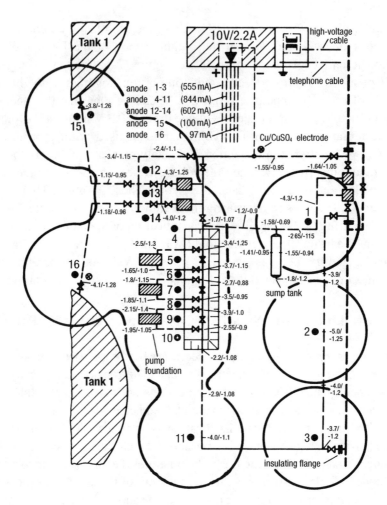

Fig. 12-6 Local cathodic protection of a tank farm in high-resistance soil using the anodic voltage cones of distributed anodes; the lines indicate soil potential values for an increase of 0.5 V relative to a remote ground; numerical pairs U_{on}/U_{off} in volts.

12.6 Tank Farms

Tank installations with underground storage tanks and station piping should, if possible, be provided with conventional cathodic protection [3]. This is sometimes not possible because electrical separation cannot be achieved between the protected installation and other parts of the plant (see Section 11.4). The necessity for cathodic protection can be tested as in Ref. 13. In tank farms, a distinction should be made between coated, buried storage tanks and aboveground, flat-bottomed tanks in which the base contacts the soil.

Local Cathodic Protection 319

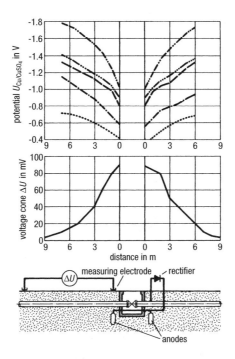

Fig. 12-7 Voltage cone and pipe/soil potential near a concrete pit.
——————— ΔU without external polarization,
· · · · · · · · · free corrosion potential,
— — — — U_{on} of the water pipeline,
— · — · — · — U_{off} of the water pipeline,
— ·· — ·· — ·· — U_{on} of the water pipeline with additional protection station,
— ··· — ··· — U_{off} of the water pipeline with additional protection station.

The same conditions as described in Section 12.1 apply for storage tanks that are coated and buried with operating station piping. The tanks are usually close together. For injection of the current, a relatively large number of anodes have to be arranged close to and up to 1 m from the protected objects (see Fig. 11-4). In Table 12-2 the number of anodes and the protection current of various tank farms with buried coated storage tanks are given [14].

In the local cathodic protection of the bottoms of flat-bottomed tanks, cell formation with steel-concrete foundations is of little importance since the surfaces are relatively small, in contrast to the installations in Sections 12.2 to 12.5. On the other hand, connected components of the installation, such as cables and grounds, take up considerable protection current. On account of the large foundations of flat-bottomed tanks, which are often bare or only poorly coated, polarization to the protection potential is only possible with very negative on potentials. In tank foundations with the

Table 12-2 Parameters of storage tanks with local cathodic protection

Tank						Soil resistivity
Number	Volume (m³)	Surface area (m²)	No. of anodes	Protection current (A)	Voltage (V)	(Ω m) [after Eq. (3-44)]
12	50	1104	39	9.3	14	$a = 1.6$ m: 7000
6	50	552	21	4.5	13	$a = 2.4$ m: 4000
10	25	554	36	7.7	14	
7	Diverse	114	13	4.0	9	$a = 1.6$ m: 9000
14	Diverse	762	28	7.4	11	$a = 2.4$ m: 4000
16	Diverse	360	25	6.0	8	
6	Diverse	445	16	2.8	9	

following construction, cathodic protection is possible with local cathodic protection [3]:

- concrete slab foundation with a covering of sand or bitumen;
- concrete ring foundation with the interior filled with gravel, grit, and sand, as well as with or without a bitumen covering layer;
- flat foundation of gravel, grit, and sand with or without a bitumen covering layer.

The installation of electrically insulating foils under the base of the tank or in its vicinity interferes with cathodic protection.

An accurate estimate of the protection current requirement is only possible with a drainage test on the filled container, in which the currents flowing to other grounded installations through connections can be taken into account. In particular, with tanks of large diameter, the base of the tank bulges upward when it is empty and is no longer in complete contact with the earth. Values obtained from drainage tests are then too low.

The arrangement of anodes for introducing a protection current with tank farms consisting of flat-bottomed tanks can be carried out as follows [15]:

- individual anodes below the tank base;
- individual anodes around the circumference of the tank;
- deep or horizontal anodes for several tank bases.

Covering the foundations of the tank with bitumen or a mixture of bitumen, sand and gravel ("oiled-sand") reduces the protection current requirement considerably. Figure 12-8 shows the protection current requirement of the base of some flat-bottomed tanks. The protection current densities of tank bases 1, 3 and 4 lie between 0.5 and 2 mA m^{-2}. The protection current density of tank No. 2 is very much greater.

Local Cathodic Protection 321

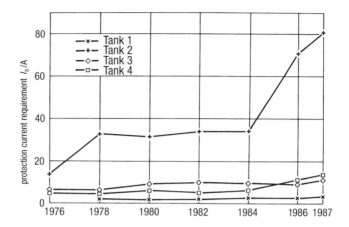

Fig. 12-8 Protection current requirement for a flat-bottomed tank.

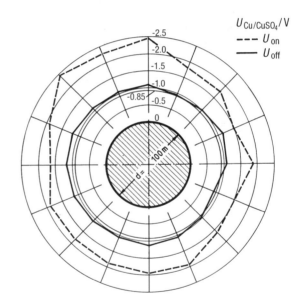

Fig. 12-9 Tank/soil potential around a flat-bottomed tank with $J_s = 1$ mA m^{-2}.

After a few years the tank base can no longer be cathodically protected by the initial protection current density of 3 mA m^{-2}. Only after installing additional anodes can the tank base be again cathodically protected with protection current densities of 10 mA m^{-2} with a total protection current of 80 A. The reason for the high protection current requirement is unsatisfactory insulation of the tank foundation.

Determination of tank/soil potentials is only possible at the outer edge of the tank. For monitoring the local cathodic protection, the distance between test points should not exceed 2 m [3]. The measurement of the tank/soil potential by the current interruption method is less prone to error than with the installation in Sections 12.2 to 12.4, due to the smaller equalizing currents.

Figure 12-9 shows on and off potentials that were measured around the circumference of a flat-bottomed tank 100 m in diameter. These values, however, give no information on the tank/soil potentials at the center of the container or at points away from the edge of the tank. In new tank constructions, long-life reference electrodes are therefore installed in the center of the base where the most positive potentials are found [15].

12.7 References

[1] M. E. Parker, Rohrkorrosion und kathodischer Schutz, Vulkan-Verlag, Essen 1963.
[2] AfK-Empfehlung Nr. 9, ZfGW-Verlag, Frankfurt 1979.
[3] TRbF 522, Carl Heymanns Verlag, Köln 1987.
[4] W. Schwenk, gwf gas/erdgas *127*, 304 (1986).
[5] H. Hildebrand und W. Schwenk, Werkstoffe und Korrosion *37*, 163 (1986).
[6] H. Hildebrand, C.-L. Kruse und W. Schwenk, Werkstoffe und Korrosion *38*, 696 (1987).
[7] R. A. King, H. Nabuzaueh u. T. K. Ross, Corr. Prev. Contr. *24*, H. 4, 11 (1977).
[8] H. Hildebrand, M. Schulze und W. Schwenk, Werkstoffe und Korrosion *34*, 281 (1983).
[9] W. Prinz, 3R intern. *17*, 466 (1978).
[10] DIN VDE 0151, Beuth-Verlag, Berlin 1986.
[11] W. v. Baeckmann u. K. Klein, Industrie-Anzeiger *99*, 419 (1976).
[12] W. v. Baeckmann, 3R intern. *26*, 310 (1987).
[13] TRbF 521, Carl Heymanns Verlag, Köln 1983.
[14] W. Behringer, HDT-Essen, Vortrag vom 26. 10. 1987.
[15] K. C. Garrity u. M. J. Urbas, NACE-Conference, San Francisco 1987, paper 320.

13

Telephone Cables

C. Gey

Telephone cables need different corrosion protection measures than pipelines because of their particular construction and operating conditions. All telephone cables have either a completely solid metal sheathing around the core or in the case of completely plastic cables, a metal band as an electric shield [1-3]. With cables that have a protective wrapping of jute and a viscous material around the cable sheathing, the leakage load G' is much higher than in cables with a protective sheathing of plastic. The cable sheathing or the shield is connected with the operational ground in the transmission or booster stations in order to improve the shielding action of the sheathing or shield. Glass fiber cables are the exception in which the shield is not grounded.

Formerly, cables with a metal sheathing were predominantly used. These cables should be cathodically protected in the case of corrosion risk (see Chapter 4). Modern cables (such as multiple sheathing or glass fiber cables) have a protective wrapping of plastic. Generally cathodic protection is not necessary in this case. In exceptional cases multiply sheathed cables may need cathodic protection if grounded metallic sleeves are used at splicing points. With the increased use of cables with plastic protective wrapping, corrosion problems only arise in grounding installations of these cables, entry structures into buildings, and cable attachments (booster and sleeves), as well as lightning protection systems.

13.1 Laying Cables

Telephone cables are either laid directly in the ground or drawn through ducts. Formerly, cable duct-shaped blocks (1000 mm long) were used for these ducts. Usually several of the cable duct blocks were laid connected to each other and the junctions between them cemented. Such ducts gave no electrical insulation against the soil. The junctions were not watertight so that constituents of the soil got into the duct with penetrating water. In the past, lead cables without coatings and cables with a sheathing material other than lead were laid in these ducts. Today cables with a plastic coating (or sheath) are generally used.

For several years now, cable ducts have been manufactured from plastic pipes, which are watertight and form a continuous run of piping. In laying the ducts, low points can occur in which condensed water or water penetrating from the ends can collect. In many cases this water has led to corrosion damage in lead-sheathed cables. Lead-sheathed cables must therefore only be used in such ducts with an additional PE sheath of type A-PM2Y. Cathodic protection of these cables is not possible because of their complete insulation by the plastic pipe.

13.2 Passive Corrosion Protection

Passive corrosion protection can be achieved by coating the metallic cable sheathing (see Chapter 5). In the past, lead was mainly used for the cable sheathing. Today steel, copper, and aluminum are being increasingly used as the sheathing material or shield. It can be seen from Table 2-1 that for the same anodic current density, the corrosion rate is 2.5 times higher for lead than for iron. It should be remembered that telephone cables have a much thinner wall than pipelines so that lead-sheathed cables are much more prone to damage than pipelines. Lead-sheathed cables without additional coating are therefore no longer used.

For direct burial, the lead sheathing is protected against corrosion by alternating layers of impregnated paper and viscous bitumen. Cables of small diameter are mechanically protected with closely wound round wires, and those of larger diameter with flat tape reinforcement. A layer of impregnated jute is applied over the reinforcement, which gives some corrosion protection but does not insulate the cable sheathing from the soil. In contrast, PE sheathing, free of junctions and pores, 1.6 mm to 4.0 mm thick, is also used. Cathodic protection is thus mainly applied to jute-covered, lead-sheathed cables and plastic-sheathed cables that have external reinforcement (e.g., lightning protection cable). Cables with other metallic sheathing can be included in cathodic protection, but special arrangements have to be made [4].

With corrugated steel-sheathed cables, the cable tube is enclosed with a band of carbon steel welded longitudinally without overlap. The sheathing core produced in this way is corrugated to make it easier to bend. The undulations are filled with a viscous material that adheres to metal and plastic and is wrapped around with a plastic tape. Then a plastic casing of PE is extruded above this layer. This plastic sheathing is virtually free of pores and therefore gives good corrosion protection. Only sleeves and mechanically damaged areas are exposed to the soil.

Cables with a copper sheathing are used only seldom. The protective cover is the same as with a corrugated steel-sheathed cable. If a cable with copper sheathing is connected to a lead-sheathed cable (A-PMbc) (see Table 13-1), the copper sheathing acts as a cathode in a galvanic cell and is therefore cathodically protected.

Table 13-1 Rest and protection potentials for telephone cable in soil ($U_{Cu\text{-}CuSO_4}$ in V)

Cable type	Construction and material	Rest potential	Protection potentials U_s		U'_s
			Aerobic	Anaerobic	
A-PM	Lead sheath, not coated	−0.48 to −0.53	−0.65	−0.65	−1.7
A-PM2Y	Lead sheath, PE coated	−0.50 to −0.52	−0.65	−0.65	−1.7
A-PMbc	Lead sheath, steel strip armoring	−0.40 to −0.52	−0.85	−0.95	−1.7
A-Mbc	Lead sheath, galvanized steel wire armoring	−0.45 to −0.65	−0.85	−0.95	−1.7
A-PWE2Y	Corrugated steel sheath, PE protective coating	−0.65 to −0.75	−0.85	−0.95	—
A-PLDE2Y	Aluminum sheathing, PE protective coating	−0.90 to −1.20	−0.62	−0.62	−1.3
	PE-cable with copper screen	+0.10 to −0.10	−0.16	−0.20	—
	Galvanized steel pipe, uncoated	−0.90 to −1.10	−1.28	−1.28	−1.7

Since copper-sheathed cables are also coated with plastic, the ratio of cathode/anode area (S_c/S_a) is very small so that there is not an increased risk of corrosion of the lead-sheathed cable by the electrical connection between the cable sheathings according to Eq. (2-44).

Aluminum-sheathed cables should not be connected to other cables because aluminum has the most negative rest potential of all applicable cable sheathing materials. Every defect in the protective sheath is therefore anodically endangered (see Fig. 2-5). The very high surface ratio S_c/S_a leads to rapid destruction of the aluminum sheathing according to Eq. (2-44). Aluminum can also suffer cathodic corrosion (see Fig. 2-11). The cathodic protection of aluminum is therefore a problem. Care must be taken that the protection criterion of Eq. (2-48) with the data in Section 2.4 is fulfilled (see also Table 13-1). Aluminum-sheathed cables are used only in exceptional cases. They should not be laid in stray current areas or in soils with a high concentration of salt.

Cables with multiple-layer sheathing have plastic-insulated cores. Solid PE or sintered PE is used as the plastic. Sintered PE is a foamed polyethylene material that has different electrical properties than solid PE. Under certain circumstances the core region is filled with a petrolatum material to give protection against condensation and to make it longitudinally watertight. Plastic tape and a metal band as a shield are wound around it. The metal band consists of aluminum or copper and is coated with plastic. In addition, a plastic coating of PE is extruded over the metal band. Public telephone and telegraph cables with a plastic coating are stamped with a prominent symbol that is only allowed on PE and not on PVC, since deep stamping can lead to internal stresses and the material can suffer stress corrosion.

13.3 Cathodic Protection

Telephone cables and their lightning protection systems can be protected against corrosion in the same way as pipelines. According to the type of cable and cable construction, different protected lengths are achieved. The range of cathodic protection is considerably greater with well-coated cables than with bitumen-jute coatings. Equation (24-75) applies in calculating the length of the protected range, which is that much greater

- the better the protected object is coated (small J_s),
- the lower the resistance load (R') is.

In contrast to pipelines, cables and their lightning protection have only poor sheathing or none and therefore a greater leakage as well as a considerably higher resistance load, so that the length of the protected range in cathodic protection is small. This necessarily means that limits are imposed on corrosion protection. Despite these disadvantages, cathodic protection of telephone cables can be applied economically by accurate calculation and careful planning.

Whereas cathodic protection—apart from a few exceptions—can be relatively easily applied outside built-up areas, in urban areas difficulties arise due to the numerous metallic installations below ground. Interference with nearby installations has to be considered, on account of the relatively high protection currents (see Section 9.2).

The switching-off method for IR-free potential measurement is, according to the data in Fig. 3-5, subject to error with lead-sheathed cables. For a rough survey, measurements of potential can be used to set up and control the cathodic protection. This means that no information can be gathered on the complete corrosion protection, but only on the protection current entry and the elimination of cell activity from contacts with foreign cathodic structures. The reverse switching method in Section 3.3.1 can be used to obtain an accurate potential measurement. Rest and protection potentials for buried cables are listed in Table 13-1 as an appendix to Section 2.4. The protection potential region lies within $U'_s < U < U_s$.

13.3.1 Stray Current Protection

In very built up areas, telephone cables are laid in ducts that mostly consist of shaped duct blocks. This means that cables running parallel to tramway tracks cannot be avoided. Metal-sheathed cables with poor coating, or with none at all, are then heavily exposed to stray currents from the tramway [5,6].

Measurement of the cable sheathing/soil potential can be used to assess the corrosion danger from stray current interference (see Section 15.5.1). Since the measured values vary widely and the stray currents cannot be switched off, IR-free potential measurements are only possible with great effort. In order to keep the IR term of the potential measurement low, the reference electrode must be placed as close as possible to the measured object. With measurements in cable ducts (e.g., underneath tramway tracks), the reference electrodes can be introduced in an open duct.

The magnitude and direction of stray currents flowing in the cable casing can be determined from measurements of the sheathing current. All formerly known measurement methods for determining entering and exiting currents are too imprecise to quantitatively determine the emerging stray currents [6,7]. The cable sheathing resistance necessary for calculating the sheathing current from the voltage drop along a definite length of cable without soldered sleeves can be seen in Fig. 13-1.

Direct current installations that are grounded in several places cause stray currents in the soil which can interfere with other installations (see Section 9.2). All dc railways are sources of stray currents. Protection methods that can be applied in the same way to cables are described in Chapter 15.

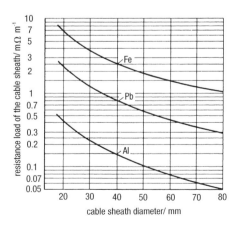

Fig. 13-1 Resistance load R' of cable sheathing as a function of cable diameter d and sheathing material.

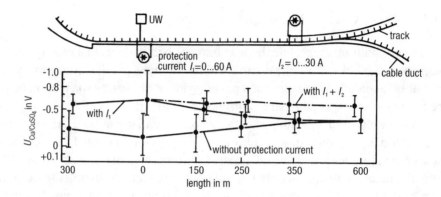

Fig. 13-2 Potential variation at a telephone cable in an area affected by stray currents in a large town with two stray current drainages.

In stray current drainage (see Fig. 15-5) in urban areas, 10 to 15% of the supply current of a tramway transformer substation is often drained. Sometimes up to 300 A can flow from cable sheathing via stray current drainage back to the source of the stray current. The lowering of potential by stray current drainage often acts for a few hundred meters with unsheathed lead cables due to their high leakage load and the proportionately high cable sheathing resistance load [8].

The active length of the protection system can be increased by installing forced stray current drainage (see Fig. 15-7). These protection measures are used when the dc installation at the point of connection to the stray current drainage is not sufficiently negative or if a much greater voltage drop arises due to the resistance of the stray current return conductor, so that the required protection potential can no longer be attained.

Potential-regulating instruments are predominantly used for forced stray current drainage (see Chapter 8). Potential control lowers current consumption and reduces the interference of foreign installations, because by it the potential of the protected object is lowered only enough for the required protection of the installation. Several smaller protection current devices are advantageous in the protection of wider cable areas since this gives uniform potential lowering at the protected object and there is less interference with foreign installations. Figure 13-2 shows the potential distribution for a cable with stray current drainage. The total current drained amounts to about 70 A.

All protection measures must be checked and monitored at regular intervals to guarantee their effectiveness. To keep the labor cost as low as possible, the time interval between inspections should be as long as possible. This requires that the components of the installation, such as transformer-rectifiers, which can often fail

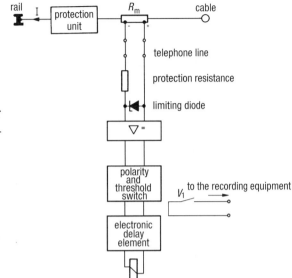

Fig. 13-3 Circuit diagram for a corrosion protection monitoring unit.

due to overloading or overvoltages, be monitored remotely [9]. Equipment for remote monitoring of forced stray current drainage works according to the principle in Fig. 13-3. The equipment should signal any disturbance under the following conditions:

- a current greater than 2 A flows from the rails to the cable. The recording apparatus should respond after a delay of 8 s. This delay is necessary so that no alarms are set off by current impulses along the telephone cables;
- the drainage current is smaller than $0.3\,I_{max}$ for longer than 8 h. This time delay is necessary since during operational shutdown of the source of the stray current, the current does not reach $0.3\,I_{max}$ for long periods of time.

The signal is indicated on a central control board.

13.3.2 Cathodic Protection with Impressed Current Anodes

Cathodic protection with impressed current anodes is used predominantly with cables or steel casing in which the cable is inserted, outside built-up areas where it is possible to build large anode installations without damaging interference with other lines. In densely populated areas, protection with impressed current anodes is often only possible with deep anodes, with surface anodes or locally at individual problem points (local cathodic protection, see Chapter 12).

Anodes of small impressed current protection installations can be installed close to the cable or the cable duct. They can, however, also be inserted as an

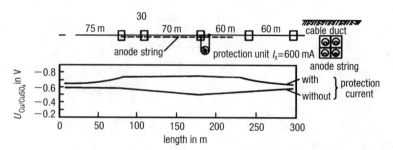

Fig. 13-4 Potential variation of telephone cables in a damp cable duct constructed of cable channel-shaped stone (without stray currents).

anode string in an open channel of a cable duct of duct-shaped blocks. This reduces interference with other lines. Figure 13-4 shows the potential distribution for three telephone cables protected with an anode string. Long-distance cables with metal sheathings should always be cathodically protected since extended corrosion cells are formed from factors arising from the soil (see Section 4.2.1).

Lead-sheathed cable near cathodically protected pipelines must be connected to the cathodic protection of the pipeline since they will otherwise suffer interference from the protection current (see Section 9.2). Figure 13-5 shows the current requirement for a 10-m length of cable and the potential distribution of such cables.

The same applies to crossings with cathodically protected pipelines to combat interference (see Section 9.2). An untenable interference by the protection current of the pipeline can nearly always be avoided if:

- the cable is coated with PE in the region of the voltage cone of the anode installations (about 100 m radius around the anode installation),
- the cable is coated with PE free of defects for at least 3 m on either side of the crossing point with the pipeline [4].

Telephone cables that lead to telephone towers are very prone to lightning damage. To avoid lightning damage, these cables should either be enclosed in gal-

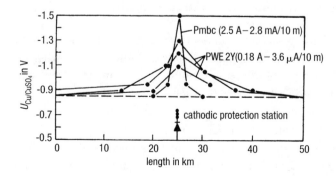

Fig. 13-5 Potential variation at a telephone cable adjacent to a cathodically protected pipeline.

vanized steel casings or laid with lightning protection. With galvanized steel casings, the individual lengths of pipe (6 m) are connected with screwed sockets that are watertight. The individual sockets must be bonded, since otherwise the resistance load of the pipeline is increased so much that cathodic protection of the pipe duct is no longer possible. Both ends of the casing duct must also be sealed watertight [2,4]. If the casing duct does not lie in frost-free soil, then cathodic protection must be applied, depending on the soil conditions. If water penetrates corrosion defects in the pipe, the water freezes in heavy frost and the cable can be damaged by pressure from the ice.

Cables with lightning protection have galvanized steel wire reinforcement without additional coating over the plastic coating to improve the leakage load G' of the cable sheathing. This outer reinforcement usually consists of 2-mm thick flat wire which is particularly endangered by corrosion in the soil. In order for the lightning protection to last the entire life of the cable, this reinforcement should always be cathodically protected.

The protection current density for uncoated galvanized steel casings and for the galvanized flat wire reinforcement is relatively high. It lies between 20 and 30 mA m^{-2}. Due to the high specific soil resistivity in rocky ground, it is often difficult to construct a correspondingly low-resistance anode installation. If the output voltage of the rectifier of the protection unit exceeds 70 V, the corresponding overvoltage lightning arrestor will not work in the event of a lightning strike and the rectifier unit will be disabled. These problems can be overcome by using a new type of flexible plastic cable anode (see Section 7.2.4). With these polymer cable anodes, low-resistance anode installations can be created, even in very high-resistance soils. These polymer cable anodes can be laid parallel to the protected object at a distance >1 m. With long lengths of the anode cables, there is a somewhat lower grounding resistance than with more conventional installations. This type of construction is proportionately more expensive but it has the following advantages:

- low output voltage and therefore lower capacity transformer-rectifiers,
- uniform potential lowering along the protected object,
- little interference with other installations.

Figure 13-6 shows the potential distribution along a galvanized steel casing pipe duct protected by a flexible polymer cable anode that leads to a transmitter. This pipe duct is 1000 m long. The specific soil resistivity along the track of the pipe is between 800 Ω m and 1600 Ω m. The flexible polymer cable anode is laid in fine coke at a distance of 1 m to one side of the pipe. The grounding resistance of the anode installation amounts to 1.2 Ω. For a protection current of 3.5 A, an output voltage from the rectifier of 11.5 V is required, and the electric power for the protection is 40 W.

332 Handbook of Cathodic Corrosion Protection

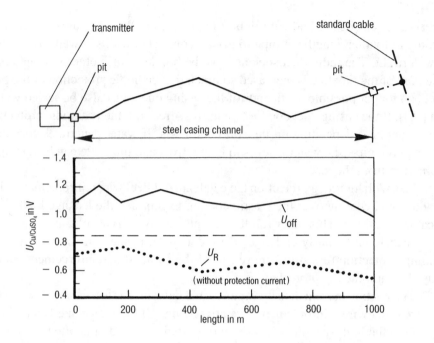

Fig. 13-6 Potential variation of a galvanized steel casing pipe channel cathodically protected with a flexible polymer cable anode.

Steel casings and lightning protection reinforcement of a cable must only be connected to the operational ground via spark gap isolators. With very high voltages (e.g., with a lightning strike), the spark gaps respond and conductively connect the installations to avoid dangerous contact voltages and damage to the telephone equipment. As soon as the voltage falls, the installations are again electrically separated. By this method separation is achieved:

- the corrosion cell "steel-reinforced concrete (foundation ground)/steel pipe" or lightning reinforcement is interrupted,
- the protection current requirement of the protected object is kept low,
- the protected range is limited.

So that the function of the spark gap isolators is not cancelled, cables with lightning protection must have insulated sleeves installed at their ends, in addition to the spark gap isolators, so that the metal sheathing or the shield that is connected at soldered joints to the lightning reinforcement is also separated from the operational ground of the building (foundation ground) and from the ground at the remote end.

Fig. 13-7 Potential of a cathodically protected special cable with galvanized steel armoring (zinc coating already corroded).

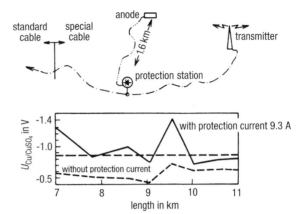

If insulated sleeves and spark gap isolators lie directly in soil, they constitute danger spots. Currents can exit and then reenter. This danger is reduced if they are installed in a dry pit and the first meter of the cable on either side is insulated.

Figure 13-7 shows the potential plan for the cathodic protection of a special cable (cable with lightning protection) that leads to an antenna installation and has unsheathed galvanized steel armoring. A low-resistance site for the anode installation could only be found at a distance of 1.6 km due to the rocky ground. The transformer-rectifier unit is protected against lightning strikes with overvoltage arrestors. The operation of the protection station is remotely monitored by a cable lead and shut down by remote control when there is warning of a thunderstorm.

13.4 References

[1] DIN VDE 0816, Beuth Verlag, Berlin 1979.
[2] Fernmeldebauordnung der Deutschen Bundespost, Teil 10 A, Bundesminister für das Post- und Fernmeldewesen, Bonn 1983.
[3] Fernmeldebauordnung der Deutschen Bundespost, Teil 12 A, Bundesminister für das Post- und Fernmeldewesen, Bonn 1973.
[4] Fernmeldebauordnung der Deutschen Bundespost, Teil 16 B, Bundesminister für das Post- und Fernmeldewesen, Bonn 1987.
[5] Fernmeldemeßordnung der Deutschen Bundespost, Teil IV B, Bundesminister für das Post- und Fernmeldewesen, Bonn 1988.
[6] A. Reinhard, VDI-Forschungsheft 482, Ausgabe B, S. 26, Düsseldorf 1960.
[7] Chr. Gey, Unterrichtsblätter der Deutschen Bundespost, Ausgabe B, S. 20, 40. Jahrgang, Nr. 1, 10. 1. 1987, Oberpostdirektion Hamburg.
[8] W. v. Baeckmann, in: Handbuch des Fernmeldewesens, Springer-Verlag, Berlin 1969, S. 360.
[9] Chr. Gey, Unterrichtsblätter der Deutschen Bundespost, Ausgabe B, S. 371, 40. Jahrgang, Nr. 10, 10. 10. 1987, Oberpostdirektion Hamburg.

14

Power Cables

H. U. PAUL AND W. PRINZ

14.1 Properties of Buried Power Cables

Cables for domestic and industrial electricity for low-voltage (220-380 V) medium-voltage (1-30 kV) and high-voltage supplies (chiefly 110 kV) are commonly buried. For low- and medium-voltage grids, plastic cable is used (e.g., for low-voltage types NYY and NAYY, which require no corrosion protection). Cables with copper shielding and plastic sheathing (e.g., types NYCY and NYCWY) are sufficiently corrosion resistant. Danger of corrosion exists in cables with lead sheathing and steel armoring, which are only coated with a layer of jute impregnated with bitumen, and also in cables sheathed with aluminum and corrugated steel with a plastic coating if the coating is damaged. This also applies to 110-kV cable, which is usually buried in steel pipes with bitumen or plastic coating.

To prevent unacceptably high contact voltages [1,2], metal sheathing of power cables in transformer and switching stations and in the distribution network is connected to grounded installations, which have a low grounding resistance. This increases the danger of corrosion and makes corrosion protection more difficult if:

(a) there is danger of corrosion due to cell formation with foreign cathodic structures (e.g., steel-reinforced concrete foundations or grounding installations) [3];
(b) there is danger of corrosion from stray currents (e.g., from installations for dc railways);
(c) stray currents can be picked up by grounds;
(d) there is an inability to apply the normal cathodic protection since the grounding is in conflict with the assumption (b) in Section 10.2.

The corrosion danger in (a) and (b) cannot be avoided by providing an improved coating as recommended in Section 4.3, if absolute freedom from defects cannot be guaranteed. From experience, areas of damage to the coating of steel conduits carrying high-voltage cables cannot be avoided even with careful construction practices. Avoidance of the corrosion danger is only possible in this case by cathodic protection and stray current protection. According to Fig. 2-12 in Section 2.4, a limitation to negative potentials has to be taken into account with lead sheathing. Since aluminum can corrode cathodically, as well as anodically, there is a corresponding limitation due to the small critical potential range (see Section 2.4); in particular with stray current interference, the process is hardly possible from a technical standpoint. The plastic coating of aluminum sheathing must be absolutely free of defects [4,5].

The absence of defects is determined by r_u measurements with dc voltage according to Eq. (5-11). Defects can be detected according to Eq. (5-10) only if the specific coating resistance r_u^0 (see Table 5-2) is greater than the specific defect resistance r_g in Eq. (5-5). From this it follows, for example, that for a very small defect of diameter $d = 0.1$ mm in a 2-mm thick coating with $\rho = 100$ Ω m, $r_g = 2.6 \times 10^7$ Ω. For a plastic coating with $r_u^0 = 10^{10}$ Ω m², this resistance would be reached with a surface area $S = 385$ m². Therefore only defects with $d > 0.1$ mm can be detected. To measure the very high defect resistance of 10^7 Ω, voltages of between 0.1 and 1 kV must be measured [4,5]. Erroneous measurements must be expected here as a result of gas formation. Such r_u measurements are only possible with recently laid and unconnected cable sheathings. The later occurrence of defects or pores is not detected on inspection.

The difficulty mentioned in (c) can be overcome by local cathodic protection according to Chapter 12. This is possible in industrial installations, but not in urban areas. Effective local cathodic protection can be achieved by strategically placed anodes (see Section 12.2) and by a limited protection range for objects with high longitudinal resistance. This is the case with lead-sheathed cables (see Fig. 13-1). Conventional cathodic protection is possible if the protected object is isolated from the grounding installation by dc decoupling devices. This is done for the steel conduits for high-voltage cables.

14.2 Cathodic Protection of the Steel Conduits for Power Cables

In 110-kV high-voltage grids, gas pressure cables (external and internal gas pressure cables) are mainly laid in steel conduit. In service, the steel pipe is filled with nitrogen at a gas pressure of 15 to 16 bar. Figure 14-1 shows the cross-section of such an external gas pressure cable.

The same consideration applies in the cathodic protection of this steel conduit as for pipelines (see Section 10.2). The electrical isolation of all other metallic structures in contact with grounds from the steel conduits carrying high-voltage cables is achieved by insulating the sealing ends of the steel conduits from the grounding installation. To exclude impermissible contact voltages with failures in the electrical grid, they must be connected to the station ground via dc decoupling devices (see Fig. 14-2) [6].

Power Cables   337

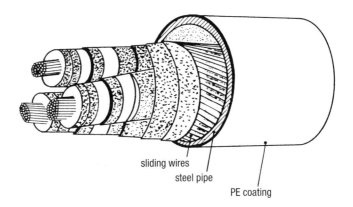

Fig. 14-1 Construction of an external gas-pressurized cable.

sliding wires
steel pipe
PE coating

14.2.1 Requirements for dc Decoupling Devices (between Casing and Ground)

The following requirements of dc decoupling devices must be met in normal operation and with grid failures:

- they must not allow any impermissibly dangerous voltages to arise;
- they must maintain currents that are thermally and mechanically constant;
- they must be designed to cope with expected transient overvoltages;
- the effectiveness of cathodic protection must be ensured;
- they should have as low a resistance as possible so that the reducing action of the pipe is maintained in the event of a short circuit.

Transient overvoltages are caused by switching operations in high-voltage installations. Thus, for example, in grounding conductors, momentary high voltages and currents can occur. In transient overvoltages, returning oscillations

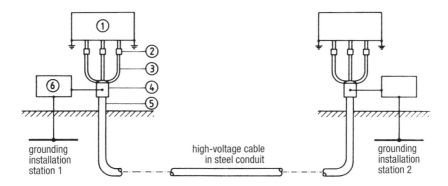

Fig. 14-2 High-voltage cable grounded via dc decoupling devices: (1) high-voltage installation, (2) insulating flange, (3) cable branching, (4) cable end sealing, (5) steel conduit, (6) dc decoupling devices.

with frequencies between a few kilohertz and several megahertz are frequently involved. Higher transient overvoltages related to frequency and amplitude occur more often in SF6 gas-insulated installations than in open-air installations.

Mechanical and thermal loading has to be taken into account when considering the restriction in the permissible contact voltages [2]. The permissible contact voltages are shown in Fig. 14-3 as a function of the length of time of a current failure. The connections from the steel pipe and the grounding installation to the dc decoupling device should be as short and as inductively low as possible to reduce the arcing tendency at the insulating units at the cable sealing ends.

14.2.2 Types and Circuits of dc Decoupling Devices

14.2.2.1 Low Ohmic Resistances

Figure 14-4 shows the circuit for a dc decoupling device with low ohmic resistances of about 0.01 Ω. With this decoupling device, no impermissible contact voltage arises, even with high failure currents. For a failure lasting up to 0.5 s with grounding short-circuit currents of up to 15 kA, the resulting contact voltages of 150 V are under the allowable limit. For cathodic protection, the necessary dc voltage separation of the steel conduit from the ground installation is achieved by means of a dc voltage source that causes a voltage drop of about 1 V at the resistance or a combination of resistances (see Section 14.2.3) [6].

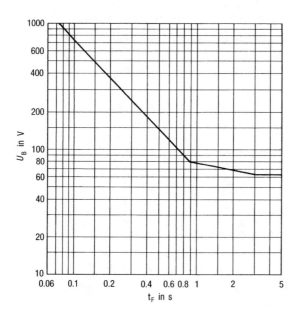

Fig. 14-3 Allowable contact voltages U_B as a function of the period t_F of the failure current.

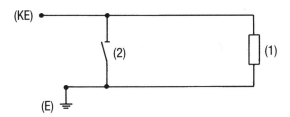

Fig. 14-4 Circuit diagram for a low-resistance dc decoupling devices. (KE) insulated cable end sealing, (E) grounding installation, (1) resistance (~10 mΩ), (2) grounding side bar or short-circuit-resistant grounding switch.

14.2.2.2 Higher Ohmic Resistances

In this context, higher ohmic resistances of about 100 mΩ are understood. These resistances cannot be loaded as high as the 10-mΩ resistances. To protect these higher ohmic resistances against overvoltages, a breakdown fuse (voltage fuse) is connected in parallel. The fuse melts as soon as the voltage drop at the resistance reaches the nominal voltage of the breakdown fuse. Figure 14-5 shows the circuit of such an arrester.

In grids with resonant grounding, in the event of a current failure the residual ground fault current or the coil current, or parts thereof, can flow for several hours through the resistance. The coil current can be up to 400 A, depending on the size of the grid. Failure currents and voltages on cathodically protected pipes are described in detail in Ref. 8.

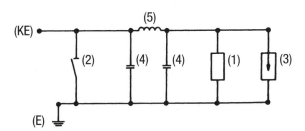

Fig. 14-5 Circuit diagram for a higher resistance dc decoupling devices. (KE) insulated cable end seal, (E) grounding installation, (1) resistance (~100 mΩ), (2) grounding side bar or short-circuit-resistant grounding switch, (3) breakdown fuse, (4) capacitors (ca. 60 μF), (5) choke.

340 Handbook of Cathodic Corrosion Protection

Unnecessary loading of the breakdown fuse through transient overvoltages can be avoided by connection to a π element which consists of a length choke and transverse capacitors. So-called iron core chokes are most conveniently used for series chokes, which are usual in power electronics. A damping element with a 67-μF capacitor is advised at the input and output of the π element.

14.2.2.3 dc Coupling Devices with Nickel-Cadmium Cell

Nickel-cadmium cells have a very low ac resistance of 1 mΩ. The charge state of the cells is of secondary importance. Nickel-cadmium cells must have sufficient current capacity and have current stability. They can be used directly as a dc decoupling device (Fig. 14-6) [6].

For normal operation, the breakdown fuse is not connected because if it is called upon to act, the cell would be short circuited, which could lead to its destruction. When there is work being done on the cable and the cathodic protection station is switched off, the fuse is connected by closing the switch. The cell (2) can be isolated by removing the connection (4). Finally, by closing the connection (1), direct grounding is established. In installing the cell, the process is carried out in the reverse direction.

The charge state of the cell must be maintained in operation to have a cell voltage of 0.9 to 1.2 V [6]. Overcharging the cell is to be avoided due to electrolytic decomposition of water and evolution of gas. The cell voltage should therefore not exceed 1.4 V. Cathodic protection stations should be operated so that the cell voltage lies in the desired range.

14.2.2.4 Polarization Cell

The polarization cell is an electrochemical component in which nickel or stainless steel electrodes are immersed in 50% KOH solution [9]. With ac, the

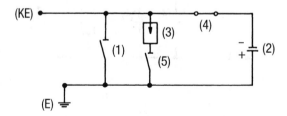

Fig. 14-6 Circuit diagram for a dc decoupling device with nickel-cadmium cell. (KE) insulated cable end sealing, (E) grounding installation, (1) grounding side bar; (2) NiCd cell, 1.2 V; (3) breakdown fuse; (4,5) isolating links.

polarization cell has a resistance on the order of a few milliohms because of the good conductivity of the electrolyte and the low polarization impedance of passive materials with electronic conductive passive layers. The relatively high dc resistance of several tens of ohms corresponds to the high polarization resistance of passive electrodes in redox-free media. This resistance breaks down if redox reactions according to Eq. (2-9) can take place, or if the voltage of the polarization cell exceeds the decomposition voltage of water (1.23 V according to Fig. 2-2), because the electrochemical reactions of Eqs. (2-17) and (2-18) take place.

In operation, small diffusion currents of up to 1 mA can flow because of unavoidable traces of redox substances (e.g., oxygen from the air). When subjected to ac, there is usually a charge reversal of the electrode capacity. However, with sufficiently high voltage amplitudes, a Faraday current flows because of the electrolysis of water. In order to keep this sufficiently small, the long-term ac should not be greater than 0.1% of the maximum permissible short-circuit current. Otherwise the polarization breaks down and the dc is no longer shut off. Figure 14-7 shows the circuit of the dc decoupling device.

The polarization cell must be inspected at regular intervals (e.g., twice a year) to check water loss caused by electrolysis. If necessary, the correct level must be restored with deionized water. In addition, the electrolyte should be renewed every 4 years. It is recommended that the dc decoupling device be designed so that the maximum expected failure current flows through the smallest possible polarization cell in order to load the cathodic protection as little as possible.

14.2.2.5 dc Decoupling Devices with Silicon Diodes

Silicon diodes have a very high internal resistance on loading in the transmission direction up to their threshold voltage of about +0.7 V. If the threshold voltage is exceeded, the internal resistance decreases. With a nonsymmetrical

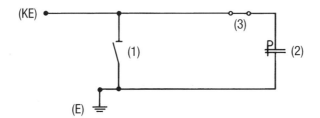

Fig. 14-7 Circuit diagram for a dc decoupling device with polarization cell. (KE) insulated cable end sealing, (E) grounding installation, (1) grounding side bar, (2) polarization cell, (3) disconnecting shackle.

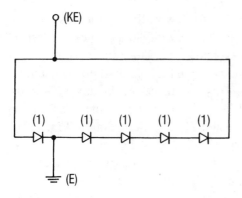

Fig. 14-8 Circuit diagram for a dc decoupling device with silicon diodes. (KE) insulated cable end sealing, (E) grounding installation, (1) silicon power diodes.

antiparallel circuit as in Fig. 14-8, the voltage over the dc decoupling device can increase during a period to values between about −2.8 V (four times the threshold voltage) and about +0.7 V (the threshold voltage).

It is necessary for cathodic protection of steel conduit that the conduit be approximately 1 V more negative than the grounding installation. The diodes comprise a high-resistance connection if the threshold voltage of the diodes is not exceeded. The threshold voltage can be exceeded by voltages induced in the steel conduit, due to a symmetric loading of the cable, by asymmetry in the construction of the cable, or by grid failures. The asymmetric diode arrangement then acts so that, with any high ac interference, the steel conduit remains about 1 V more negative than the grounding installation. This involves an average dc voltage that results in the dc decoupling device becoming low ohmic over a branch during the positive half-cycle of the ac voltage. The voltage between the steel conduit and the grounding installation then amounts to about +0.7 V. During the negative half-cycle of the ac

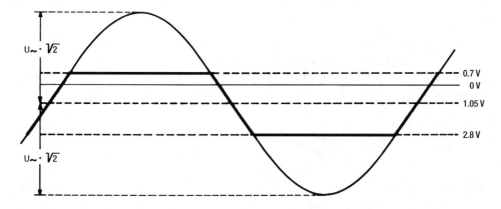

Fig. 14-9 Voltage across a dc decoupling device with 4:1 asymmetry (see Fig. 14-8).

voltage, the voltage is about −2.8 V. The arithmetic mean is therefore −1.05 V (see Fig. 14-9) [10]. To achieve other values of the voltage, the branch with only one diode remains unchanged while the number of diodes in the other branch can be correspondingly varied [see Eq. (23-45)].

A dc decoupling device with a symmetric antiparallel circuit is not logical because a 50-Hz ac voltage usually exists between the steel conduit and the grounding installation. With a symmetric diode circuit, so many diodes must be connected in series in each branch that the ac voltage between the steel conduit and grounding installation remains below the sum of the threshold voltage of the diodes connected in series because otherwise the average voltage would be zero. Cathodic protection of the steel conduit would then no longer be possible.

14.2.3 Installation of Cathodic Protection Station

If the cathodic protection of the steel conduit requires only a small protection current, up to 10 mA, the positive pole of the transformer-rectifier can be connected to the station ground if there is no danger of any marked anodic damage to the ground or installations connected to it. This is the case when the potential of the grounding installation does not change by more than 10 mV in the positive direction on switching on the protection station [6]. With greater current requirements, additional rectifiers can be provided in the stations so that the anodic loading of the station grounds is eliminated. The impressed current anodes are most conveniently installed as deep anodes (see Sections 9.1.3 and 12.3).

The current required with low-resistance dc decoupling devices is always large. For direct injection via the resistance in the station, a rectifier with low output voltage and high current is necessary. With low-resistance dc decoupling devices with Ni-Cd cells or with silicon diodes, the corrosion protection stations can be arranged along the route of the cable. With dc decoupling devices with Ni-Cd cells, also with a low protection current requirement, rectifiers are installed to maintain the cell voltage; in the case of dc decoupling devices with polarization cells or silicon diodes, galvanic anodes are also involved in the cathodic protection.

For adjustment and for monitoring the cathodic protection, test points are necessary along the cable (see Section 10.3.2). It is convenient to install these test points at the cable sleeves. This gives distances between test points of about 0.5 km. The installation of test points to measure conduit current is also convenient for locating accidental contacts.

14.2.4 Control and Maintenance of Cathodic Protection

Off potentials cannot be used to control cathodic protection. Such a measurement is only correct if the dc decoupling devices are isolated from the grounding

344 Handbook of Cathodic Corrosion Protection

installation. This, however, is not permissible with installations in operation. If the dc decoupling devices are not isolated, the off potentials are falsified by equalizing currents, electrochemical capacitance currents or by rectified alternating currents (see Section 23.5.4.2). To measure the *IR*-free potentials during operation, only methods with external measuring probes as in Section 3.3.3.2 should be considered. These should therefore be built in immediately after setting up an installation.

Off potentials can be measured on external gas pressure cables before commissioning and on temporary shutdown. As a rule, carrying out U_{off} and U_{on} measurements gives comparative information on the effectiveness of the cathodic protection. Deviations from these reference values can usually be traced to foreign contacts. In this respect, foreign line crossings, casings and failures in the dc decoupling devices are possible sources.

To evaluate the cathodic protection—with the exception of very high-resistance soils—from experience, an average value of the on potential of $U_{Cu\text{-}CuSO4} = -1.5$ V is to be used. With this value, no danger from stray currents should be experienced [6].

14.3 Stray Current Protection

In urban areas with dc railway lines/tramways, power cables are usually in much greater danger from stray currents (see Chapter 15). The metal sheathing of low- and medium-voltage cables should be included in stray current protection installations in the neighborhood of rectifier substations. With three-core cable in medium-voltage grids, the thermal loading of the cable can be exceeded due to

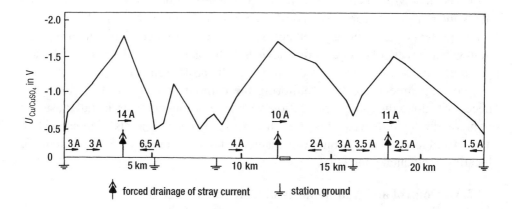

Fig. 14-10 Forced drainage of stray currents and partial cathodic protection of a 110-kV pressurized cable with a low-resistance connection to the station grounds.

additional stray currents. These can necessitate limiting the stray current drainage with resistances. Stray current protection measures are similar to those for pipelines and are described in Section 15.5. In spite of the low-resistance grounding of the cable sheathing, cathodic protection can be achieved over wide areas of the length of the cable by forced stray-current drainage (see Fig. 14-10). When laying power cables in steel conduit and connections to low-resistance grounding installations, the stray current behavior can be considerably altered, which makes the checking of foreign protection systems absolutely necessary.

14.4 References

[1] DIN VDE 0100, Beuth-Verlag, Berlin 1973.
[2] DIN VDE 0141, Beuth-Verlag, Berlin 1976.
[3] DIN VDE 0151, Beuth-Verlag, Berlin 1986.
[4] H. Sondermann u. J. Baur, ÖZE *31*, 161 (1978).
[5] E. Jäckle ETZ-A *99*, 356 (1978).
[6] AfK-Empfehlung Nr. 8, ZfGW-Verlag, Frankfurt 1983.
[7] W. v. Baeckmann u. J. Matuszczak, ETZ-A *96*, 335 (1975).
[8] A. Kohlmeyer, ETZ-A *96*, 328 (1975).
[9] J. B. Prime, jr., Materials Perform. *16*, H. 9, 33 (1977).
[10] J. Pestka, 3R intern. *22*, 228 (1983).

15

Stray Current Interference and Stray Current Protection

W. v. BAECKMANN AND W. PRINZ

15.1 Causes of Stray Current Interference

Stray current is a current flowing in an electrolyte (soil, water) that arises from metal conductors in these media and is produced by electrical installations [1]. It can be the result of dc or ac, predominantly with a frequency of 50 Hz (public electricity supply)[1] or $16\frac{2}{3}$ Hz (traction power supply). In its course through the soil, the stray current can also flow in metallic conductors (e.g., pipelines and cable casings). Direct current causes anodic corrosion at exit points on these conductors into the surrounding electrolyte (see Section 4.3). In a similar manner, ac produces anodic corrosion in the anodic phase. On account of the very high capacity of the steel/electrolyte interface, anodic corrosion is very frequency-dependent and at $16\frac{2}{3}$ Hz or 50 Hz only occurs with high current densities (see Section 4.4). Usually the ratio of corrosion current to ac is also dependent on the electrolyte and the type of metal, so that steel, lead and aluminum all behave differently. In this chapter only stray current corrosion resulting from dc is dealt with, along with interference from telluric currents.

15.1.1 dc Equipment

An electrical installation can only produce stray current if a conductor in the operating current circuit or part of the installation is grounded in more than one place. Such installations are:

(a) dc-operated railways where the rails are used for conducting the current;
(b) overhead tram installations in which more than one conducting connection consists of a pole from the current supply with a ground or a return conductor of the railway line;
(c) high-voltage dc transmission;
(d) electrolysis installations;
(e) dc operations in outer and inner harbors, dc welding equipment, particularly in shipyards;

---

[1] In the United States this is 60 Hz.

(f) dc telephone network and traffic light equipment [2];
(g) cathodic protection installations (impressed current equipment, stray current drainage and forced drainage).

Installations with only one operating ground (e.g., welding equipment, electrolysis grounds and dc cranes) can also produce stray currents if additional grounding at another place (e.g., a ground fault) is present. Stray currents can only occur with grounded equipment if two ground faults are present in different places at the same time.

15.1.2 General Measures at dc Equipment

Measures for avoiding or reducing stray currents are given in Ref. 1. The ground in operation should not be used to carry current. An exception is made with small, intermittently flowing currents from telephone equipment, currents from dc railways, high-voltage dc transmission lines, and cathodic protection equipment. Special requirements are set out for these installations. All current-carrying conductors and parts of the installation belonging to the operating current circuit must be insulated. Ground fault monitoring is recommended for extended and ungrounded dc installations with high operating currents. By this means a ground fault can immediately be recognized and failure avoided before a second ground occurs. Several simultaneous grounding connections must be reliably reduced over a short period of time. If grounding of a conductor or part of the equipment belonging to the operational current circuit is necessary for operational reasons or for prevention of too high contact voltages, the installation must only be grounded at one point. Direct current grids with PEN conductors as protective measures are not allowed [3]. Welding equipment, crane rails and other dc equipment with high operating currents should have connecting leads as short as possible. Grounded metal components such as industrial railway lines, crane rails, pipe bridges, and pipelines should not be used to carry current. Rectifiers with large output for supplying several users should be avoided. An ac supply should be aimed for, with the production of dc with small rectifiers at the actual point of use (e.g., for welding in shipyards).

15.2 Stray Currents from dc Railways

15.2.1 Regulations for dc Railways

Nearly all dc railways use the rails to return the operating current. The rails are mounted on wood or concrete sleepers (ties) and have a reasonably good contact with the soil in surface railway installations. The electrolytically conducting soil is

Stray Current Interference and Stray Current Protection

connected in parallel with the rails. The rail network is regarded as being grounded over its entire length. This relationship and the resulting risk of corrosion was recognized early on (see Section 1.4). Stray currents from railway installations must be minimized by suitable design and corresponding monitoring [1,4]. Since stray currents, however, cannot be completely avoided, additional measures for pipelines and cables are advisable and in many cases necessary.

The most important measures to reduce stray currents are:

- a railway line network completely interconnected (low resistance load R'_s);
- good electrical insulation of the rails from the soil (low leakage load G');
- small feed areas (as many transformer stations as possible).

Table 15-1 lists the important parameters for tramway tracks (see Section 24.4.2). Welded rails have a much smaller resistance load compared with rails formerly joined by tongue-and-groove connections. Bridging at switch points and crossings with longitudinal connections is, however, always necessary. Rails of a single track are provided with cross link, usually every 125 m, and rails of double or multiple tracks, every 250 m. Exceptions are permitted for insulated rails and track-current circuits of signalling equipment. The effect of rail fractures is reduced by cross links. Links between double and multiple track systems result at the same time in a considerable reduction in the voltage drop in the rail network since the return current of a train can be distributed over parallel connected rails.

Tracks on wooden sleepers in ballast have a relatively low leakage load that is a hundred times smaller for a ballast bed than for tracks in the street. Good electrical insulation is, however, only ensured if the rails have no electrical contact with other installations with low grounding resistance. Overhead conductor poles should in principle not be connected to the rails. Exceptions are overhead conductor poles with electrical equipment that is connected to the rails to avoid inadmissibly high contact voltages in the event of a failure. Such overhead conductor poles should be

Table 15-1 Nominal parameters for tramway rails with a resistance load of $R'_s = 10$ mΩ km^{-1}

Parameter	Rails in the street				Rails on gravel ballast			Rails in tunnel	Unit
G'	50	33	20	10	5	2	1	0.1	S km^{-1}
$\alpha = \sqrt{R'_s G'}$	0.71	0.57	0.45	0.32	0.22	0.14	0.1	0.032	km^{-1}
$l_k = 1/\alpha$	1.41	1.74	2.24	3.16	4.47	7.1	10	31.6	km

350 Handbook of Cathodic Corrosion Protection

given a sufficiently high grounding resistance by electrically insulating the foundations [4].

The railway lines on bridges are often electrically connected with steel or reinforced steel structures which usually have a very low grounding resistance. In new installations, an electrical separation of the rails from the bridge structure is required according to the grounding resistance of the structure and the type of rail bed. Independent of this, pipelines and metal sheathing of cables are always electrically separated from the structure in order to exclude direct transmission of stray current from the rails in these conductors.

Bus bars of a transformer substation must not be directly grounded. They must be connected with rails by at least two insulated cables. Metal sheathing of feeder and return current cables must only then be connected with the rails or bus bar if an increase in anodic corrosion on other buried installations is absolutely excluded. The insulation of all return cables must therefore be monitored regularly.

To reduce stray currents, voltage drops in the rail network should not exceed a certain limiting value over an average time (see Table 1 in Ref. 1). A distinction should be made here between the average voltage drop in the core of the network and the voltage drop in outlying stretches of the network.

The core of the network is the region of the rail network in which the branch line of two transformer substations or of any other branch line is less than 2 km distant in a direct line. All branch lines outside the core of the network are termed outlying lines. In a branched rail network, not only the stretches of line within a circle with a radius of 2 km from the most negative return current point of a transformer substation belong to the core of the network, but also connecting branch lines that are less than 2 km from each other [1]. The area of a network core can be simply determined on a track plan with the help of a circular template as in Fig. 15-1.

External lines are generally found only in outlying districts and between different cities. Of considerable importance in reducing stray currents and keeping

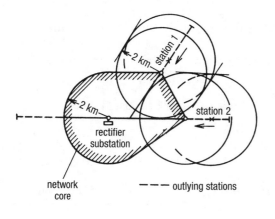

Fig. 15-1 Schematic diagram of a network core in a branched rail network.

within the limiting value in a total network core is the requirement that the voltage between the return conductor points and various transformer substations be as low as possible over an average time period. This can be achieved by favorable connecting points for the return conductor cable, variable resistors and suitable choice of feeder areas. In the case of parallel operation of several substations, a small change in the feeding voltage of individual substations is often sufficient.

The permitted limiting value applies both in the planning as well as for the measured values in operation. In calculating the voltage drop at the rails, the parallel current passing through the soil is neglected. The resistance of the track is considered to be the highest permitted level. Uniform current loading is assumed within each section of the track. When checking during operation, the timed average value of the voltage drop should be measured over a period of at least 3 h [1]. A measurement during operation alone does not give sufficient information on the state of the rail network. It could even lead to false conclusions if low voltage drops are ascribed to a high leakage load value of the rail network which then allows particularly high stray currents. Assessment of the rail network is only possible by comparing the calculated values with the measured values.

In the Verband Deutscher Elektrotechniker (VDE) regulations [1,4], no demands are made on the accuracy of the measured or calculated voltage drops in a rail network. An inaccuracy of ±10% and, in difficult cases, up to ±20%, should be permitted. A calculation of the annual mean values is required. If the necessary equipment is not available, a calculation is permitted over a shorter period (e.g., an average day). Voltage drops in the rail network only indicate the trend of the interference of buried installations. Assessment of the risk of corrosion of an installation can only be made by measuring the object/soil potential. A change in potential of 0.1 V can be taken as an indication of an inadmissible corrosion risk [5].

In almost all dc railways in Germany, the positive pole of the rectifier is connected to the tram wire of the current rail and the negative pole with the running rails. A connection of the positive pole with the running rails is technically possible and was previously convenient when mercury vapor rectifiers were in use for contact protection; however, this resulted in difficulties with corrosion protection measures which involved stray current drainage or forced drainage. For this reason, connection of the negative pole with running rails is strongly recommended [4].

The resistance of a section of line with n rails of cross-section S per rail is given from the rail resistance load R'_s and for length, l, by:

$$R = R'_s l = \frac{\rho_{ST}}{nS} Kl \qquad (15\text{-}1)$$

Here $K = 1$ for welded rail joints or $K = 1.15$ for butted junctions, which increase the resistance and $\rho_{st} = 13$ to $25 \ \mu\Omega$ cm, the specific resistivity of the steel, depending on its strength. The usual resistance figure for a track with welded joints with $n = 2$, $S = 74$ cm² and $\rho_{st} = 22$ $\mu\Omega$ cm is $R'_s = 15$ mΩ km⁻¹.

15.2.2 Tunnels for dc Railways

Underground railways in large cities are being extended or newly constructed as are tunnels for tramways. Based on the permitted anodic potential change of up to 0.1 V in installations affected by stray currents [1], the limiting value for the voltage drop in a tunnel within an individual feeder section and over the whole of the tunnel length is specified as 0.1 V. In all tunnels with walls of reinforced concrete, or steel or cast iron, as well as combinations of steel and reinforced concrete (e.g., steel sheet linings and steel tubings), the following requirements given in Ref. 4 must be met:

- continuous electrical connection of individual tunnel joiners or structural components over the whole length of the tunnel;
- good electrical insulation of the driving rails from the tunnel structure (e.g., by ballast bed);
- electrical separation of all metal installations outside the tunnel both from the tunnel itself as well as from the rails.

The first two requirements ensure that the voltage drop of 0.1 V is maintained in the tunnel and current leakage into the soil is limited. The third requirement prevents direct stray current leakage to foreign installations. Special conditions for the electrical insulation of tunnels from the soil are not regulated. Experiments on existing installations or those in construction have shown that coatings of economically viable materials scarcely lead to a measurable reduction in the leakage load of the tunnel.

The maximum voltage drop in the tunnel is approximately (neglecting current flow from the tunnel structure into the surrounding soil) [6]:

$$U_{T_{max}} = \frac{l^3}{12} R'_S R'_T G'_{ST} I_{max} \qquad (15\text{-}2)$$

Here l is the feed section in a tunnel length, R'_s the resistance load of the rails, R'_T the resistance of the tunnel structure, G'_{ST} the leakage load of the rails to the tunnel structure, and I_{max} the highest possible operating current in the rails for the theoretically worst case of current injection at the end of the feeding section l. The current should be considered here that leads to shutdown of the section circuit or the substation transformer due to overloading. Equation (15-2) indicates the great effect of the length of the feeder section due to its being raised to the third power.

Stray Current Interference and Stray Current Protection

In choosing the spacing of substations, it should be remembered that the permitted limiting value for the voltage drop in the rails in aboveground railways applies to the whole rail network if the underground and aboveground railways have a common connected rail network. With a given rail profile with a given resistance load of the rails, the voltage drop in the tunnel can only be influenced by small leakage load values of the leakage load of the rails G'_{ST}, and the resistance load of the tunnel R'_T. Values for the leakage load $G'_{ST} < 0.1$ S km^{-1} can be achieved according to measurements on new and well-drained tunnel installations with the rails in the usual ballast bed. This value will in the course of time become greater but should not exceed 0.1 S km^{-1} even under the least favorable conditions.

To reduce stray currents, foreign installations (e.g., building foundations, bridges, pipelines, sheathings of cables, grounding installations and grounding conductors) should not make electrical contact with the rails or the tunnel structure. In the interior of the tunnel, plastic pipes and cables with plastic sheathing should be used. All supply lines should be insulated at the point of entry in the tunnel (e.g., in stations). An insulating joint is built into pipelines outside the tunnel. Current supply from the public grid must be via transformers with separate windings. Insulating couplings are only necessary in pipelines and metal shielding of cables that cross the tunnel if there are accidental connections with the tunnel structure that cannot be avoided.

15.3 Stray Currents from High-Voltage dc Power Lines

High-voltage dc power lines are economical for distances exceeding 1000 km and can be operated as reaction-free couplings between different ac grids with regard to frequency and phase. At the end of a high-voltage dc power line with two systems, a converter is attached to each of them. The dc flows normally over two-stranded conductors. The midpoint is grounded, but because the systems work symmetrically to ground, no current flows over them. The operating voltage can be about 600 kV and the currents can be up to 1.2 kA. When one of the systems breaks down, a current can flow for a few hours to ground, so that the second system can remain in operation. In contrast, in Sweden and the United States, there are high-voltage power lines with permanent current return lines via the ground. This method of operation is not likely to be found in densely populated areas (e.g., in middle Europe).

Figure 15-2a shows the stray current interference by a bipolar high-voltage dc power line [7]. When the system breaks down, large voltage cones occur in the soil at the grounding installation. A few kilometers away, the current density in the soil is relatively low.

The principal circuit diagram for a bipolar system is shown in Fig. 15-2a and the current distribution in the case of breakdown of a monopolar system in Fig. 15-2b. I and II are the respective ac systems.

The current density, J, in the soil in the middle region between the two grounding installations $2L$ can be derived as follows from Eq. (24-39) with $r = L + x$ and $R = L - x$

$$\phi(x) = \frac{I\rho}{2\pi}\left(\frac{1}{L+x} - \frac{1}{L-X}\right) \qquad (15\text{-}3)$$

$$-\frac{d\phi}{dx} = \frac{I\rho}{2\pi}\frac{(L+x)^2 + (L-x)}{(L+x)^2(L-x)^2} \qquad (15\text{-}4)$$

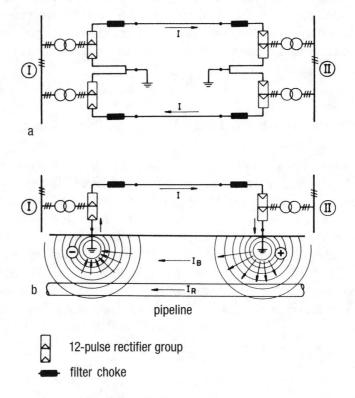

Fig. 15-2 Stray current interference from high-voltage dc transmission installations: (a) bipolar system, (b) monopolar system.

Stray Current Interference and Stray Current Protection

Using Eqs. (24-1) and (24-2), it follows from Eq. (15-4) for $x = 0$

$$J = \frac{I}{\pi L^2} \tag{15-5}$$

For $I = 1000$ A and $L = 100$ km, the current density is calculated $J = 3 \times 10^{-8}$ A m^{-2}. The field strength, E, in soil of specific resistivity, $\rho = 100$ Ω m becomes 3 mV km^{-1}.

The field strength is lower than that due to telluric currents. With different operating currents, the equalizing currents flowing to ground are even lower. No damage is caused by this. Damaging effects on pipelines or cables are only to be expected in the neighborhood of the grounding installations.

In experiments on high-voltage power lines over a straight-line distance of 6 km in the Mannheim/Schwetzingen area, feeding 100 A to the ground of a mast resulted in a voltage of 150 V. The pipe/soil potential of a gas pipeline that was 150 m from the mast's ground was changed by 1 V. A calculated change of 5 V had been expected, assuming homogeneous soil. The reason for the much lower effect was that the soil had a lower resistance at greater depths. The effects of dc railways at a distance of 10 km were greater than that of the test current. Damaging effects can be alleviated by installing cathodic protection stations.

15.4 Stray Currents Due to Telluric Currents

The magnetic field of the earth is strongly affected by solar wind (protons and electrons radiating from the sun). The earth's magnetic field is deformed according to the strength of the solar wind from sunspots. This results in changes over time in the earth's magnetic field; these can be measured as magnetic variation by geophysical observatories using magnetometers. Due to induction by the change in magnetic field, electrical changes in field strength occur in the soil (not in the pipeline), depending on the geology and the specific resistivity of the soil, which can be as high as 0.1 V km^{-1}, particularly in conducting layers on high-resistance rocky ground. A well-coated pipeline represents an equipotential line in this electric field at whose ends differences in the pipe/soil potential are measurable (see Fig. 15-3). Polarization currents arise from these at defects in the pipe coating. The greater the leakage load value of the pipeline, the greater the polarization current resulting from magnetic variations in the pipeline. However, with high leakage load values, the polarization current densities and the changes in pipe/soil potential are small. In cathodically protected pipelines, this current is superimposed on the protection current, which it can either increase or decrease [8].

356   Handbook of Cathodic Corrosion Protection

It can be seen in Fig. 15-3 how telluric currents with different soil resistivities over a long pipeline can emerge or enter and thereby cause changes in potential. The pipeline enters the particular region at point 1 and leaves it at point 8. Because of the low resistance load, part of the telluric current flows through the pipeline from point 1 to point 8. The current enters the pipeline in the neighborhood of point 1. The greater the voltage drop between points C and D, the greater this current. In the absence of a pipe coating, current reaches a value so that the voltage drop between A and B is the same as between C and D. Next, if the pipeline lies in a region of high soil resistivity in which the current density decreases, the gradient in the soil, EF, rises. The pipeline absorbs current from the soil at point 2, which exits it again at point 3. In the region with low specific soil resistivity, the current

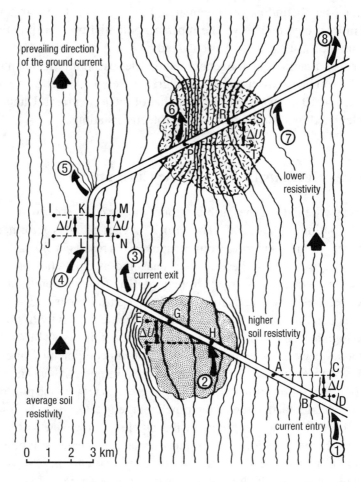

Fig. 15-3 Stray current interference of pipelines by telluric currents.

density in the soil is higher. In order for the voltage drop, *PR*, to equal the voltage drop, *TS*, in the soil, telluric current has to leave the pipeline at point 6 and reenter in the area of point 7. Because the telluric currents vary slowly and can even reverse direction from the preferred direction of the ground current, continual changes in potential occur at defects in the pipe coating.

The risk of anodic corrosion on a pipeline due to magnetic variations can be accessed with the help of the planetary earth magnetic index, K_p. These indices give the degree of the maximum change in field strength for a time period of 3 h, each derived from observations at geophysical observatories. The K_p indices are published monthly in tables and diagrams.

The relation between the change in the earth's magnetic field and the variation in pipe/soil potential, $U_{Cu\text{-}CuSO_4}$, can be seen in Fig. 15-4. H, D and Z are the changes of the earth's magnetic field components, north-south, east-west and vertical.

Extensive investigations show [9,10] that in 1981 the pipe/soil potential was more positive than the protection potential for only 20 h. Therefore, no danger of corrosion should be expected from magnetic variation. Because it cannot be foreseen when the earth's magnetic field might change with different intensity, leading to changes in pipe/soil potential, considerable difficulties arise in monitoring cathodic protection with regard to *IR*-free measurements of pipe/soil potentials. *IR*-free potentials using a switching-off technique can only be made in periods when there is no magnetic variation. In principle, the only possible way of

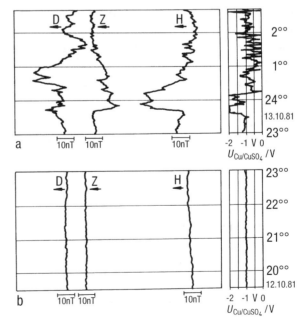

Fig. 15-4 Pipe/soil potentials for a pipeline (a) with and (b) without interference due to magnetic variation.

15.5 Protective Measures

15.5.1 Stray Current Protection for Individual Pipelines

Frequently, measures for dc installations are not sufficient to limit stray currents. This applies particularly to dc railways. In many cases, additional protective measures for the affected installations are advisable [1] or even necessary [11].

Protective measures restrict the outflow of stray current from the particular installations into the surrounding soil. This can be done in the simplest case by a connecting cable between the installation to be protected and the railway lines with a sufficiently negative potential (see Fig. 15-5). The stray current that formerly flowed through the soil can then flow back via the cable connection to the rails with no risk. Protection against stray current corrosion is achieved if the potential of the installation to be protected–with the exception of momentary peak values–is equal to or more negative than the free corrosion potential. Lowering the potential gives partial cathodic protection.

The same conditions as for cathodic protection apply for reliable stray current drainage or forced drainage for rails of dc railways (see Section 10.2). The objects to be protected must be electrically connected. Individual insulating sockets (e.g., lead-caulked or rubber screw sockets) must be bridged. The installations to be protected must have no electrical contact to the rails as these can possibly be at bridges. Even in cases where stray current protective measures are applied, bridges are a source of danger due to carrying over of stray current. Figure 15-5 shows the effect of a pipeline laid parallel to tramway rails with and without stray current drainage.

The figure shows the current and potential distribution for a single tram which feeds a current, I, into the rails at the end of the parallel stretch. In the vicinity of the tram, stray current flows from the driving rails through the soil to the pipeline; in the absence of stray current drainage in the region of the transformer substation, the current flows back again from the pipeline through the soil to the rails and here causes anodic corrosion of the pipeline. In the area of the positive rails, from $l/2$ to l, there is cathodic polarization of the pipeline and in the area of the negative rails from 0 to $l/2$ anodic polarization. The pipeline assumes the potential of the rails as a result of the low-resistance stray current drainage to the rails before the transformer substation.

The pipe/soil potential can therefore become very negative. The pipe currents become very large with poorly coated pipelines. The pipe current can be limited by

Stray Current Interference and Stray Current Protection   359

including a resistor, R, in the stray current drainage and the potential of the pipeline lowered so far, corresponding to curve in Fig. 15-5, that the pipe/soil potential becomes only slightly less negative than the free corrosion potential.

The curves in Fig. 15-5 are greatly simplified and only apply to one instant of time. In practice there are always several trams with varying current draw on a stretch of track. Currents and potentials are then subjected to wide time-dependent variations. Without protective measures, the pipe/soil potentials are usually always more negative in the outer regions of a tramway system and more positive in the vicinity of the transformer substation. In a wide intermediate region, potential changes occur in both directions. The current entrance and exit areas can be deter-

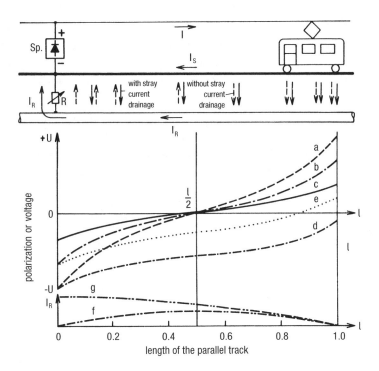

Fig. 15-5 Stray current interference in the region of dc railway: (a) Polarization of the railway lines, (b) voltage between the soil in the vicinity of the rails against a remote ground. Polarization of the pipeline: (c) without stray current drainage, (d) with stray current drainage without a resistor, (e) with stray current drainage via a resistor R. Current in the pipeline: (f) without stray current drainage, (g) with stray current drainage.

mined by measuring the pipe/soil potential. The results of the measurement give no quantitative measure of the risk of stray current corrosion because they include a considerable *IR* component. A line without stray current interference usually has a free corrosion potential $U_{Cu-CuSO_4} = -0.5$ to -0.6 V. This value can be measured when the tramway is not operating.

Direct stray current drainages (with or without resistor R) to limit the current are provided through cable connections (stray current return conductor) to the negative regions of the installation producing the stray current. In the neighborhood of a stray current drainage, the rails can at times be more positive than the pipeline if the tramway network is fed by several transformer substations. In the stray current return conductor, current reversal occurs, which leads to an increased risk of stray current in remote regions of the installation to be protected. Stray current reversal must be avoided by a rectified stray current drainage (e.g., in Fig. 15-6) by using a rectifying cell (diode) or an electrically operated relay.

The disadvantage of rectifier cells is the relatively high cutoff voltage amounting to 0.4 V for selenium and 0.7 V for silicon. The advantage is a rectified stray current drainage to the negative bus bar in the transformed substation since its potential on account of the voltage drop at the return connector cable in operation is more negative than that of the rails. The effectiveness can be increased by an additional resistor in the return conductor cable. Stray current drainage is not sufficient for complete cathodic protection so that in most cases transformers rectifiers are preferred over forced stray current drainage (see Fig. 15-7).

Figure 15-8 shows synchronous recordings of the voltage between the pipeline and the rails, U_{R-S}, of the pipe/soil potential $U_{Cu-CuSO_4}$ for a drained current in the region of a tramway transformer substation with and without various protective measures. Figure 15-8a records values without protective measures. If the rails are negative with respect to the pipeline ($U_{R-S} > 0$), the pipe/soil potential becomes more positive. Stray current exit exists. From time to time, however, $U_{R-S} < 0$.

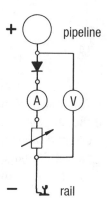

Fig. 15-6 Rectified stray current drainage with rectifier and adjustable resistance.

Stray Current Interference and Stray Current Protection 361

Then stray current enters the pipeline and the pipe/soil potential becomes more negative. The recording in Fig. 15-8b shows the behavior with a direct stray current drainage to the rails. With $U_{R-S} > 0$, a current flows off the pipeline via the stray current return conductor back to the rails so that there is no anodic polarization of the pipeline. With $U_{R-S} < 0$, a current flows over the connection in the pipeline and anodically polarizes it. Direct stray current drainage is therefore not possible in this case. Figure 15-8c shows the result of a rectified stray current drainage to the rails. Now the pipeline is always cathodically polarized. Cathodic protection is, however, also not fully attained.

With forced stray current drainage, the current is returned from the pipeline to the rails by means of a grid-fed rectifier. The transformer-rectifier is connected into the stray current return conductor, the negative pole is connected with the installation to be protected and the positive pole is connected to the rails or the negative side of the bus bar in the transformer substation.

Various arrangements for the protection rectifier and its possible applications are described in Chapter 8. Large fluctuations of the protection current can be reduced by a current-reducing resistor, R (see Fig. 15-7). This means that the pipe/soil potential in the middle becomes less negative. A current smoothing action can be achieved with rectifiers connected to the network or by reducing losses by an inductive resistance between the transformer and bridge rectifier (see Section 8.4).

Stray current drainage and forced drainage with uncontrolled protection current devices often require expensive drainage tests in setting up the protection drainage test station. The use of controlled protection current devices can reduce this expense (see Section 8.6). Also, with controlled protection current devices, potential fluctuations occur in the adjoining sections of the pipeline which are governed by voltage drops caused by stray currents in the soil.

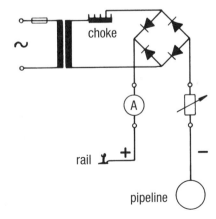

Fig. 15-7 Forced current drainage with resistor and/or choke to limit current.

Very positive tramway rails can produce such a negative pipe/soil potential by high stray current entry that it becomes more negative than the imposed nominal potential. The protection current device must be designed so that in this case the dc output is controlled at zero. If no further protection stations are installed on the pipeline, the protection current device must be arranged so that an imposed minimum protection current is not undercut.

With increasing distance from the crossing or the proximity of the tramway rails, the stray currents absorbed in the pipeline emerge again. The current exit occurs mainly at crossings with other tramway tracks. In addition to potentiostatic control, a basic current adjustment is necessary.

In Fig. 15-9 two potentiostatically controlled protection rectifiers and an additional diode are included to drain peak currents. At pipeline crossings with an external rail network (e.g., in regions outside the urban area), the forced stray current drainage should be installed as close as possible to the rails that display negative potentials for the longest operation time. The currents absorbed from the positive rails continue to flow also in the region outside the rail crossings. Here the use of potentiostatically controlled rectifiers is recommended; these should be connected not only to the rails but also to impressed current anodes.

15.5.2 Combined Stray Current Protective Measures in Urban Areas

Old gas and water supply networks in urban areas often have only bitumen coatings and a high leakage load. This also applies to electricity supply cables and telephone cables. Stray current protective measures for these installations alone are not usually possible because of the large number of connections in old delivery stations and accidental connections at underground crossings. All pipelines and cables in the neighborhood of tramway transformer substations are at risk. Therefore it is expedient to lead connecting cables from the individual pipelines and cables to be protected into a box close to the substation and to connect them via a variable resistor to a common stray current bus bar. From here a connection can be

---

Fig. 15-8 Synchronous current, voltage and potential recording with stray current interference from dc railways: (a) Without protective measures, (b) direct stray current drainage to the rails, (c) rectified stray current drainage to the rails, (d) forced stray current drainage with uncontrolled protection rectifier, (e) forced stray current drainage with galvanostatically controlled protection rectifier (constant current), (f) forced stray current drainage with potentiostatically controlled protection rectifier (constant potential), (g) forced stray current drainage with potentiostatically controlled protection rectifier and superimposed constant current.

Stray Current Interference and Stray Current Protection 363

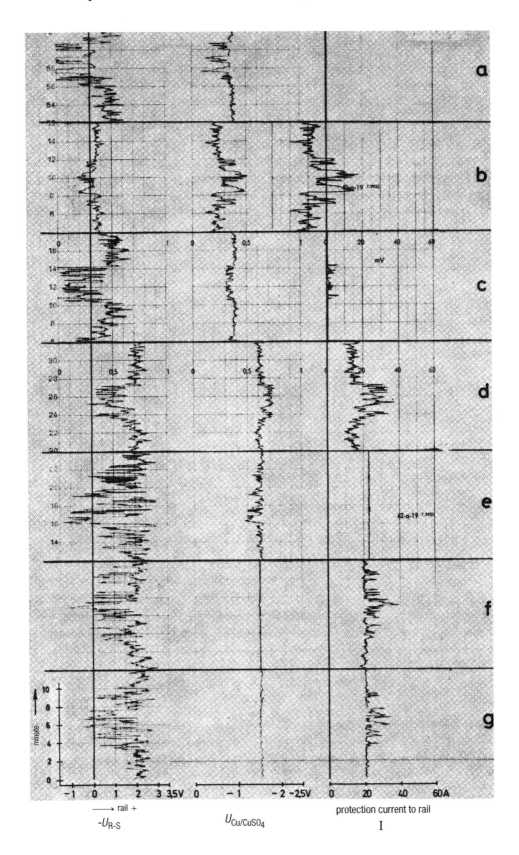

made to the rails and the negative bus bar in the substation by a cable of sufficient cross-section directly via diodes, relays or grid-fed rectifiers. Protection currents of up to several hundred amperes are to be expected in large substations. For forced stray current drainage rectifier, output voltages of only a few volts are usually necessary.

Tramway networks are fed from a large number of transformer substations. Stray current drainage and forced drainage should be installed as close to the substation as possible. With high-capacity substations in distribution networks, direct stray current drainage to the negative bus bar can be sufficient. Measurements on such installations should be carried out simultaneously by all the participants. The breakdown of the cost is described in Ref. 12. All interested parties should be involved in setting up and costing common stray current protective measures; measurements should be made simultaneously as far as possible [12].

15.6 Stray Current Protection in Harbor Areas

In harbors, central dc supply equipment is still occasionally used as welding equipment on wharves, travelling dc cranes, dc supply for ships in dock, etc. In Ref. 1 proposals for avoiding stray currents are detailed in which separate dc supply equipment is stipulated. In ac distribution networks in which the subsequent dc network is grounded in only one place, there is no possibility that the dc currents will occur as stray currents.

Considerable stray currents can, of course, be caused by dc-driven cranes that load and unload ships where the rails act as the return conductor for the current. The rails run parallel to the harbor basin, quay walls of steel-reinforced concrete or steel piling walls. These can take up a large part of the stray current and conduct it further because of their small longitudinal resistance. Noticeable stray current inter-

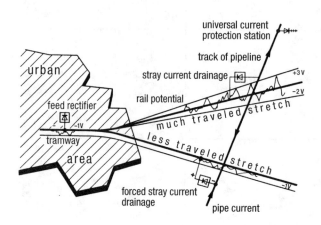

Fig. 15-9 Stray current interference for a pipeline crossing.

Stray Current Interference and Stray Current Protection 365

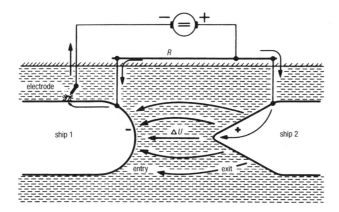

Fig. 15-10 Stray current interference of ships at the supply quay from welding operations with a central welding transformer.

ference on ships is of course to be expected only in exceptional cases. On the other hand, pipelines and cables on the dock are often at great risk from corrosion. Here stray current drainage or forced drainage should be installed to protect the equipment at risk.

Today welding current is supplied mostly from individual rectifiers on wharves. However, stray current effects on ships are possible with central welding equipment. Figure 15-10 shows two ships lying at a fitting-out quay supplied by a central welding transformer. Ships 1 and 2 are firmly connected to the return conductor collecting cable and the positive pole of the welding transformer. On ship 1, welding is carried out with the negative pole of the welding electrode. This welding current flows through the common collecting cable and results in the voltage drop ΔU over its resistance, R. This voltage also exists between ships 1 and 2 and causes stray current exit at ship 2 and therefore considerable risk of anodic corrosion. Remedial measures are sufficiently large zinc plates for low voltages, or potential-controlled impressed current protection stations for larger voltages. In both cases the anodes must have grounding resistance in water that is as low as possible.

15.7 References

[1] DIN VDE 0180, Beuth-Verlag, Berlin 1983.
[2] AfK-Empfehlung Nr. 7, ZfGW-Verlag, Frankfurt 1974.
[3] DIN VDE 0100 Teil 410, Beuth-Verlag, Berlin 1983.
[4] DIN VDE 0115 Teil 3, Beuth-Verlag, Berlin 1982.
[5] vorgesehene Änderung von [1].
[6] VÖV-Empfehlung 04.740.5, Verband öffentlicher Verkehrsbetriebe, Köln 1975.
[7] W. J. Mitchel, NACE-Konferenz, paper No. 31, Toronto 1981.
[8] A. L. Smart, IEEE Transactions on Industry Applications, Vol. *IA-18*, Nr. 5, 557 (1982).
[9] W. v. Baeckmann, gwf gas/erdgas *123*, 530 (1982).
[10] H. Brasse u. A. Junge, gwf gas/erdgas *125*, 194 (1984).
[11] W. v. Baeckmann u. W. Heim, gwf gas/erdgas *110*, 393 (1969).
[12] AfK-Empfehlung Nr. 4, ZfGW-Verlag, Frankfurt 1970.

16

Marine Structures and Offshore Pipelines

W. V. BAECKMANN AND B. RICHTER

The cathodic protection of marine structures (i.e., drilling and oil production platforms, steel piling walls, locks and other harbor installations) is increasingly used today and is always prescribed for subsea pipelines. Thin coatings (see Section 5.1.1) have only a limited life under water; in general, the thin coatings listed above cannot be repaired. Therefore, such coatings are rarely used and almost never in offshore applications. Regulations based on comprehensive investigations govern the setting up of cathodic protection [1-3].

In the tidal zone and the spray zone (known as the splash zone), cathodic protection is generally not very effective. Here thick coatings or sheathing with corrosion-resistance materials (e.g., based on NiCu) are necessary to prevent corrosion attack [4]. The coatings are severely mechanically stressed and must be so formed that repair is possible even under spray conditions. Their stability against cathodic polarization (see Section 17.2), marine growths, UV rays and seawater must be ensured [4,5].

In contrast to pipelines and harbor installations, platforms are dynamically loaded. Therefore in the choice of steels, in addition to strength and types of machinability, the risk of corrosion fatigue and strain-induced stress corrosion must be taken into account in combination with cathodic protection (see Sections 2.3.3 to 2.3.5).

16.1 Cathodic Protection Measures

Protection with impressed current, with galvanic anodes, and a combination of both processes is used for marine structures and offshore pipelines. Their properties, as well as their advantages and disadvantages, are given in Table 16-1. The protective measures must be optimized for every structure. In the impressed current protection of offshore platforms, for example, the difficulties of maintenance and repair will be of major importance, whereas in harbor installations these problems can be

Table 16-1 Comparison of cathodic protection systems for marine structures

Property	Galvanic anodes	Impressed current
Maintenance	Practically not	Yes
Cost of installation	Medium	High
Anode weight	High	Small
Anode number	Large	Small
Life	Limited	High
Current output	Limited, self-regulating	Controlled (manual or automatic)
Current distribution	With many anodes — good	Less good with few anodes
Damage to proper	Usually none coatings near anodes	Yes, special protection measures necessary
Running costs	Favorable with small objects	Favorable with large objects
Usual life	Over 10 years	>20 years

ignored. Cathodic protection always demands an accurate knowledge of the nature of the corrosive medium.

16.1.1 Design Criteria

A large number of parameters are involved in the choice of the corrosion protection system and the provision of the protection current; these are described elsewhere (see Chapters 6 and 17). In particular, for new locations of fixed production platforms, a knowledge of, for example, water temperature, oxygen content, conductivity, flow rate, chemical composition, biological activity, and abrasion by sand is useful. Measurements must be carried out at the sea location over a long period, so that an increased margin of safety can be calculated.

The protection potentials for seawater are described in Section 2.4. In pipelines and harbor installations, there is no limiting negative potential U'_s for uncoated carbon steel or for steel provided with thick coatings over 1 mm, with yield points up to 800 N mm^{-2}. With dynamically highly loaded structures, the protection potential ranges in Table 16-2 should be adhered to as in the regulations [1-3] because of the risk of hydrogen-induced stress corrosion (see Section 2.3.4).

The reference potential of the Ag-AgCl electrode in brackish water must be corrected for chloride ion content (i.e., a change in chloride ion concentration by a factor of 10 shifts the reference potential by about 50 mV in the positive direction

Table 16-2 Protection potential regions for plain carbon and low-alloy steels (YP $\leqq$ 800 N mm^{-2}) for marine structures

Marine structure	Reference electrode	$U_{\text{Cu-CuSO}_4}$ (V)	$U_{\text{Ag-AgCl}}$ in seawater (V)	U_{Zn} (V)
Seawater				
	Protection potential U_s	−0.85	−0.80	+0.25
	Limiting potential U'_s	−1.10	−1.05	0
Anaerobic conditions				
	Protection potential U_s	−0.95	−0.90	+0.15
	Limiting potential U'_s	−1.10	−1.05	0

[6]). In the cathodic protection of high-strength steels with yield points above 800 N mm^{-2}, the limiting potential U'_s is assumed to be at least 0.1 V more positive; this does not exclude adjustment for individual cases (see Section 2.3.4). The underwater area of marine structures also contains machined components of stainless steel or nonferrous metals. These structural components must be attached in a suitable form so that the protection potential ranges are observed (see Section 2.4). This can be taken care of in a situation involving several materials by a suitable arrangement of galvanic anodes.

In determining the protection current required, the surfaces of the objects to be protected in the water and on the seabed, as well as those of foreign constructions that are electrically connected to the object to be protected, should be isolated. The protection current densities derived from experience and measurements for various sea areas are given in Table 16-3. In exceptional cases measurements must be carried out beforehand at the location of the installation. Such investigations, however, provide little information on the long-term development of the protection current. By using a suitable coating [4], the protection current density in the early years of service will be only about 10% of the values in Table 16-3. For a planned operational lifetime of 30 years, about 50% of these values is necessary.

Steel constructions and pipelines must either be electrically connected to the reinforcement of reinforced concrete structures or electrically separated. If they are connected, a current density of about 5 mA m^{-2} should be applied to the external reinforcement and calculated on the total area of the concrete surface.

For structures in brackish water, harbor water and fresh water, the conditions in each case should be considered and addressed on the basis of experience gained from other installations. Since harbor installations are usually very accessible, the cathodic protection installation can be extended if necessary.

Table 16-3 Protection current densities for the cathodic protection of uncoated steel [1-3]

Sea region	Protection current density (mA m^{-2})
North Sea (southern)	100 to 140
North Sea (northern)	110 to 160
Baltic (west)	90 to 120
Gulf of Mexico	65
U.S. West Coast	85
Cook Inlet	440
Persian Gulf	108
Indonesia	65
Seabed	20

In the corrosion protection of marine structures, it is often found that the corrosion rate decreases strongly with increasing depth of water, and protection at these depths can be ignored. Investigations in the Pacific Ocean are often the source of these assumptions [7]. However, they do not apply in the North Sea and other sea areas with oil and gas platforms. Figure 16-1 is an example of measurements in the North Sea. It can be seen that flow velocity and with it, oxygen access, is responsible for the level of protection current density. Increased flow velocity raises the transport of oxygen to the uncoated steel surface and therefore determines the

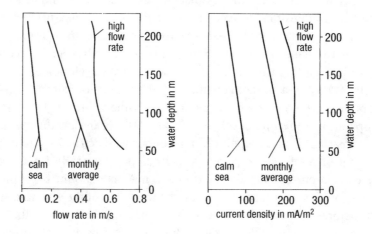

Fig. 16-1 Protection current densities as a function of depth and flow rates in the North Sea [8].

Marine Structures and Offshore Pipelines 371

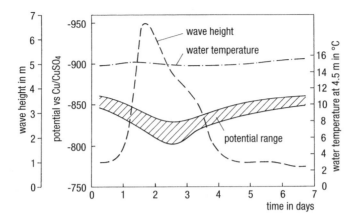

Fig. 16-2 Potential variation as a function of wave height [9].

corrosion rate (see Section 4.1). Figure 16-2 shows the relationship between wave height—as an indirect function of flow rate—and potentials on the research platform Nordsee, 80 km northwest of Helgoland. During relatively short stormy periods, a clear shift of potential to more positive values is observed. The platform legs with different degrees of marine growths can be measured separately and the relation between the changing oxygen access and the potentials can be determined [9]. Figure 16-3 shows synchronous changes in wave height and potential. Even small variations in wave height show potential changes associated with them.

Practically all other parameters were kept constant in recording these curves so that the potential changes are indirectly related to wave height. The potentials also become more positive than the protection potential of $U_{Cu-CuSO_4} = -0.85$ V. This is, however, harmless and need not affect the values in Table 16-2.

The possible influence of marine growths and sulfate-reducing bacteria on a planned service life of 30 or more years has not been sufficiently researched. Marine growths change considerably with the passage of time and can change the conditions for corrosion protection and, under certain circumstances, even improve them. Calcareous deposits are formed which are not due to cathodic polarization but are the deposits of marine fauna. These growths must be partly removed, particularly at nodal joints, for inspection purposes and also if they become too thick. However, this only applies to small areas compared with the total surface. There is no general agreement at present on the interaction of marine growths, sulfate-reducing bacteria and cathodic protection. It is not known whether corrosion due to sulfate-reducing bacteria under hard deposits can be sufficiently neutralized by cathodic protection. Their behavior on the external surfaces of platforms can be investigated with random tests. In inaccessible parts of the platform (e.g., in the

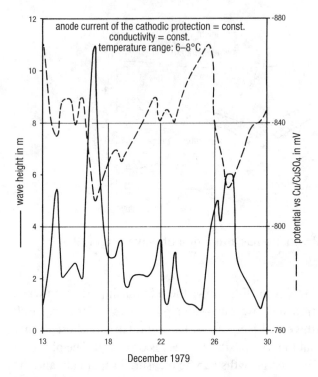

Fig. 16-3 Synchronous variation in wave height and off potentials [9].

interior of the platform legs containing standing water or between piles and legs), inspection is hardly possible. Corrosion protection by coatings is necessary in such endangered areas.

16.1.2 Protection with Galvanic Anodes

The composition and driving voltage of galvanic anodes as well as their material, size and application are dealt with in Chapter 6 and are also laid down in the regulations for offshore structures [1-3]. Galvanic anodes are always welded by the protruding iron core or for pipelines, connected via a cable. Screw connections are not allowed even for repairs. Magnesium anodes are not used on account of their high self-corrosion in seawater, but aluminum anodes are installed on platforms and zinc anodes on ships, harbor installations and tanks.

Chemical analyses should be provided for all anodes used in the offshore and harbor area, together with results for current content in A h kg^{-1} and current output in amperes [2,3]. The geometric shape and the number of anodes required is determined by these parameters. Expensive calculations for design based on grounding resistances are made only in exceptional cases because in practice there are too many uncertainties and the number and mass of the anodes have to be quoted with a corresponding safety factor.

16.1.3 Impressed Current Protection

Materials for impressed current anodes and their consumption and service life are described in Chapter 7. Chapter 17 contains information on protection current equipment for ships that would also apply to platforms. The use of impressed current systems for these structures requires very great care at the planning stage since later repairs can be carried out only at a disproportionately high cost. In particular, cable mountings under water must be designed so that mechanical damage by wave and ice movement and shipping traffic is absolutely prevented. The protection current equipment should be installed as close as possible to the anodes; however, this is not always possible. For this reason the protection of platforms with impressed current has not yet been successfully achieved.

16.2 Platforms

Production platforms for extraction of oil and gas are planned to have service lives of several decades. These structures—there are about 6000 of them worldwide—can only be inspected and monitored *in situ* by divers or diving equipment. On the other hand, semisubmersible floating platforms and drilling rigs—there are about 500 in operation worldwide—are treated in the same way as ships (see Chapter 17) since they are movable and can be maintained and repaired in dock. The docking interval of these installations is normally 5 years, which must be taken account of in the design. Aluminum anodes are almost exclusively used as galvanic anodes because of weight considerations. Titanium and niobium coated with platinum are used as impressed current anodes (see Section 7.2.2). The negative limiting critical potential U'_s of the steel is particularly important at the design stage, since support legs and floats are increasingly fabricated from high-strength steels with yield points of over 800 MPa (see Section 16.1.1).

16.2.1 Steel Structures

Production platforms are coated only in exceptional cases or for the purposes of investigation because the life of the structure is greater than the life of the coating. Therefore in the design of the cathodic protection, only the protection potential U_s of the steel need be considered. Steels with an ultimate tensile strength of up to 350 N mm^{-2} are used for these structures, which are weldable even in thick sections, and the hardness of the welded material can be kept to 350 HV (see Section 2.3.4 [2,10]). Aluminum anodes with the same protection effect and life as zinc anodes have much less weight. This is a very important advantage for

374   Handbook of Cathodic Corrosion Protection

the uncoated surface that is to be protected. Several thousands of tons aluminum anodes are used on platforms at greater depths, which must be taken into account of construction and transport to the installation site.

The anode mountings are welded to lap joints in the yard, and the anodes are installed at a minimum distance of 30 cm from the structure to achieve the most uniform current distribution [1-3]. Nonuniform potential distribution occurs even with this distance.

Nodal points of the platform require special attention for corrosion protection. Therefore the anodes have to be installed in the vicinity of these points, as indicated in Fig. 16-4. The spacing must be sufficiently large that the welded joints of the nodes do not lie in the area of the lap joints. The effort for calculating the optimal distribution with the lowest weight of anodes is considerable and has led to computer programs by which the anode distribution can be estimated [11].

The production platforms are equipped with a large number of production tubes and risers, which are connected to the flow lines. The corrosion protection of these installations must be carried out particularly reliably since any damage can result in the escape of gas or oil and cause great danger to personnel and the environment. These pipes are usually provided with heavy-duty thick coatings; in the splash zone the thickness is over 10 mm. These coatings must be resistant against cathodic polarization (see Sections 5.2.1 and 17.3). In choosing galvanic anodes, the temperature of the material to be conveyed must be taken into account. In some areas and according to national regulations, the risers must be electrically separated (from the platform) if different corrosion protection systems are used, e.g., platforms with impressed current protection and risers with galvanic anodes. To date, impressed current protection for fixed platforms has only been installed

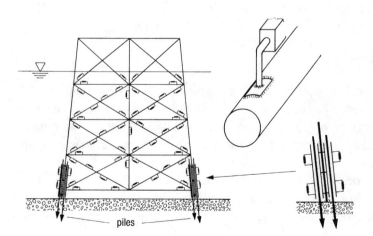

Fig. 16-4
Arrangement of galvanic anodes on platforms.

on about ten structures. The problems are with maintenance and repair. In the underwater area, recabling of fixed anodes is almost impossible, especially as it must be carried out through the highly stressed splash zone.

The difficulties of such operations on the research platform Nordsee are described in Ref. 9. The Murchison platform was provided with a combination of impressed current protection and galvanic anodes because there was a limit to the load to be transported [12]. The anodes for platforms are installed and provided with cables at the yard. They are installed with redundancy and excess capacity so that no repairs are necessary if there is a breakdown. The lower part of the platform up to the splash zone is usually placed in position in the designated location at least 1 year before the erection of the deck structure so that impressed current protection cannot initially be put in operation. This requires cathodic protection with galvanic anodes for this period. This also means that the impressed current protection is more expensive than the galvanic anodes.

New types of anodes have been developed and tested as shown in Fig. 16-5 to improve the possibility of maintenance and repair. They can be lifted onto a ship and repaired. The connecting cables are also replaceable. In shallow water, the anchorage must be accurately calculated because considerable dynamic stressing can occur in heavy seas. The ocean floor must be suitable for long-term anchorage. No supply ships must anchor in the area around the platform. This requirement alone often prevents the installation of impressed current anodes since the operator does not wish or is not able to restrict himself to these conditions.

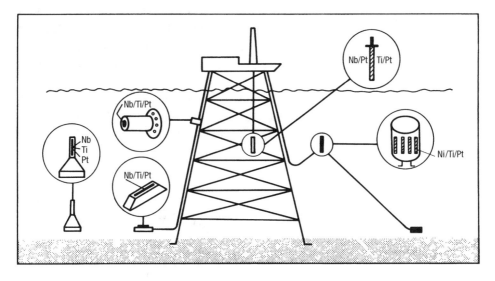

Fig. 16-5 Types of impressed current anodes and their installation [13].

The distance between the structure and fixed impressed current anodes is an important factor. The number of anodes has to be small so the anodes need to be relatively large, which will result in too negative a potential if the distance is not sufficiently great. A minimum distance of 1.5 m is prescribed [1-3], but this involves considerable construction effort due to the effects of heavy seas. Besides the so-called restriction on impressed current installations, there is the requirement that the corrosion protection be switched off when diving work is being carried out [14]. This regulation is not justifiable. Work on the underwater region of production platforms takes place continuously, as far as the weather allows; if the protection must be switched off each time, the impressed current protection becomes very limited.

The protection current equipment must be installed on the deck, so great lengths of cable with a corresponding cross-section are required. Only potential-controlling protection current equipment, as in the case of ships, should be employed since the necessary current densities are continually changing due to the changing heavy seas (Figs. 16-2 and 16-3).

16.2.2 Concrete Structures

A series of platforms of reinforced concrete have been installed in the North Sea and the Baltic. The concrete covering in the underwater region of 60 mm for weak reinforcement and 75 mm for prestressed concrete as well as 75 mm and 100 mm in the splash zone should prevent corrosion of the reinforcing steel [15]. A large number of steel components either built in or attached are always present on these structures. The passive reinforcing steel acts as a foreign cathodic structure to these attachments; therefore, they must be cathodically protected. In designing protection with galvanic anodes, current densities according to Table 16-3 must be allowed for, together with the surface area of nearby electrically conducted reinforcement, at about 5 mA m^{-2} of concrete surface. The connection fittings for pipelines protected with galvanic anodes can be seen in Fig. 16-6. The large vertically attached anodes serve to protect the steel casing wall, which is driven into the seabed. The anode is connected to the wall by flat iron.

16.3 Harbor Structures

In planning cathodic protection, the specific resistivity of the water, the size of the surfaces to be protected and the required protection current densities have to be determined. The protection current density depends on the type and quality of the coating. Thermosetting resins (e.g., tar-epoxy resin coatings) are particularly effective and are mostly used today on coastal structures. They are chemically

Fig. 16-6
Galvanic anodes on concrete structures.

resistant in various types of water and are not destroyed even by marine growths. Thicknesses of 0.4 to 0.6 mm have a high electrical resistance, high resistance to cathodic disbondment and very good wear resistance (see Section 5.2.1.4).

The protection current densities for structures near the sea can amount to 60 to 100 mA m^{-2} for uncoated surfaces in the area in contact with water and 20% of that for parts driven into the soil. The land sides of retaining walls take so little current that they do not have to be taken into account in the calculation. With coated objects, the protection current density lies between 5 and 20 mA m^{-2}, depending on the quality of the coating. About half this value must be expected for the part in the soil because either the coating is absent or is damaged by the driving.

16.3.1 Impressed Current Equipment

Impressed current installations should always be designed with a large excess capacity [16]. The cost of the extra expenditure is insignificant in relation to the total objective, especially if the life of the anodes is lengthened by calling on the excess capacity. A larger protection installation provides the possibility of carrying out prepolarization on uncoated components. Damage or degradation of the protective coating in the case of coated surfaces can be compensated for. Impressed current anode materials are detailed in Chapter 7.

The anodes can either be laid on the bottom, hung between the pillars or attached to the object to be protected. In every case they should be mounted so they can be easily exchanged. Anodes laid on the seabed are installed on concrete sleds or concrete slabs (see Fig. 16-7) so that they do not sink into the mud or become covered with sand. The current output in sand is considerably reduced due

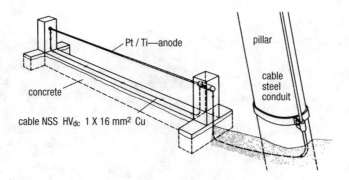

Fig. 16-7 Fixture for PtTi anodes for laying on the sea bed.

to the high specific resistivity. Anodes cannot be laid on the seabed in harbors because they can cause interference with ships lying there or can be damaged by dredging operations. A proper design for fixing the impressed current anodes directly onto retaining walls, dolphins, lock gates, etc., is necessary. There is a limited range (5 to 8 m) and considerable overprotection in the vicinity of the anodes, which frequently cannot be avoided in harbors.

Anode supports, cable insulators and the coating on the object to be protected can be destroyed by anodic evolution of chlorine (see Section 7.1). Only chorine-resistant materials should be used. Anodes on retaining walls or between pile foundations can be installed in perforated or fabricated plastic tubes (half-shells) (see Fig. 16-8). They must naturally be provided with very many holes to avoid uneven removal of anode material. Filter tubes of a chlorine-resistant special material or

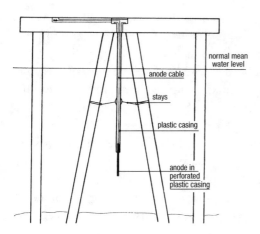

Fig. 16-8 Impressed current anodes for protecting tube supports of a loading bridge.

Marine Structures and Offshore Pipelines

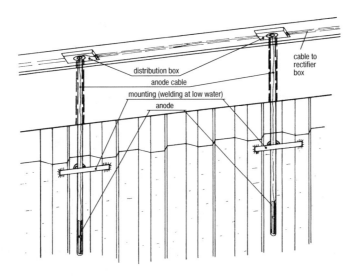

Fig. 16-9 Cathodic protection of a steel piling with impressed current.

polyethylene, polyester or polypropylene in which the removal of anode material is practically uniform have proved successful.

On steel pilings, the anodes are inserted loosely in a perforated plastic tube, or in the anode area, in fabricated plastic tubes and secured at the head of the anode. The holders can be constructed to be replaceable and the anodes can be attached via casings to the superstructure of steel pilings (see Fig. 16-9). If possible, attachments should be provided before piles are driven, otherwise expensive underwater welding will be necessary. The advantage is a technically simple mounting with low repair costs and the possibility of using all types of anodes. The disadvantage is the relatively high investment costs and the somewhat uneven removal of anode material.

16.3.2 Protection with Galvanic Anodes

Galvanic anode materials are described in Chapter 6. Zinc anodes are predominantly used in marine harbor installations since aluminum anodes tend to passivate in such water and magnesium anodes are a danger to thin coatings. In harbors with brackish or fresh water, on the other hand, only magnesium anodes are employed. Galvanic anodes are chiefly used for movable structures such as locks and sluices, small steel pilings, coated steel retaining walls and in inaccessible areas. Combinations with impressed current equipment are also used, as illustrated by the Columbus quay in Bremerhaven [17]. On the other hand, there are harbors or quays where the laying of cables and the installation of impressed current protection equipment appears impractical.

16.4 Steel Sheet Piling

With steel sheet pilings, every joint must be electrically bridged with a weld or with a welded cover plate to avoid an ohmic voltage drop in the return conduction of the protection current since joints, even with apparently excellent clamping, cannot be assumed to have a low-resistance connection. Testing of many sheet steel piling joints has given values of over 0.1 mΩ. Subsequent connection of the panels is very uneconomical since the tops of the panels are mostly covered with concrete or with a concrete apron reaching down to the water. The electrical bonding of the steel sheet must be carried out well before its final installation [18]. Table 16-4 contains the data on cathodic protection systems for marine structures.

Very often steel sheet pilings exist in conjunction with steel-reinforced concrete structures in harbors or locks. If cathodic protection is not necessary for the reinforced concrete structure, there is no hindrance to the ingress of the protection current due to the connection with the steel surfaces to be protected. The concrete surface has to be partly considered at the design stage. An example is the base of the ferry harbor at Puttgarden, which consists of reinforced concrete and is electrically connected to the uncoated steel sheet piling.

16.5 Piling Foundations

Loading bridges and piers are generally built on piles with reinforced concrete superstructures. They consist of an access bridge with an extended pier for mooring several ships. Zinc anodes operate without maintenance and independent of current supply as well as operational requirements, as for example, in the ore loading pier in Monrovia, Libya. Between the piles there are 186 slab-shaped zinc anodes, each of 100 kg, which are joined by 82 chains. The chains are connected to the steel reinforcement and via this to the piles and the cross-beams by a copper cable about 1 mm long and with a 16 mm^2 cross-section; this can be used to measure the current output of the anodes. The potential of the bridge piles is $U_{Cu\text{-}CuSO_4} = -1.0$ V; the average protection current density was initially 25 mA m^{-2}; after 2 years it was 11 mA m^{-2} and after 15 years, 6.5 mA m^{-2}. The relatively low current densities are due to these installations being coated in the underwater areas. The current per anode fluctuates initially between 0.9 and 1.2 A and decreases at the rate of 0.2 A per year. According to more recent measurements, the life expectancy of the anodes is 25 years for 85% consumption. Table 16-4 contains further data on loading piers [19].

Table 16-4 Cathodic protection systems for coastal and harbor structures

Installation	Coating	Splash zone (m²)	Current density (mA m⁻²)	Type of anode	No. of anodes	Anode weight (t)	Rectifier output (A)	Life (years)
Loading pier, Liberia	Tar pitch	5400	25→6.5	Zn	190	14	–	25
Loading pier, San Salvador	Tar pitch	27,000	70	C	120	–	7 × 300	15
Tanker pier, North Sea	Tar pitch	39,000	30	FeSi-Mo	210	–	65 × 20	15
Ore pier, Malaysia	Tar pitch-epoxy	35,000	15→5	PtTi	30	–	4 × 100	>10
Steel piling, Elbe	Tar pitch-epoxy	25,000	16	FeSi-Cr	380	4	20 × 100	>20
Drawbridge, Wilhelmshaven	Tar pitch-epoxy	22,000	10	FeSi-Mo	160	8	18 × 150	>25
Loading quay, Lomé-Togo	Tar pitch-epoxy	70,000	18	PtTi	71	–	2 × 250 2 × 150	>25
Ferry harbor, Puttgarden	None	8500 and ca. 5500 steel-reinforced concrete	160	PtTi	360	–	20 × 100	>10
Tonasa II Indonesia	None	11,250 and 5140 soil	70→30	PtTi	45	–	1 × 600 2 × 120	>20

382   Handbook of Cathodic Corrosion Protection

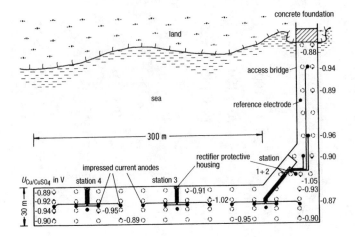

Fig. 16-10 Cathodic protection of an ore loading bridge.

Today loading piers are mostly cathodically protected with impressed current. At moorings for tankers, cathodic protection rectifiers are installed on extinguisher bridges as far as possible from the hazardous area. Otherwise, they must be of an explosion-proof type.

Figure 16-10 shows a plan of a bridge structure for supplying the Krakatau steel works in Indonesia [20]. The tubular piles are between 25 and 30 m in length. In designing the cathodic protection, a surface area of 4×10^4 m^2 was calculated. The piles

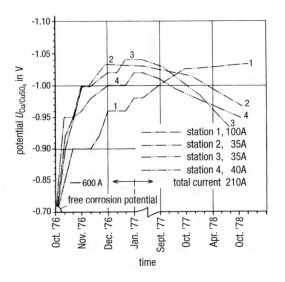

Fig. 16-11 Variation in cathodic polarization with time.

are coated with 300-μm-thick tar-epoxy resin, giving a protection current density of 10 mA m^{-2}. To take account of expected damage, the equipment was designed for 1200 A, which is provided by four rectifiers each with an output of 10 V/300 A. The equipment is regulated by stepless transformers and the rectifiers are oil cooled. A three-phase grid supply with full-wave rectification was chosen on account of the high capacity and the residual ripple of 5% required for the anodes. Thirty-two PtTi anodes 2.7 m long and 10 mm thick were installed (see Fig. 16-7). The anodes were inserted in holes in the concrete support, grouted in with epoxy resin, and lowered onto the seabed. The current loading for individual anodes was between 10 and 15 A for a rectifier output voltage of 6 V.

The free corrosion potential of $U_{Cu\text{-}CuSO_4} = -0.7$ V was lowered to -0.9 V with a protection current totalling 600 A. Figure 16-11 shows that after a polarization period of 4 months, an average of -1.0 V was reached. The protection current was then reduced to 210 A. This corresponds to an average protection current density of 5.3 mA m^{-2}. Long-life zinc reference electrodes were installed for potential control whose measurements over a constant voltage source were converted to the Cu-CuSO$_4$ scale. In the summer of 1978, the pier on the other side was lengthened to 300 m so that two more cathodic protection installations were set up, with a total current output of 200 A. The average protection current density amounted to only 3 mA m^{-2} as the result of a thicker coating. Installing, equipping and checking the cathodic protection took about 3 months.

16.6 Offshore Pipelines

Every year several thousand kilometers of pipeline are laid in the sea in the offshore regions of the world. The total at present is about 35,000 km. Whereas the early offshore pipelines were limited to shallow water and short pipelines of small diameter, today pipelines of several hundred kilometers and up to DN 1000 are laid. With short pipelines, impressed current cathodic protection is also possible, but is seldom used [21], e.g., for the 10-km-long pipeline of the German platforms Schwedeneck in the Baltic and Emshörn Z1A in the North Sea. According to Eq. (24-75), the possible protection range for a pipeline DN 300 with $s = 16$ mm and good insulating coating can amount to 100 km. For offshore applications, the coating resistances r_u will not be as good for onshore pipelines (see Section 5.2.1.2). For this reason, the protection current densities J_s are sharply increased [see Eq. (5-11')] and the length of the protected region is relatively short [see Eq. (24-75)]. It is inefficient to increase the protected length by raising the anode voltage. Therefore, the valid area of Eq. (24-75) has to be abandoned. According to the information in Section 24.4.4, only hydrogen evolution in the vicinity of the anodes has to be taken into account with cathodic overprotection so that the length of the protected region can be extended only insignificantly, by some tens of percents. For this reason, only galvanic anodes need to be considered for the protection of longer offshore pipelines. The usual installation is zinc anodes [1-3, 22-26].

Pipelines in seawater are usually covered with a thick coating. For weighting and for mechanical protection, a 5-cm-thick concrete casing is applied which is reinforced with 2- to 3- mm-diameter galvanized wire mesh. This wire mesh should not be in electrical contact with either the pipe or the anode. The pipes should be partly water jetted into the seabed to protect them from movement and damage from deep drag nets or anchors. The soil removed by water jetting is used to fill the trench or the trench is filled with rubble. The pipelines have to be anchored where the seabed is stony or rocky. Zinc anodes for offshore pipelines can consist of two half-shells. These are cast into sheet iron that protrudes at the ends; these are welded together and connected to the pipeline by copper cables. There are also bracelets made from individual blocks of anodes which are welded onto sheet iron around the pipelines (see Section 6.5.4).

Table 16-5 gives details of pipelines in the North Sea that are protected with zinc anodes. Complete cathodic protection is achieved even with about 5% of free pipe surface arising from damage to the coating [27-30]. The surface area of the expected damage, the specific resistivity of the surroundings and the length of unburied pipeline are unknown, so that an accurate estimate of the current output of the zinc anodes cannot be made. In the Ekofisk pipeline, for example, zinc anodes each weighing 450 kg are attached at 134-m intervals. With a maximum current output in seawater of 2 A and in the seabed of 0.2 A, it follows that the service life should be over 20 years [23]. A protection current density of 2 mA m^{-2} is calculated assuming an average surface area of defects of 2%.

Table 16-5 Cathodic protection of pipelines in the North Sea with zinc anodes

Pipeline	Construction date	Diameter rating	Product	Length (km)	Anode weight (t)
Ekofisk – Teesside	1973-1975	850	Oil	330	1200
Ekofisk – Emden	1974-1977	900	Gas	415	1600
Frigg – St. Fergus	1975-1977	800	Gas	350	1100
Ekofisk Flowline	1971-1977	100/750	Gas/oil	180	500

16.7 Control and Maintenance of Cathodic Protection

The potential of the object to be protected is measured in the electrolytic solution at certain distances, depending on the structure, to control the effectiveness of cathodic protection. The anodes and measuring electrodes are observed with a TV camera. In seawater, it is not necessary to interrupt the cathodic protection current when measuring the off potential. The ohmic voltage drop is negligible due to the good conductivity of seawater so that only the on potential need be considered for comparison with the protection potential [see Eq. (2-34)]. The silver-silver chloride-saturated KCl reference electrode is used for measurements in seawater [6]. The Cu-$CuSO_4$ reference electrode, which is otherwise usual for measurements in soil, is sensitive to diffusion of chloride ions and reacts with them, thus altering the reference potential.

If only a Cu-$CuSO_4$ electrode is available, the housing should always be filled with saturated copper sulfate solution, be free from bubbles, and be emptied before reuse; the copper rod must be cleaned down to bare metal and refilled.

16.7.1 Production Platforms

Potential measurements have been carried out at suitable times on platforms with galvanic anodes after the structures have been commissioned. Where impressed current protection was installed, the potential as well as the anode current was measured with fixed, built-in measuring electrodes during the commissioning period.

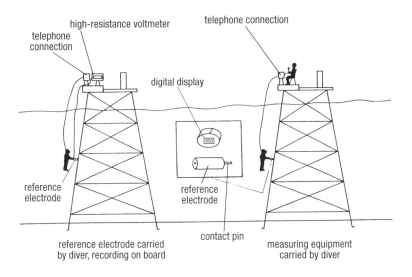

Fig. 16-12 Monitoring potential of cathodic protection on platforms.

386 Handbook of Cathodic Corrosion Protection

Two possibilities for monitoring potential with the aid of divers are shown in Fig. 16-12 [31]. In Fig. 16-12A, the diver holds the reference electrode as close as possible to the structure and gives the position by radio telephone to those on board, where the measurement is recorded. This is the only method suitable for coated structures. An instrument that contains a reference electrode and is equipped with a digital display is carried by the diver shown in Fig. 16-12B; electrical connection to the structure is made with a contact probe [11]. The diver relays the position and potential over the radio telephone. It is not always possible to produce good electrical contact because the structure is covered with marine growths that can only be removed at considerable expense and then only partially [32].

The technician on board relies entirely on the information transmitted by the diver, which is not always reliable and reproducible. Therefore, such measurements are carried out with the additional aid of a television camera so that the technician on board can record the position and measurement on videotape. The arrangement shown in Fig. 16-12A is advantageous because the reference electrode can be coupled with the TV camera. The state of the anodes, their possible passivation and material loss can be investigated at the same time as the potential measurements, and the marine growths can be removed if they threaten to smother the anodes. Platforms are given an annual visual examination. Potential measurements are also carried out on these occasions. Impressed current installations are also subject to continual monitoring so that defects can be detected at an early stage and measures to repair them instituted.

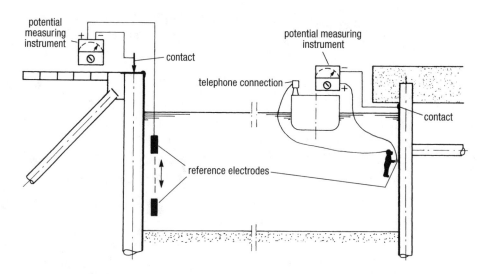

Fig. 16-13 Measurements of potential on harbor structures.

16.7.2 Harbor Structures

Harbor structures are very accessible and can be investigated without the effects of wave motion. Grounding of steel pilings presents no problems and the work can be carried out from the quay (see the left-hand side of Fig. 16-13). With steel-reinforced concrete structures, measurements have to be made from a boat if no reliable contact has been provided in their construction (see the right-hand side of Fig. 16-13).

In protection with impressed current installations, it depends on the conductivity of the harbor water whether off potentials can be measured and also whether the *IR* component can be taken into account. Only very few inland harbors require such expenditure; usually the conductivity is sufficiently high.

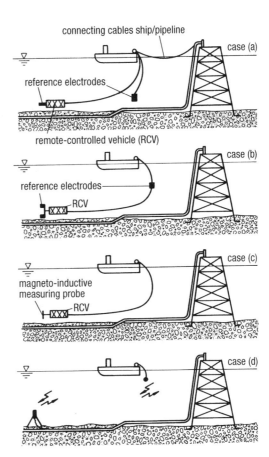

Fig. 16-14 Monitoring cathodic protection of offshore pipelines.

16.7.3 Offshore Pipelines

It is very expensive to check cathodic protection on long pipelines since protection is afforded by galvanic anodes. The few short pipelines with impressed current protection are not considered here since built-in measuring electrodes are provided and therefore no problems are expected in monitoring.

For offshore pipelines, there are four possibilities for monitoring the efficiency of the cathodic protection [11,31,33-36]. These methods are represented schematically in Fig. 16-14.

(a) Cable connection ship-pipeline
The ship is connected by cable to the platform and the reference electrode is steered by a remote-operated vehicle (ROV) or—in shallow waters—directly from the ship over the pipeline. In the case of long pipelines of more than 20 km, there can be considerable difficulty in running the cables.

(b) Potential difference measurements between reference electrodes
In this process, a reference electrode is attached to the supply cable of the ROV. There are two further electrodes on the ROV. As it travels along the pipeline, the potential differences between the electrodes provide information on the effectiveness of the galvanic anodes, on the location and current uptake at defect areas and on the potentials. Section 3.6.2 provides information on the evaluation of the potential differences. From the known potential at the initial point on the platform, the potentials given for the method described in Section 3.7.2 can be determined if ohmic voltage drops in the pipeline do not have to be considered.

All measurement data are recorded on tape—with a plausibility check during the measurements—and later evaluated with the computer. This method has been tested in practice and is sufficiently accurate.

(c) Magnetic inductive method
A probe is passed over the pipeline, which enables the flowing protection current to be determined by magnetic inductive methods, the efficiency of the actual anodes to be measured and areas of defects to be detected. An increased current flow is recorded at the defect areas as well as at the anodes installed at definite distances. This method is now frequently used, particularly in combination with method (b).

(d) Electrodes with data teletransmission
Potential probes have been developed that are anchored near the pipeline with buoyancy floats and electrically connected to the pipeline. These probes are switched on by coded signals from the ship; they read the potential and

send the information back to the ship acoustically. The batteries of the probes have to be renewed at definite intervals. A very large number of probes is required, and these can be at risk of mechanical damage by drag nets.

Methods (a) to (c) are principally only applicable to pipelines that are not buried; otherwise, the probes cannot be accurately located over the pipeline. With buried pipelines, in addition to the methods described, it is necessary to register their position according to depth and distance from the ROV.

16.8 References

[1] NACE, Standard RP-01-76 (Rev. 83), Houston Texas 1976.
[2] Germanischer Lloyd, Vorschriften für meerestechnische Einrichtungen, Hamburg 1985.
[3] Det Norske Veritas, Technical Note TNA 701-703, 1981.
[4] Schiffbautechnische Gesellschaft, Richtlinien Nr. 2215 u. Nr. 2220, Hamburg 1987.
[5] B. Richter, Schiff u. Hafen *33*, H. 8, 78 (1981).
[6] VG 81259 Teile 1-3, Kathodischer Korrosionsschutz von Schiffen, Beuth Verlag, Köln 1986.
[7] F.J. Kievits und H. Hebos, Werkstoffe u. Korrosion *23*, 1975 (1972).
[8] Fischer, Bue, Battes, und Steensland, OTC Houston 1980, Paper No. 3889.
[9] B. Richter, Schiff u. Hafen *32*, H. 10, 102 u. H 11, 81 (1980).
[10] DIN 50929 Teil 3, Beuth Verlag, Berlin 1985.
[11] V. Ashworth u. C. J. L. Booker, Cathodic Protection, Ellis Horwood LTD, Chichester 1986, S. 36.
[12] R. Vennelt, R. Saeger u. M. Warne, Materials Performance *22*, H. 2, 22 (1983).
[13] B. Richter, Schiff u. Hafen, Sonderheft Meerwasserkorrosion, Februar 1983, S. 88.
[14] Hauptverband der gewerblichen Berufsgenossenschaften, Taucherarbeiten, VBG 39, Köln 1984.
[15] Recommendations for the design and construction of concrete sea structures, Federation Internationale de la Precontrainte, Paris.
[16] Hafenbautechnische Gesellschaft, Kathodischer Korrosionsschutz im Wasserbau, Hamburg 1981.
[17] H.-P. Karger, Korrosion und Korrosionsschutz im Stahlwasserbau, T.A. Wuppertal 1978.
[18] H. Determann u. E. Hargarter, Schiff u. Hafen *20*, 533 (1968).
[19] G. Hoppmann, 3 R-intern. *16*, 306 (1977).
[20] K. Klein, 3 R-intern. *16*, 421 (1977).
[21] K. Klein, 3 R-intern. *15*, 716 (1976).
[22] NACE Standard, RP-06-75, Houston, Texas, 1975.
[23] W. v. Baeckmann, Erdöl/Erdgas *93*, 34 (1977).
[24] B.S. Wyatt, Anti-Corrosion H. 6, 4 u. H. 7, 8 u. H. 8, 7 (1985).
[25] H. Hinrichsen u. H. G. Payer, Hansa *119*, 661 (1982).
[26] C. Schmidt, Erdöl – Erdgas *100*, 404 (1984).
[27] M. Reuter u. J. Knieß, 3 R-intern. *15*, 674 (1976).
[28] T. Michinshita u.a., 2nd Int. Conf. Int. Ext. Prot. Pipes, Paper C2, Canterbury 1977.
[29] F.Q. Jensen, A. Rigg u. O. Saetre, NACE Int. Corr. Forum Pap. 217, Houston (Texas) 1978.
[30] A. Cozens, Offshore Technology Conf., Houston 1977.
[31] B. Richter, Schiff u. Hafen *34*, H. 3, 24 (1982).
[32] B. Richter, Meerestechnik *14*, 105 (1982).
[33] T. Sydberger, Material Performance *22*, H. 5, 56 (1983).
[34] C. Weldon, A. Schultz u. M. Ling, Material Performance 22, H. 8, 43 (1983).
[35] G. H. Backhouse, NACE Corrosion, Paper 108, Toronto 1981.
[36] R. Strømmen u. A. Rødland, NACE Corrosion, Paper 111, Toronto 1981.

17

Cathodic Protection of Ships

H. BOHNES AND B. RICHTER

Cathodic protection of ships extends from the external protection of the underwater area, including all attachments and openings (e.g., propeller, rudder, propeller bracket, sea chests, buoyancy tanks, scoops thruster), to the internal protection of various tanks (ballast and fresh water, fuel storage), pipework (condensers and heat exchangers) and bilges. Advice on dimensions and distribution of anodes is given in regulations [1-6]. Ships are exposed to water of very different compositions, unlike all the other objects to be protected which are described in this handbook. The salt content and conductivity are particularly important because they have a profound influence on the action of corrosion cells (see Section 4.1) and current distribution (see Sections 2.2.5 and 24.5). In addition, on ships, problems of dissimilar metals must be considered. Protective measures against stray currents are dealt with in Section 15.6.

17.1 Water Parameters

17.1.1 Dissolved Salts and Solid Particles

The waters through which ships travel are categorized by their salt content. The following are approximate values: seawater, 3.0 to 4.0% salt; coastal brackish water, 1.0 to 3.0%; river brackish water, 0.5 to 1.8%; salty river water, 0.05 to 0.5%; river water, <0.05%. Seawater mainly contains NaCl. The salt content is approximately 1.8 times the chloride ion content. The salt content of the world's oceans is almost the same. Different salt contents can occur in more enclosed seas [e.g., the Adriatic (3.9%), Red Sea (4.1%) and the Baltic (1.0%)]. Table 17-1 gives as an example average analyses for seawater and the Rhine River.

The salt content determines the specific electrical conductivity of the water (see Section 2.2.2). In coastal areas this varies according to tide and time of year. The following average values in ohms per centimeter serve as a guide: Narvik roadstead, 33 [7]; Helgoland, 27 [7]; North Sea, 30; Elbe/Cuxhaven, 100 [7]; Elbe/Brunsbüttelkoog, 580; Elbe/Altona, 1200; Lübeck wharf, 75; Antwerp (Quay 271), 120; Rotterdam Botlek, 240; Tokyo Gulf, 25 [8].

From these figures it can be seen that when a ship enters Hamburg harbor the conductivity of the water falls by a factor of 40. The range of the protection current is correspondingly decreased [see Eq. (24-111)]. In addition, the formation of cathodic layers is made more difficult by the low Ca^{2+} ion content. This leads through mechanical abrasion to lower layer resistances, increased current requirements, and a reduced protection range according to Eq. (2-45). It is therefore understandable that the risk of corrosion in harbor increases; moreover, the action of corrosion cells is greater when the ship is at rest than when it is in transit (see Section 4.1); pitting corrosion has to be considered.

Coastal waters and particularly stagnant water can be contaminated with effluent containing partly inhibitors and passivating materials (e.g., phosphates) and partly reducing components (e.g., sulfides and organic matter). Such components bring about incomplete inhibition [9] and anaerobic corrosion. In both cases pitting corrosion will occur in the absence of coating and cathodic protection. Effluent contains mostly ammonium salts and amines that can attack cuprous materials. Local corrosion due to cell formation on pipework has to be taken into account in ships that remain in harbor for long periods if the pipes are not filled with clean seawater or fresh water.

Layers and coatings that are sensitive can be destroyed by solid matter in water, such as sand, mud and ice. It is correspondingly necessary to design a higher

Table 17-1 Composition and average concentrations of chemicals in seawater and Rhine River water (Duisburg)

Component X_i	Seawater		Rhine River	
	c_i (mg L^{-1})	$c_i |z_i|$ (mol m^{-3})	c_i (mg L^{-1})	$c_i |z_i|$ (mol m^{-3})
Cl^-	18,980	535	119	3.36
SO_4^{2-}	2650	55	90	1.87
HCO_3^-	140	2	60	0.98
Na^+	10,556	459	93	4.04
K^+	–	–	8	0.21
Mg^{2+}	1272	105	21	1.73
Ca^{2+}	400	20	43	2.15
O_2	8 to 20		2 to 6	
pH	7.6 to 8.3		6.6 to 8.1	

protection current output for ships that are subject to such conditions. On the other hand, marine growths have both positive and negative importance for corrosion and cathodic protection. On one hand they raise the diffusion resistance for oxygen access and on the other they can destroy coatings that do not meet the regulations [10,11].

17.1.2 Aeration and Oxygen Content

The solubility of oxygen in water with a salt content up to 1 mol L^{-1} is only dependent on the temperature. The oxygen concentrations in equilibrium with air amount to (in mg L^{-1}): 0°C, 14; 10°C, 11; 20°C, 9; and 30°C, 7. The depth of water has no effect in the case of ships. In Hamburg harbor in summer, 7.3 mg L^{-1} are measured in depths up to 7 m. The value can be much lower in polluted harbors and even fall to zero [8]. In the open sea, constant values are found at depths of up to 20 m. With increasing depth, the O$_2$ content in oceans with low flow rates decreases [12] but hardly changes at all with depth in the North Sea [13].

Oxygen is essential as a component in the cathodic partial reaction. In the case of differential aeration, heterogeneous film formation occurs together with local corrosion (see Section 4.1). Just as with steel pilings, local corrosion is observed on ships in the splash zone [9] due to differential aeration, particularly at welded joints [14]. On uncoated steel, the corrosion usually leads to shallow pitting. The oxygen supply depends on flow rate according to Eq. (4-6). The quantity K_w depends on the flow rate. In the course of some years K_w can increase due to film formation. According to Eq. (2-40), equivalence can be assumed between the minimum protection current density J_s and the corrosion rate. The observed corrosion rate of 0.1 to 1.0 mm a^{-1} corresponds to protection current densities of 0.1 to 1 A m^{-2}. The protection current density decreases markedly with time due to cathodic film formation (see Fig. 17-1), so that an almost constant protection current density of 50 mA m^{-2} can be assumed in still seawater. In ballast water tanks, corrosion rates

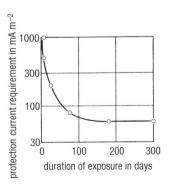

Fig. 17-1 Protection current vs. time for uncoated steel in a calm sea.

of 3 mm a^{-1} are observed on horizontal surfaces, particularly on the base; these require corresponding provision of a cathodic protection current density of 3 A m^{-2}.

Oxygen solubility decreases almost linearly with increasing temperature but the diffusion rate increases exponentially. This leads to a slight increase in corrosion rate with increasing temperature although in Eq. (4-6) the factor is assumed to be greater. For this reason an increase in corrosion rate of about 1.5 times is considered in tropical waters compared with the North Atlantic.

17.1.3 Flow Rate in the Case of a Moving Ship

Flow rate not only raises the rate of oxygen transport by lowering the K_w value in Eq. (2-6), but also adversely affects surface film formation. Figure 17-2 shows the relation between protection current density and steaming velocity. Factor F_1 relates to undisturbed film formation. The influence of flow is not very great in this case. Factor F_2 represents the real case where surface films are damaged by abrasion [15]. The protection current density can rise to about 0.4 A m^{-2} at uncoated areas.

17.1.4 Variations in Temperature and Concentration

Differences in temperature and concentration can in principle lead to corrosion cell formation, but have little effect below the water line. On the other hand, they have to be taken into account in the interior corrosion of containers and tanks in relation to their service operation (see Section 2.2.4.2). Generally the action of corrosion cells can be reduced or eliminated by cathodic protection.

In strong sunlight, water can evaporate at defects in coatings and surface films, and lead to concentration and crystallization of salts (e.g., in the upper decks of the ship). This can damage surface films, giving rise to local anodes. This is the case when a ship slowly rises in the water on unloading and is later reimmersed on loading.

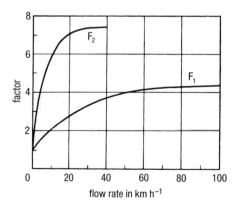

Fig. 17-2 Protection current densities of uncoated steel as a function of flow rate in flowing seawater, F_1 with undisturbed film formation; F_2 with disturbed film formation due to erosion.

17.2 Effect of Materials and Coating Parameters

Almost all common metals and structural steels are liable to corrode in seawater. Regulations have to be followed in the proper choice of materials [16]. In addition, there is a greater risk of corrosion in mixed constructions consisting of different metals on account of the good conductivity of seawater. The electrochemical series in seawater (see Table 2-4), the surface area rule [Eq. (2-44)] and the geometrical arrangement of the structural components serve to assess the possibility of bimetallic corrosion (see Section 2.2.4.2 and Ref. 17). Moreover the polarization resistances have considerable influence [see Eq. (2-43)]. The standards on bimetallic corrosion provide a survey [16,17].

Different microstructural regions in a material which has an almost uniform composition can also lead to the formation of corrosion cells (e.g., in the vicinity of welds). Basically, corrosion cells can be successfully overcome by cathodic protection. However, in practice, care has to be taken to avoid electrical shielding by large current-consuming cathode surfaces by keeping the area as small as possible. In general, with mixed installations of different metals, it must be remembered that the protection potentials and the protection range depend on the materials (Section 2.4). This can restrict the use of cathodic protection or make special potential control necessary.

The protection of aluminum ships demands particular attention since electrical contact with steel and copper materials can seldom be avoided and a whole range of aluminum alloys are unsuitable for cathodic protection (see Section 2.4 and Fig. 2-11). Later protective measures must therefore be observed during the construction stage since even good coatings in combination with cathodic protection are frequently not adequate to protect gaps or openings.

Coatings assume the function of passive protection on ships and are indispensable as carriers for antifouling substances. By combining them with cathodic protection, the aim is to substantially reduce the protection current requirement and increase the protection range by raising the polarization parameter (see Section 5.1). Besides chemical and mechanical durability, electrical coating resistivity as well as porosity and degree of damage determine the quality of the coating. With good resin coatings, the coating resistivity on pore-free specimens can be over 10^5 Ω m^2. After soaking in water, the resistivity can fall by several tens of percents and even fall as low as 30 Ω m^2 [18,19]. This corresponds from Eq. (5-11) to a protection current density of $J_s = 10$ mA m^{-2}. The factors that affect the coating resistivity are thickness, type of coating and quality of the substrate preparation. For a practical protection current requirement, the current uptake at uncoated surfaces and defects also has to be considered (see Section 5.2.1.2).

In contrast to thick coatings, thin coatings combined with cathodic protection present certain risks on ships. Blisters filled with highly alkaline liquids resulting from electro-osmotic processes, ion migration, and (depending on the alkali ion concentration) potential, temperature and the properties of the coating system have to be considered (see Section 5.2.1.4). To avoid blistering, the cathodic protection should be limited to negative potentials (e.g., $U'_{Hs} = -0.65$ V). Even closed blisters are undesirable on ships because they increase the resistance to motion. One of the objects of cathodic protection is to reduce the resistance to motion by preventing the formation of rust pustules. In general, this resistance is due 70% to friction and 30% to shape and wave resistance. The latter is constant for a given ship but the frictional resistance can be raised up to 20% by corrosion. It is considerably reduced by having hull surfaces as smooth as possible, which are not damaged by local corrosion products. A further factor increasing resistance to motion is marine growths, which can be countered by antifouling coatings. The loss in efficiency caused by roughness can correspond to an increase of 12% in fuel consumption. The loss due to marine growths can be three times greater.

Coatings for application to the hull below the waterline and for tanks must be suitable for combining with cathodic protection. The type of protection—impressed current or galvanic anodes—is not important since the potential alone is the governing factor. The requirements in DIN 50928 [20] (see Section 5.2.1) must be observed in the case of continual action of an electrolyte and electrochemical polarization. In this respect, the Ship Building Technical Society (Schiffbautechnische Gesellschaff, STG) has developed a standard [11] in which the testing and evaluation of coatings for application under water are laid down. $U_H = -0.73$ V is prescribed as the test potential in an experiment lasting at least 9 months in operating conditions or a similar period in which marine growths take place in natural experiments. Blister formation, efflorescence, and adhesion are assessed on damaged as well as undamaged surfaces. The regulation is based on comprehensive investigations and has been internationally endorsed [19,21].

The economic use of ships requires lengthening the intervals between dry dockings and the associated maintenance costs. For this reason, newly developed high-grade coating materials are used that require greater expense in their application. The STG has enacted a standard [10] governing coating materials for use on ships and offshore structures. The standards include their application and maintenance, which is related to DIN 55928 Part 5 [22]. These high-grade coating materials today demand better surface preparation; the standard value for the degree of purity for the outer surface is Sa 2½ [10], while for special coatings in tanks SA3 is demanded. The use of other preparation methods (e.g., flame cleaning) has decreased and vapor cleaning does not give the desired surface quality.

Welds represent particularly weak points. The sheet itself is mechanically shot blasted in the factory or in the shipyard and given a shop primer. The installation welds can usually only be derusted mechanically and coatings combined with cathodic protection are particularly prone to blistering and loss of adhesion.

In addition to careful surface preparation, these high-grade coatings require specific climatic conditions, such as humidity and temperature, as well as specific repainting intervals and control of the film thickness. In open docks even in climatically favorable zones, such as central Europe, these conditions can be observed only at considerable expense. Damage arising from unfavorable application conditions is often attributed to the cathodic protection although it is not responsible if the correct coating material has been chosen.

Blistering is influenced by the total film thickness as well as by the application conditions and the coating material. A minimum dry film thickness of over 250 μm is required, and this does not take account of the antifouling component [10].

17.3 Cathodic Protection Below the Waterline

Cathodic protection of an uncoated ship is practically not possible or is uneconomic due to the protection current requirement and current distribution. In addition, there must be an electrically insulating layer between the steel wall and the antifouling coating in order to stifle the electrochemical reduction of toxic metal compounds. Products of cathodic electrolysis cannot prevent marine growths. On the contrary, in free corrosion, growths on inert copper can occur if cathodic protection is applied [23].

A distinction is made between complete and partial protection of the underwater area, depending on the extent of the protected region. In partial protection, only the stern is protected; it is particularly endangered because of the high flow rate and aeration as well as the formation of cells on attachments, such as the propeller and rudder. Partial protection can also be extended to the bow, which also experiences high rates of flow. The complete protection of the ship with galvanic anodes or impressed current is becoming increasingly important since defects in the coating due to mechanical damage are more frequent at the bow and amidships. Installation of galvanic anodes on the bilge keel presents no problems. The extent to which attachments such as propellers and rudders can be covered by the protection or be given individual cathodic protection depends on the design of the ship and the method of protection.

In all cases partial or total hulls of aluminum or stainless steel must be provided with cathodic protection. This also applies to high-alloy steels with over 20% chromium and 3% molybdenum since they are prone to crevice corrosion underneath the coatings. The design of cathodic protection must involve the particular conditions and is not gone into further here.

17.3.1 Calculations of the Protection Current Requirement

If the surface area cannot be obtained from structural drawings, the underwater surface area can be calculated from the following relation [1]:

$$\frac{S_0}{\mathrm{m}^2} = \frac{L_{\mathrm{CWL}}}{\mathrm{m}}\left(\frac{B_{\mathrm{CWL}}}{\mathrm{m}} + 2\frac{T_{\mathrm{CWL}}}{\mathrm{m}}\right)\delta \qquad (17\text{-}1)$$

where L_{CWL} is the length of the construction waterline, B_{CWL} is the breadth of the waterline at moulding edge (machined at 0.5 L_{CWL}), T_{CWL} is the depth of construction on 0.5 L_{CWL} (calculated as basis), and δ is the block coefficient.

In addition, the surface area of all attachments and openings must be separately determined from construction drawings. The sum gives the total area, S. Attachments include rudders and propellers as well as propeller brackets and shaft bosses as in multiscrew ships. Further, special driving gear (Kort nozzles and Voith-Schneider perpendicular propellers) must be paid particular attention. Openings include sea chests, which sometimes have an uncoated bronze grid, as well as scoop openings and bow thrusters [1]. The attachments and openings require considerably larger current densities so that the current requirement cannot be determined merely from the sum of the surface areas. From the protection current density, J_{si}, of the individual surfaces, S_i, the total protection current requirement, I_s, is calculated as:

$$I_s = \sum_i J_{\mathrm{si}} S_i \qquad (17\text{-}2)$$

The protection current density for steel ships. J_{si}, depends on the quality of coating, the flow behavior and the type of components to be protected (see Sections 17.1 and 17.2). For example, a propeller that is assembled with a slip ring requires protection current densities of up to 0.5 A m^{-2}. Experience has to be relied on for the service behavior of coated surfaces (e.g., possible damage from ice or sand abrasion). Protection current densities are usually a few mA m^{-2} for typical ships' coatings. They increase somewhat with time. After a year, average values of between 15 and 20 mA m^{-2} can be assumed. It is usual in designing with galvanic anodes for 15 mA m^{-2} to include a mass reserve of 20%. For steel merchant ships, impressed current equipment giving 30 mA m^{-2} is designed so that it can eventually deliver more current to cope with damage to the coating. This value has to be increased for ice breakers and ice-going ships according to the area and time of travel (e.g., in the Antarctic at least 60 mA m^{-2} is required). The additional expenditure is negligible with this system compared with galvanic anodes.

Cathodic Protection of Ships

The protection current requirement for aluminum ships is considerably less because of the dense adherent oxide films. The necessary protection current requirement is being clarified in current investigations [24] but good results have been obtained by assuming a figure of 10% of that for steel. With aluminum there is only a very narrow permissible potential range [25] (see Section 2.4) so that impressed current protection cannot be used because of the anodic voltage cone and only selected anode materials can be considered.

17.3.2 Protection by Galvanic Anodes

17.3.2.1 Size and Number of Anodes

The required anode mass is given by Eq. (6-7) (see also data in Ref. 1). With $J_s = 15$ mA m^{-2}, total surface, S, and 2 years of life, the required anode mass (m) comes to:

$$\frac{m_{Zn}}{kg} = 0.337 \frac{S}{m^2} \tag{17-3a}$$

$$\frac{m_{Al}}{kg} = 0.12 \frac{S}{m^2} \tag{17-3b}$$

Since the current per unit volume is approximately the same for both types of galvanic anode, the same dimensions can be assumed for each of them. Flat elongated anodes or groups of anodes are used almost exclusively for the underwater zone. These are cast onto supports of ship construction steel, aluminum or stainless steel. The supports on the bilge keel or the so-called doubling of the ship's side are most conveniently welded, which accommodates the dimensions of the supports. In this way the wall of the ship is not impaired when anodes are exchanged. This type of fixing is prescribed by the German navy [1]. Today magnesium is only used in the interior of ships due to the high driving voltage, the low current output and the severe self-corrosion. In contrast to the protection of offshore structures (see Chapter 16), zinc anodes maintain their preeminent role. The feared increase in resistance to motion has not been apparent either experimentally or in service at speeds up to 18 knots. Anodes 20 mm thick were not a disadvantage, even on speedboats.

Cathodic protection, complete or partial (stern and bow), is arranged by the distribution of the anodes so that the desired current distribution is maintained correctly in the relevant areas. Galvanic anodes, depending on their dimensions and current output, deliver a certain maximum current which depends on the conductivity. The calculated maximum current from Eq. (6-12) based on the driving voltage and grounding resistance is reduced in practice on working anodes due to film formation and polarization resistances which depend on the anode material, the electrolyte and the time (i.e., the operating conditions). It is understandable therefore that the maximum currents given by manufacturers for a particular anode material can be upset in practice by such changes. In the design it is important to consider not only the total current but also to see that the necessary current densities and protection ranges are achieved. Initially the coatings have a high resistance and a low degree of damage. The range according to Eq. (24-111) is then large and the protection current requirement is low. In service, the coating resistance decreases so that the current requirement not only increases but the protected range decreases. In particular, attention has to be paid to the fact that a reduction in conductivity (e.g., in harbor) also restricts the range as in Eq. (2-45). If the protection potential range is temporarily not reached at the total surface to be protected, there is nevertheless not a great danger of corrosion since the cathodic protection usually encourages the action of corrosion cells. Figure 2-9 shows the relation between corrosion rates and potential.

The required number, n, of anodes can be calculated using Eq. (17-2) from the current requirement, together with the maximum current output I_{max} of the anodes. The arrangement of the anodes is dealt with in Section 17.3.2.2. Galvanic protection systems are usually designed to give protection for 2-4 years. After this period, a maximum of up to 80% of the anodes should be consumed.

$$n_i\, I_{max} = J_{si}\, S_i \quad \text{and} \quad n = \sum_i \frac{J_{si}\, S_i}{J_{max}} \qquad (17\text{-}4)$$

Here, n_i is the number of anodes for the ith surface area region.

The equipping of a steel ship with a below-waterline area of 4500 m² is given as an example. With $J_s = 15$ mA m⁻², the current requirement is 67.5 A; according to Eq. (17-3a), this would require 1517 kg of zinc for a life of 2 years. This would require 96 anodes each of 15.7 kg of zinc (16.8 kg gross). Such anodes have a current output of 0.92 A.

The total current therefore amounts to 88 A. This covers the required protection current density. On the other hand, by choosing a larger anode size of 25.9 kg of zinc and with a current output of 1.2 A, 58 anodes would provide the necessary protection current. For a service life of 4 years, the amount of zinc is doubled to 3033 kg. Here the possibility for groups of anodes could be, for example, 82 two-anode groups, each anode weighing 18 kg. With a current output per group of 1.3 A, this gives a total current output of 107 A. This level of current supply could also be given, if practical considerations demanded it, by 54 three-anode groups.

Equipping a ship with aluminum anodes to last 2 years would require, according to Eq. (17-3b), 540 kg. Ninety-six anodes of the same size as those of zinc, each weighing 6.2 kg of aluminum (7.3 kg gross) give a total of 595 kg and overprovision. The current output of the anodes is practically the same as with zinc. With aluminum, therefore, there is a greater reserve. For a service life of 4 years, 1080 kg would be required. Eighty-two two-anode groups with each anode weighing 7.3 kg gives a total of 1197 kg. Since there is overprovision with 106 A also, even in this case with three-anode groups, anodes can be saved.

Ten percent of the anode mass is calculated for aluminum ships. The anode supports must also be of aluminum in order to allow them to be welded and to avoid bimetallic corrosion.

17.3.2.2 Arrangement of Anodes

The anodes should be uniformly distributed over the underwater surfaces to achieve good current distribution [1]. The following basic principles should be observed: about 25% of the total weight of anodes should be used to protect the stern. The remaining anodes should be distributed on the bow and amidship. They should be installed on the bilge radius so that they are protected against being pulled off after they are attached. In the area of the bilge keel they should be alternately attached on the upper and lower sides. The spacing of the anodes on the bilge in the middle of the ship should not be greater than 6 to 8 m to ensure an overlap of the protected zones. In water with high protection current densities (e.g., the tropics) and with low resistivity (e.g., the Baltic) the range is smaller. For such ships, a spacing of 5 m is chosen. It is reduced even further for ships subjected to high mechanical damage, e.g., ice-going ships in Arctic waters.

The foremost anodes in the bow area should be installed obliquely because of the water flow. Care has to be taken that they cannot be damaged by the anchor chain. Because of the high loading, the anodes should be installed not only on the bilge but also in the vicinity of the central throughplate keel. At the stern the anodes should be predominantly sited in the area of the stern tube, the sole piece, the propeller well and perhaps the heel. In distributing the anodes, care must be taken that eddies caused by the anodes do not impinge on the propeller. For this reason no anodes should be installed in the region 0.4 to 1.1 r (r = radius of the propeller)— a forbidden region. In more recent times, it has also been required that the anodes in the area of the stern tube be at least a distance of $2\,r$ from the propeller. The anodes attached over the propeller well are often not immersed when the ship is travelling under ballast so they should be fixed obliquely on the stern profile. This also applies to anodes above and below the forbidden region (see Fig. 17-3 [1]).

The rudder is provided with anodes on both sides; these should be fixed either at the level of the propeller hub or as far as possible above and below on the rudder blade. There are specially shaped rudder anodes which are welded to the front edge of the rudder. Sea chests and scoop openings should be specially considered and provided with anodes because of their increased current requirement.

For partial protection of the stern, 33% of the anodes used for complete protection should be installed instead of the usual 25%. Of these, 25% serve as actual protection for the stern and 8% as shielding for the stern area against the remainder of the current-consuming body of the ship. These anodes are known as gathering anodes and are fixed in front of the anodes protecting the stern.

Propeller brackets in multipropeller ships must be particularly protected. In small ships the anodes are attached on both sides at the base of the propeller bracket. With large ships the anodes are welded onto the propeller brackets (see Fig. 17-4).

Shafts and screws should also be included in cathodic protection with galvanic anodes and shaft slip rings. The transmission voltages should be below 40 mV (see Section 17.3.3.3).

Special propulsion also requires relevant calculations and distribution of the anodes. For Kort nozzles, the total surface area of the rudder is determined and a basic protection current density of 25 mA m^{-2} imposed. The anodes are attached on the external surface at a spacing of 0.1 r to 0.25 r at the region of greatest diameter. Internally the anodes are fixed to the strengthening struts. With Voith-Schneider propellers, the anodes are arranged around the edge of the base of the propeller.

Ships with nonmetallic hulls frequently have metallic attachments which can be cathodically protected. Here the anodes are screwed onto the timber or plastic hull and electrically connected with low resistance via the interior of the ship to the objects to be protected. The metallic foundation serves for flotation and copper bands.

17.3.2.3 Control and Maintenance of Cathodic Protection

To measure the potential, reference electrodes are lowered on unbreakable ropes tensioned with 20 kg of lead as near as possible to the ship's side. *IR* errors can be neglected because of the good conductivity of seawater [see Eq. (2-34)]. In contrast to fresh water, the switching method in seawater is not necessary (see Section 3.3.1).

Ag-AgCl electrodes are usually chosen (see Section 16.7 and Table 3-1). Care has to be taken that the electrical connection to the ship is sufficiently low resistance

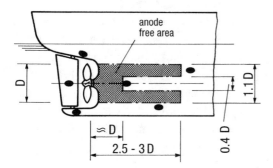

Fig. 17-3 Arrangement of galvanic anodes on the stern according to VG 81256 [1].

Cathodic Protection of Ships

and dry. Clamps on the drums are usually used. The information on protection potentials is given in Section 2.4.

17.3.3 Protection with Impressed Current

The number of galvanic anodes is approximately proportional to the area of the surfaces to be protected due to their limited current output and range. With large numbers of anodes, galvanic protection is economically inferior to impressed current because costs of material and mountings are proportional to the number of galvanic anodes whereas the cost of impressed current protection increases less than proportionally with the surface area. The boundary lies with ships of about 100 m in length. Additional advantages of impressed current protection are the control-

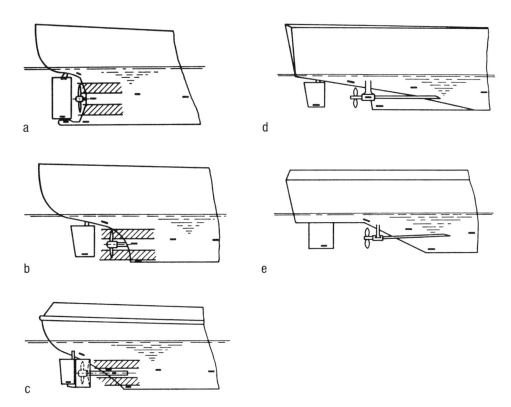

Fig. 17-4 Distribution of galvanic anodes for different shapes of stern. (a) Steamer stern: one propeller, one balance rudder. (b) Steamer stern: one propeller, a suspension rudder. (c) Tugboat stern: two propellers, a Kort nozzle rudder. (d) Transom stern: two propellers, a suspension rudder. (e) Transom stern: two propellers, two suspension rudders.

lable current output and the use of long-life anodes according to Table 7-1. In comparison with galvanic protection, impressed current protection involves higher driving voltages and fewer anodes. An elevated current density of 30 mA m^{-2} is applied to coated surfaces because of the average greater reduction in potential; however, the potential $U'_H = -0.65$ mV must not be undershot to avoid damage to the coating. For the example given in Section 17.3.2.1, 4500 m^2 would require 113 A, which could easily be provided by a central supply. Four anodes serve for current injection with an output of 30 A each.

17.3.3.1 Current Supply and Rectifiers

In contrast to fixed objects, ships require potential regulating equipment because the protection current requirement varies with environmental and operational demands. Chapter 8 contains more detailed information on rectifiers. The protection current equipment on ships must be particularly robust and resistant to vibration [3]. The regulation is met with phase-changing control using thyristors. In contrast to instruments for protecting pipelines against stray current interference, the controlled time constant can be very large because the protection current requirement changes only very slowly. The supply equipment contains additional current and potential measuring instruments for the individual impressed current anodes and measuring electrodes. The most important data are recorded in large installations.

Silicon rectifiers are predominantly used because of the proportionately higher power. An automatic current restriction must be provided to protect against overloading in low-resistance contacts to well-grounded installations (e.g., in harbors). Any breakdown must be indicated by a visible or audible warning signal. Correspondingly, voltage limitation can also be provided if the impressed current anodes require it (see Section 7.2.2). Figure 17-5 shows the circuit diagram and the components of a protection installation for ships as an example.

Current supply for medium-sized ships is provided by an instrument in the engine control room or the engine room. With large ships, the rectifier should be installed near the anodes so that only cables of small cross-section are required. Formerly, in such cases two protection units independent of each other were used in the engine room and the forecastle. Meanwhile, the anodes were installed in the rear quarter of the ship, even with large tankers, and the poor current distribution had to be tolerated.

Another possibility is to decentralize the dc supply. Here the ac supply for the individual protection rectifiers comes from a potential-controlled central supply that is situated in the engine control room. The protection rectifiers can then be situated amidships with relatively short dc cables to the anodes. The dimension of the dc cable should maintain a voltage drop below 2 V.

Cathodic Protection of Ships 405

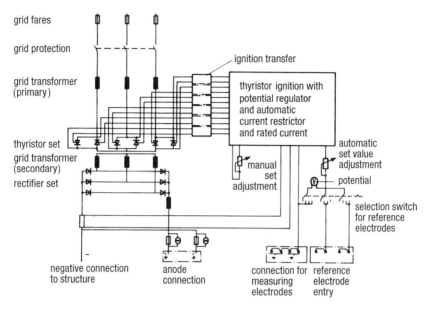

Fig. 17-5 Circuit diagram for an impressed current protection installation.

17.3.3.2 *Impressed Current Anodes and Reference Electrodes*

Two types of impressed current anodes are mainly used on ships: attached and recessed anodes, which are fixed flush with the ship's side. Several years ago, towing anodes of aluminum or platinized silver were also used. These anodes were not used later on because of insufficient protection of the bow region. With attached anodes [27,28] as in Fig. 17-6, which mainly applies to lead-silver anodes, the holder consists of a long trapezoid-shaped plastic (usually fiberglass-reinforced polyester) anode body into whose sides the active anode surfaces are set. The plastic body is screwed to staybolts welded to the ship's side and containing female inserts. The plastic body is provided with a protective shield as a support which guards the ship's side in the immediate vicinity of the anode surfaces and protects the coating against the cathodic effects of very negative potentials. This protection shield is fixed to the hull of the ship by welded support brackets. The current is connected inside the plastic body to a cap at the end of the body. From there the cables are either led directly through the ship's side or they emerge at the end of the anode body and are installed on the hull under a protective cover.

Attached anodes provide the particular advantage of proportionately low grounding resistance for the protection current because the actual anode body is set into both sloping sides of the plastic body as long narrow strips. The disadvantage is that the body is exposed on the ship's side and can easily be damaged mechanically.

406 Handbook of Cathodic Corrosion Protection

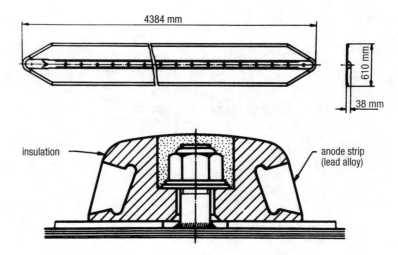

Fig. 17-6 Externally attached PbAg impressed current anodes.

Therefore, two new anode shapes have been developed which are more compact and have an almost rectangular or round active surface, but have a less favorable grounding resistance [29]. They lie flat on the ship's side (see Fig. 17-7) or are set flush with the ship's side (recessed) (see Fig. 17-8). The actual anode body is a flat plate embedded in the plastic supporting body which is surrounded by a metal ring. The anode is welded to the ship's hull by this metal ring. Such anodes have active surfaces of lead-silver alloys or of platinized titanium and are up to 100 mm wide and up to 2000 mm long. Recessed anodes are mainly used on high-duty ships such as ice breakers and are installed in the bow region. Usually in such cases additional

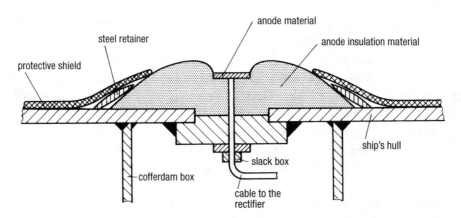

Fig. 17-7 Externally attached impressed current anode.

Cathodic Protection of Ships 407

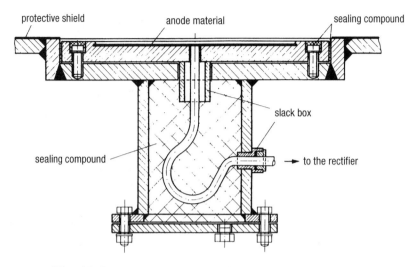

Fig. 17-8 Recessed anode with slack box and casing.

guards are provided for protection. Detailed data on anode materials and reference electrodes for ships are available in VG 81259 [3].

The choice of anodes depends on the expected severity of operating conditions together with cost and durability. The current output is inversely proportional to the grounding resistance for a given voltage. This is proportional to the conductivity of the water (see Table 24-1 and Section 24.1) and can vary by a factor of 100 (see Section 17.1.1). The anode voltages must be correspondingly raised in poorly conducting waters to achieve the required protection current densities. It is not unlikely that the voltage limit of the protection installation and the permissible driving voltages of the particular anodes can be exceeded. In view of this, the protection has to be designed for the ship in question and the particular type of water. Short periods in harbor can be ignored. However, in high-resistance waters, underprotection can occur as a result of voltage limitation.

The surroundings of each flat impressed current anode must be protected against the anodically produced oxygen and chlorine and its reaction products HCl and HOCl [see Eqs. (7-1) and (7-2a,b)] as well as against the strong alkaline environment resulting from the large drop in potential and cathodic conditions (see Section 5.2.1). Protection shields are fitted for this. A special coating, fiberglass-reinforced plastic or mastic, at least 1 mm thick, is applied to the shot-peened substrate (standard surface profile Sa 2½) up to 3 m around the anode, which must not become brittle, have sufficient ductility, and must not change over long periods in dock. Coating thicknesses of up to 2.5 mm have to be considered for particularly severe conditions (e.g., those that can be expected in exposed places with respect to water flow).

The size of the protection shield depends on the protection current, the voltage and the shape of the anode [3].

Measuring electrodes for impressed current protection are robust reference electrodes (see Section 3.2 and Table 3-1) which are permanently exposed to seawater and remain unpolarized when a small control current is taken. The otherwise usual silver-silver chloride and calomel reference electrodes are used only for checking (see Section 16.7). All reference electrodes with electrolytes and diaphragms are unsuitable as long-term electrodes for potential-controlled rectifiers. Only metal-medium electrodes which have a sufficiently constant potential can be considered as measuring electrodes. The silver-silver chloride electrode has a potential that depends on the chloride content of the water [see Eq. (2-29)]. This potential deviation can usually be tolerated [3]. The most reliable electrodes are those of pure zinc [3]. They have a constant rest potential, are slightly polarizable and in case of film formation can be regenerated by an anodic current pulse. They last at least 5 years.

Cables for anodes and measuring electrodes are laid in pipes or cable banks. The anode cables must be sufficiently low resistance (see Section 7.4). Occasionally axial watertight quality is demanded for anode cables. Previously the current connection was made with a short cable and a socket bolt under the anode, rarely with a lateral lead. Today the cable socket to the anodes below the waterline is connected via a cofferdam box (see Fig. 17-8) and an iron bushing [5]. The wall thickness of these boxes must correspond to that of the wall of the ship [4-6]. The anode cable leads through the cofferdam box to the current supply if the connection with the current supply cable in the cofferdam box is not made with a connecting bolt. Anode cables and connecting bolts are led in through a watertight bulkhead slack box. The supply cable to the rectifier is led through two slack boxes. The cofferdam boxes can be also packed with viscous insulating material (Fig. 17-8).

17.3.3.3 *Arrangement of Anodes and Reference Electrodes*

The same observations as those for galvanic anodes apply to impressed current anodes concerning forbidden areas and their arrangement to guard against mechanical damage (see Section 17.3.2.2). The essential differences are the smaller number of anodes required and difficulties arising from their connections. In tankers even current leads in armored tubes are forbidden in the area of the loading tank. The current lead must only be exposed with a suitable cover. The absence of anodes in this area will impair current distribution. Figure 17-9 shows the ideal arrangement of anodes and reference electrodes allowed for in the ship's construction [3]. The ideal arrangement is that in which current distribution is the deciding factor in the design. This is the case with ice breakers and ice-going ships where large surface areas in the bow region can occur where there is no coating. Their placement in the region of the engine room is therefore preferred on economic grounds and accessibility.

Anodes are not attached to the rudder but are situated between the rudder shaft and the ship's wall and connected via a copper strip. The propeller is protected via a slip ring on the shaft. To achieve a low-resistance contact, the divided copper or bronze ring has a rolled silver-bearing surface on which metal graphite brushes slide. The transmission voltage should be below 40 mV.

In contrast to galvanic anodes, there is no fixed rule on the spacing of impressed current anodes because the current output and range can be regulated. In large ships at least 150 m in length the stern anodes should be at least 15 m from the propeller. This distance can be reduced to 5 m on small ships. The reference electrodes should be sited where the lowest drop in potential is expected (i.e., distant from the anodes). In large ships they should be at least 15 to 20 m from the anodes and proportionally nearer in small ships. The bow thruster rudder, scoops and sea chests are equipped with galvanic anodes.

Impressed current anodes and reference electrodes must be very carefully mounted. Insulation damage can arise (e.g., due to welding) and must be immediately repaired. Anode and electrode surfaces should be covered with paper attached by a water-soluble adhesive to protect them against coating materials and dirt after being mounted in dry situations. If after mounting, blast cleaning and coating for corrosion protection are applied, then the covering must be correspondingly more robust.

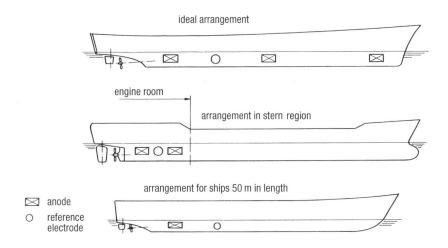

Fig. 17-9 Example of the arrangement of anodes and reference electrodes from Ref. 13.

17.4 Internal Cathodic Protection of Tanks and Containers

The general advice given in DIN 50927 applies to internal cathodic protection [30]. The interior of tanks is protected by galvanic anodes. Impressed current protection is not allowed due to the danger of ignition by sparking and short circuits. Objects to be protected include ballast tanks, loading tanks and water tanks (see also Chapters 20 and 21). In some types of ships (e.g., deep-sea tugs and navy units) fuel consumed is replaced by seawater in the tanks for reasons of stability. In such cases the fuel tanks also have to be included in the cathodic protection system. The regulations in Ref. 5 cover the design of the protection system and the choice of anodes, which include possibilities in new ship building to reduce the thickness of materials covered by the class designation KORR. The extent of corrosion protection can be seen in Table 17-2. The following criteria are distinguished:

(a) ballast tanks that travel either empty or containing water;
(b) tanks for dirty material.

Magnesium anodes must not be used in tanks. Aluminum anodes may be installed in all tanks according to the agreements of the International Association of Classification Societies, which are included in the individual regulations [5,6], but in tanks (b) in the event of the anode falling off, the kinetic energy must not exceed 275 J, i.e., a 10 kg anode must not be fixed more than 2.8 m above the bottom of the tank. There are no restrictions on the use of zinc anodes. The restrictions on the use of aluminum anodes are due to the possible danger of sparks if the anode falls off.

Anodes are designed to last 4 years so that ballast times for (a) are 40% of the service time and for (b) 25%, which corresponds to 146 and 91 days per year, respectively. The shortest filling time should amount to 5 days per ballast journey. Protection current densities are given in Table 17-3.

The surfaces to be protected should be the total surface, including inserts, spars and pipes. The upper 1.5 m of the side walls and the covers should be provided with a coating of recognized quality [10] to protect against corrosion.

The horizontal surfaces should be coated because there is residual water in the ballast and there are water-oil mixtures in the crude oil tanks when ships travel empty and these can cause severe corrosion attack. In the lower part of the tank, up to about 1.5 m from the base, a combination of coating and cathodic protection with special anodes is chosen. Basically the anodes could take over the exclusive protection in this area, but with empty ballast tanks containing residual water or empty crude oil tanks with aggressive oil-water mixtures containing sulfur compounds, they do not prevent corrosion.

Suitable anode shapes have been developed for different types and constructions of tanks. For example, there are long flat anodes which when installed on the base of the tank can cope with the protection against residual water. All anodes must be arranged so that they can be cleaned by tank cleaning equipment.

Coatings for tanks have to be chosen so that they can withstand the particular conditions [10]. The high-grade modern systems demand very good surface preparation of Sa 2½, which cannot always be achieved in practice, because of their ribbed construction and confinement. The increased frequency of defects has to be reckoned with.

Tanks for the transport of chemicals cannot usually be cathodically protected because of the danger of impurities contaminating the cargo. Particular emphasis is placed on the quality of the coating to avoid contamination by corrosion products.

The advice given in Section 17.3.1 and Eqs. (17-2) to (17-4) applies in determining the current requirement and number and weight of anodes, taking into account the different current requirements of individual surface areas S_i. The number of anodes is derived from Eq. (17-4), taking into account the maximum current output I_{max} of the galvanic anodes, which cannot be given with certainty because of problems with internal protection and interaction with dirty electrolyte.

In place of Eq. (17-3a,b) the expression for the anode mass, m, taking into account the ballast time, t_B, as a percentage of the service life and time of protec-

Table 17-2 Corrosion protection of tanks

Surfaces	Ballast tanks	Crude oil tanks
Upper tank area to 1.5 m below deck	Coating	
Middle area	Coating or coated horizontal surfaces in combination with protection by anodes	Horizontal surfaces coated
Base to 1 m high	Coating and protection with anodes	

Table 17-3 Protection current densities for tanks

Surface treatment	Ballast tanks (mA m^{-2})	Crude oil tanks (mA m^{-2})
Coated surfaces	20	20
Uncoated surfaces	120	150

tion, t_s; and the current content Q'_{pr} [see Eq. (6-9)] is given by:

$$\frac{m}{\text{kg}} = 8.76 \times 10^{-2} \frac{J_{si}}{\text{mA m}^{-2}} \frac{S_i}{\text{m}^2} \frac{\text{A h kg}^{-1}}{Q'_{pr}} \frac{t_s}{a} \qquad (17\text{-}5)$$

Practical values for Q'_{pr} are 780 for zinc anodes and 2250 to 2800 for aluminum anodes, depending on the particular alloy. The weight of the anodes is increased by 20% so that at the end of the proposed life the remains of the anode are still active. The number of anodes given by Eq. (17-4) and their weight calculated from Eq. (17-5) for individual anodes must agree with the actual anodes so that for the object in question the special data concerning geometrically endangered areas and cell formation (e.g., in mixed material installations) are taken into account.

In contrast to external protection, the anodes in internal protection are usually more heavily covered with corrosion products and oil residues because the electrolyte is stagnant and contaminated. The impression can be given that the anodes are no longer functional. Usually the surface films are porous and spongy and can be removed easily. This is achieved by spraying during tank cleaning. In their unaltered state they have in practice little effect on the current output in ballast seawater. In water low in salt, the anodes can passivate and are then inactive.

17.5 Cathodic Protection of Heat Exchangers, Condensers and Tubing

Galvanic or impressed current anodes are used to protect these components. The anode material is determined by the electrolyte: zinc and aluminum for seawater, magnesium for freshwater circuits. Platinized titanium is used for the anode material in impressed current protection. Potential-regulating systems working independently of each other should be used for the inlet and outlet feeds of heat exchangers on account of the different temperature behavior. The protection current densities depend on the material and the medium.

The internal cathodic protection of pipes is only economic for pipes with a nominal width greater than DIN 400 due to the limit on range. Internal protection can be achieved in individual cases by inserting local platinized titanium wire anodes (see Section 7.2.2).

17.6 Cathodic Protection of Bilges

The layer of dirty water standing in bilges is usually so small that protection with anodes of the usual dimensions is not possible. Experiments with very flat anodes attached to polished surfaces with conducting adhesive have shown that the method is not sufficiently effective. Better results have been achieved with anode wires of aluminum or zinc alloys with steel cores. Such wire anodes have diameters of 6 to 10 mm and are laid directly on the deck in long loops that pass over standing structural items and are soldered. The wire loops can be laid in bunches, depending on the corrosion intensity. Additional protection can also be given by scattered zinc powder with activating additions. The zinc particles (10 to 100 μm) sink into corrosion craters and pits and give good local protection. This addition must, of course, be repeated frequently (e.g., every 3 to 6 months). Anode wires last about 2 years.

17.7 Cathodic Protection of Docks

Floating docks are protected with galvanic and impressed current anodes, that is, on the outside, the gates and the base. Since docks are stationary, types of anodes and arrangements that would otherwise be in danger of mechanical damage can be chosen, as for marine structures. In the past, graphite, magnetite and silicon-iron anodes (see Chapter 16) in the form of round rods or plates were hung on the outside sufficiently far into the water so that uniform current distribution was ensured. Chambers can also be provided in which the anodes are drawn up when the docks are flooded. In every case care has to be taken for traction relief of the anode cable. Finally, the anodes can be fixed to racks and anchored to the ground. Often the cathodic protection of docks is managed in the same way as the cathodic protection of ships so that every dockside has a potential-controlled protection installation or one protection rectifier with a transfer cable (to the other side).

Since docks are usually situated in brackish water, the anodes must have large surface areas to keep the grounding resistance and the driving voltage low. Cathodic protection of the interior is not necessary because the dock is only flooded for a short time and can be otherwise maintained.

17.8 References

[1] VG 81 255, Beuth Verlag, Köln 1983.
VG 81 256 Teile 1 bis 3, Beuth Verlag, Köln 1979 und 1980.
VG 81 257 Teile 1 und 3, Beuth Verlag, Köln 1982.
[2] VG 81 258, Beuth Verlag, Köln 1984.
[3] VG 81 259 Teile 1 bis 3, Beuth Verlag, Köln 1985 und 1986.
[4] Brit. Stand. Inst., Code of Practice for Cathodic Protection, London 1973.
[5] Germanischer Lloyd, Vorschriften für Konstruktion und Bau von stählernen Seeschiffen, Hamburg.
[6] Det Noske Veritas, Rules for Classification of Steel Ships, Oslo.
[7] H. Determann u. E. Hargarter, Schiff u. Hafen 20, 533 (1968).
[8] R. Tanaka, Sumitomo Light Met. Techn. Rep. 3, 225 (1962).
[9] P. Drodten, G. Lennartz u. W. Schwenk, Schiff u. Hafen 30, 643 (1978).
[10] STG-Richtlinie Nr. 2215, Hamburg 1987.
[11] STG-Richtlinie Nr. 2220, Hamburg 1988.
[12] H. Kubota u.a., Nippon Kokan techn. Rep. Overs. 10, 27 (1970).
[13] Fischer, Bue, Brattes, Steensland, OTC, Houston 1980, Paper No. 388.
[14] B. Richter, Schiff u. Hafen, Sonderheft Meerwasser-Korrosion, Februar 1983, S. 88.
[15] A. Bäumel, Werkstoffe u. Korrosion 20, 391 (1969).
[16] VG 81 249 Teile 1 und 2, Beuth Verlag, Köln 1977 u. 1983.
VG 81 250, Beuth Verlag, Köln 1988.
[17] DIN 50919, Beuth Verlag, Berlin 1984.
[18] K. Meyer u. W. Schwenk, Schiff u. Hafen 26, 1062 (1974).
[19] H. Hildebrand u. W. Schwenk, Werkstoffe u. Korrosion 30, 542 (1979) und 33, 653 (1982).
[20] DIN 50928, Beuth Verlag, Berlin 1985.
[21] H. Determann, E. Hargarter, H. Sass, Schiff u. Hafen 32, 89 (1980);
H. Determann, E. Hargarter, H. Sass, Schiff u. Hafen 28, 729 (1976);
W. Bahlmann, E. Hargarter, H. Sass, D. Schwarz, Schiff u. Hafen 33, H. 7, 50 (1981);
W. Bahlmann, E. Hargarter, Schiff u. Hafen, Sonderheft Meerwasser-Korrosion, Februar 1983, S. 98.
[22] DIN 55928 Teil 5, Beuth-Verlag, Berlin 1980.
[23] H. Pircher, B. Ruhland u. G. Sussek, Thyssen Technische Berichte 18, H. 1, 69 (1986).
[24] J. H. Gobrecht, E. Hargarter, W. Huppatz u. F. J. Reker, Schiff u. Hafen, Sonderheft Meerwasser-Korrosion, Februar 1983, S. 24.
[25] V. Brücken, H. Dahmen u.a., Int. Leichtmetalltagung, Leoben-Wien, 22–26. 6. 1987.
[26] U. Neumann, Techn. Rundschau 16, H. 4, 65 (1971).
[27] J. H. Morgan, Corrosion 13, 198 (1975).
[28] J. H. Morgan, Cathodic Protection, Leonhard Hill Ltd, London 1959.
[29] H. Bohnes, Kathodischer Korrosionsschutz mit galvanischen Anoden und Fremdstrom, Grillo-Ampak, Duisburg 1979.
[30] DIN 50927, Beuth Verlag, Berlin 1985.

18

Cathodic Protection of Well Casings

W. Prinz and B. Leutner

18.1 Description of the Object to be Protected

In oil and natural gas fields and reservoirs, the boreholes are cased to stabilize the wells. Depending on the depth and the operating conditions, several pipes may be fitted inside each other in the area near the surface (telescope casing) (see Fig. 18-1).

The annular space between the outer pipe and the surrounding rock is filled with cement over the whole depth up to the ground in new wells. The purpose of this is to seal the deposits at the top and to keep the fresh water and salt water zones separate. In addition, it serves as a protection against pressure from the rock and as corrosion protection which, however, is only effective so long as there is no current exit caused by extended corrosion of cells or due to foreign anodic influences. The cement filling of the borehole casing is usually not uniformly spread over the pipe surface. It has to be remembered that there can be sections which are either not covered or are only thinly covered with cement.

With old borehole casings, it is often only the oil- or gas-producing depths of the deposit and the upper regions that are cemented. The noncemented depths contain residue from drilling mud and various-sized particles of rock. The circulation during drilling is carried out with a suspension of $BaSO_4$ and water with a high density, which usually contains salt. This promotes the effectiveness of corrosion cells.

18.2 Causes of Corrosion Danger

18.2.1 Formation of Corrosion Cells

The usual practice in old wells of only partially cementing the outer pipe can lead to cell formation (steel in the cement-steel in the soil) in the transition regions to the uncoated sections (see Sections 4.2 and 4.3). In contrast to the well-known cathode "steel-soil" in the vicinity of the ground surface, the cathodic activity of the

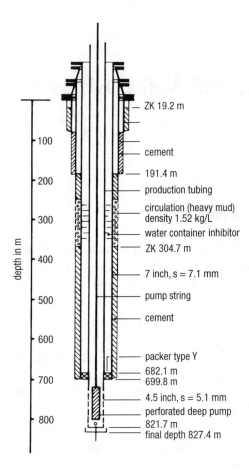

Fig. 18-1 Profile of a telescopic well casing.

cement filling is restricted because the access of oxygen is hindered in the wet state (see Section 5.3.2). For this reason, corrosion by cell formation with the "steel-cement" as cathode is to be expected mainly in the soil near the surface (see Fig. 4-3d).

Further cell currents flow between the wells as a result of electrical connections established between them by the flow lines and of the different free corrosion potentials, thereby allowing them to behave as anode or cathode. The currents can amount to a few amps so that considerable corrosion damage can arise. The action of these cells can be prevented by building in insulators between the drilling and the field cable.

Sections of the borehole casing threatened by corrosion can be located with the help of the profile measurement technique described in Section 18.3.1. In general, the profile measurement cannot identify which factors are the main cause of corrosion danger.

18.2.2 Free Corrosion in Different Soil Layers

Throughout its depth the borehole casing passes through several geological regions which can range from layers with saline water to inert bedrock. In addition, corrosive gases (CO_2, H_2S) and temperature differences up to 50°C may occur, depending on the type of deposit. These factors lead to an increased corrosivity in deeper layers of soil where film formation is restricted by salt and concentration cells may be formed (see Figs. 4-3b to 4-3c).

18.2.3 Conditions for the Occurrence of Stress Corrosion

Since borehole casings are subjected to increased temperatures at increasing depths and since additional tensile stresses are acting, there is the possible danger of stress corrosion. From the data in Section 2.3.3, there exists a danger, in the case of cathodically protected casings, of intergranular stress corrosion arising from conditions e, g and h where NaOH is generated according to Eqs. (2-17) and (2-19) and the remaining components are generated by the reaction of NaOH with CO_2 from the soil. The critical potential and protection potential ranges are given in Section 2.4. Under the prevailing service conditions it can be assumed that stress corrosion by $NaHCO_3$ does not occur because of adequate cathodic polarization, nor does stress corrosion by Na_2CO_3 and NaOH due to their low concentration. Furthermore, the high-carbon, high-strength casing steels have increased resistance to this type of stress corrosion [1], especially as the necessary critical stress level is not likely to be reached. From previous experience, no damage by stress corrosion has been observed in cathodically protected casings.

18.2.4 Corrosion by Anodic Interference (Cell Formation, Stray Currents)

In oilfields with high paraffin content, the production tubing has to be cleaned by heating every week to remove paraffin deposits. The heating of the pipe takes place while a dc of about 1.2 kA is fed into the production tubing from the well head. The current is fed back via the casing to the dc installation by contacts between the production tubing and the casing in the lower part of the borehole. Part of the return current in the casing is shunted through the rock to the well head. This has the consequence that during the heating period the lower part of the casing is anodically polarized. This can interfere with wells within a radius of 1000 m.

Since in oilfields or reservoirs borehole casings are frequently sited close to one another, they can be affected by the voltage cone of the anode bed and the cathodic voltage cone of other casings. Thus, in the case of large oilfields and reservoirs, it is appropriate to protect all wells cathodically.

18.3 Measurements for Assessing Corrosion Protection of Well Casings

With buried pipelines, the degree of corrosion danger from cell formation and the effectiveness of cathodic protection can be determined by pipe/soil potential measurements along the pipeline (see Sections 3.6.2 and 3.7). This is not possible with well casings since the only point available for a measuring point is at the well head. Therefore, other methods are required to identify any corrosion risk or the effectiveness of corrosion protection.

18.3.1 Investigations for Corrosion Damage

Various electromagnetic and mechanical measuring methods are used to investigate the old well casings to determine whether there is external or internal corrosion.

The individual methods provide information on:

(a) exact depth of the screw socket connections;
(b) reduction from the average wall thickness irrespective of whether the reduction in wall thickness is on the inside or outside of the pipe;
(c) reduction in wall thickness or penetration with the distinction as to whether the areas of attack are due to external or internal corrosion;
(d) attack by internal corrosion and penetration.

Methods for (a) to (c) are electromagnetic or use eddy currents. The method for (d) is a mechanical system similar to the caliper pig, in which the probe runs over the inner wall of the pipe over pits, shallow pits and penetration holes.

18.3.2 Measurement of ΔU Profiles

When measuring ΔU profiles, the voltage drop is measured along the interior of the casing [2]. The voltage drop is caused by cell currents as in Section 18.2.1 or by cathodic protection currents. The anodic and cathodic sections of the casing and also the effect of the cathodic protection can be determined from the ΔU profiles.

A ΔU profile measurement is carried out with a measuring probe which is introduced into the borehole. Two measuring contacts arranged one above the other act as the measuring probe (see Fig. 18-2); they are pressed against the wall of the casing under great pressure. The contacts, which are about 8 m apart, must be very well insulated from each other. To avoid measuring errors, the internal wall of the casing must be practically free of contamination (e.g., cement residues, paraffin, oil deposits and corrosion products). Cleaning with a scraper may be necessary. In addition, the casing must be filled with a nonconducting or poorly conducting

Cathodic Protection of Well Casings

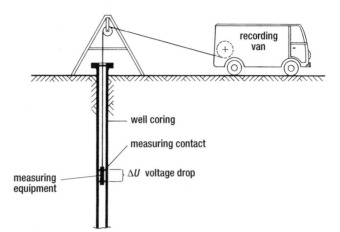

Fig. 18-2
Principle of ΔU profile measurement.

medium (e.g., diesel oil or desalinated water) since otherwise galvanic voltages can arise at the contacts and falsify the measurements.

The voltage drop ΔU is measured over the distance between the measuring contacts in the casing. Continuous measurement during the insertion and removal of the probe is not possible since the voltage drops are so small, being in the range of a few microvolts [3]. Thermovoltages arising at the contact surfaces are of the same magnitude as the actual voltage drops between the measuring probes and therefore lead to errors in the measured values. The measurement may only be made when the probe is stationary. The final reading must be reproducible after three or four insertions of the probe. If reproducible results cannot be obtained, the reason must be investigated (e.g., contact or insulation problems) and overcome. The measuring distance must be decided for a given length of pipe (i.e., over every piece of pipe but not including a socket connection). The depth of the sockets and the individual lengths of pipe are determined by electromagnetic measurements [e.g., casing collar locator (CCL)]. These measurements are transmitted to the measuring van via a steel armored cable which also acts as a haulage rope.

In the profile measurements, the sign of the ΔU value is related to the lower measuring contact. A positive ΔU value corresponds to the ohmic voltage drop for a current flowing to the well head. In the area of decreasing ΔU values, which can become even more negative, current flows from the casing into the rock. This indicates local anodic activity and points at risk from corrosion. In areas where ΔU values are increasing, current is flowing into the casing. These are the cathodic regions of the casing.

Before a drainage test is carried out, a so-called zero profile is measured. This involves the indication of corrosion currents, which, according to whether ΔU values are increasing or decreasing, locate the anodic or cathodic regions (see Fig. 18-3).

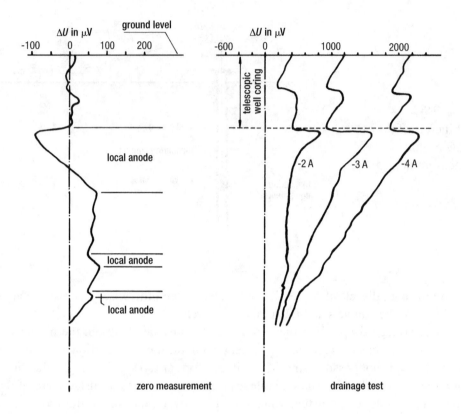

Fig. 18-3 ΔU profiles with and without protection current injection.

After measuring the zero profile, ΔU measurements are carried out with the injection of a cathodic protection current. In contrast to the zero profile measurements, the distance between the individual measurements is 25 to 50 m. Shorter distances between the measuring points are used only at depths where there are unusual ΔU profiles. Current should be injected at at least three different levels. The protection current density of about 12 mA m^{-2} obtained from experience should be the basis for determining the maximum required protection current. As shown by the results in Fig. 18-3, the ΔU profiles are greater with increasing protection current. The action of local cells is suppressed when the ΔU values no longer decrease in the direction of the well head. This is the case in Fig. 18-3 with a protection current I = 4A.

In the region of the double casing, part of the return current flows from the inner to the outer pipe, depending on the resistance of the pipe. This is shown by a sharp decrease in the ΔU values. The current that flows results from undefined electrical connections between the pipes. For this reason, evaluating the profile measurements in

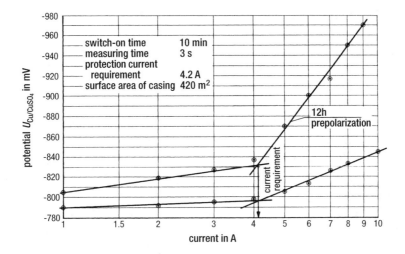

Fig. 18-4 U_{off} (log I) curves for determining the Tafel potential.

the area of double or multiple casing is pointless. Electrolytic corrosion does not occur in the annular region because of the current flow through electrical connections.

18.3.3 Measurement of the Tafel Potential

The Tafel potential is given by a bend in the U_{off} (log I) curve.[1] According to criterion No. 4 in Table 3-3, under the conditions given in Section 3.3.3.1, it corresponds to the protection potential.

The average protection current requirement can be conveniently determined for individual wells in an oilfield by measuring the Tafel potential. In contrast to profile measurements, internal measurements on the casing are not necessary. These measurements cannot be used to predict polarization behavior at greater depths.

To determine the Tafel potential, the casing is cathodically polarized over a fixed time period of about 10 minutes with increasing levels of current. The results in Fig. 18-4 show how the Tafel potential can be determined from the intersection of two straight lines plotted on a logarithmic scale [4]. Casing potentials can also be calculated from the electrostatic fields and the ohmic voltage drop in the casing [5]. This assumes homogeneous conductivity and polarization.

---

[1] In the United States, these are E log *I* tests.

18.4 Design and Construction of Cathodic Protection Stations

The most important parameters in the cathodic protection of a well casing are the depth of penetration of the protection current and its magnitude. These depend on the geological formations and the specific electrical resistance of the individual layers. Insulating layers are a disadvantage because they limit the extent of the protected region. Borehole profiles are not adequate to make an assessment [6]. Depending on the type of field or reservoir, it is recommended that measurements be carried out on the casing as detailed in Section 18.3.2. The results can be applied to other wells [7]. With well casings for oil and gas deposits in salt domes, previous experience has shown that information from measurement of Tafel potentials is sufficient for planning the cathodic protection of casings. The protection current requirement of the flow lines is determined by a drainage test or from experience (see Section 10.3.3).

In planning the cathodic protection of oil and gas fields or reservoirs, depending on the subsoil of the pipelines, the required technical data on the casing include diameter, wall thickness, depth, cement filling and the geological situation in its vicinity. Other factors that have to be considered are operational data such as temperature of the deposits, well head temperature, electrical heating of oil in oilfields and the conductivity of the transported media. While the pipelines for gas and oil reservoirs do not carry electrolytic media, media in pipelines of oilfields—governed by the high saline content of the extracted oil—are extremely electrolytically conductive. In this case, an anodic corrosion danger behind the insulating unit has to be considered (see Sections 10.3.5 and 24.4.6).

It follows from this that the casing and the flow lines in oilfield installations should not be separated from one another by insulting units. The cathodic cable is always connected to the well to be protected. The flow lines are all equipped with pipe current measuring points in order to calculate the current uptake of the individual well casing. The current flowing from the electrical drive of the oil well pump to the grounding system is not important for the current balance of the casing and, from experience, can in most cases be neglected. Protective measures against internal corrosion have to be taken at the connecting point with the flow lines (see Section 10.3.5). Setting up local cathodic protection as described in Chapter 12 can also be advantageous in oilfields because here there is no interference from buried installations due to the voltage cone of the anodic system or the voltage cone of the casing. Measurement of *IR*-free potentials by the switching method is not possible because of the very different polarization of the casings and the flow lines. The comparative measurement and control of polarization potentials of long-distance pipelines can be performed with measuring probes as in Section 3.3.3.2.

Cathodic Protection of Well Casings   423

With pipelines that do not carry electrolytically conducting media, the casings and flow lines can be electrically separated from each other by insulating units. For flow lines with a high protection current requirement, it is usually advisable to construct special anode beds for well casings and flow lines. For flow lines with low protection current requirements, these can be connected to the protection current unit for the casing via diodes and balancing resistors (see Section 10.4) or via a potential connection to the insulating unit at the cathodically protected casing. In both cases the switching method can be used to determine the *IR*-free pipe/soil potentials of the flow lines. As to the potential connection, the bridging of the insulating unit can then be switched on and off.

The construction of the anode bed is determined by the details of the field or reservoir. The anode bed should have a spacing of >100 m to give an even potential distribution at the well casing. The distance from other well casings not

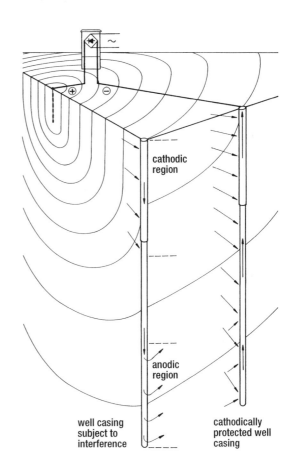

Fig. 18-5 Current distribution in the interference of a borehole well casing by an anodic voltage cone.

cathodically protected should, if possible, be greater in order to avoid detrimental interference (see Section 9.2).

With interference from the voltage cones of the anodes, the interfering currents flow in the vicinity of the anodes (i.e., in the upper soil layers into the well casing) and leave it in the deeper soil layers before the end of the casing or the point at which it enters a nonconductive salt dome in which the leakage load of the pipe goes to zero (see Fig. 18-5). In order to avoid detrimental interference in the section of the casing near the surface, anodes should be installed in this area. Deep anodes should have little covering.

The anodic and cathodic current densities are inversely proportional to the length of the anode bed or the well casing. Since the length of the casing is usually greater by a factor of 10 than the length of the anode bed, the current density of the

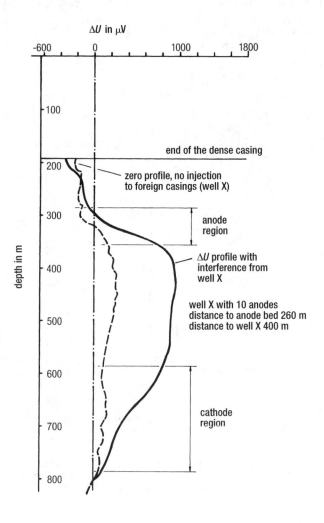

Fig. 18-6 ΔU profile of a borehole well casing with interference from a cathodic voltage cone.

well casings is smaller by this factor. But even here interference with cathodically unprotected well casings is possible. This is the case when an unprotected casing is sunk in the immediate vicinity of a cathodically protected casing. If the electric field of the cathodic voltage cone is stronger than that of the anodic voltage cone at the place of interference, current could enter the lower region of the cathodically unprotected casing and leave it for the cathodically protected casing, causing anodic corrosion of this upper part of the unprotected casing (see Fig. 18-6).

18.5 Commissioning, Maintenance and Control

When commissioning the cathodic protection station for the well casing, a protection current about 10% higher than that determined in the measurements as in Section 18.3 is supplied. In the case of a separate protection current supply for the flow lines, the pipe/soil potential should be set at a U_{on} value of $U_{Cu\text{-}CuSO_4} = -1.5$ V. After a sufficient polarization period, which for the well casing is about a year, for flow lines with a high protection current requirement about half a year, and for flow lines with good coating about a month, the cathodic protection for the flow lines can be verified as in Section 10.4. If the cathodic protection is set up as local cathodic protection, checking is only necessary after a year.

It is very important to measure interference. The measurements can be made as described in Section 9.2. In general, raising of the potential cannot be measured except in the case of interference from the cathodic voltage cone of a well casing since it occurs at a distance from the surface. With anodic interference, the 0.5-V limit must be observed (see Section 9.2.1). Damaging interference can only be suppressed by an equal potential bond to the cathodically protected object.

The protection station must be carefully maintained (see Section 10.5). The function of the rectifier should be monitored at monthly intervals. The pipe/soil potentials of the pipelines should be measured at least once a year. The *IR*-free potentials should be determined as far as possible by the switching method, especially when new pipework is installed and connected to the protection system.

18.6 References

[1] G. Herbsleb, R. Pöpperling und W. Schwenk, 3R internat. *13*, 259 (1974).
[2] H. Kampermann u. W. Harms, Erdöl Erdgas *101*, H. 2, S. 45 (1985).
[3] R. Graf u. B. Leutner, 3R internat. *13*, 247 (1971).
[4] E. W. Haycock, 13. NACE-Jahrestagung, April 1957, S. 769 t.
[5] F. W. Schremp u. L. E. Newton, NACE-Jahrestagung, März 1979, Paper 63.
[6] W. F. Gast, Oil and Gas Journal 23.4.1973, S. 79.
[7] E. P. Doremus u. F. B. Thorn, Oil and Gas Journal 4.8.1969, S. 127.

19

Cathodic Protection of Reinforcing Steel in Concrete Structures

B. ISECKE

19.1 The Corrosion System Steel-Concrete

The corrosion protection of reinforcing steel in steel-reinforced concrete and prestressed concrete structures acts through the alkalinity of the porous concrete water content, since the steel passivates under these conditions [1-5] (see also Section 5.3). This corrosion protection is long term if the construction work has been carried out according to the state of the art [6,7] and if no changes likely to impair the passivity occur in use. Deficiencies in the construction process in the thickness and density of the concrete covering as well as the action of chloride-containing electrolytes (deicing salts, seawater, PVC combustion, waste incinerators) can lead to depassivation [8-12], requiring additional corrosion protection measures. These measures must either give the concrete greater resistance to the corrosive environments or, if depassivation has already set in after severe attack on the concrete, they must involve additional direct protective action. Coating of the concrete to prevent penetration of chloride ions [10], coating the reinforcing steel with epoxy resin [13-18], or hot dip galvanizing [18-20] are used for this purpose.

Cathodic protection of reinforcing steel with impressed current is a relatively new protection method. It was used experimentally at the end of the 1950s [21,22] for renovating steel-reinforced concrete structures damaged by corrosion, but not pursued further because of a lack of suitable anode materials so that driving voltages of 15 to 200 V had to be applied. Also, from previous experience [23-26], loss of adhesion between the steel and concrete due to cathodic alkalinity [see Eqs. (2-17) and (2-19)] was feared, which discouraged further technical development.

Cathodic protection as a restoration measure was revived in 1974 [27] as a result of the increasing damage to roadwork structures from Cl^--induced corrosion of the reinforcing steel, which resulted in high repair costs in the United States [18]. This development was promoted by considerable negative experience with other methods. Today impressed current is used to protect reinforcing steel on road bridges, retaining walls, marine structures, multistory car parks, salt tanks and waste incinerator plants [28-37].

Cathodic protection cannot work with prestressed concrete structures that have electrically insulated, coated pipes. There is positive experience in the case of a direct connection without coated pipes; this is protection of buried prestressed concrete pipelines by zinc anodes [38]. Stability against H-induced stress corrosion in high-strength steels with impressed current has to be tested (see Section 2.3.4).

19.2 Causes of Corrosion of Steel in Concrete

The passivating action of an aqueous solution within porous concrete can be changed by various factors (see Section 5.3.2). The passive film can be destroyed by penetration of chloride ions to the reinforcing steel if a critical concentration of ions is reached. In damp concrete, local corrosion can occur even in the presence of the alkaline water absorbed in the porous concrete (see Section 2.3.2). The Cl^- content is limited to 0.4% of the cement mass in steel-concrete structures [6] and to 0.2% in prestressed concrete structures [7].

A further cause of depassivation is a reduction in the alkalinity of the concrete as in Fig. 2-2 (i.e., a reduction in the pH of the absorbed water). This occurs with carbonizing of the concrete by reaction with CO_2 in the atmosphere. In structures with a sufficiently thick concrete covering over the steel inserts, especially with dense, low-porosity concrete of good quality, carbonization is unimportant. With poorer quality concrete and/or with too little concrete covering, carbonization penetrates to the steel reinforcement, which then loses its passivity. With depassivation due to chloride ions or carbonization, there is a danger of corrosion in damp concrete only if there is access to oxygen. If the concrete is thoroughly soaked on all sides, access is severely restricted, so that the cathodic partial reaction according to Eq. (2-17) cannot take place at any point on the reinforcing steel (see Section 5.3.2). Then, however, the anodic reaction according to Eq. (2-8) also cannot occur, i.e., the depassivated steel will not corrode.

However, if part of the reinforcing steel is aerated, a cell is formed as in Section 2.2.4.2. With a high surface area ratio S_c/S_a and with well-aerated cathodes, very high corrosion rates can occur at anodic regions.

19.3 Electrolytic Properties of Concrete

The specific electrical resistance of concrete can be measured by the method described in Section 3.5. Its value depends on the water/cement value, the type of cement (blast furnace, portland cement), the cement content, additives (flue ash), additional materials (polymers), the moisture content, salt content (chloride), the temperature and the age of the concrete. Comparisons are only meaningful for the water-saturated state. In wet portland cement concrete, specific resistivities can lie between 2 and 6 kΩ cm [5,39,40] and values between 10 and 200 kΩ cm have been determined for blast furnace concrete and for dry concretes [5,31,40]. Usually the resistivity shows wide variation, depending on the temperature and moisture content [41,42].

19.4 Criteria for Cathodic Protection

The information in Sections 2.2, 2.4 and 3.3 is relevant for protection criteria. Investigations [43] with steel-concrete test bodies have shown that even in unfavorable conditions with aerated large-area cathodes and small-area damp anodes in Cl$^-$-rich alkaline environments, or in decalcified (neutral) surroundings with additions of Cl$^-$ at test potentials of $U_{Cu\text{-}CuSO_4} = -0.75$ and -0.85 V, cell formation is suppressed. After the experiments had proceeded for 6 months, the demounted specimens showed no recognizable corrosive attack.

Figure 19-1 shows the experimental setup with the position of the steel test pieces and the anodes. The anodes were oxide-coated titanium wires and polymer cable anodes (see Sections 7.2.3 and 7.2.4). The mixed-metal experimental details are given in Table 19-1. The experiments were carried out galvanostatically with reference electrodes equipped to measure the potential once a day. Thus, contamination of the concrete by the electrolytes of the reference electrodes was excluded. The potentials of the protected steel test pieces are shown in Table 19-1. The potentials of the anodes were between $U_{Cu\text{-}CuSO_4} = -1.15$ and -1.35 V.

At the beginning of the experiment, the measured free corrosion potential of the reinforcing steel in the Cl$^-$-rich environment was $U_{Cu\text{-}CuSO_4} = -0.58$ to -0.63 V; in the neutral environment it was -0.46 to -0.55 V, and in straight concrete it was -0.16 V. The demounted test pieces at the end of the experiment are shown in Fig. 19-2. After 6 months, no corrosive attack on the cathodically protected test pieces was detected.

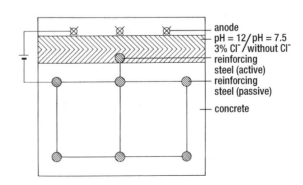

Fig. 19-1 Experimental set-up for the cathodic protection of an "active steel concrete-passive steel" cell.

Table 19-1 Protection current densities and potentials of the cathodically protected steel test pieces in Fig. 19-1

Test potential $U_{Cu-CuSO_4}$ (V)	Medium at the anode	Protection current density (mA m^{-2})	
		Anode 1	Anode 2
−0.75 to −0.80	Dealkalized without	38	36
−0.80 to −0.95	addition of Cl$^-$	39	50
−0.75 to −0.80	Alkaline, pH 12,	51	44
−0.80 to −0.95	3 weight% Cl$^-$/cement	61	58

Anode 1 = plastic cable anode; anode 2 = mixed metal oxide-titanium.

With unprotected comparison test pieces, the corrosion rate was 4 mm a^{-1}, which from cell current measurements indicated that the self-corrosion was 50%.

In practice, the current densities for protecting concrete structures are generally lower than the values in Table 19-1. The reason is that the cathode surfaces are not well aerated and areas of the anodes are dry. Practical experience and still-incomplete investigations [43] indicate that at even more positive potentials than those given in Table 19-1 with $U_H = -0.35$ V, noticeable protection can be achieved so that $U_{Hs} = -0.4$ V can be regarded as the protection potential. In DIN 30676, $U_{Hs} = -0.43$ V is given [44] (see also Section 2.4).

The information given in Section 5.3.2 [40,44] applies to the limits of the protection potential range. With the types of concrete used in practice, there is no danger of cathodic corrosion. Similarly, damage from cathodically evolved hydrogen due to the porous structure of the usual types of concrete can be ruled out. Also, the feared reduction in the steel-concrete bond could not be confirmed. In this respect, investigations [45] had shown that the bonding was unaffected after 2.5 years at non-*IR*-free potentials of $U_{Cu-CuSO_4} = -1.62$ V. At $U_{Cu-CuSO_4} = -1.43$ V, the bond actually increased. Considerable reduction in the migration of chloride ions was observed with this degree of polarization [35], which agrees with Eq. (2-23).

Since cathodic protection of concrete structures in the United States has been very much advanced, protection criteria have been developed [46]. They correspond to the pragmatic criteria Nos. 3 and 4 in Table 3-3 (see Section 3.3.3.1). It is assumed that the protective effect is adequate if, upon switching off the protection current, the potential becomes more than 0.1 V more positive within 4 hours. The measurements are carried out in various parts of the protected object with built-in Ag-AgCl reference electrodes or with any electrodes on the external surface.

Cathodic Protection of Reinforcing Steel in Concrete Structures

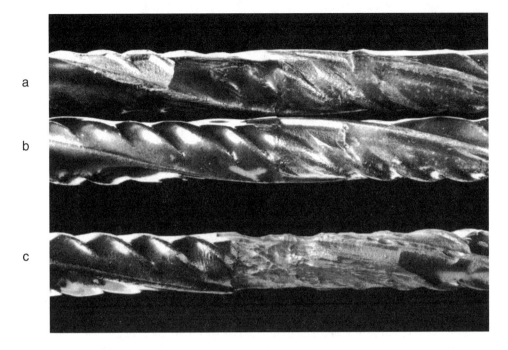

Fig. 19-2 Appearance of the active test pieces in a 6-month experiment according to Fig. 19-1 and Table 19-1. (a) $U_{Cu\text{-}CuSO_4} = -0.80$ to -0.95 V, (b) $U_{Cu\text{-}CuSO_4} = -0.75$ to -0.80 V, (c) without cathodic protection, in electrical contact with passive steel.

Drainage tests and initial measurements should not be made before 28 days have elapsed after the anodes are embedded in the artificial concrete system in order to allow the hydration of the concrete and to ensure moisture equilibrium, which can affect the potentials. The protection current density is limited to 20 mA m^{-2} (at the steel surface) to avoid possible reduction in the steel-concrete bond. Usual current densities lie in the range 1 to 15 mA m^{-2} [29-33].

19.5 Application of Cathodic Protection to Reinforced Concrete Structures

19.5.1 Design and Installation

The decision to cathodically protect reinforced concrete structures depends on technical and economic considerations. Cathodic protection is not an economic process for small area displacements of the concrete due to corrosion of the reinforcing steel arising from insufficient concrete covering. On the other hand, the main areas of application of such reinforced concrete structures must be looked at, particularly where there are high chloride ion concentrations at considerable depth in the concrete. To repair this, the Cl⁻-contaminated old concrete can only be removed down to the uppermost layer of reinforcing steel for technical reasons. If a high Cl⁻ content remains in the concrete at greater depths, then corrosion will continue there. Since the repaired upper layer represents a new cathode surface, cell action is increased and corrosion in the deeper layer is promoted. Negative experience with repairs of this kind is unfortunately very widespread [47].

19.5.2 Determination of the State of Corrosion of the Reinforcing Steel

A diagnosis of possible damage should be made before beginning repairs with other construction measures [48,49]. There should be a checklist [48] of the important corrosion parameters and the types of corrosion effects to be expected. Of special importance are investigations of the quality of the concrete (strength, type of cement, water/cement ratio, cement content), the depth of carbonization, concentration profile of chloride ions, moisture distribution, and the situation regarding cracks and displacements. The extent of corrosion attack is determined visually. Later the likelihood of corrosion can be assessed using the above data.

The extent of the corrosion shows itself by, for example, cracks and displacements arising from growing solid corrosion products (rust). A little-used method to find these is to tap the concrete surface with a hammer; this allows hollow regions and cracks to be clearly distinguished acoustically from undamaged regions. These simple methods cannot, however, answer the question of whether the reinforcing steel is already depassivated. Corrosion damage in concrete is attributed to solid corrosion products arising from Fe^{2+} ions from the anode according to Eq. (2-21) and subsequent reaction with oxygen. Since the porous structure of the concrete can absorb Fe^{2+} ions, the site of this oxidation is not necessarily the surface of the anode. The reaction site is determined mainly by the diffusion currents of the partners to the reaction—Fe^{2+} ions and O_2 in the concrete (see the explanation in Section 4.1). When oxidation occurs at some distance from the anode the corroding anode can remain uncovered. At a later stage, rust stains can appear on the concrete surface. If the oxidation takes place in the vicinity of the anode, where the highest concentration of corrosion products is to be found, internal stresses arise due to the growing solid corrosion products of hydrated ferric oxide (rust) with cracks, voids and displacement as a consequence.

Two-dimensional potential measurements on the concrete surface serve to determine the corrosion state of the reinforcing steel. This method has been proved for one-dimensional systems (pipelines), according to the explanation for Fig. 3-24 in Section 3.6.2.1 on the detection of anodic areas.

In the United States, measurements of this kind have been carried out since the beginning of the 1970s on road bridges damaged by chloride ions [50]. Statistical evaluation finally led to a regulation [51] giving an assessment rating of the likelihood on corrosion of reinforcing steel (see Table 19-2). This methodology was tested at different places [52-54].

Since the rest potential of steel in concrete depends on many parameters [5,41,42], the data in Table 19-2 are only a rough approximation. Of major influence are the type of cement (blast furnace or portland), aeration (concrete covering and moisture) and age (film formation). Even more negative potentials than $U_{Cu\text{-}CuSO_4} = -0.4$ V can be found without there being any risk of corrosion. On the other hand, anodic endangered areas can show a relatively positive potential due to polarization by nearby cathodes that would be judged as harmless according to Table 19-2. Apparently the rest potential is largely determined by the actual aeration conditions [41-43].

However, if the interpretation of the potentials measured for regions with a covering as uniform as possible and aeration or moisture is extended to estimate the potential gradients corresponding to the explanation for Fig. 3-24, there follows the possibility of classifying the state of corrosion [52-54]. Furthermore, the sensitivity of the estimate can be raised by anodic polarization according to the explanation given for Fig. 2-7, because the depassivated steel is less polarizable than the passive steel in concrete [43].

19.5.3 Reinforcement Continuity

The cathodic protection of reinforcing steel and stray current protection measures assume an extended electrical continuity through the reinforcing steel. This is mostly the case with rod-reinforced concrete structures; however it should be verified by resistance measurements of the reinforcing network. To accomplish this, measuring cables should be connected to the reinforcing steel after removal of the concrete at different points widely separated from each other. To avoid contact resistances, the steel must be completely cleaned of rust at the contact points.

Table 19-2 Assessment of susceptibility to corrosion from Ref. 15

Corrosion likelihood (depassivation)		Steel/concrete potential $U_{Cu\text{-}CuSO_4}$ (V)
10%:	Passivation likely	< -0.25
	Prediction not possible	-0.25 to -0.35
90%:	Depassivation likely	< -0.35

434 Handbook of Cathodic Corrosion Protection

Measured values of >1 Ω show that the continuity is insufficient. The reinforcing steel should then be short circuited with the rest of the reinforcement, which has a sufficient electrical continuity.

19.5.4 Installation and Types of Anode System

Loose old concrete must be removed before installing the anode system. This can be achieved with a high-pressure water jet or by sand blasting. This produces a roughening of the surfaces outside the anodes to be attached, which ensures good adhesion of the sprayed concrete. Cracks in the concrete must be chiseled out to the base. The anodes can be attached by two different methods according to the degree of damage to the covering concrete.

With damage of a large area, the concrete is removed down to the uppermost layer of the reinforcement; then the first layer of sprayed concrete is applied (see Fig. 19-3a). The anode is fixed to this layer and is followed by a second layer of sprayed concrete. With large areas of old concrete which are still solid, the anodes can be attached to this and finally embedded with sprayed concrete (see Fig. 19-3b). The anodes must supply the required protection current and be structurally robust. The bond between the old concrete and sprayed concrete must not be damaged during the spraying process or in operation.

In the first reconstruction [27] of road slabs contaminated with Cl⁻, silicon iron anodes were embedded in a layer of coke breeze as shown in Fig. 19-4a or the current connection was achieved with noble metal wires in a conducting mineral bedding material. Slots were ground into the concrete surface for this purpose at spacings of about 0.3 m (see Fig. 19-4b). This system is not suitable for vertical structures.

The anode systems used today consist of a fine-meshed, oxide-covered titanium network [55,56] (see Section 7.2.3), polymer cable anodes of high flexibility

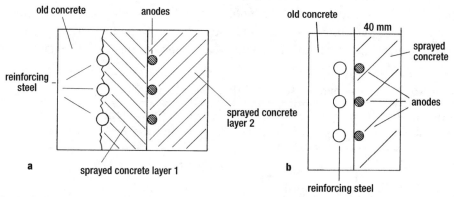

Fig. 19-3 Methods of installing anodes: (a) Old loose concrete, two sprayed concrete layers, (b) solid concrete, one layer of sprayed concrete.

Cathodic Protection of Reinforcing Steel in Concrete Structures

with a Cu core [28,29,37] (see Section 7.2.4) or conducting coatings with noble metal leads [28,29]. Investigations of their long-term behavior in damp soil and on the expected acidification in the vicinity of the anode (Eq. 5-24), depending on the anode current density, have not yet been completed. Accelerated tests with very high current densities of 100 mA m^{-2} lead to a false assessment because no account is taken of neutralization by diffusion of OH$^-$ ions from the surroundings. With graphite anodes in areas containing Cl$^-$ ions, reactions according to Eqs. (7-3) and (7-4) can lead to weight loss of conducting polymers. These processes and the acidification are being quantitatively investigated under service conditions [57]. Drilling cores are being taken at 3-year intervals to measure the loss in alkalinity. Anodic acidification could also limit the cathodic protection in carbonized concrete because of the lack of alkalinity. Therefore the anode system in Fig. 19-3b cannot be used. In addition, an enrichment of chloride ions is to be expected in the anode area due to migration according to Eq. (2-23) [35,58].

19.5.5 Concrete Replacement Systems for Cathodic Protection

There are different concrete replacement systems available for renovating reinforced concrete structures. They range from sprayed concrete without polymer additions to systems containing conducting polymers (PCC-mortar). Since with the latter alkalinity is lower, more rapid carbonization occurs on weathering [59] and the increased electrical resistivity has to be taken into account, so that with cathodic protection only sprayed concrete should be used as a repair mortar.

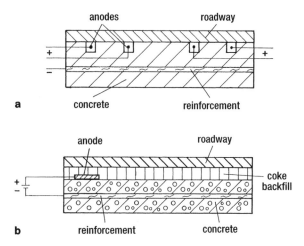

Fig. 19-4 Older anode system for roadway plates: (a) Silicon iron anodes in coke breeze bed, (b) noble metal lead in conducting bedding in a trench.

19.5.6 Commissioning, Maintenance and Control

In designing cathodic protection for concrete structures, it is convenient to divide the surfaces to be protected into separate areas that take account of the local steel area per volume of concrete as well as the distribution of moisture. Figure 19-5 shows schematically the division of a reinforced concrete support wall which is to be protected in four areas. The object to be protected had a concrete surface area of 200 m², and the component areas ranged from 20 to 90 m². The reinforcing index (steel surface/concrete surface) was 1 and had to be known in each case in order to determine the protection current density. Plastic cable anodes were used for the anode system. These were loop shaped and dowelled into the concrete surface in areas of low reinforcing index in straight lines. Ag-AgCl reference electrodes were used for the control system. These were built in at a distance of 2 cm from the reinforcing steel and had to be protected against mechanical influences. In the object considered in Fig. 19-5, eight reference electrodes were used with separate anode connections to each component area.

Control of the cathodic protection was carried out by potential measurements with the built-in reference electrodes and with mobile electrodes on the surface. In spite of there being little information on the long-term stability of built-in reference electrodes, they should not be abandoned for this reason, because the *IR* error according to Eq. (2-34) is small due to the small distance from the reinforcing steel. Special reference electrodes are being developed [59]. Mobile reference electrodes have the advantage that potential control can be carried out at any desired location. *IR* errors have to be considered, particularly near the anodes, due to the large distance from the reinforcing steel.

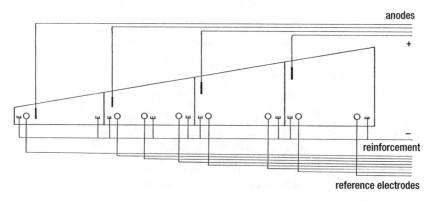

Fig. 19-5 Renovation of a retaining wall with cathodic protection, division into sections.

Cathodic Protection of Reinforcing Steel in Concrete Structures 437

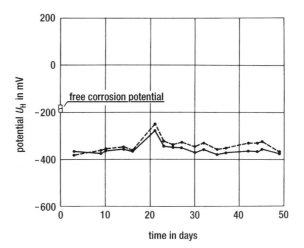

Fig. 19-6 Steel/concrete potentials before and after interrupting the protection current.

Since the object to be protected represents a cell consisting of active and passive steel, considerable *IR* errors in the cell current must be expected in measuring the off potential. The considerations in Section 3.3.1 with reference to Eqs. (3-27) and (3-28) are relevant here. Since upon switching off the protection current, I_s, the nearby cathodes lead to anodic polarization of a region at risk from corrosion, the cell currents I_e and I_s, have opposite signs. It follows from Eqs. (3-27) and (3-28) that the *IR*-free potential must be more negative than the off potential. Therefore, there is greater certainty of the potential criterion in Eq. (2-39).

It has to be remembered that hydration processes and moisture exchange occur with old concrete when sprayed concrete is applied. Both processes can affect the potentials so that the protection current should only be switched on 4 weeks after

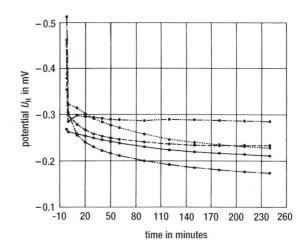

Fig. 19-7 Potential variation with time at five test points after switching off the protection current.

438 Handbook of Cathodic Corrosion Protection

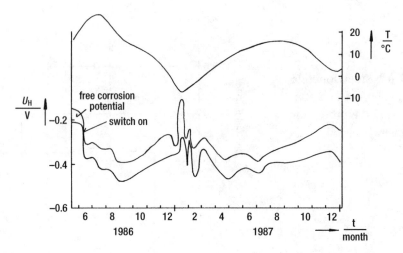

Fig. 19-8 Steel/concrete potentials (two records) and temperature as a function of the time of year. (The protection installation was switched on in June 1986.)

the last application of sprayed concrete skin. Figure 19-6 shows steel/concrete potential measurements versus time made with built-in reference electrodes before and after switching on the protection current. The noticeable jump in potential after 20 days is due to a defective interruption of the protection current. The potential criterion of $U_{Hs} = -0.4$ V according to Eq. (2-39) is still not attained.

Figure 19-7 shows off potential measurements as an example, in which the 100-mV criterion, No. 3 in Table 3-3, as well as the potential criterion $U_{off} < U_s$ is fulfilled. It has to be remembered with off potential measurements that according to the data in Fig. 3-6, depolarization is slower with age, so that the 100 mV criterion must lead to errors with a measuring time of 4 hours. Off potential measurements should be carried out after commissioning at 1-, 2-, 6- and 12-month intervals and then annually.

Depolarization times greater than 48 hours should be allowed when applying the 100-mV criterion. Comparison of measured potentials is only possible where conditions of temperature and moisture are similar. Figure 19-8 shows the dependence on season and temperature which can probably be ascribed to different aeration. In repair work, potential probes should be provided at different places on the object to reduce IR errors in off potential measurements arising from equalizing currents.

19.6 Stray Current Effects and Protective Measures

Stray current effects with reinforced concrete are not likely from the usual causes, but are possible with roadways and bridges over which dc railways pass. In the Federal Republic of Germany, the railways are mainly ac. Usually, the rails are not grounded and are well insulated with high-resistance ballast. If, however, stray currents escape, the protective measures in Chapter 15 are applicable so that all other lines that run parallel to the reinforcing steel and the rails must be included in the protection system.

19.7 References

[1] A. Bäumel, Zement, Kalk, Gips *12*, 284 (1959).
[2] H. Kaesche, Zement, Kalk, Gips *12*, 289 (1959).
[3] A. Bäumel u. H.J. Engell, Arch. Eisenhüttenwes. *30*, 417 (1959).
[4] H. Kaesche, Arch. Eisenhüttenwes. *36*, 911 (1965).
[5] W. Schwenk, Betonwerk + Fertigteil-Technik, *51*, 216 (1985).
[6] DIN 1045, Beuth-Verlag, Berlin (1988).
[7] DIN 4277, Beuth-Verlag, Berlin (1988).
[8] B. Isecke, Werkst, Korros. *37*, 322 (1986).
[9] G. Ruffert, Schäden an Betonbauwerken, Verlagsgesellschaft Rudolf Müller, Köln-Braunsfeld 1982.
[10] D. Jungwirth, E. Beyer, P. Grübel, Dauerhafte Betonbauwerke, Beton-Verlag, Düsseldorf, 1986.
[11] H.J. Lasse u. W. Stichel, Die Bautechnik *60*, 124 (1983).
[12] D. Weber, Amts- und Mitteilungsblat BAM *12*, 107 (1982).
[13] P. Schießl, Bautenschutz u. Bausanierung *10*, 62 (1987).
[14] B.D. Mayer, Bautenschutz u. Bausanierung *10*, 66 (1987).
[15] B. Isecke, Bautenschutz u. Bausanierung *10*, 72 (1987).
[16] G. Rehm u. E. Fielker, Bautenschutz und Bausanierung *10*, 79 (1987).
[17] G. Thielen, Bautenschutz und Bausanierung *10*, 87 (1987).
[18] J.E. Slater, Corrosion Metals in Association with Concrete, ASTM STP 818, Philadephia 1983.
[19] U. Nürnberger, Werkst. Korros. *37*, 302 (1986).
[20] D.E. Tonini u. S.W. Dean (ed.), Chloride Corrosion of Steel in Concrete, ASTM STP 629, Philadephia 1977.
[21] R.F. Stratfull, Corrosion *13*, 174t (1957).
[22] R.F. Stratfull, Corrosion *15*, 331t (1959).
[23] A.E. Archambault, Corrosion *3*, 57 (1947).
[24] E.B. Rasa, B. McCollom, O.S. Peters, Electrolysis in Concrete, U.S. Bureau of Standards, Technological Paper No. 18, Washington 1913.
[25] O.L. Eltingle, Engineering News *63*, 327 (1910).
[26] G. Mole, Engineering *166*, H. 11, 5 (1948).
[27] R.F. Stratfull, Materials Performance *13*, H 4, 24 (1974).
[28] Proc. "Corrosion in Concrete B Practical Aspects of Control by Cathodic Protection," Seminar London, published by Global Corrosion Consultants, Telford, England, 1987.

[29] Proc. 2nd International Conference on the deterioration and repair of reinforced concrete in the Arabian Gulf, Bahrain 1987.
[30] Cathodic Protection of Reinforced Concrete Decks, NACE, 20 papers, 1985.
[31] D. Whiting u. D. Stark, Cathodic Protection for Reinforced Concrete Bridge Decks B Field Evaluation, Final Report, Construction Technology Laboratories, Portland Cement Association, Skokie, Illinois, NCHRP 12-13A (1981).
[32] Proc. NACA Corrosion 87, San Francisco, papers 122 bis 147, 1987.
[33] B. Heuzé, Materials Performance, *19*, H. 5, 24 (1980).
[34] O. Saetre u. F. Jensen, Materials Performance, *21*, H. 5, 30 (1982).
[35] O.E. Gjorv u. Ø. Vennesland, Materials Performance, *19*, H. 5, 49 (1980).
[36] B. Isecke, Beton, *37*, 277 (1987).
[37] K.B. Pithouse, Corrosion Prevention and Control, *15*, H. 10, 113 (1986).
[38] J.T. Gourley u. F.E. Moresco, NACE Corrosion 87, paper 318, San Francisco 1987.
[39] B.P. Hughes, A.K.O. Soleit, R.W. Brierly, Magazine of Concrete Research *37*, 243 (1985).
[40] H. Hildebrand, M. Schulze u. W. Schwenk, Werkst, Korros. *34*, 281 (1983).
[41] H. Hildebrand u. W. Schwenk, Werkst. Korros. *37*, 163 (1986).
[42] H. Hildebrand, C.-L. Kruse u. W. Schwenk, Werkst. Korros. *38*, 696 (1987).
[43] J. Fischer, B. Isecke, B. Jonas, Werkst. Korros., demnächst.
[44] DIN 30676, Beuth-Verlag, Berlin 1985.
[45] J.A. Shaw, Civil Engineering (ASCE), H. 6, 39 (1965).
[46] Cathodic Protection of Reinforcing Steel in Concrete Structures, Proposed NACE-Standard, Committee T-3K-2 NACE, Houston 1985.
[47] P. Vassie, Corrosion Prevention and Control, *32*, H. 6, 43 (1985).
[48] H.U. Aeschlimann, Schweizer Ingenieur und Architekt, *15*, 867 (1985).
[49] V. Herrmann, Beton- und Stahlbetonbau, *82*, 334 (1987).
[50] J.R. van Daveer, ACI Journal *72*, 697 (1975).
[51] ASTM C 876, Philadelphia 1980.
[52] J. Tritthardt u. H. Geymayer, Beton, *31*, 237 (1981).
[53] B. Elsener u. H. Böhni, Schweizer Ingenieur und Architekt *14*, 264 (1984).
[54] R. Müller, Proc, "Werkstoffwissenschaften und Bausanierung," TA Esslingen, 689 (1987).
[55] J.E. Bennett, Beitrag in [28] und [29].
[56] S. Kotowski, B. Busse u. R. Bedel, METALL *42*, 133 (1988).
[57] Forschungsprojekt 102D der EG im BRITE-Forschungsprogramm "Electrochemically Based Techniques for Assessing and Preventing Corrosion of Steel in Concrete," Bericht erhältlich 1990.
[58] G. Mussinelli, M. Tettamanti u. P. Pedeferri in [29], S. 99.
[59] R. Kwasny, A. Roosen u. M. Maultzsch, Betonwerk u. Fertigteil-Technik 52, 797 (1986).

20

Internal Cathodic Protection of Water Tanks and Boilers

G. Franke and U. Heinzelmann

Internal cathodic protection of water tanks and boilers is most economical if it is taken care of at the design stage. It can, however, be installed at a later stage as a rehabilitation measure to halt the progress of corrosion. Tanks and boilers in ships were described in Section 17.4. Further applications of internal protection are dealt with in Chapter 21.

20.1 Description and Function of Objects to be Protected

Water tanks are used for storing hot and cold drinking water, service water, cooling water, condensate and sewage. Filter tanks are used for purifying water, and reaction vessels are used for physical and chemical reactions (e.g., removal of ozone). The tanks may be open, with variable water levels, filled as pressure vessels, or partly filled as compression tanks; they may also be vacuum tanks. They can be equipped with inserts such as heating or cooling surfaces, separating partitions, probes, connected tubes or distribution tubing.

All these components can be included in the cathodic protection if certain features are considered in the design [e.g., bundles of heating tubes in a square array (see Fig. 20-1)], and the electrodes are arranged so that all the surfaces receive sufficient protection current i.e., so that the criterion in Eq. (2-39) is fulfilled.

In enamelled tanks with protection electrodes of low current output, fittings [e.g., heating surfaces (cathodic components)] must be electrically isolated from the tank and the ground. Figure 20-2 shows such a bushing. Smaller cathodic components which take up only negligible protection current (e.g., temperature probes) do not need to be insulated.

In uncoated cathodically protected tanks, inserts must have an electrical connection to the tank to avoid damage due to anodic interference in the cathodic voltage cone of the surfaces of the object to be protected. These processes are also

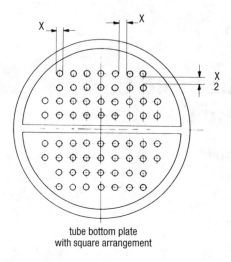

Fig. 20-1 Heating element tube bundle on a rectangular pattern.

tube bottom plate with square arrangement

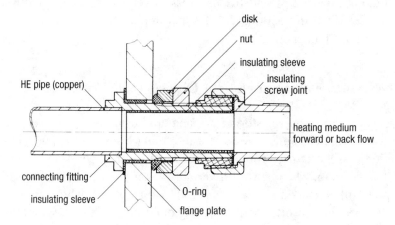

Fig. 20-2 Insulated installation of heating pipes in enamelled containers with electrical separation of container and ground.

well known in the case of buried pipelines and are described in Section 24.3.4. This anodic danger in the internal protection of tanks is known as "current exit corrosion" and is discussed further in Section 20.1.4.

20.1.1 Materials for Objects to be Protected and Installation Components

Materials for metal tanks and installations include plain carbon steel, hot-dipped galvanized steel, stainless steel [e.g., steel No. 1.4571 (AISI 316Ti)], copper and its alloys. The corrosion resistance of these materials in water is very variable and can be assessed with the aid of Ref. 1. Furthermore, the rest potentials are very different (see Table 2-4) as are the protection potentials (see Section 2.4). Plain carbon steel that is uncoated is not corrosion resistant. Stainless steel can suffer chloride-induced localized corrosion. The rest of the materials are corrosion resistant due to the formation of protective films but can be attacked if the films have defects locally. In all these cases, cathodic protection is applicable.

20.1.2 Types of Linings and Coatings

Different linings and coatings are used to provide corrosion protection for the tank which, according to Section 5.1, often makes economical cathodic protection possible. The properties of these materials are extensively described in Chapter 5.

Organic coatings include phenol formaldehyde, epoxy, polyacryl and polyacryl acid resins, polyamide, polyolefin, bitumen and rubber. The required properties of lining materials in combination with cathodic protection are given in Section 5.2.1. In addition, adequate resistance to water vapor diffusion is required (see Section 5.2.2). These properties are discussed in the basic standard for internal cathodic protection [2], which also reports how these properties can be assured by testing with the aid of Ref. 3. It is convenient to limit the protection potential region to $U'_{Hs} = -0.8$ V in the presence of resin coatings, which makes potential-controlled protection current equipment necessary.

Cathodic blistering is not expected with high-resistance polyolefins and rubberized coatings; however, where there are defects, local reduction in adhesion (cathodic disbonding) is apparent. Enamel linings have the best resistance against cathodic influence (see Section 5.4).

20.1.3 Preconditions for Internal Cathodic Protection

The preconditions for internal cathodic protection are assembled in Ref. 2 and discussed in Section 21.1. The following directions are applicable to water tanks:

A protection potential range must exist for the materials of the internal surfaces (tank and components) (see Section 2.4). The electrolytic conductivity of water in uncoated tanks should be >100 μS cm^{-1}. For lower conductivities, a coating must be applied to maintain adequate current distribution that has a sufficiently high coating resistance r_u as in Eq. (5-2), in order to raise the polarization parameter k in Eq. (2-45). For this the specific polarization resistance r_p and the specific coating resistance r_u are considered equal. Usually adequate current distribution has to be ensured by structural design (see Fig. 20-1) as well as by the arrangement and number of anodes.

With water of pH < 5, the data for Eq. (6-4) should be used to check whether the galvanic anodes have a self-corrosion that is too high (see Section 6.1.1). The data and test information in Section 6.2.4 should be used for magnesium anodes. Impressed current protection can lead to too high hydrogen evolution (see the protection measures in Section 20.1.5).

20.1.4 Measures to Prevent Anodic Interference

Current exit corrosion arising from anodic interference of the surfaces of the object to be protected in the cathodic voltage cone and at insulating units can be seen in Fig. 20-3. The object to be protected is an internally coated boiler with an uncoated heating element (e.g., copper pipe) that is electrically insulated from the object to be protected according to Fig. 20-2. To avoid electrical shorting through the ground, an additional insulating unit must be built into the pipe outside the tank. Defects are present in the tank coating near the insulator, which produces a cathodic voltage cone. The arrows show the path of the protection current; the hatched region represents the location of current exit corrosion.

The anodic danger in the interior of the heating element is dealt with theoretically in Section 24.4.6. Assuming high polarization resistances in the system pipe material-heating water, a current density distribution is set up in the interior of the pipe according to Eq. (24-91) with constants according to Eqs. (24-101) and (24-102). Out of these, the maximum current density J_0 according to Eq. (24-102) is of particular interest. Quantitative application of this relationship is not possible

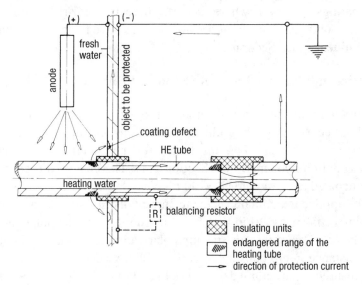

Fig. 20-3 Heating pipe (HE) in a boiler with a high-resistance coating with electrical separation from the boiler and from the ground by insulating units and a balancing resistor.

because of the idealized assumption and the unknown parameters involved. However, it gives an indication of the trend of reduced danger of corrosion. This decreases with the length of the insulator, L, and with the square root of the polarization parameter, $\sqrt{r_p/k}$; it increases, however, with the square root of the pipe diameter, $\sqrt{d}$.

A circular defect in the coating of diameter d is assumed to indicate the anodic danger in the region of the cathodic voltage cone. With current density J_s, and grounding resistance according to Eq. (24-17), the cathode voltage U_0 is given by

$$U_0 = \frac{\pi}{8} J_s d \rho \qquad (20\text{-}1)$$

The electric field is given by the equations in Table 24-1, column 2. U_r and U_t give the voltage distribution in the radial and depth directions respectively, where $2r_0 = d$. For small arguments of the arc sin and arc tan functions (i.e., for $r_0 \ll r$ and $r_0 \ll t$), it follows that at a distance $x = r$ or $x = t$ from the defect:

$$U_x = \frac{2}{\pi} U_0 \frac{r_0}{x} = \frac{U_0 d}{\pi x} \qquad (20\text{-}2)$$

Finally, it follows from Eqs. (20-1) and (20-2):

$$U_x = \frac{1}{8} J_s d^2 \rho \frac{1}{x} \qquad (20\text{-}3)$$

If the object interfered with (heating element) lies in the region between x_1 and x_2, then the interfering voltage due to the voltage cone from Eq. (20-3) is given by:

$$\Delta U_x = \frac{1}{8} J_s d^2 \rho \frac{\Delta x}{x_1 x_2} \qquad (20\text{-}4)$$

The size of the defects in the coating has an important influence.

It should be clearly pointed out that with anodic interference according to the data in Fig. 2-6 in Section 2.2.4.1, the corrosivity of the electrolyte for the particular material has no influence on the current exit corrosion. On the other hand, the conductivity of the electrolyte has an effect according to Eqs. (24-102) and (20-4). Chemical parameters have a further influence that determines the formation of surface films and the polarization resistance.

To avoid the described corrosion risk from current exit corrosion, heating elements should be electrically connected to the object to be protected via a balancing resistor. Figure 20-3 shows a correctly placed balancing resistor R (dashed line).

The heat exchanger is brought into the cathodic protection circuit with a correctly rated balancing resistor, thus eliminating the anodic danger according to

446 Handbook of Cathodic Corrosion Protection

Eq. (20-4). In addition, the voltage at the insulating unit ΔU in Eq. (24-102) is substantially reduced. Increasing the length, L, of the insulating unit can also be helpful.

20.1.5 Measures to Prevent Danger from Hydrogen Evolution

Hydrogen can be generated in internal corrosion protection with impressed current and also with protection by galvanic anodes by self-corrosion, which can form an explosive gas mixture with oxygen from water, air, or the anode reaction. With cathodically protected tanks, this is always a possibility. The construction and operation of the protection installation affects this reaction. Since protection electrodes of noble metals or valve metals with noble metal coatings can catalyze the ignition of gas mixtures containing hydrogen and oxygen, their use in protecting uncoated closed water tanks should be avoided on safety grounds. Ignitable gas mixtures must be removed from the uppermost part of the tank with automatic exhausters of sufficiently wide diameter (at least DN15). The exhaust tailpipe protrudes into the tank (about 50 mm) in order to prevent gas bubbles from being carried over into the water pipes. The drop conduit attached to the exhauster must drain off water down-

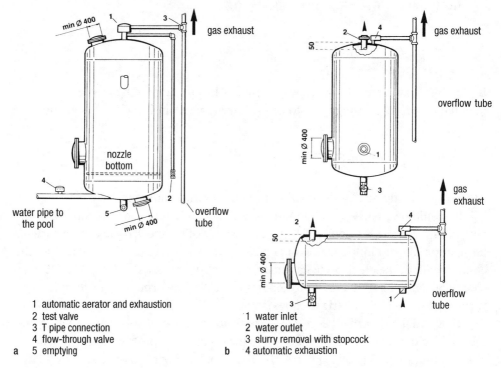

Fig. 20-4 Safety precautions for water tanks (a) and filter tanks (b) to prevent danger of explosion from hydrogen.

ward and gas upward via a T piece (see Fig. 20-4a). The exhaust tailpipe must be led at a constant slope into the open if the water tank is in an insufficiently ventilated room. Before a cathodically protected uncoated tank is emptied, the protection equipment must be switched off and the tank completely filled with water.

In filter tanks, a pipe (at least DN15) with a test valve must be placed beneath the exhausters leading to the floor level through which water must be taken to the open air every day until the water is free from gas bubbles (see Fig. 20-4b). The partly filled cathodically protected pressure booster tank cannot be provided with automatic gas exhaustors. Therefore, nozzles (at least DN15) with manually operated valves must be provided at the highest point of the tank. Before emptying, the protection system must be switched off and the tank filled with water through the manual valve.

Tanks that are enamelled or that have other high-resistance coatings with volumes below 4 m^3 do not require automatic exhausts if they are equipped with galvanic anodes in which the surface area does not exceed 5 dm^2 per square meter of surface to be protected, or with controlled impressed current anodes with a low current output of 150 mA.

20.2 Protection with Galvanic Anodes

Only magnesium anodes are practical for the cathodic protection of water tanks because their driving voltage is sufficiently high (see Section 6.2.4). Zinc and aluminum anodes are not suitable for fresh water because their driving voltages are too low and they passivate too easily. Aluminum anodes containing Hg, which have a sufficiently negative rest potential and low polarizability, cannot be used for drinking water (see Section 20.5).

The range of applications for magnesium anodes includes the internal protection of boilers, feedwater tanks, filter tanks, coolers, pipe heat exchangers and condensers. They are mainly used in conjunction with coatings and where impressed current equipment is too expensive or cannot be installed.

Anode consumption is high with uncoated objects. Care therefore has to be taken that the $Mg(OH)_2$ slurry does not interfere with the function of the system. Regular deslurrying has to be undertaken.

Magnesium anodes are frequently used as an additional protection measure at a later stage for stainless steel tanks. In this case the anodes are connected through a 5- to 10-Ω resistor to the tank to avoid an unnecessarily high current for the cathodic protection of the tank and simultaneous high consumption of the anodes.

Magnesium anodes are widely used in conjunction with enamel coatings. This type of corrosion protection is particularly economical and convenient in small- and medium-sized boilers. The anode only has to ensure protection of small defects in the enamel, which are unavoidable when enamelling in mass production, and are detailed in Ref. 4. The life of magnesium anodes designed with 0.2 kg m^{-2} of enamelled surface is at least 2 years and, on average, over 5 years.

20.3 Protection with Impressed Current

The impressed current protection method is used mainly for the internal protection of large objects and particularly where high initial current densities have to be achieved (e.g., in activated charcoal filter tanks and in uncoated steel tanks). There are basically two types of equipment: those with potential control, and those with current control.

20.3.1 Equipment with Potential Control

Protection current devices with potential control are described in Section 8.6 (see Figs. 8.5 and 8.6); information on potentiostatic internal protection is given in Section 21.4.2.1. In these installations the reference electrode is sited in the most unfavorable location in the protected object. If the protection criterion according to Eq. (2-39) is reached there, it can be assumed that the remainder of the surface of the object to be protected is cathodically protected.

Since usually the reference electrode is not equipped with a capillary probe (see Fig. 2-3), there is an error η_Ω in the potential measurement given by Eq. (2-34); in this connection see the data in Section 3.3.1 on *IR*-free potential measurement. The switching method described there can also be applied in a modified form to potential-controlled protection current devices. Interrupter potentiostats are used that periodically switch off the protection current for short intervals [5]. The switch-off phase is for a few tens of microseconds and the switch-on phase lasts several hundred microseconds.

During the switching-off phase, the *IR*-free potential measured by the protection current equipment is the actual potential and is compared with the nominal potential. The protection current delivered by the device is adjusted so that the *IR*-free potential meets the nominal value. As a simplification, the impressed current anode can also be used in the switching-off phase as the reference electrode in the event one of them indicates a constant potential in this phase. This applies to noble metal coatings on valve metals (e.g., TiPt) since O_2 is evolved in the switching-on phase according to Eq. (7-1) and a definite redox system is established. Figure 20-5 shows potential-time curves for switching-on and switching-off phases of a TiPt electrode in tapwater. Immediately after switching off and switching on, there is an ohmic voltage drop corresponding to Eq. (2-34) and represented by a dashed line in the figure. During the first 50 μs after switching off, a noticeable depolarization

Internal Cathodic Protection of Water Tanks and Boilers 449

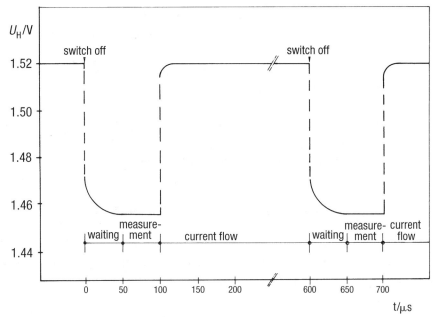

Fig. 20-5 Potential-time curves for a TiPt anode in the switching-off and switching-on phase (schematic) in the measurements in Ref. 7.

of the electrode can be seen. In the following 50 μs, the potential, however, is constant. The electrode can be used as a reference electrode during this time period [6]. According to the data in Section 3.3.1, 100 μs after switching off one does not have to expect an inadmissibly high depolarization for the object to be protected (steel-water system). Therefore, the time schedule for operation of the interrupter potentiostat shown in Fig. 20-5 is suitable for internal cathodic protection [7].

By using only a single reference electrode in the object to be protected, the potential can be determined only in the vicinity of this electrode and not in more remote areas. Section 3.3.1 together with Eq. (3-27) provides further explanation of this. To improve the current and potential distribution, the number and location of the anodes must suit the geometry of the object to be protected. Occasionally, additional reference electrodes are required for potential control [2]. The optimum nominal potential for potential control can be found by this method by considering remote IR errors.

Nonuniform current and potential distribution is usually to be expected with uncoated objects to be protected. The distribution can be considerably improved by coatings (see Section 20.1.3). In enamelled tanks, the current and potential distribution of cathodic protection is very good. By arranging the anode centrally, IR errors from equalizing currents in the switching-off phase can be ignored. The anode potential in the switching-off phase can be evaluated from the information

450   Handbook of Cathodic Corrosion Protection

in Fig. 20-5 for potential control so that the whole surface of the protected object can be adequately protected. A suitable arrangement of several anodes is necessary only for larger tanks of over 2 m^3 in volume.

In general, potential control according to this technique allows the most accurate and stable maintenance generally possible of the potential of the object to be protected. Changes in the protection current requirement (e.g., through formation of surface films or enlargement of defects in coatings) can be readjusted. Therefore the overprotection associated with increased hydrogen evolution according to Eq. (2-19) is not possible. This is particularly important in internal protection (see Section 20.1.5).

20.3.2 Equipment with Current Control Based on Water Consumption

In some cases it is convenient to control the protection current according to the water consumption. This mainly concerns tanks that supply fresh water in small amounts at irregular intervals (e.g., storage tanks for fire-fighting installations). Hydrogen evolution occurs with otherwise constant protection current in a stagnation period after cathodic oxygen reduction as in Eq. (2-19). In this case, current control with a contact water meter acting as a pulse generator to limit hydrogen evolution is advisable (see Section 20.4.2).

20.4 Description of Objects to be Protected

20.4.1 Boilers with Enamel Linings

Cathodic protection of enamelled tanks with Mg anodes has long been the state of the art, with potential-controlled equipment being used with increasing frequency in recent years. A high-resistance coating with limited defects according to Ref. 4 enables uniform current distribution to be maintained over the whole tank.

A danger of corrosion exists at defects in the enamel due to cell formation with cathodic components (i.e., inserts with a more positive free corrosion potential). These are mainly heating elements with a somewhat larger surface area than the enamel defects. With uninsulated inserts, they consume protection current so that defects lying in the voltage cone of the inserts not only receive no protection current but, under certain circumstances, can become anodic. For this reason, heating elements must be electrically isolated from the boiler (see Figs. 20-2 and 20-3). The inserts must not be grounded in order to prevent a short circuit of the heating element with the object to be protected. In this case, safety regulations [8] must be observed. These assume that the steel tank has a protective lead connection and

Internal Cathodic Protection of Water Tanks and Boilers 451

that external ungrounded components of the heater are covered to prevent accidental contact.

Heat exchangers with flowing cold electrolyte in fresh-water heat pump reservoirs must be treated similarly [9]. Here the cold electrolyte circuit need not have expensive insulated bushings installed as shown in Fig. 20-3, and the entire water pump equipment does not have to be grounded but must be covered to prevent contact. Electrical separation as in Fig. 20-3 is only undertaken with hot-water heat exchangers.

Magnesium anodes are usually built into the object to be protected through isolating sockets or holes (see Fig. 20-6) and joined to them with cables. They must be readily accessible and easily exchangeable for convenient control [4]. The directions for use have to indicate the necessity for an inspection after 2 years of operation. During operation, control can be exercised by electrical measurements (current, resistance). In addition, acoustic and optical methods exist to determine the amount of anode consumption [4]. The life of the anodes is usually more than 5 years (see Section 6.6).

Impressed current anodes are also built in and insulated through sockets or holes in the object to be protected as in Fig. 20-6. With coated valve metals, a region of about 200 mm from the connection is uncoated. In the case of an inter-

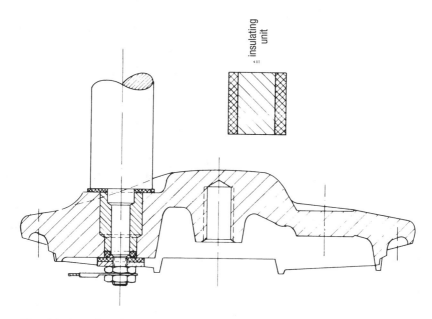

Fig. 20-6 Electrically insulated wall passage of a magnesium anode.

rupter potentiostat, its functioning can be optically indicated with the aid of current-dependent diode lamps.

The periods for inspection of cathodically protected plants are:

- installations with galvanic anodes: 2 years.
- installations with uncontrolled protection current equipment: 1 year.
- installations with controlled protection current equipment: 0.5 year.

The performance of the protection current equipment and the protection current should be checked at these frequencies and, if necessary, potential measurements carried out. In addition, maintenance periods specific to the installation and a control program should be provided by the installer.

As an example, Fig. 20-7 shows potential and protection currents of two parallel-connected 750-liter tanks as a function of service life. The protection equipment consists of a potential-controlled protection current rectifier, a 0.4-m long impressed current anode built into the manhole cover, and an Ag-AgCl electrode built into the same manhole [10,11]. A second reference electrode serves to control the tank potential; this is attached separately to the opposite wall of the tank. During the whole of the control period, cathodic protection is ensured on the basis of the potential measurement. The sharp decrease in protection current in the first few months is due to the formation of calcareous deposits.

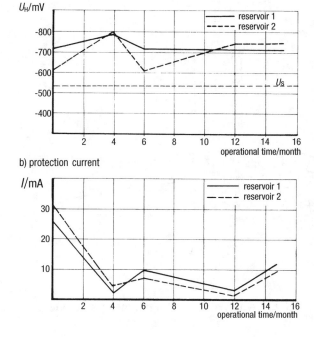

Fig. 20-7 Potential and protection current measurements on two 750-liter tanks; $\varkappa\,(20°C) = 730\ \mu S\ cm^{-1}$; operating temperature 70°C.

Internal Cathodic Protection of Water Tanks and Boilers

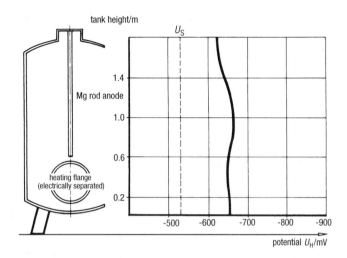

Fig. 20-8 Vertical potential profile in a 300-liter electrically heated boiler with Mg anodes; $I_s = 0.4$ mA; $\varkappa\,(20°C) = 30\ \mu S\ cm^{-1}$.

As an example of potential distribution, Fig. 20-8 shows the potential on the vertical axis in a 300-liter electric storage reservoir. The water had an extremely low conductivity of $\varkappa\,(20°C) = 30\ \mu S\ cm^{-1}$. A Mg rod anode served for cathodic protection; it reached to just above the built-in heating element to give uniform current distribution. This was confirmed by the measurements.

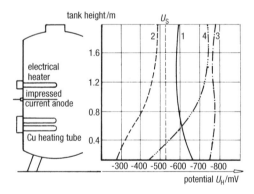

Fig. 20-9 Vertical potential profile in a 250-liter boiler

Curve	Cu heat exchanger	$\varkappa(20°C)/\mu S\ cm^{-1}$	I_s/mA
1	Insulated	130	16.2
2	Not insulated	130	30.2
3	Insulated	1040	36.8
4	Not insulated	1040	107

Figure 20-9 shows the negative effect of uninsulated heating elements on corrosion protection. In a 250-liter tank, an electric tube heating element with a 0.05-m² surface area was screwed into the upper third without electrical separation, and in the lower third a tinned copper tube heat exchanger with a 0.61-m² surface area was built in. The Cu heat exchanger was short-circuited for measurements, as required. For cathodic protection, a potential-controlled protection system with impressed current anodes was installed between the two heating elements. The measurements were carried out with two different samples of water with different conductivities.

As the measurements show, the small heater without an electrical separation (from the boiler) is not detrimental to cathodic protection. However, with the uninsulated built-in Cu heat exchanger without an electrical separation, cathodic protection was not achieved. As expected, the polarization increased with increasing conductivity of the water. It should be pointed out that the Cu tube was tinned and that the tin could act as a weak cathodic component. Apart from the unknown long-term stability of such a coating, the apparent raising of the cathodic polarization resistance of tin is not sufficient to provide cathodic protection with such a large fixture. This applies also to other metal coatings (e.g., nickel).

The anodic interference of built-in inserts from the cathodic voltage cone described in Section 20.1.4 is clearly seen in the measured values in Fig. 20-10. The

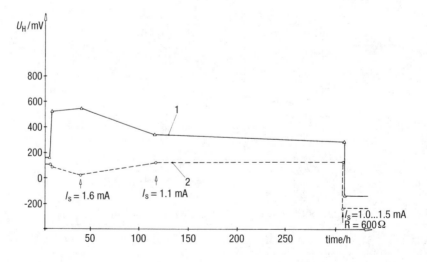

Fig. 20-10 Potential-time curves for a copper tube heating element in an enamelled container with brass screws. Curve 1: potential in the voltage cone of the brass screws. Curve 2: potential outside the voltage cone. Tapwater with $\varkappa$ (20°C) = 100 μS cm^{-1}; 50°C.

object to be measured is an electrical heater copper tube with an insulating bushing in a brass screw joint. The brass threaded part represents a relatively large defect and produces a voltage cone right at the fitting. Curves 1 and 2 in the figure show potential-time curves at the copper insert inside and outside the cathodic voltage cone respectively. The differences in potential depend, as expected, on the protection current. As a result of the different surface areas, the cathodic polarization at current entry (curve 2) is smaller than the anodic polarization at current exit (curve 1). After installing a balancing resistor of 600 Ω between the copper insert and the tank, the insert was brought into cathodic protection, which can be seen immediately by the change in potential [12].

The size of the balancing resistors must usually be decided empirically. A resistance of 500 Ω is usually sufficient for heating elements of up to 2.5 m² in surface area. With larger surface areas, it must be raised to 1000 Ω because otherwise cathodic protection at defects in the enamel is endangered.

The action of different balancing resistors can be seen in Fig. 20-11. The enamelled water heater had an electrically isolated heating element of stainless steel with a 2.5 m² surface area. With $R = 600$ Ω, the tank can no longer be cathodically protected; at $R = 800$ Ω, on the other hand, all is well [13]. The potential of the

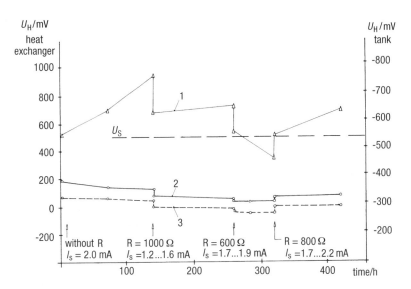

Fig. 20-11 Potential-time curves of an enamelled container with built-in stainless steel heat exchanger as a function of equalizing resistance, R. Curve 1: container potential in the region of the heat exchanger. Curve 2: heat exchanger potential in the voltage cone of defects in the enamelling. Curve 3: heat exchanger potential outside the voltage cone of the defects.

456 Handbook of Cathodic Corrosion Protection

stainless steel [material no. 1.4435 (AISI 316)] is clearly more negative than the expected pitting potential of this material [14].

20.4.2 Boilers with Electrolytically Treated Water

The electrolysis protection process using impressed current aluminum anodes allows uncoated and hot-dipped galvanized ferrous materials in domestic installations to be protected from corrosion. If impressed current aluminum anodes are installed in water tanks, the pipework is protected by the formation of a film without affecting the potability of the water. With domestic galvanized steel pipes, a marked retardation of the cathodic partial reaction occurs [15]. Electrolytic treatment alters the electrolytic characteristics of the water, as well as internal cathodic protection of the tank and its inserts (e.g., heating elements). The pipe protection relies on colloidal chemical processes and is applied only to new installations and not to old ones already attacked by corrosion.

In applying electrolytic protection, galvanized tubes can be installed downstream from copper components in water boilers without danger of Cu^{2+}-induced pitting corrosion. The protection process extends the application range for galvanized tubes with respect to water parameters, temperature and material quality beyond that in the technical regulations [16, 17].

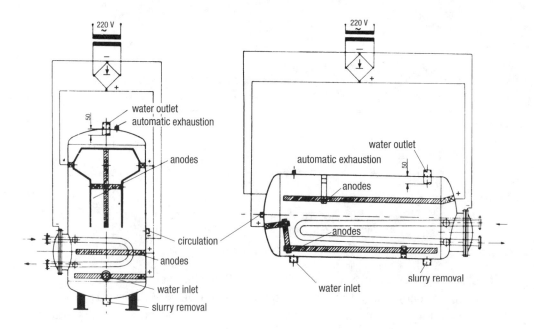

Fig. 20-12   Boiler protected by electrolysis protection process (schematic).

Figure 20-12 shows schematically the arrangement of aluminum anodes in vertical and horizontal boilers with heating tubes. More than 33% of the aluminum anodes must be sited in the upper third of the tank for the formation of protective films in the tubing [18]. Cathodic currents needed to protect the tubing may be three times as high as the current requirement for cathodic protection [19].

The treated water contains sufficient concentration of surface film-forming agents if cold water spends about 12 min and warm water at least 20 min in the tank [19]. Sudden temperature variations over 10°C must be prevented because the active form of $Al(OH)_3$ is sensitive to them [20]. If mixing with cold water or subsequent warming cannot be avoided, a short-term electrolytic aftertreatment must be provided in a small reaction tank. The development of undisturbed protective films in the tubing assumes continuous water flow with forced circulation by pumps [20].

The main part of the anode slurry, which can contain $CaCO_3$, depending on the hardness of the water, is chemically inert from the corrosion point of view, and settles to the bottom of the tank. The slurry must be removed once or twice a week through a stub with a stopcock; this takes place in short bursts of opening and closing of the tap at full water pressure. The deslurrying outlet is at the end opposite to the water inlet in a horizontal tank.

A higher content of Al_2O_3 and SiO_2 is critical for the composition of the protective films in the tubing, assuming the water contains silicates or silicic acid. The protective films have a maximum thickness of 1.5 mm and cannot grow further. The corrosion process can be stopped even in copper pipe networks with type I pitting [21] by providing a reaction tank with impressed current aluminum anodes.

Some construction points for tanks when using the electrolysis protection method should be noted:

(a) automatic exhausters must be provided to remove reaction gases (hydrogen) [2,22] (see Section 20.1.5).
(b) the water outlet must be inserted about 50 mm into the tank to promote effective exhaustion.
(c) stubs with stopcocks must be provided in the bottom of the tank for deslurrying.
(d) heating pipe bundles must be arranged on a square pattern with consideration of values for a minimum spacing of the pipe (see Section 20.1).
(e) the sealing materials for the current lead sockets and anode fixtures must be suitable for use with drinking water.

The current yield of aluminum depends on the composition of the water and the operating conditions; it usually lies between $\alpha = 0.8$ and 0.9 (see Section 6.2.3). Self-corrosion occurs, as with Mg, with hydrogen evolution.

Pure aluminum is used in the electrolysis protection process, which does not passivate in the presence of chloride and sulfate ions. In water very low in salt with a conductivity of $\varkappa < 40$ μS cm^{-1}, the polarization can increase greatly, so that the necessary protection current density can no longer be reached. Further limits to its application exist at pH values < 6.0 and > 8.5 because there the solubility of Al(OH)$_3$ becomes too high and its film-forming action is lost [19]. The aluminum anodes are designed for a life of 2 to 3 years. After that they must be renewed. The protection currents are indicated by means of an ammeter and/or a current-operated light diode. In addition to the normal monitoring by service personnel, a qualified firm should inspect the rectifier equipment annually.

In water containing sulfate, the use of the electrolysis protection process with low water consumption can sometimes result in the formation of small amounts of H$_2$S, which is detectable by the smell. Sulfate reduction occurs through the action of bacteria in anaerobic areas (e.g., in the slurry zone of the tank).

There are heat-resistant sulfate-reducing bacteria with high activity at 70°C [23]. On the other hand, electrochemical sulfate reduction has not been confirmed [24]. To avoid microbial sulfate reduction and the formation of excessive aluminum compounds, it is advisable to adjust the electrolysis currents to the water consumption. Flow rate or differential pressure switches in the inlet tube allow the current to be controlled according to the water consumption. The electrolysis currents can also be regulated for alternating water consumption by installing a contact counter in the inlet tube. The contact water counter governs the protection current equipment electronically and smoothly controls the current strength. Regularly varying water consumption can be catered for with a time switch. The activity of anaerobic bacteria can be avoided by installing metal oxide-covered titanium anodes in the slurry area, on which anodic oxygen is evolved [25].

20.4.3 Water Storage Tanks

The protection current requirement of a water storage tank depends mainly on the presence and quality of an internal coating. The requirement increases with rising temperature, increasing salt content and lowering pH of the water. It is recommended that an experimental estimate of the protection current requirement be made in such cases. A drainage test may also be necessary with organic coat-ings since the extent of defects or mechanical damage is unknown. Such testing is not necessary with inorganic coatings (enamelling) because of the limited number and size of defects, according to Ref. 4, as well as the higher degree of stability of these coatings.

To achieve satisfactory current distribution and to avoid cathodic damage to the lining (see Section 5.2.1), the distance between the anodes and the object to be protected should not be too close and the driving voltage should not be too high. In

Internal Cathodic Protection of Water Tanks and Boilers 459

the vicinity of the anodes, the potential should be limited to $U_H = -0.8$ to -0.9 V. A large number of anodes have to be installed in uncoated steel tanks to give satisfactory protection current distributiom. The specific conductivity determines the minimum distance of the anodes from the object to be protected [26].

A tank with a fixed cover of plain carbon steel for storing 60°C warm, softened boiler feed water that had a tar-pitch epoxy resin coating showed pits up to 2.5 mm deep after 10 years of service without cathodic protection. Two separate protection systems were built into the tank because the water level varied as a result of service conditions. A ring anode attached to plastic supports was installed near the bottom of the tank and was connected to a potential-controlled protection rectifier. The side walls were protected by three vertical anodes with fixed adjustable protection current equipment.

Partially platinized titanium impressed current anodes were chosen because contamination of the feed water by anodic decomposition products had to be avoided. Four pure zinc reference electrodes were installed in the tank to control and regulate the potential. The supports for the anodes were of polypropylene, which can operate for short periods up to 100°C, in contrast to the usual PVC supports used in cold water.

The measured grounding resistances of the anodes were 6.5 Ω for the 24-m-long ring anode and 8 Ω for the three 7-m-long vertical anodes and were within the range of the calculated design values. By gradual adjustment of U_{off} values, the protection potential of the initial protection current density amounted to about 450 $\mu A\ m^{-2}$.

Potential control with zinc reference electrodes presented a problem because deposits of corrosion products are formed on zinc in hot water. This caused changes in the potential of the electrode which could not be tolerated. Other reference electrodes (e.g., calomel and Ag-AgCl reference electrodes) were not yet available for this application. Since then, Ag-AgCl electrodes have been developed which successfully operate at temperatures up to 100°C. The solution in the previous case was the imposition of a fixed current level after reaching stationary operating conditions [27].

After 10 years of service, a crown-shaped drinking water tank of 1500 m^3 capacity showed pitting 3 mm deep at defects in the chlorinated rubber coating. After a thorough overhaul and a new coating with a two-component zinc dust primer and two top coats of chlorinated rubber, impressed current protection was installed [27]. Taking into account a protection current requirement of 150 mA m^{-2} for uncoated steel at pores covering 1% of the surface, protection current equipment delivering 4 A was installed. Two protection current circuits were provided to cope with the calculated varying protection current requirement resulting from the water level. One serves the bottom anode and is permanently fixed. The other serves the wall electrodes and is potential controlled. Partially platinized titanium wire with copper leads was used as the anode material.

The bottom ring anode was 45 m long. The vertical wall anodes were fixed 1.8 m above the bottom and had lengths of 30 and 57 m for the inner and outer walls respectively. High-grade zinc reference electrodes which have a stable rest potential in drinking water acted as potential control. The supporting bolts for the anodes and reference electrodes were plastic.

After commissioning the protection installation, an on potential of $U_H = -0.82$ V was imposed. In the course of a few years of operation, the protection current rose from 100 to 130 mA. The mean U_{on} values were $U_H = -0.63$ V and the U_{off} values were $U_H = -0.50$ V. An inspection carried out after 5 years showed that cathodic blisters, whose contents were alkaline, had developed in the coating. The steel surface at these places was rust free and not attacked. Uncoated steel at the pores was covered with $CaCO_3$ deposits as a result of cathodic polarization.

Figure 20-13 shows current and potential time curves for a stainless steel 500-liter tank with cathodic protection by impressed current and interrupter potentiostat.

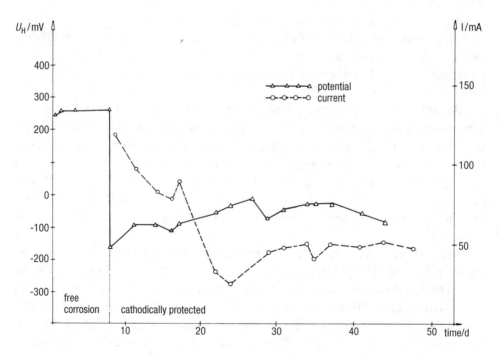

Fig. 20-13 Current and potential-time curves for a 500-liter stainless steel water tank. Impressed current protection with an interrupter potentiostat $\varkappa$ (20°C) = 2250 μS cm^{-1} c (Cl$^-$) = 0.02 mol L^{-1}; 60°C.

Internal Cathodic Protection of Water Tanks and Boilers

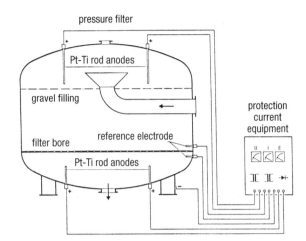

Fig. 20-14 Internal cathodic protection of pressure filter boiler with impressed current.

On switching on the protection current, there was an immediate shift of the potential into the protection region $U < U_{Hs} = 0.0$ V [2] (see group I in Section 2.4). The initial high current requirement of about 120 mA fell in the course of operation to below 50 mA. If protection is effected potentiostatically, there is no danger from hydrogen evolution because the negative potentials necessary for electrolysis will not be reached [28].

20.4.4 Filter Tanks

The arrangement and distribution of anodes in gravel and activated charcoal filters is different. Cathodic protection of activated charcoal filters is basically feasible but requires a large number of electrodes and high protection current densities that are twice those for gravel bed filters, so that an electrically insulating layer can be deposited on the steel wall.

Figure 20-14 shows the protection of a gravel filter in the treatment of raw water. The 200-m² internal surface was coated with 300 μm of tar-pitch epoxy resin. Long-term experiments indicated that at $U_H = -0.83$ V, no cathodic blisters were formed, but were formed at more negative potentials (see Section 5.2.1.4). TiPt anodes 400 and 1100 mm in length, 12 mm in diameter, and with an active surface of 0.11 m² were installed [29].

After several years of operation, protection current densities between 50 and 450 μA m^{-2} were measured on several vessels. The filter tanks and pure water tanks were provided with protection current by separate current circuits. This was necessary because the nozzle bottom out of the filter tank made it impossible to supply the pure water tank with sufficient protection current.

Monitoring the cathodic protection of filter tanks extends to the control of protection current levels with the use of ammeters and/or current-operated light diodes. The internal condition of tanks should be inspected visually every 2 years.

20.5 Requirements for Drinking Water

Only those anodes can be used in the internal protection of drinking water tanks whose reaction products in water are beyond suspicion from the health point of view regarding type and concentration. This excludes galvanic anodes containing toxic elements (e.g., Hg-activated aluminum anodes or PbAg anodes). Only magnesium and impressed current aluminum can be considered for galvanic anodes in drinking water because their reaction products are absolutely harmless to health. Noble metal anodes, titanium anodes coated with noble metals or oxides, or magnetite anodes can also be used without hesitation in drinking water since they do not release reaction products into the water. The reaction products resulting from corrosion protection with aluminum anodes in drinking water tanks used for heating or storage are "additives" and cannot be avoided technically [30]. They can be used if the products are not detrimental to the customer and are unobjectionable concerning health, taste, and smell, which applies to $Al(OH)_3$ [31].

Cathodic protection with impressed current, aluminum or magnesium anodes does not lead to any promotion of germs in the water. There is also no multiplication of bacteria and fungi in the anode slime [32,33]. Unhygienic contamination of the water only arises if anaerobic conditions develop in the slurry deposits, giving rise to bacterial reduction of sulfate. If this is the case, H_2S can be detected by smell in amounts which cannot be detected analytically or by taste. Remedial measures are dealt with in Section 20.4.2.

20.6 References

[1] DIN 50930 Teile 2 bis 5, Beuth-Verlag, Berlin 1980.
[2] DIN 50927, Beuth-Verlag, Berlin 1985.
[3] DIN 50928, Beuth-Verlag, Berlin 1985.
[4] DIN 4753 Teile 3 und 6, Beuth-Verlag, Berlin 1987 u. 1986.
[5] Patentschrift DE 2928998, 1982.
[6] Patentschrift DE 3707791, 1988.
[7] H. Rickert, G. Holzäpfel u. Ch. Fianda, Werkstoffe u. Korrosion *38*, 691 (1987).
[8] DIN VDE 0720 Teil 2 Eb, Beuth-Verlag, Berlin 1978.
[9] G. Hitzblech, Heizungsjournal, 2. Themenheft NT-Heizung, 1981.
[10] G. Franke, Der Elektromeister und Deutsches Elektrohandwerk *13*, 942 (1980).
[11] C. L. Kruse u. G. Hitzblech, IKZ *10*, 42 (1980).
[12] G. Franke, in: "Korrosion in Kalt- u. Warmwasser-Systemen der Hausinstallation", S. 127–144, Deutsche Gesellsch. Metallkd., Oberursel 1984.
[13] G. Franke, SHT *3*, 205 (1987).
[14] G. Herbsleb u. W. Schwenk, Werkstoffe u. Korrosion *18*, 685 (1967); *24*, 763 (1973).
[15] C.-L. Kruse, Werkstoffe u. Korrosion *34*, 539 (1983).
[16] Teil 3 aus [1].
[17] Merkblatt 405 der Stahlberatungsstelle "Das Stahlrohr in der Hausinstallation – Vermeidung von Korrosionsschäden", Düsseldorf 1981.
[18] Deutsches Bundespatent 2144514.
[19] U. Heinzelmann, in: "Korrosion in Kalt- und Warmwassersystemen der Hausinstallation", S. 46–55, Deutsche Ges. Metallkd., Oberursel 1974.
[20] G. Burgmann, W. Friehe u. K. Welbers, Heizung-Lüftung-Haustechn. *23*, Nr. 3, 85 (1972).
[21] Teil 5 aus [1].
[22] DIN 4753 Teil 10 (Entwurf), Beuth-Verlag, Berlin 1988.
[23] Hygiene-Inst. d. Ruhrgebietes, Gelsenkirchen, Tgb. Nr. 5984/72 v. 5. 10. 1972, unveröfftl. Untersuchung.
[24] W. Schwenk, unveröffentl. Untersuchungen 1972–1975.
[25] Deutsches Bundespatent 2445903.
[26] Deutsches Bundespatent 2946901.
[27] A. Baltes, HdT Vortragsveröffentlichungen 402, 15 (1978).
[28] P. Forchhammer u. H.-J. Engell, Werkstoffe u. Korrosion *20*, 1 (1969).
[29] F. Paulekat, HdT Veröffentlichungen 402, 8 (1978).
[30] Lebensmittel- u. Bedarfsgegenständegesetz v. 15. 8. 1974, Bundesgesetzbl. I, S. 1945–1966.
[31] K. Redeker, Bonn, unveröfftl. Rechtsgutachten v. 8. 11. 1978.
[32] R. Schweisfurth u. U. Heinzelmann, Forum Städte-Hygiene *36*, 162 (1988).
[33] R. Schweisfurth, Homburg/Saar, unveröffl. Gutachten 1987.

21

Internal Electrochemical Corrosion Protection of Processing Equipment, Vessels, and Tubes

H. GRÄFEN AND F. PAULEKAT

Measures a and c in Section 2.2 are directly relevant for internal electrochemical protection. In the previous chapter examples of the application of not only cathodic protection but also anodic protection were dealt with; in this connection see the basic explanation in Sections 2.2 and 2.3 and particularly in Section 2.3.1.2.

Besides the use of anodic polarization with impressed current to achieve passivation, raising the cathodic partial current density by special alloying elements and the use of oxidizing inhibitors (and/or passivators) to assist the formation of passive films can be included in the anodic protection method [1-3].

Electrochemical corrosion protection of the internal surfaces of reaction vessels, tanks, pipes and conveyor equipment in the chemical, power and petroleum industries is usually carried out in the presence of strongly corrosive media. The range stretches from drinking water through more or less contaminated river, brackish and seawater frequently used for cooling, to reactive solutions such as caustic soda, acids and salt solutions.

21.1 Special Features of Internal Protection

In comparison with the external cathodic protection of pipelines, tanks, etc., internal protection has limitations [4] which have already been indicated in Section 20.1 but are fully listed here:

(a) The free corrosion potentials and protection ranges must be determined for every protection system and for all the operational phases, because of the multiplicity of materials and media. Section 2.4 gives a summary of important protection potentials or potential ranges.

(b) The ratio of the volume of the electrolyte to the surface area of the object to be protected is considerably lower than with external protection. For this reason the current distribution, especially on uncoated surfaces, is limited.

Corrosion Protection of Processing Equipment, Vessels, and Tubes

Particular difficulties can arise with complicated objects with inserts (stirrers, heaters, etc.) Uniform current distribution thus can only be achieved by an increased number and proper location of several protective electrodes.

(c) The electrochemical reactions and consecutive reactions at the object to be protected and at the impressed current electrodes must be considered in internal protection. The type of electrochemical protection and the material of the impressed current anodes must be chosen with regard to the electrolyte and the requirements for its purity. It must be ascertained which electrolytic reactions are occurring and to what extent these reactions can lead to detrimental effects (e.g., contamination of the medium by reaction products from the electrode material or electrochemical reactions with constituents of the medium).

(d) With internal coatings, either applied or planned, it must be determined whether they have the necessary stability against electrochemical effects of the protection current (see Section 5.2.1) [5].

(e) Hydrogen can evolve from the effects of the protection current or self-corrosion of galvanic anodes, and can form explosive gas mixtures. In this case the protective measures described in Section 20.1.5 must be applied [4]. Catalytically active noble metal anodes and anodes coated with noble metals cannot be used if they penetrate the gas space. The use of catalytically active noble metals should be avoided for safety reasons and the protection potential region limited to $U'_{Hs} = -0.9$ V.

(f) With inserts electrically insulated from the object to be protected and not included in the electrochemical protection, it should be determined whether the protection current could lead to damaging interference and which protective measures are to be adopted [4] (see Section 20.1.4).

(g) It should be determined whether special measures have to be instituted when the operation is nonsteady (e.g., with erratic or interrupted throughput) and when changes in the temperature and concentration of the electrolyte occur.

(h) Test points must be provided for measuring the object and medium potential, depending on the requirements of the protective system. This involves finding the sites where the protective effect is expected to be at its lowest because of the protection current distribution.

Reference electrodes at the test points may only be needed part of the time, depending on the mode of operation of the protective systems (e.g., for monitoring or for permanent control of potential-controlled protection current equipment). Potentiostatic control is always preferred to galvanostatic systems where operational parameters are changing.

Half-cells and simple metal electrodes with nearly constant rest potential serve as reference electrodes (see Tables 3-1 and 3-2 in Ref. 4). They have to be tested to see whether their potential is sufficiently constant and their stability adequate under the operating conditions (temperature and measuring circuit loading).

The following information is required for planning internal protective systems:

- Construction plans with accurate information on the materials to be used and additional welding materials.
- Chemical composition of the electrolyte with information on possible variations.
- Information on the operating conditions, such as height of filling, temperature, pressure, and flow velocity, as well as range of variations of these parameters.

From this it is apparent that the choice of protection method and protection criteria should always be properly tailored. Coatings are usually applied to reduce the protection current requirement as well as to improve the protection current distribution in cathodic internal protection systems, because they raise the polarization parameter (see Sections 2.2.5 and 5.1). If the polarization parameter lies in the same range as the protected object, the current distribution is sufficiently good.

21.2 Cathodic Protection with Galvanic Anodes

Galvanic anodes (see Chapter 6) are used for smaller objects to be protected (e.g., water heaters [see Chapter 20], feedwater tanks, coolers and pipe heat exchangers). Large objects (power station condensers, inlet structures, ducts, sluices, water turbines and large pumps) are mainly protected with impressed current on economic grounds. Galvanic anodes are, however, also installed if account has to be taken of protection against explosion and safety regulations regarding products of electrolysis. Areas of application include tanks for ballast, loading, petrol, water, and exchange tanks of ships (see Section 17.4) as well as crude oil tanks.

Aluminum anodes are used for the internal cathodic protection of large crude oil tanks which are susceptible to damage from corrosive salt-rich deposits. In an earlier example [6] 71 anodes were equally spaced in the base area. The base region up to 1 m in height in the region of the water/oil interface had, including the inserts, an area of 2120 m^2 and was protected with 17 A. The protection current density was 8 mA m^{-2}. With a total anode weight of 1370 kg and with Q'_{pr} = 2600 A h kg^{-1}, the service life was calculated to be 24 years.

The type of anode used consisted of two parts, each 1 m long, which were welded to specially prepared plates. Potential measurements were carried out for controlling the protection. This was achieved with floating reference electrodes in the water sump.

Corrosion Protection of Processing Equipment, Vessels, and Tubes 467

Today for this kind of object, the aluminum anodes are usually insulated and connected via cables outside the tank. By this means it is possible to purify and activate the anodes by applying anodic current pulses from an external voltage source. This is necessary during the course of operation since the anode surfaces can be easily passivated by oil films [7].

21.3 Cathodic Protection with Impressed Current

Cathodic protection by means of impressed current is very adaptable and economic because of the long durability of anodes and the large number of anode materials and shapes. Some examples are described here. Internal cathodic protection of fuel oil tanks has already been dealt with in Section 11.7. The internal protection of water tanks is described in detail in Chapter 20.

21.3.1 Internal Cathodic Protection of Wet Oil Tanks

The output from oil fields, after secondary and tertiary processing [7], is considered "wet oil." It consists of oil field brine, crude oil and natural gases containing a high CO_2 content and in certain circumstances it may also contain small amounts of H_2S. Wet oil can also contain solid particles from the bedrock.

Wet oil is collected and treated in large steel tanks (knockout tanks) between 10^2 and 10^4 m^3 in size, depending on the application. The aqueous phase is drained off in the process. The corrosivity of the aqueous phase is determined by the high salt content (1 mol L^{-1}), high CO_2 content (corresponding to a partial pressure of 1 bar), and after buffering with $CaCO_3$, a pH of 6; as well as small amounts of H_2S and traces of O_2 [7]. Investigations have shown that a protection potential of $U_{Cu\text{-}CuSO_4} = -0.95$ V must be applied under these conditions and for operating temperatures up to 70°C. In unbuffered media, the pH is 4; in this case the protection potential must be reduced by an additional 0.1 V. The steep total current density-potential curve means that one has to contend with a correspondingly high protection current requirement. The lowest current densities are up to 10 A m^{-2}. They can, however, be much smaller if scales are formed [7].

Corrosion protection by coatings alone demands freedom from pores and defects during the entire service time. Since this is difficult to guarantee and there is the danger of cell formation (see Section 5.1), a combination with cathodic protection is advisable. The coatings involved must be compatible with the cathodic protection and the operating conditions (see Section 5.2.1.4). Reference 5 gives the basis for such decisions. By neglecting the danger of cathodic damage to coatings, considerable damage occurred to the coatings in the past, which caused a marked increase in protection current densities. Internal cathodic protection without causing damage to coatings is only possible by applying resin coatings having a sufficiently high electrical resistance and a thickness of >800 μm. The stability against cathodic blistering should be determined by long-term and accelerated testing and verified under operating conditions.

The impressed current method with metal oxide-coated niobium anodes is usually employed for internal protection (see Section 7.2.3). In smaller tanks, galvanic anodes of zinc can also be used. Potential control should be provided to avoid unacceptably negative potentials. Pure zinc electrodes serve as monitoring and control electrodes in exposed areas which have to be anodically cleaned in the course of operation. Ag-AgCl electrodes are used to check these reference electrodes.

Good current distribution can be expected because of the good conductivity of the electrolyte. However, if a large area of the coating is damaged, local underprotection cannot be ruled out due to the low polarization resistance. For this reason, control with several reference electrodes is advisable.

21.3.2 Internal Cathodic Protection of a Wet Gasometer

Figure 21-1 shows the object to be protected and the arrangement of impressed current anodes and reference electrodes. A central anode and two ring anodes of platinized titanium wire 3 mm in diameter provided with additional copper wire conductors are installed here. It is worth noting that the central anode is suspended from a float, whereas the ring anode is mounted on plastic supports. The zinc reference electrodes are also on floats near the inner side of the bell, while the 17 reference electrodes are mounted on plastic rods on the bottom of the cup and in the ring

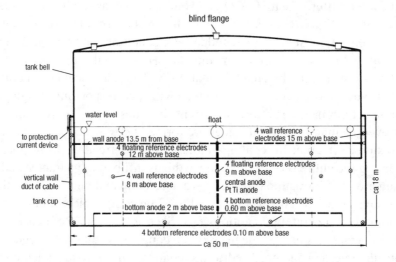

Fig. 21-1 Internal cathodic protection of a wet gasometer.

Corrosion Protection of Processing Equipment, Vessels, and Tubes 469

space between the cup and bell. The three anodes are supplied by separate protection current devices. The current output of both ring anodes is regulated potentiostatically so that the protection current can adapt to the changing surface area [8].

21.3.3 Internal Cathodic Protection of a Power Plant Condenser Cooled by Seawater

Internal cathodic protection of power plant condensers cooled by seawater with impressed current has been the state of the art for some years. An example of its application to the internal protection of large condensers with tubes of copper and aluminum alloys is described in the following paragraphs. It concerns an undivided one-way tube-bundle heat exchanger with box headers. Two such main condensers are necessary for every power station. The coated steel box headers have surface areas 65 m^2 to be protected. The pipes and inlet and outlet areas of the tubes account for 20 m^2 of surface. The cooling water flows through the condensers at about 2 m s^{-1} at water temperatures of 30 to 40°C. Cathodic protection was not only necessary for the internal protection of the coated steel box headers, but also to fight against bimetallic corrosion of the steel and pipework materials.

In addition, with high solid content of the cooling water and at high flow velocities, severe corrosive conditions exist which continuously destroy surface films. Cathodic protection alone is not sufficient. Additional measures must be undertaken to promote the formation of a surface film. This is possible with iron anodes because the anodically produced hydrated iron oxide promotes surface film formation on copper.

Six iron anodes are required for corrosion protection of each condenser, each weighing 13 kg. Every outflow chamber contains 14 titanium rod anodes, with a platinum coating 5 μm thick and weighing 0.73 g. The mass loss rate for the anodes is 10 kg A^{-1} a^{-1} for Fe (see Table 7-1) and 10 mg A^{-1} a^{-1} for Pt (see Table 7-3). A protection current density of 0.1 A m^{-2} is assumed for the coated condenser surfaces and 1 A m^{-2} for the copper alloy tubes. This corresponds to a protection current of 27 A. An automatic potential-control transformer-rectifier with a capacity of 125 A/10 V is installed for each main condenser. Potential control and monitoring are provided by fixed zinc reference electrodes. Figure 21-2 shows the anode arrangement in the inlet chamber [9].

21.3.4 Internal Cathodic Protection of a Water Turbine

In Europe, the first internal cathodic protection installation was put into operation in 1965 for 24 water-powered Kaplan turbines with a propeller diameter of 7.6 m. These were in the tidal power station at La Rance in France. The protected object consisted of plain carbon and high-alloy stainless steels. Each turbine was

Fig. 21-2 Arrangement of anodes in the entry chamber of a power station condenser.

protected in the stator and impeller rotor region as well as the suction hose with 36 platinized niobium anodes coated with 50 μm of Pt. Each of the three turbines was connected to an unregulated transformer-rectifier with an output of 120 A/24 V, so that the required potential equalization was achieved at the individual anodes with the use of variable resistors. One hundred fixed Ag-AgCl reference electrodes were installed for control [10,11].

In 1987 at the Werra River, four Kaplan turbines of 2.65 m diameter in two power stations were cathodically protected. The turbines were of mixed construction with high-alloy CrNi steels and nonalloyed ferrous materials with tar-EP coating. Considerable corrosion damage occurred prior to the introduction of cathodic protection, which was attributed to bimetallic corrosion and the river's high salt content of $c(Cl^-) = 0.4$ to 20 g L^{-1}.

Plate anodes were used for corrosion protection in order to avoid damage due to erosion and cavitation. These consisted of enamelled steel bodies in which a metal oxide-coated titanium anode 1 dm^2 in surface area was fitted. The enamel

Corrosion Protection of Processing Equipment, Vessels, and Tubes

coating served as further insulation between anode and protected object. The current loading of these anodes amounted to 50 A for a useful life of 6 years in the Cl$^-$-rich water. Fixed Ag-AgCl reference electrodes were used for potential regulation and monitoring. Potential control was needed because the current requirement fluctuated considerably, depending on the operation of the turbine and variations in the conductivity of the water.

Figure 21-3 shows a schematic diagram of the turbine in which the black areas are CrNi steel. The plate anodes are situated in the region of the suction manifold (A1 to A3) in a triangular arrangement in the impeller ring (A4 to A6) and distributed in the vicinity of the segmented water inlet ring (A7 to A10) (see Fig. 21-4). These four anodes were individually connected to the transformer-rectifier with chloride-resistant pressure-resistant cable.

Four magnetite rod anodes (A11 to A14), each weighing 9 kg, in perforated polypropylene protection tubes were used to protect the box headers, the in-flow hood and inlet housing as well as the CrNi steel oil cooler. Protection tubes keep out solid matter carried in the river water. The oil cooler was connected to the protection equipment with its own cathode conductor.

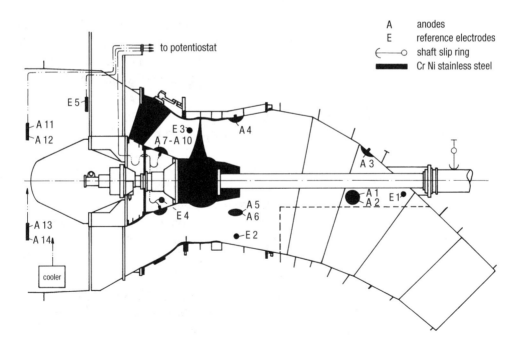

Fig. 21-3 Diagram of the protected Kaplan turbine with position of anodes (A1 to A14) and reference electrodes (E1 to E5).

Fig. 21-4 Mounted plate anode (see text) on the CrNi-stainless steel water inlet ring between stator and rotor.

A reference electrode is fixed in each protection region in the most unfavorable place for current distribution and serves to regulate the potential. The Ag-AgCl reference electrodes in the turbine sections (E1 to E4) have a threaded connection that will withstand up to 60 bars of pressure. The electrode E5 in the box header is pure zinc.

The protection current requirement is determined mainly by the uncoated surfaces of the stainless steel whose protection potential is a few tenths of a volt more positive than that of the plain carbon steel, to avoid pitting (see Section 2.4). The protection current requirement for the turbine section is about 10 A so that the plate anodes are only loaded to about 1 A.

The protection current equipment slowly regulates the potential in a three-phase ac bridge circuit. A motor-driven regulating column-type transformer is used as a servocomponent. A millivolt-limiting value controller with max/min limits controls the transformer. The 10 and the 4 anode outlets in each protected area are provided with 3 A safety cutouts and with a fuse monitoring system. The signals from this monitor for thermal overloading and potential control warn of a disturbance at each rectifier. A circular chart recorder serves for continuous monitoring of the potential. An additional monitoring system is built in that provides an early warning of a drop in the water level in the turbine for both running and idling conditions. Table 21-1 shows the measured potentials before and after switching in the protective installation. According to these figures, the polarization is sufficient for cathodic protection.

Corrosion Protection of Processing Equipment, Vessels, and Tubes

Table 21-1 Structure/electrolyte potentials in a Kaplan turbine as in Fig. 21-3 before and after switching on the cathodic protection system.

Location of reference electrode	E1	E2	E3	E4	E5
Material of object to be protected (A or B)[a]	A	B	B	B	A
Free corrosion potential $U_{\text{Cu-CuSO}_4}$ in V	−0.33	−0.30	−0.50	−0.49	−0.31
On potential $U_{\text{Cu-CuSO}_4}$ in V	−0.95	−0.50	−0.70	−0.68	−0.95
Protection potential U_s from Section 2.4, $U_{\text{Cu-CuSO}_4}$ in V	−0.85	−0.30	−0.30	−0.30	−0.85

[a] Material A: tar-EP coated plain carbon steel; material B: CrNi stainless steel.

It was established by feeding measurements that there was a potential difference of 0.45 V between the turbine shaft and the housing. A shaft slip-ring served for the necessary potential equalization (see Fig. 21-5). After this system was installed, the potential difference was only <5 mV. According to the operating conditions of the turbine, a current up to 1.5 A flows through this system.

Fig. 21-5 Shaft slip-ring system with silver-graphite brushes for potential equalization.

Cathodic protection of water power turbines is characterized by wide variations in protection current requirements. This is due to the operating conditions (flow velocity, water level) and in the case of the Werra River, the salt content. For this reason potential-controlled rectifiers must be used. This is also necessary to avoid overprotection and thereby damage to the coating (see Sections 5.2.1.4 and 5.2.1.5 as well as Refs. 4 and 5). Safety measures must be addressed for the reasons stated in Section 20.1.5. Notices were fixed to the turbine and the external access to the box headers which warned of the danger of explosion from hydrogen and included the regulations for the avoidance of accidents (see Ref. 4).

21.4 Anodic Protection of Chemical Plant

21.4.1 Special Features of Anodic Protection

The fundamentals of this method of protection are dealt with in Section 2.3 and illustrated in Fig. 2-15. Corrosion protection for the stable-passive state is unnecessary because the material is sufficiently corrosion resistant for free corrosion conditions. If activation occurs due to a temporary disturbance, the material immediately returns to the stable passive state. This does not apply to the metastable passive state. In this case anodic protection is necessary to impose the return to the passive state. Anodic protection is also effective in the unstable passive state of the material but it must be permanently switched on, in contrast to the metastable passive state.

The protection potential ranges are commonly limited with anodic protection: group IV in Sections 2.3 and 2.4. For this reason potential-regulating protection current devices must be used. The protection potential range can be severely limited by certain corrosion processes (e.g., chloride-induced pitting corrosion in stainless steels). Often anodic corrosion protection is no longer applicable under these circumstances. In such cases, partial cathodic protection (to prevent pitting corrosion) was investigated by lowering the potential slightly below that of the pitting potential. Local corrosion susceptibility caused by the material's condition can also render anodic protection ineffective. This includes, for example, the susceptibility to intergranular corrosion of high chromium-bearing stainless steels and nickel-based alloys.

Three types of anodic protection can be distinguished: (1) impressed current, (2) formation of local cathodes on the material surface and (3) application of passivating inhibitors. For impressed current methods, the protection potential ranges must be determined by experiment (see information in Section 2.3). Anodic protection with impressed current has many applications. It fails if there is restricted current access (e.g., in wet gas spaces) with a lack of electrolyte and/or in the

Corrosion Protection of Processing Equipment, Vessels, and Tubes

transition region between the electrolyte and the gas phase. It is recommended that the filling height of the protected tank be changed to achieve electrochemical protection in the otherwise scarcely wetted but endangered transition surface area. Problems of a lack of current distribution, as in cathodic protection, do not exist in anodic protection because of the high polarization resistances [see Eq. (2-45)].

The formation of local cathodes is used primarily on materials with high hydrogen overvoltage to avoid acid corrosion. Figure 21-6 shows the anodic potential current-potential curve a, for a passivatable metal in an electrolyte with passivating current I_p and the corresponding cathodic partial current potential curve b for hydrogen evolution. The passivating current is not reached because of the high hydrogen overvoltage. An active state exists with the rest potential U_{Ra}. If the material is brought into contact with a metal of lower hydrogen overvoltage corresponding to the cathodic partial current potential curve c, the cathodic partial current is sufficient for passivation. A rest potential U_{Rp} in the passive state is now attained in free corrosion. Local cathodes which are introduced into the material by alloying behave in the same way. By the same electrochemical principles, the overvoltage can also be reduced for the reduction of other oxidizing agents (e.g., oxygen) so that unstable-passive materials become stable-passive materials as in Fig. 2-15.

Passivating inhibitors act in two ways. First they can reduce the passivating current density by encouraging passive film formation, and second they raise the cathodic partial current density by their reduction. Inhibitors can have either both or only one of these properties. Passivating inhibitors belong to the group of so-called dangerous inhibitors because with incomplete inhibition, severe local active corrosion occurs. In this case, passivated cathodic surfaces are close to noninhibited anodic surfaces.

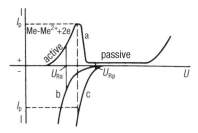

Fig. 21-6 The dependence of the passivation process on the shape of the cathodic partial current potential curve: (a) Anodic partial current potential curve, (b) cathodic partial current-potential curve without local cathode rest potential U_{Ra}, (c) cathodic partial current potential curve with local cathode rest potential U_{Rp}.

21.4.2 Anodic Protection with Impressed Current

21.4.2.1 Preparatory Investigations

Practical application demands laboratory investigations of the protection potential range, the passivating current density and the current requirement in the passive range. In this the parameters of practical interest such as temperature, flow velocity and concentration of corrosive agents in the medium must be considered [12-14]. Chronopotentiostatic investigations according to DIN 50918 are generally applicable.

In addition, the reactions occurring at the impressed current cathode should be heeded. As an example, Fig. 21-7 shows the electrochemical behavior of a stainless steel in flowing 98% H_2SO_4 at various temperatures. The passivating current density and the protection current requirement increase with increased temperature, while the passive range narrows. Preliminary assessments for a potential-controlled installation can be deduced from such curves.

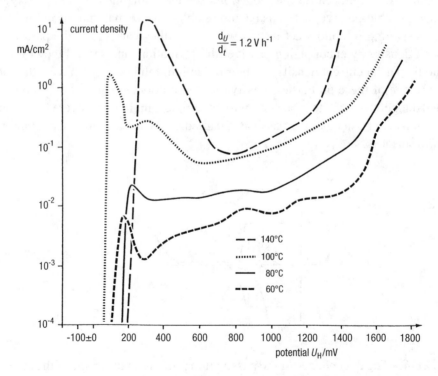

Fig. 21-7 Effect of temperature on the passivation of a stainless steel [material No. 1.4571 (AISI 316 Tr)] in flowing 98% H_2SO_4 (0.8 m s^{-1}), potentiodynamic measurements at 1.2 V/hr.

Corrosion Protection of Processing Equipment, Vessels, and Tubes

The determination and evaluation of potentiodynamic curves can only be used as a preliminary assessment of corrosion behavior. The protection current requirement and the limiting value for the potential control can only be determined from so-called chronopotentiostatic experiments as in DIN 50918. In systems that react with spontaneous activation after the protection current is switched off or there is a change in the operating conditions, quick-acting protection current devices must be used. Figure 8-6 shows the circuit diagram for such a potentiostat.

In most cases the time dependence of passivating and activating processes permits the use of slow-acting protection current devices (see Fig. 8-7) if activation occurs only after a lengthy period when a drop in protection current occurs. Usually a recording instrument is used as a two-point controller for monitoring the set potential-limiting value. The dc input signal from the reference electrode is continuously recorded. The set-limiting value contacts activate upon under- or overshooting of the limiting potential values imposed for the control process of the protection current device, giving optical and acoustic warning signals for serious deviations.

Half-cells based on Ag-AgCl, Hg-HgO, Hg-HgSO$_4$ and other systems can be used as reference electrodes (see Table 3-1) [4]. Electrodes have been developed to operate up to 100 bar and 250°C (see Fig. 21-8).

Materials that are corrosion resistant to the expected cathodic polarization qualify as impressed current cathodes. Austenitic CrNi steels are used with strong acids. The oleum (i.e., fuming sulfuric acid) and concentrated sulfuric acid tanks used in sulfonating alkanes and in the neutralization of sulfonic acids are anodically protected using platinized brass as cathodes [15]. Lead cathodes are used to protect titanium heat exchangers in rayon spinning baths [16].

A particular problem arises in the anodic protection of the gas space because the anodic protection does not act here and there is a danger of active corrosion. Thus these endangered areas have to be taken account of in the design of chemical

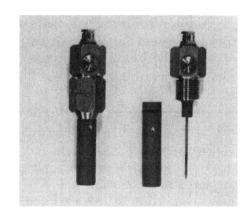

Fig. 21-8 Reference electrodes (Pt to redox system) for high pressures and temperatures (max. 250°C and 100 bar).

installations. If gas spaces cannot be avoided in the construction, they must be clad with corrosion-resistant materials. Tanks and storage containers should be kept completely full wherever possible and the time between emptying and filling should be kept as short as possible so that activation does not occur spontaneously.

21.4.2.2 Protection Against Acids

Anodic protection against acids has been used in a number of processes in the chemical industry, as well as during storage and transport. It is also successful in geometrically complicated containers and tubings [12]. Carbon steel can be protected from nitric and sulfuric acids. In the latter case, temperature and concentration set application limits [17]. At temperatures of up to 120°C, efficient protection can only be achieved with concentrations over 90% [18]. At concentrations between 67 and 90%, anodic protection can be used at up to 140°C with CrNi steels [19].

In sulfuric acid production involving heat recovery and recovery of waste sulfuric acid, acids of various concentrations at high temperatures can be dealt with. Corrosion damage has been observed, for example, in sulfuric acid coolers, which seriously impairs the availability of such installations. The use of anodic protection can prevent such damage.

It is known that the common austenitic stainless steels have sufficient corrosion resistance in sulfuric acid of lower concentrations (<20%) and higher concentrations (>70%) below a critical temperature. If with higher concentrations of sulfuric acid (>90%) a temperature of 70°C is exceeded, depending on their composition, austenitic stainless steels can exhibit more or less pronounced corrosion phenomena in which the steels can fluctuate between the active and passive state [19].

Anodic protection allows the use of materials under unfavorable conditions if they are also passivatable in sulfuric acid. CrNi steels [material Nos. 1.4541 (AISI 321) and 1.4571 (AISI 316 Ti)] can be used in handling sulfuric acid of 93 to 99% at temperatures up to 160°C. This enables a temperature of 120 to 160°C to be reached, which is very suitable for heat recovery.

Anodic protection today allows safe and efficient protection of air coolers and banks of tubes in sulfuric acid plants. In 1966 the air cooler in a sulfuric acid plant in Germany was anodically protected. Since then more than 10,000 m^2 of cooling surfaces in air- and water-cooled sulfuric acid plants worldwide have been protected. The dc output supply of the potentiostats amounts to >25 kW, corresponding to an energy requirement of 2.5 W per m^2 of protected surface. As an example, Fig. 21-9 shows two parallel-connected sulfuric acid smooth tube exchangers in a production plant in Spain.

Two heat exchangers, each 1.2 m in diameter and 9 m in length, with 1500 heat exchanger tubes of stainless CrNiMo steel [material No. 1.4571 (AISI 316 Ti)] were subjected to 98 to 99% sulfuric acid. The flow rate around the tubes was

Corrosion Protection of Processing Equipment, Vessels, and Tubes

Fig. 21-9 Bank of tube heat exchangers for sulfuric acid with plug-in cathodes.

1 m s^{-1} as cooling of the acid with water took place. The input temperature of the acid was 140°C and the output temperature about 90°C. The cooling water outlet temperature of 90°C was suitable for heat recovery. The protection current for the acid-contacting surfaces of the tubes and the cylindrical surfaces of the smooth tube exchanger of 1775 m^2 area was provided by a system of 300 A/6 V capacity. Electrical power of 1 W m^{-2} was necessary, corresponding to an average protection current density of 0.16 A m^{-2}. The interchangeable insertion cathodes were made of the same material and installed and insulated at four positions on removed cooling tubes. The centrally or concentrically arranged cathodes inside the bank of tubes have the advantage of homogeneous current distribution with little voltage drop. Hg-Hg$_2$SO$_4$ reference electrodes were used as screwed-in electrodes in the outer shell of the smooth tube exchanger in both the hot inlet and cold outlet tubes for potentiostatic control and monitoring of the installation. The protective equipment was monitored and operated automatically from a control room. Test cou-

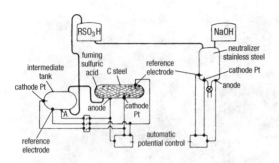

Fig. 21-10 Diagram for the internal anodic protection of a sulfonation plant.

pons had been built in for monitoring and the corrosion rate of these coupons was found to be less than 0.1 mm a^{-1}.

Economic heat recovery could be achieved in certain cases by the further development of anodic protection at higher temperatures. High temperatures allow the production of steam and lead to a considerable increase in efficiency [20].

Anodic protection is particularly suitable for stainless steels in acids. Protection potential ranges are given in Section 2.4. Besides sulfuric acid, other media such as phosphoric acid can be considered [13, 21-24]. These materials are usually stable-passive in nitric acid. On the other hand, they are not passivatable in hydrochloric acid. Titanium is also a suitable material for anodic protection due to its good passivatability.

21.4.2.3 Protection Against Media of Different Composition

Figure 21-10 shows the circuit diagram for a sulfonation plant [25]. In the oleum tank and the intermediate tank, carbon steel is anodically protected. In the neutralizer, CrNi steel is exposed to NaOH or RSO$_3$H (sulfonic acid), depending on the operational phase. Here the anodic protection must be designed to give passivity in both media. The protection potential ranges overlap in a narrow range 0.25 V wide. The limiting potentials were fixed at U_H = 0.34 and 0.38 V. The polarization parameters according to Eq. (2-45) are extremely high due to the high polarization resistances in the passive range and the good conductivity, so that the protection range is sufficiently high for the protection of the tubings.

Carbon steels can be anodically protected in certain salt solutions. This involves mainly products of the fertilizer industry such as NH$_3$, NH$_4$NO$_3$ and urea. Anodic protection is effective up to 90°C [26]. Corrosion in the gas space is suppressed by control of pH and maintenance of a surplus of NH$_3$.

21.4.2.4 Protection Against Alkaline Solutions

According to the current density-potential curves in Figs. 2-18 and 21-11, carbon steels can be passivated in caustic soda [27-32]. In the active range of the

Corrosion Protection of Processing Equipment, Vessels, and Tubes 481

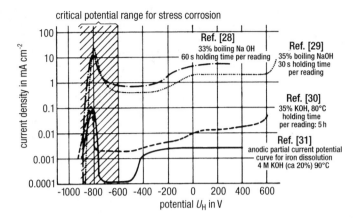

Fig. 21-11 Current density-potential curves for plain carbon steel in hot caustic soda from Refs. 28-31.

curves, corrosion occurs with the formation of FeO_2^{2-} ions (see field IV in Fig. 2-2). The steel can be stable or metastable passive, depending on the temperature, the NaOH concentration, and the presence of oxidizing agents (e.g., O_2). A well-known example is the protection of some cellulose digesters in Canada, which are operated with alkali digesting solutions [32]. The passivatability must be checked with NaOH concentrations of over 50% at high temperatures.

In spite of the possibility of cathodic corrosion discussed in relation to Eq. (2-56), practical experience has shown that carbon steel is a suitable material for impressed current cathodes. Stress corrosion of the cathode material does not have to be considered because of the strong cathodic polarization as shown in Fig. 2-18.

Stress corrosion can arise in plain carbon and low-alloy steels if critical conditions of temperature, concentration and potential in hot alkali solutions are present (see Section 2.3.3). The critical potential range for stress corrosion is shown in Fig. 2-18. This potential range corresponds to the active/passive transition. Theoretically, anodic protection as well as cathodic protection would be possible (see Section 2.4); however, in the active condition, noticeable negligible dissolution of the steel occurs due to the formation of FeO_2^{2-} ions. Therefore, the anodic protection method was chosen for protecting a water electrolysis plant operating with caustic potash solution against stress corrosion [30]. The protection current was provided by the electrolytic cells of the plant.

Six caustic soda evaporators were anodically protected against stress corrosion in the aluminum industry in Germany in 1965 [27]. Each evaporator had an internal surface area of 2400 m². The transformer-rectifier had a capacity of 300 A/ 5 V and was operated intermittently for many years. Automatic switching on of the protection current only took place in case of need when the drop in potential reached

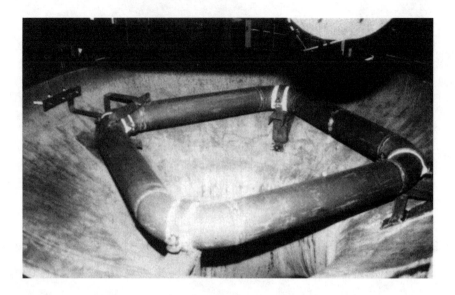

Fig. 21-12 Steel ring-shaped cathode in the heating chamber of a caustic evaporator.

the critical potential range. This method of operation was economic from a point of view of the apparatus and the expense of installation. The plant was rebuilt because of the high number of new passivating processes involved, which always result in increased mass loss and which had endangered operational safety over the long run. A start-up rectifier (500 A/12 V) is now provided for the initial passivation of the respective evaporators so that the required passivation can be achieved in a shorter time. Afterward the evaporators are either manually or automatically switched over by a bus-bar distribution system to the normal maintenance rectifier (150 A/10 V), which supplies the permanent current required to maintain the passivity. The protection current equipment is housed in large corrosion-resistant plastic housing equipped with air conditioning.

The external heating boxes for the caustic soda evaporator with forced circulation must be anodically protected separately. Ring-shaped impressed current electrodes of carbon steel are mounted and insulated on supporting brackets (see Fig. 21-12).

Cathodes made of steel tube rings 10 m in diameter are used in the internal protection of several 30-m-high caustic soda stirring tanks, each with a surface area of 3100 m^3. The cathodes were insulated and mounted on supporting brackets. The current supply was via armored parallel connected cables. One-inch bolts were used for the entrance in the tank wall as shown in Fig. 21-13 for a feed current of 500 A. The size of the cathodes has to be such that they can maintain sufficient cathodic protection current. The surface ratio of the object to be protected and the

Corrosion Protection of Processing Equipment, Vessels, and Tubes 483

Fig. 21-13 Current duct bolts (1 in.) for 500-A protection current.

cathode must be >100 : 1. This can be achieved with several cathodes with a small surface area or one long central cathode of small diameter.

21.4.2.5 Combined Protection by Impressed Current and Inhibitors

In the same way that corrosion stimulators and tensile stresses in the case of stress corrosion can narrow the protection range or even render electrochemical protection impossible (see Sections 2.3 and 2.4), inhibitors can widen the protective region and even create regions. Characteristic examples occur with stainless steels in electrolyte, producing pitting with chloride ions and inhibiting sulfate or nitrate ions. The critical potentials are substantially displaced or, in the case of nitrate ions, newly created (see Fig. 2-16). In this case, the critical pitting potential range is limited by a second pitting potential at more positive potentials. This critical upper limiting potential is also called an inhibition potential and can be used with anodic protection [33]. Perchlorate ions can also inhibit pitting corrosion [34].

21.4.3 Protective Effect of Local Cathodes due to Alloying

The protective action of alloying has been demonstrated by a series of investigations on stainless steels [35-37], titanium [38-40], lead [41,42], and tantalum [43].

Figure 21-6 shows the possibility of reducing the overvoltage of cathodic hydrogen evolution. One can also reduce restrictions in the O_2 reduction by using copper in lead alloys. Such alloying elements can be very effective because they result in more positive potentials. For example, Pt acts in this manner, Pd is somewhat less effective, and Au, on the other hand, is practically ineffective [36].

In the construction of plants, titanium with 0.2% Pd is mainly used. It can be employed with advantage in nonoxidizing acid media and also has increased resistance to pitting and crevice corrosion because of its more favorable pitting potential [40].

The addition of cathodically active elements to pure lead was the main objective of investigations to improve its corrosion resistance to H_2SO_4 [42,44]. Best known is copper-lead with 0.04 to 0.08% Cu. By adding combinations of alloying elements, it was possible to produce lead alloys that not only had much better corrosion resistance, but also had greater high-temperature strength. Lead alloy with 0.1% Sn, 0.1% Cu and 0.1% Pd is an example [45].

An interesting field of application is the protection of tantalum against hydrogen embrittlement by electrical connection to platinum metals. The reduction in hydrogen overvoltage and the shift of the free corrosion potential to more positive values apparently leads to a reduced coverage by adsorbed hydrogen and thereby lower absorption [43] (see Sections 2.1 and 2.3.4).

21.4.4 Protective Action of Inhibitors

Inhibitors are materials that reduce either one or both of the partial corrosion reactions as in Fig. 2-5. Anodic or cathodic inhibitors inhibit the anodic or cathodic reaction respectively so that the rest potential becomes either more positive or more negative. Most inhibitors, however, inhibit the anodic partial reaction. This is because the transfer of metal ions can be more easily restricted than that of electrons.

Anodically acting inhibitors facilitate passivation in passivatable systems and also reduce the passivating current density by covering the surface well. With non-passivating systems, corrosion protection only occurs with complete coverage of the surface. If the coverage is incomplete or damaged, there is a danger of local attack.

The inhibitive effect can be detected by a high increase in polarization resistance. It can be determined from current density potential curves which are used successfully to check it [46,47]. Raising the cathodic partial current density is also favorable for passivation as well as the reduction of the passivating current density. For this reason oxidizing agents are also termed passivators with the anodically acting inhibitors. For example, nickel passivates in 0.5 M H_2SO_4 with the addition of $Fe_2(SO_4)_3$, H_2O_2, $KMnO_4$, $Ce(SO_4)_2$ and $K_2Cr_2O_7$ [48].

Alkalizing and film-forming inhibitors are used to prevent corrosion in water containing O_2 or salt. These include Na_2CO_3, NaOH, Na_3PO_4, Na_2SiO_3, $NaNO_2$ and Na_2CrO_4. The number of organic compounds that effectively inhibit metal

Corrosion Protection of Processing Equipment, Vessels, and Tubes 485

corrosion in various media is small. Typical materials that act as inhibitors are organic sulfur and nitrogen compounds, higher alcohols and fatty acids. These materials are described in the comprehensive references [49-51] on their type and effect.

21.5 Trends in the Application of Internal Electrochemical Protection

The use of electrochemical protection in the chemical industry started about 20 years ago, which is somewhat recent, compared with its use for buried pipelines 40 years ago. Adoption was slow because the internal protection has to be tailored to the individual plant, which is not the case with the external protection of buried objects. Interest in internal protection came from the increasing need for greater safety for operating plants, increased demands for corrosion resistance, and larger plant components. While questions of its economy cannot generally be answered (see Section 22.6), the costs of electrochemical protection are generally less than the cost of equivalent and reliable coatings or corrosion-resistant materials.

The safety, availability and capacity of production plants are predetermined by the quality of the materials and the corrosion protection measures in the essential areas; both are major considerations in the initial planning. Even today, damage to equipment and tanks is often assumed to be unavoidable and the damaged components are routinely replaced. By carrying out damage analysis, which points the way to knowledge of prevention of damage, the availability and life of plants can be increased considerably. This particularly applies to the use of anodic protection.

It was demonstrated that the anodic protection of chemical plants constructed of plain carbon or low-alloy steels in contact with caustic solutions is technically possible and can be used without disrupting the operation of the plant. Anodic protection reduces mass loss from corrosion and prevents stress corrosion. The protective system can be interrupted without spontaneous activation of the protected surfaces. The range of the protection is increased through the homogeneously formed passive film and the consequent increase in the polarization resistance. It should be specially mentioned that the formation of sodium aluminum silicate in aluminate solutions from kaolinite and the difficult removal of these encrustations, particularly on heating elements, are reduced as a result of the smooth anodically protected surface. In this respect it should be mentioned that safe operation was made possible by anodic protection, because annealing adequate to relieve stress in apparatuses susceptible to stress corrosion, particularly large tanks, is practically impossible and constructional or operational stresses are unavoidable. The desired and, from a cost point of view, expected operational safety achieved by using very high-cost corrosion-resistant materials cannot be reliably achieved in many cases. This is particularly so in chemical installations.

In recent years, a number of protective installations have come into operation, especially where new installations must be maintained, or where older and already damaged installations have to be saved and operating costs have to be lowered. Worldwide, equipment, tanks and evaporators in the aluminum industry and industries using caustic alkalis with a capacity of 60,000 m^3 and a surface area of 47,000 m^2 are being anodically protected. Equipment for electrochemical protection has been installed with a total rating of 125 kW and 12 kA.

In spite of considerable advances in recent years, only 20% of the endangered plants and tanks in German industry urgently requiring protection are anodically protected against pitting and stress corrosion; worldwide, aluminum oxide producers have protected only 1.5% of their plants against corrosion. Over the past 20 years, experience has shown that 40 to 50% of surfaces wetted with caustic solutions exhibit corrosion attack and that in almost all installations the electrochemical and mechanical conditions are those causing the most severe damage. Of the worldwide annual production of 23,500 t of alumina, only about 20% is produced in anodically protected installations. It is therefore to be expected that internal electrochemical protection is going to show a continuing increase in use. DIN 50927 [4], published in 1985, should be an important factor in this development.

21.6 References

[1] H. Gräfen, Z. Werkstofftechnik 2, 406 (1971).
[2] H. Gräfen, G. Herbsleb, P. Paulekat u. W. Schwenk, Werkst. Korros. 22, 16 (1971).
[3] M. J. Pryor u. M. Cohen, J. Electrochem. Soc. 100, 203 (1953).
[4] DIN 50927, Beuth-Verlag, Berlin 1985.
[5] DIN 50928, Beuth-Verlag, Berlin 1985.
[6] F. Paulekat, HdT Vortragsveröffentlichung 402, 8 (1978).
[7] B. Leutner, BEB Erdgas u. Erdöl GmbH, persönliche Mitteilung 1987.
[8] W. v. Baeckmann, A. Baltes u. G. Löken, Blech Rohr Profile 22, 409 (1975).
[9] F. Paulekat, Werkst. Korros. 38, 439 (1987).
[10] M. Faral, La Houille Blanche, Nr. 2/3, S. 247–250 (1973).
[11] J. Weber, 23rd Ann. Conf. Metallurg., Quebec 19.–22. 8. 1984, Advance in Met. Techn. S. 1–30.
[12] J. D. Sudbury, O. L. Riggs jr. u. D. A. Shock, Corrosion 16, 47t (1960).
[13] D. A. Shock, O. L. Riggs u. J. D. Sudbury, Corrosion 16, 55t (1960).
[14] W. A. Mueller, Corrosion 18, 359t (1962).
[15] C. E. Locke, M. Hutchinson u. N. L. Conger, Chem. Engng. Progr. 56, 50 (1960).
[16] B. H. Hanson, Titanium Progress Nr. 8, Hrsg. Imperial Metal Industries (Kynoch) Ltd., Birmingham 1969.
[17] W. P. Banks u. J. D. Sudbury, Corrosion 19, 300t (1963).
[18] J. E. Stammen, Mat. Protection 7, H. 12, 33 (1968).
[19] D. Kuron, F. Paulekat, H. Gräfen, E.-M. Horn, Werkst. Korros. 36, 489 (1985).
[20] H. Gräfen, in: DECHEMA-Monografie Bd. 93 "Elektrochemie der Metalle, Gewinnung, Verarbeitung und Korrosion", S. 253–265, VCH-Verlag, Weinheim 1983.
[21] O. L. Riggs, M. Hutchinson u. N. L. Conger, Corrosion 16, 58t (1960).
[22] C. E. Locke, Mat. Protection 4, H. 3, 59 (1965).
[23] Z. A. Foroulis, Ind. Engng. Chem. Process Design. 4, H. 12, 23 (1965).
[24] W. P. Banks u. E. C. French, Mat. Protection 6, H. 6, 48 (1967).
[25] W. P. Banks u. M. Hutchinson, Mat. Protection 8, H. 2, 31 (1969).
[26] J. D. Sudbury, W. P. Banks u. C. E. Locke, Mat. Protection 4, H. 6, 81 (1965).
[27] H. Gräfen, G. Herbsleb, F. Paulekat u. W. Schwenk, Werkst. Korros. 22, 16 (1971).
[28] K. Bohnenkamp, Arch. Eisenhüttenwes. 39, 361 (1968).
[29] M. H. Humphries u. R. N. Parkins, Corr. Science 7, 747 (1967).
[30] H. Gräfen u. D. Kuron, Arch. Eisenhüttenwes. 36, 285 (1965).
[31] W. Schwarz u. W. Simons, Ber. Bunsenges. Phys. Chem. 67, 108 (1963).
[32] T. R. B. Watson, Pulp Paper Mag. Canada 63, T-247 (1962).
[33] G. Herbsleb, Werkst. u. Korrosion 16, 929 (1965).
[34] I. G. Trabanelli u. F. Zucchi, Corrosion Anticorrosion 14, 255 (1966).
[35] N. D. u. G. P. Tschernowa, Verl. d. Akad. d. Wiss. UdSSR, 135 (1956).
[36] G. Bianchi, A. Barosi u. S. Trasatti, Electrochem. Acta 10, 83 (1965).
[37] N. D. Tomaschow, Corr. Science 3, 315 (1963).
[38] H. Nishimura u. T. Hiramatsu, Nippon Konzeku Gakkai-Si 21, 465 (1957).
[39] M. Stern u. H. Wissenberg, J. elektrochem. Soc. 106, 759 (1959).
[40] W. R. Fischer, Techn. Mitt. Krupp 27, 19 (1969).

[41] M. Werner, Z. Metallkunde *24*, 85 (1932).
[42] E. Pelzel, Metall *20*, 846 (1966); 21, 23 (1967).
[43] C. R. Bishop u. M. Stern, Corrosion *17*, 379 t (1961).
[44] H. Weißbach, Werkst. u. Korrosion *15*, 555 (1964).
[45] H. Gräfen u. D. Kuron, Werkst. u. Korrosion *20*, 749 (1969).
[46] K. Risch, Werkst. u. Korrosion *18*, 1023 (1967).
[47] F. Hovemann u. H. Gräfen, Werkst. u. Korrosion *20*, 221 (1969).
[48] J. M. Kolotyrkin, 1st Intern. Congr. on Met. Cor. Butterworths, London 1962, S. 10.
[49] J. I. Bregman, Corrosion Inhibitors, The Macmillan Comp., N.Y. Collier – Macimillan Limit., London 1963.
[50] C. C. Nathan, Corrosion Inhibitors, NACE, Houston 1973.
[51] H. Fischer, Werkstoffe u. Korrosion *23*, 445 u. 453 (1973); *25*, 706 (1974).

22

Safety and Economics

W. V. BAECKMANN AND W. PRINZ

Corrosion protection not only maintains the value of industrial plants and structures but also ensures safe operational conditions and the reliability of these structures; it prevents damage that can endanger people and the environment. For this reason corrosion protection of such installations forms the most important means of protecting the environment.

The use of corrosion-resistant materials and the application of corrosion protection measures are in many cases the reason that industrial plants and structures can be built at all. This is particularly so in pipeline technology. Without cathodic protection and without suitable coating as a precondition for the efficiency of cathodic protection, long-distance transport of oil and gas under high pressures would not be possible. Furthermore, anodic protection was the only protective measure to make possible the safe operation of alkali solution evaporators (see Section 21.5).

In all these applications it should be realized that the safety provided by corrosion protection depends on its efficacy. The system must not only be correctly installed at the commissioning stage but must also be monitored continually, or at regular intervals during operation [1]. In this respect the methods of measurement used in monitoring cannot be too highly emphasized.

22.1 Safety

The greatest possible safety against corrosion damage is achieved by passive protection with coatings in combination with cathodic protection. Therefore coating and cathodic protection of pipelines that have strong safety requirements are compulsory in order to protect both people and the environment [2-5].

22.1.1 Statistics of Pipeline Failures

According to Ref. 6, 70 incidents of corrosion damage for 1000 km of buried steel pipeline transporting gas were reported on average each year in Germany. For 1000 km of pipeline on lines operating at pressures up to 4 bar, 100 incidents of corrosion damage were recorded; there were 150 instances of damage to service pipes and 75 instances of damage to supply pipelines. In high-pressure gas pipelines operating at pressures of 4 to 16 bar, there were 20 incidents of corrosion damage, and in high-pressure gas pipelines over 16 bar there were 1.4 incidents per 1000 km. The highest corrosion rate occurred in gas pipelines operating below 4 bar, where cathodic protection is not compulsory. In this case service pipes are particularly endangered by corrosion.

Cathodic protection has been compulsory since 1976 for high-pressure gas pipelines operating at pressures between 4 and 16 bar. Not all such pipelines operating in this pressure range which were installed before 1976 have been subsequently provided with cathodic protection. The total number of new corrosion cases per 1000 km per year has been reduced to 20 by cathodic protection. In high-pressure gas pipelines operating at pressures greater than 16 bar, the number of corrosion failures is only 1.4 (i.e., two orders of magnitude less than for pipelines without cathodic protection). These figures clearly show the effect of cathodic protection in raising the safety level.

In the United States, all steel pipelines for gas transport and distribution must be cathodically protected. The damage statistics for pipeline safety for 1983 show that the likelihood of damage for all steel pipelines is about 1 per 1000 km of pipeline per year [7]. The number of corrosion failures in gas pipelines in the United States is an order of magnitude less than in Germany. The trend of damage statistics compared with the American statistics shows that the safety of gas pipelines can be considerably improved by cathodic protection. German damage statistics clearly show, moreover, a considerably higher risk of corrosion in pipelines in the distribution network (see Section 10.3.6). Reference 8 details how cathodic protection of gas distribution networks is carried out.

22.1.2 Measures for Control and Maintenance

At present, Ref. 9 is the best source of information on checking and monitoring cathodic protection and fulfilling the protection criterion of Eq. (2-39) at the potential measuring points along the pipeline by using the switching-off method (see Section 3.3.1). Since, however, there are *IR* errors in these potential measurements, arising from equalizing currents and foreign currents, monitoring according to Ref. 9 can result in insufficient cathodic protection, which can remain unknown and overlooked [1]. This occurs with unfavorable soil conditions (e.g., with locally widely varying values of the soil resistivity, ρ) and also, more important, through large defects in the coating between the test posts (see Section 10.4). Considerable improvement in maintenance can be achieved by reducing errors in potential measurement by applying Eq. (3-28) with the help of more intensive types of surveys (Section 3.7) and also by the use of external measuring probes (Section 3.3.3.2) [1].

On about 2500 km of pipeline laid since 1970, overline surveys showed 84 places totalling 5 km in length where the protection criterion had not been reached. In 21 exploratory excavations, 7 cases of pitting corrosion with penetration depths $l_{max} > 1$ mm were found. At three places the pipe had to be replaced or repaired with split sleeves. Seven hundred sixty-five places with a total length of 95 km in 2500 km of pipeline laid between 1928 and 1970 were found to have failed to reach the protection criterion. Thirty-two examples of pitting corrosion with $l_{max} > 1$ mm were discovered in 118 excavations. In eight cases the pipe had to be replaced or made safe with a split sleeve [10]. It can be clearly seen from these examples that weak points in the cathodic protection system can be detected with overline surveys and corrected, thus increasing the reliability of corrosion protection of pipelines.

22.2 General Comments on Economics

The cost and economics of cathodic protection depend on a variety of parameters so that general statements on costs are not really possible. In particular, the protection current requirement and the specific electrical resistance of the electrolyte in the surroundings of the object to be protected and the anodes can vary considerably and thus affect the costs. Usually electrochemical protection is particularly economical if the structure can be ensured a long service life, maintained in continuous operation, and if repair costs are very high. As a rough estimate, the installation costs of cathodic protection of uncoated metal structures are about 1 to 2% of the construction costs of the structure, and are 0.1 to 0.2% for coated surfaces.

In some cases, electrochemical protection allows older structures to be kept functional that otherwise would have to be replaced due to local corrosion damage (nonuniform corrosion attack, pitting, corrosion cracks, etc.). In certain cases the operation has been technically possible only by using economical materials protected electrochemically. The advantages of electrochemical protection are generally recognized today. Its application is constantly being extended into new areas and applications [11]. It is particularly advantageous in the following:

1. Nonuniform corrosion or pitting corrosion frequently occurs on steel structures in seawater and in soil. Nonuniform and pitting corrosion easily lead to damage in tanks, pipelines, water heaters, ships, buoys and pontoons, because these structures lose their functional efficiency when their walls are perforated (see Chapter 4).

2. Frequently wall thicknesses are increased to meet the planned service life, resulting in a considerable increase in the materials used. This not only raises installation and transport costs but also the weight, which makes construction more difficult.

3. Many tanks could be constructed with low-alloy, high-strength steels if they could be given adequate corrosion resistance. In the absence of electrochemical protection, corrosion-resistant high-alloy steels or alloys have to be used which often have less favorable mechanical properties and are much more expensive. Areas of application are heat exchangers, cooling water piping for seawater turbines, reaction vessels, and storage tanks for chemicals (see Chapter 21).

4. Cathodic protection can be used as a renovation measure for steel-reinforced concrete structures (see Chapter 19). Although material costs of from 100 DM m^{-2} (particularly with preparation, erection, and spray coating costs) up to 300 DM m^{-2} are quite high, they do not compare with the costs of demolition or partial replacement.[1]

22.3 Costs of Cathodic Protection of Buried Pipelines

Cathodic protection of uncoated objects in the soil is technically possible; however, the high current requirement, as well as measures for the necessary uniform current distribution and for *IR*-free potential measurement, result in high costs. In determining the costs of cathodic protection of pipelines, it has to be remembered that costs will increase with increases in the following factors:

- necessary current requirement,
- electrical soil resistivity at the location for installation of the anodes,
- distance of the nearest power supply for impressed current anodes.

The current requirement depends on the leakage load [Eq. (24-71)] and the dimensions of the protected surface [12]. Measures to reduce the leakage load are described in Section 10.2.2. The level of current output from galvanic anodes or the grounding resistance of impressed current anodes—and with it the necessary rectifier output voltage for supplying the protection current—depend mainly on the soil resistivity.

The higher the costs of connecting the current of an impressed current installation, the more economical galvanic anodes become. Usually the choice of one or the other protection method is made not only from an economic point of view, but also from technical considerations. Only the economic point of view is dealt with here.

---

[1] The costs here and in the table are based on deutsche mark values at the time the original text was written (the late 1980s) and therefore should be used only as relative numbers.

Table 22-1 Installation costs of a cathodic protection station with three magnesium anodes for I_{total} = 100 mA, r = 30 W m and service life of 36 years.

1	Setup and installation of test post, including material used and operational testing	DM	1200
2	Provision and installation of three 10-kg anodes	DM	4500
3	Laying and burying cable, including cable and earthworks (58 DM m^{-1})	DM	600
4	Connection of sleeves	DM	200
5	Connection of cable pipes by Thermit welding, field coating, including material and earthworks	DM	700
6	Commissioning	DM	300
7	Total cost with earthworks	DM	7500

22.3.1 Galvanic Anodes

Galvanic anodes (Chapter 6) for buried installations are predominantly of magnesium. With specific soil resistivities of $\rho < 20\ \Omega$ m, zinc anodes can also be economical because of their longer life. Table 22-1 gives the costs of cathodic protection with magnesium anodes with a total current output of 100 mA. The bare material costs of magnesium anodes with backfill amount to about 40 DM kg^{-1}. If the galvanic protection method is to be compared with the impressed current method, test points separating the anodes for off potential measurements, which are usual with long-distance pipelines, have to be taken into account, so that the costs, including those of the test points, come to about K_A = 2500 DM/anode.

On the other hand, the costs for an average cathodic protection station for 6 A come to K_{IC} = 40,000 DM according to Table 22-2. For very small installations as, for example, the external cathodic protection of a tank, the costs of an impressed current system where a current supply is already available without cost, with lower current output, can be reduced to about 4000 DM. With larger tanks and greater soil resistivity, the following considerations point to the increased suitability of an impressed current system.

The following economic considerations apply particularly to the cathodic protection of pipelines. The total cost of protection with galvanic anodes $K_{\Sigma A}$ should be less than the costs of an impressed current installation K_{IC}:

$$K_{\Sigma A} = nK_A \leq K_{IC} \tag{22-1}$$

Table 22-2 Installation costs of an impressed current cathodic protection station for 6 A with four FeSi anodes.

Fixed Costs

Rectifier housing with glass fiber-reinforced base, housing connection box, meter table, corrosion protection rectifier 10 A/20 V, terminal strips, cable terminal and wiring, completely installed	DM	8500
Pipeline connection with reference electrode and pipe current test point, including material and earthworks	DM	4500
Low-voltage supply 220 V	DM	8000
Total costs	DM	21,000

Variable Costs

Anode installation near surface 100-m cable trench, including cable and laying (58 DM m^{-1})	DM	7600
Installation of an anode bed with 4 FeSi anodes in coke backfill, cable and connecting sleeves, including earthworks	DM	6000
Commissioning	DM	3400
Total costs of the near-surface anode bed	DM	38,000
Deep anode installation 20-m cable trench, including cable and laying	DM	1600
50-m borehole, including 4 FeSi anodes and coke backfill	DM	16,000
Commissioning	DM	3400
Total costs of a deep anode installation	DM	42,000

Fig. 22-1 Economic application range for cathodic protection with magnesium anodes or with impressed current.

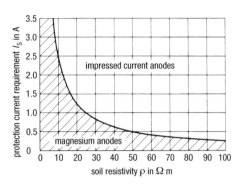

The required number of anodes n is given by the protection current requirement of the object I_s and the current output of a single magnesium anode I_{max}:

$$n = \frac{I_s}{I_{max}} \leq \frac{K_{IC}}{K_A} \tag{22-2}$$

Figure 22-1 shows the limiting curve derived with the help of Eq. (6-12) and Table 24-1 from Eq. (22-2) [13]. There is the relationship:

$$\frac{I_s}{A} \gtrless \frac{F_C}{(\rho/\Omega\,\text{m})} \quad \text{impressed current is} \quad \frac{\text{economical}}{\text{uneconomical}} \tag{22-3}$$

F_C contains the cost K_{IC} and K_A as well as the anode geometry. In the example in Fig. 22-1, it is assumed that $F_C = 25$.

The decision on whether cathodic protection with impressed current or with magnesium anodes is more economical depends on the protection current requirement and the soil resistivity. This estimate only indicates the basic influence of the different variables. In the individual case, installation costs can vary widely so that a specific cost calculation is necessary for every project.

22.3.2 Impressed Current Anodes

The effective costs of a cathodic protection station depend very much on the local conditions, particularly on the costs of the power supply and the extent of the anode bed. Table 22-2 contains the costs of a protection station with four silicon iron anodes.

With such a protection station and an assumed current density of 10 μA m^{-2}, about 100 km of pipeline can be protected. In practice, the protection range L is

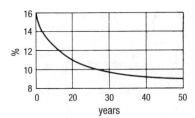

Fig. 22-2 Annuity as a function of service life for a depreciation of 8% interest plus 0.8% charges.

limited to about 80 km. The inclusive costs of providing cathodic protection for this length are about 40,000 DM, which corresponds to 500 DM km^{-1}. This means that the output from a 6-A rectifier would not be fully used. From experience, one must expect that the cathodic protection current requirement will increase in the course of time (e.g., by connections to further branch pipelines or other structures and through defects in the coating, as well as by subsequent external effects).

In order to calculate the annual cost of cathodic protection, the capital and operating costs must first be determined. Figure 22-2 gives the annuity for a service life of up to 50 years with 8% interest, including trade tax and property tax. After 50 years the curve flattens out because the annuity varies very little. In general, a service life of 30 years can probably be assumed for a cathodic protection station. In these considerations the useful life is limited to 20 years so that repair and rebuilding costs can be neglected. At 20 years, the annuity amounts to 11%, giving an annual capital cost of installing the cathodic protection of 55 DM km^{-1}. The annual capital cost for the test points amounts to about 110 DM km^{-1} after installation costs of 1000 DM km^{-1}. The capital cost therefore amounts to 165 DM km^{-1}. The annual cost, including current, upgrades, and monitoring, as well as occasional costs of overline surveys, comes in total to about 235 DM km^{-1}. The total annual cost of cathodic protection of long-distance pipelines is estimated at roughly 400 DM km^{-1}. With distribution networks in urban areas, the costs are much higher and can amount to 3000 DM km^{-1}, including the costs of insulators for the connection of service pipe [14,15]. Section 9.1.4 gives details on the design of anode equipment.

22.3.3 Prolonging the Life of Pipelines

In order to carry out a cost comparison of cathodic protection with the prolongation of service life of pipelines that it provides, the construction costs and the material costs of the pipeline have to be known. If there are no particular difficulties (e.g., having to lay the pipe in heavily built-up areas, river crossings, or rocky soil), the construction costs for a high-pressure DN 600 pipeline are about 10^6 DM km^{-1}. If it is simply assumed that a pipeline has a useful life of 25 years without cathodic protection whereas with cathodic protection it has a life of at least 50 years, the

Safety and Economics

economic advantages of cathodic protection are immediately apparent. According to Fig. 22-2, the increased life saves capital costs of 10.17% − 8.98% = 1.19% of 10^6 DM km^{-1}. That amounts to 11,900 DM km^{-1} per year. On the other hand, the annual cost of cathodic protection is only about 400 DM km^{-1}.

With frequent occurrence of pitting corrosion, individual repairs are no longer adequate. Either large stretches must be repaired and subsequently recoated or whole sections of the pipeline must be replaced. The repair costs for several lengths of pipeline are often higher than the costs of laying a new pipeline. If it is assumed that after 25 years without cathodic protection 10% of a pipeline in corrosive soil has to be replaced, then the additional costs amount to 10% of 1.19% of 10^6 DM km^{-1}. That is, 1190 DM km^{-1}. These costs can be saved with cathodic protection costing roughly 400 DM km^{-1}. This illustrates the economic benefits of cathodic protection.

These considerations are always relevant to water pipelines, which must have a very long service life. With natural gas, oil, and product pipelines, shorter periods are usually assumed for depreciation. Disregarding the fact that in this case cathodic protection is compulsory for safety reasons and protection of the environment, the costs of repairs resulting from wall perforation after a long service life can exceed the costs of cathodic protection.

Unfortunately there are very little reliable data on the frequency of wall perforation caused by corrosion; in most cases comprehensive data about wall thickness, pipe coating, type of soil, etc. are lacking. The incidence of wall perforation is usually plotted on a logarithmic scale against the service life of the pipeline (see Fig. 22-3). Cases are also known where a linear plot gives a straight line. Curve 1 in

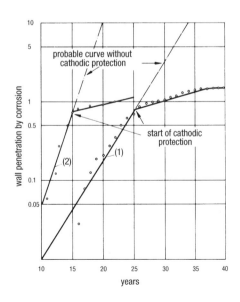

Fig. 22-3 (1) Total number of wall penetrations by corrosion per kilometer of a DN 500 pipeline as a function of service life. (2) Total number of wall penetrations per kilometer of a pipeline with severe stray current exit.

498   Handbook of Cathodic Corrosion Protection

Fig. 22-3 shows the total number of perforations per kilometer in a 180-km DN 500 long-distance gas pipeline with a wall thickness of 9 mm which was laid in 1928 in a corrosive red-marl soil. There was no influence from stray currents.

Since stray current corrosion damage can occur after only a few years, the economy of stray current protection measures is obviously not questionable [12]. In Fig. 22-3 the effect of stray currents is shown by curve 2 [14]. Without there being firm evidence, it is apparent that the shape of the corrosion damage curve in steel-reinforced concrete (see Sections 10.3.6 and 4.3) is similar to that for stray current corrosion [15].

Perforation increases with time due to pitting corrosion because two processes overlap:

1. Corrosion starts rapidly at many points simultaneously. The maximum penetration rate, however, depends on the type of soil and the existence of foreign cathodic structures [16,17].

2. The number of corrosion spots increases with time, but the maximum penetration rate remains roughly constant locally and with time. The penetration rate corresponds to a Gaussian distribution curve [18].

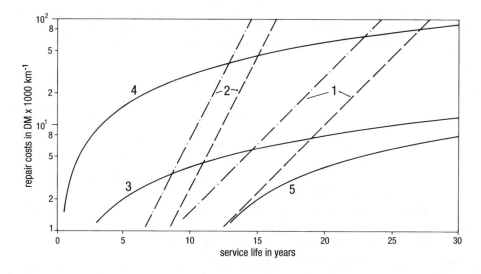

Fig. 22-4   Total repair costs (1) and (2) for wall penetrations (1) and (2) in Fig. 22-3 and comparison with total annual cost of cathodic protection at 400 DM km^{-1} (3), at 3000 DM km^{-1} (4), and with subsequent installation (400 DM km^{-1}) after 10 years (5). —·—·— DN 800, PN 64, 150,000 DM/case of damage. - - - DN 300, PN 40, 30,000 DM/case of damage.

Both cases can be understood as being an increasing growth of coating defects with time, leading to the exponential curve shown in Fig. 22-3. The assumptions that with time the coating permits more oxygen to diffuse through it and that the coated surface acts as a cathode, lead to the same dependence on time (see Section 5.2.1.3). If after the cathodic protection is commissioned, further wall penetrations occur that correspond to the curve in Fig. 22-3, this is usually due to the difficulty of fulfilling the protection criterion on old corroded spots. According to the information in Eq. (3-21), this is to be expected when too little cathodic polarization is combined with too great a resistance of the local anodes [19].

The costs for repairing one area of corrosion damage in a high-pressure gas pipeline are approximately 150,000 DM for large-diameter and 50,000 DM for small-diameter pipes. According to Fig. 22-4, the total curves 1 and 2 of the annual cost of repairing wall perforations increase exponentially with service life. This is based on the curves for the frequency of wall perforation in Fig. 22-3. The rising costs of cathodic protection related to service life are shown by curves 3 and 4 in Fig. 22-4. The cost advantage of cathodic protection occurs relatively late, as shown by the intersection of the two curves with the straight lines 1 and 2. With old pipelines, cathodic protection was only applied after commissioning (curve 5).

22.4 Corrosion Protection of Well Casings

The danger of corrosion in well casings is described in Section 18.2. The cost of repairing corrosion damage in well casings is very high. On average it is around

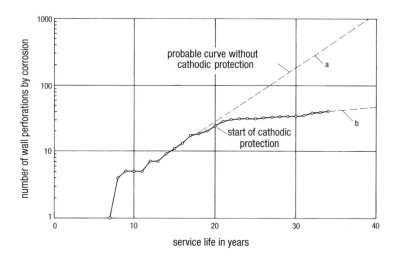

Fig. 22-5 Total number of wall penetrations in well casings in an oil field.

300,000 DM [20]. Figure 22-5 shows the number of wall perforations in an oil field in which the first corrosion damage appeared after 8 years. The number of corrosion incidents rose considerably in the course of subsequent years, as curve a in Fig. 22-5 indicates. The well casings were cathodically protected. Inhibitors were used to protect against internal corrosion. The number of fresh corrosion incidents clearly decreased, as shown by curve b. In the 12 years after the cathodic protection was commissioned, only 16 examples of corrosion occurred, whereas from curve a about 350 would have been expected. The savings in repair costs of about 63×10^6 DM are set against the annual capital costs of the cathodic protection of 231,000 DM and operational costs of 150,000 DM, amounting to a total of about 4.6×10^6 DM for the period of 12 years. This clearly shows the economic benefits of cathodic protection for well casings.

22.5 Corrosion Protection in Seawater

Shipping is the oldest area of application of cathodic protection (see Section 1.3). While galvanic anodes are predominantly used in ship construction and offshore structures, the high current requirement for harbor installations and loading bridges demands the use of impressed current anodes. Characteristic corrosion problems in offshore structures became apparent in the Gulf of Mexico in the middle 1950s. In addition to nonuniform and pitting corrosion, problems of corrosion fatigue became more prominent [21]. In contrast to dolphins and ships, no coatings were applied in most offshore installations. Corrosion protection was achieved with cathodic protection. The current output ratios of galvanic anodes of aluminum, magnesium and zinc are shown in Tables 6-2 to 6-4 as 3.3:1.4:1. On the other hand, the ratio of the costs of anodes in DM kg^{-1} is 1.8:3.2:1, so that the specific costs in DM A^{-1} hr^{-1} are 1:4.2:1.8 (i.e., very favorable for aluminum anodes). Long-term observations on the three types of anode in the Gulf of Mexico gave the cost ratio as 1:3.5:2 [21]. Magnesium anodes are therefore uneconomic in the offshore field and are also not suitable (see Section 16.1.2). Protection with zinc anodes is more costly than with aluminum anodes.

A particular advantage of impressed current systems is the ability to control the output voltage of the rectifier. Also, there are the comparatively low installation costs and relatively uniform current distribution. The costs of impressed current protection compared with aluminum anodes are 0.8:1. With ships this ratio depends on the length of the ship; with larger ships it is 1:2.5 since the calculation is made in comparison with zinc and aluminum anodes. The order of magnitude of the annual costs depends on the structure and the investment costs.

Table 22-3 Cost survey for cathodic protection of various installations.

Structure to be protected	Protective measures carried out	Cost of cathodic protection (DM m^{-2} a^{-1})
5000-m^3 saline water tank	Cathodic protection, coating	2.65
Pipeline	PE coating $J_s = 10\ \mu A\ m^{-2}$, cathodic protection	0.2
Pipeline	Bitumen coating, $J_s = 1\ mA\ m^{-2}$, cathodic protection	0.44
Well casing	Cathodic protection	1.2
Well casing and pipelines in oilfield	Cathodic protection Bitumen coating	1.35

The structural costs of impressed current protection of harbor and coastal structures are about 1.5 to 2.5% of the total cost of the object to be protected. As an example, for the installation of a cathodic protection station in a tanker discharge jetty, the construction costs amounted to 2.2% of the total costs. The annual cost of current, maintenance, testing, and repairs amounted to 5% of the construction costs of the cathodic protection [22].

22.6 Cost of Internal Protection

Information on the costs of cathodic and anodic internal protection of tanks varies very widely since the costs depend not only on the material costs but also on the special installation costs and these in turn depend on the internal geometric layout of the tank and its pipework.

22.6.1 Internal Cathodic Protection

The economic aspects of internal cathodic protection are naturally most important where there is a danger of pitting or nonuniform corrosion. Usually no

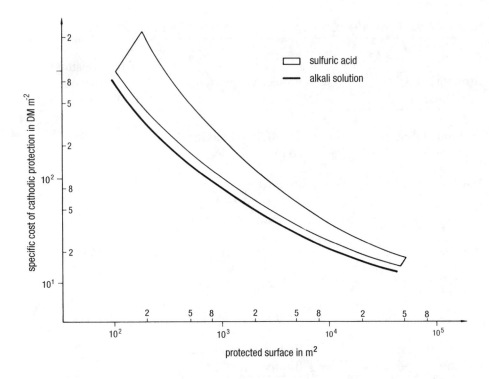

Fig. 22-6 Costs of anodic protection as a function of area to be protected.

potential measurements are made in the interior of smaller tanks, and the protection current is decided on by experience. To protect 1 m² of uncoated surface, about 1.5 kg of magnesium should be provided for a life of 4 to 5 years [23] (see Section 20.2). The costs of mounting and affixing are roughly the same as the cost of the anodes. Although protection of tanks with galvanic anodes does not involve current costs and the system runs practically without maintenance, cathodic protection of larger tanks is being carried out more and more with impressed current anodes, so that the installation costs are mostly over 20 DM m^{-2}, depending on the size of the tank [24]. Table 22-3 gives costs for the cathodic protection of various installations.

22.6.2 Internal Anodic Protection

The electrochemical protection of chemical plants is a question of customizing and therefore no universally applicable cost data can be established. Since chemical equipment and the materials used in its construction are proportionately more expensive and frequently other protection methods give no absolute guarantee of operational safety, developments for protection in this area are rapidly moving forward (see Section 21.5).

In spite of lower anodic current densities for uncoated steel surfaces, the technical applications of anodic protection are generally greater than those of cathodic protection. Naturally the costs of the plant and reaction vessels are also considerable, so that the costs of anodic protection can be estimated at about 3% of the installation costs of the object to be protected. The installation costs calculated for the surface area of anodic protection against stress corrosion by caustic soda in a large chemical plant are in the region of 60 DM m^{-2}. Stable coatings cannot be applied for this price nor can the tank be constructed of stress corrosion-resistant materials. There are many cases where, after 1 to 2 years, 20% of the tank repair costs were for repairing stress corrosion. Figure 22-6 shows the cost of anodic protection for two very different installations as a function of the surface area to be protected [25]. Current costs for anodic protection are, in general, negligible as in the case of cathodic protection.

22.7 References

[1] W. Schwenk, 3R intern. *26*, 305 (1988); Werkst. Korros. *39*, 406 (1988).
[2] TRbF 301, Carl Heymanns Verlag, Köln 1981.
[3] TRGL 141, Carl Heymanns Verlag, Köln 1977.
[4] DVGW Arbeitsblatt GW 463, ZfGW-Verlag, Frankfurt 1983.
[5] DVGW Arbeitsblatt GW 462 Teil 2, ZfGW-Verlag, Frankfurt 1985.
[6] DVGW Schaden- und Unfallstatistik, Frankfurt 1984.
[7] Pipeline & Gas Journal, Dallas/Texas, August 1985.
[8] DVGW Arbeitsblatt GW 412, ZfGW-Verlag, Frankfurt 1988.
[9] DVGW Arbeitsblatt GW 10, ZfGW-Verlag, Frankfurt 1984.
[10] W. Prinz, gwf gas/erdgas *129*, 508 (1988).
[11] H. J. Fromm, Mat. Perform. *16*, H. 11, 21 (1977).
[12] W. v. Baeckmann u. D. Funk, 3R intern. *17*, 443 (1978).
[13] C. Zimmermann, Diplomarbeit, TU Berlin 1962.
[14] W. Pickelmann, gwf gas/erdgas *114*, 254 (1973).
[15] W. Schwarzbauer, gwf gas/erdgas *121*, 419 (1980).
[16] K. F. Mewes, Stahl und Eisen *59*, 1383 (1939).
[17] S. A. Bradford, Mat. Perform. *9*, H. 7, 13 (1970).
[18] M. Arpaia, CEOCOR, interner Bericht, Kom. 1, 1978.
[19] W. v. Baeckmann u. D. Funk, Rohre, Rohrleitungsbau, Rohrleitungstransport *10*, 11 (1971).
[20] B. Leutner, pers. Mitteilung 1988.
[21] D. E. Boening, OTC Proc. Pap. 2702 (1976).
[22] G. Hoppmann, 3R intern. *16*, 306 (1977).
[23] K. Sautner, Sanitäre Technik *26*, 306 (1977).
[24] W. v. Baeckmann, A. Baltes u. G. Löken, Blech *10*, 409 (1975).
[25] F. Paulekat, pers. Mitteilung 1988.

23

Interference Effects of High-Voltage Transmission Lines on Pipelines

H.-U. PAUL AND H.G. SCHÖNEICH

Areas of dense building, a lack of corridors for erecting overhead transmission lines, and conditions imposed by regulatory authorities have made it necessary to use common "energy rights-of-way" more and more frequently. This results in frequent crossings and contiguity, and in stretches of lines running parallel for several kilometers, which can lead to ac interference. This results not only in danger to the pipeline maintenance personnel, but also in damage to the cathodic protection.

By ac interference is understood the action of a high-voltage power plant on a pipeline by capacitive, ohmic and inductive coupling. In contrast to the other chapters in this handbook, in this chapter only ac and its relevant parameters are dealt with. The theoretical bases correspond closely to the information in Section 24.4; however, analogous parameters are designated differently and must be represented as complex. The concept <u>potential</u> as used in this chapter (e.g., the pipeline potential) is the ac voltage between the pipeline and the reference ground (remote ground). The important pipe/soil potential used in corrosion protection, which is a superimposed dc component, is not considered here.

Various kinds of interference can be defined:

- Capacitive interference: the production of electrical potentials in conductors due to the influence of alternating electrical fields (e.g., by high-voltage transmission lines under power).

- Ohmic interference: the production of electrical potentials in conductors by electrical contact, by arcing or by a local voltage cone, caused by fault currents or stray currents in the soil (see Chapter 15).

- Inductive interference: the production of electrical potentials in conductors due to the induction from alternating magnetic fields arising from short-circuit currents or operational currents in high-voltage power lines.

Depending on the length of time the interference processes are active, a distinction is made between short-term and long-term interference. Short-term interference is a rare occurrence resulting from a failure in a high-voltage installation (short circuit to ground, effect of lightning, etc.) and very high interference currents (e.g., in the kiloampere range). The current failure usually lasts ≤0.5 s. Interference voltages of several kilovolts can occur. Long-term interference is a continuing action which, however, involves lower voltages than short-term interference. Such interference can arise from operational currents. However, it can also occur for a limited period in high-voltage grids with ground connection compensation if there is a phase failure. The voltages in long-term interference can amount to a few tens of volts [1].

To avoid disruption or danger to installations and people, recommendations for avoiding interference should be taken into account in planning pipelines and high-voltage installations [2]. These should include regulations for routing energy corridors, measures to be taken in the construction and operation of pipelines, and the methods used to determine pipeline potentials under long- and short-term interference and safety requirements.

23.1 Capacitive Interference

Capacitive interference of pipelines is of minor importance. It arises in the immediate vicinity of overhead power lines or railway power lines in the construction of pipelines where the pipe is laid on a foundation that is well insulated from soil (e.g., on dry wood). The pipeline picks up a voltage with respect to the soil. The value of this voltage depends on the voltage of the interfering conductor at the time as well as the capacities between the conductors and the pipeline.

With aboveground, welded lengths of pipes that are insulated from the ground and that are within 10 m of overhead power lines, measures have to be taken against inadmissible capacitive interference if the following length limits are exceeded [2]:

- 200 m alongside three-phase overhead power lines with nominal voltage ≥110 kV.
- 1000 m alongside 110-kV railway power lines and overhead conductors as well as supply lines of ac railway lines.

For distances greater than 10 m, the length limits for capacitative interference can be read off from Fig. 23-1. For 110-kV railway power lines operating at 16⅔ Hz, the limiting lengths are 5 times the values shown in Fig. 23-1. In order to avoid possible damage to pipe lengths greater than the length limit, it is recommended that these stretches of pipe be connected to a 1-m-long ground [2]. By this measure, however, on very long stretches of pipeline, the pipeline potential at the other end can be raised to inadmissibly high values by inductive interference [see Eq. (23-29)]. In this case grounds of sufficiently low resistance should be connected at both ends.

Interference Effects of High-Voltage Transmission Lines on Pipelines

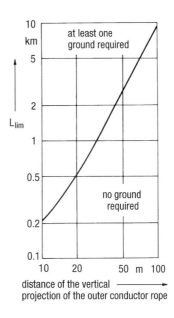

Fig. 23-1 Limiting length L_{lim} of a parallel insulated stretch of pipe with capacitive interference by a 50-Hz, three-phase overhead power line with a nominal voltage ≥110 kV [2].

23.2 Ohmic Interference

23.2.1 Contact with a Conductor under High Voltage

During construction in the vicinity of overhead power lines, there is a great danger that construction equipment (e.g., cranes) may inadvertently come too close to the voltage-carrying conductor, causing arcing or even direct contact with the conductor cable. In both cases, dangerous contact voltages occur at the construction equipment itself and its surroundings. Great care should be taken in the construction and repair of pipelines to observe sufficiently wide distances for safety (see Fig. 23-2) [2-4].

This can be achieved by limiting the height and reach of equipment working in the vicinity of high-voltage overhead power lines and by careful supervision of construction [2]. It is recommended that equipment remain at least 5 m away from high-voltage power lines with a nominal voltage of ≥110 kV, and 3 m away from those of <110 kV, including overhead conductors and supply lines.

To avoid immediate current discharge from a pylon to a pipeline due to a high-voltage failure, the distance between the pipeline and the corner legs of the pylon or the pylon ground should be greater than 2 m if possible. Reducing the distance by not less than 0.5 m is permitted only with an agreement by the parties concerned. It has been proved that, with a distance of 0.5 m between a ground and a PE-coated pipeline, even under unfavorable conditions no arcing occurs when there is a failure of the high-voltage grid [5]. If the tracks of ac railways cross under-

508   Handbook of Cathodic Corrosion Protection

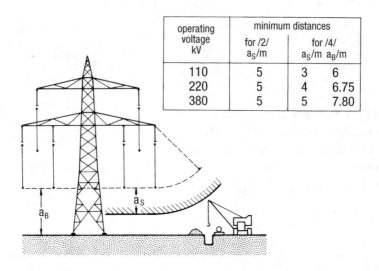

Fig. 23-2   Safe distances from high-voltage overhead power lines.

ground, a distance of 1.5 m must be maintained between the pipeline and the upper edge of the ties.

Where pipelines and high-voltage cables cross, a distance of at least 0.2 m must be observed to prevent contact between cable and pipeline (this can be achieved by interposing insulating shells or plates). Such intermediate materials can be PVC or PE. Their disposition and shape must be determined by mutual agreement [2,6].

Where pipelines and high-voltage cables run parallel to each other, spacing of at least 0.4 m should be observed to ensure adequate working space. In bottle-necks, the spacing should not be less than 0.2 m [2].

23.2.2   Voltage Cone of a Pylon Grounding Electrode

Figure 23-3 shows in principle the change in the potential at the soil surface in the vicinity of an overhead power line pylon as a function of distance. If there is a ground connection at a high-voltage pylon grid, part I_G of the ground leakage current I_F flows via the grounding impedance Z_G of the pylon into the ground. The pylon reaches the voltage $U_G = I_G \times Z_G$ against the reference ground. A pipeline with bitumen or plastic coating has the potential of the reference ground. If it is within the voltage cone of a pylon, a person touching the pipeline (e.g., when carrying out repair work) could receive a shock from the potential between the surrounding soil and the pipeline.

Interference Effects of High-Voltage Transmission Lines on Pipelines

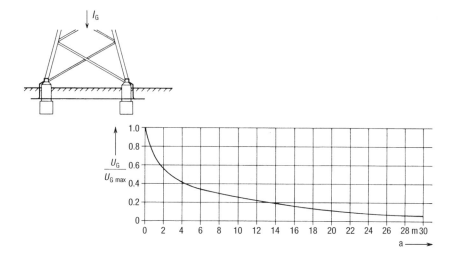

Fig. 23-3 Variation of the potential at ground level in the vicinity of a power line pylon.

If the contact voltage exceeds 65 V for long-term interference or 1000 V for short-term interference, protective measures are necessary for people working on the pipeline (e.g., rubber boots, insulated gloves, or insulated protective padding) [2]. On no account can there be an electrical connection between a pipeline and an overhead power line pylon or its ground. In the normal operation of a three-phase overhead power line, pylon grounding voltages arise from a main conductor exchange at a transposition of pylon lines [7].

The potentials from larger grounding installations on the soil surface are essentially flatter but more extended than those from overhead power line pylons (e.g., power stations, transformers and switchboard plants). The maximum expected grounding voltage in case of failure and its variation at the soil surface can be obtained from the facility's management.

The lines of an ac railway produce changes in the potential of the soil surface at right angles to the surface of the soil [8]. Since the potential differences on the soil surface are, however, much lower than the voltage between the rails and the reference ground, which, for contact protection, must not exceed 65 V, the potential changes are negligibly small.

Special conditions arise if a pipeline is laid in the region of the grounding installation of a power station or a transformer installation. If a lasting or a transitory connection with the grounding installation results during a grounding fault, the grounding voltage will be transferred to the pipeline and then appear outside the voltage cone as a contact voltage. Depending on the characteristics of the pipeline, the contact voltage decreases more or less quickly at greater distances. The variation is identical to the contact voltage U_B outside close proximity in the case of inductive interference (see Section 23.3). The grounding voltage is substituted for by U_{Bmax}. The pipeline is usually cathodically protected and electrically isolated from the grounding installation by an insulating unit at the works boundary or in the vicinity of its entry into the buildings. In the first case, the pipeline can be connected to the grounding installation in the works area. It is not possible to conduct the voltage to the outside because of the insulating unit. In the second case, additional measures may be necessary to prevent accidental connections with the grounding installation or with grounded parts of the plant, and to avoid high contact voltages inside the works area.

Contact voltages at pipelines must not exceed the limiting values in Fig. 14-3 [9] inside high-voltage plants. With long-term interference (i.e., acting for ≥3 s) the contact voltage must be $U \leq 65$ V; with short-term interference, the limiting value of the contact voltage depends on the duration of the failure current t_F (e.g., $U_B \leq 370$ V for $t_F = 0.2$ s and $U_B \leq 740$ V for $t_F = 0.1$ s).

If these values cannot be conformed to, additional measures are necessary (e.g., wearing insulating footwear and using insulating padding) [2]. There is increased danger if there is a possibility of simultaneous contact with the pipeline and a ground or grounded part of the installation. At distances closer than 2 m, the ground or grounded parts of the installation should be covered with insulating cloths or plates when work is being carried out on the pipeline.

23.3 Inductive Interference

23.3.1 Causes and Factors Involved

Inductive interference on pipelines is usually only to be expected with extended and close or parallel routing with three-phase, high-voltage overhead power lines, as well as with conductors and supply lines of ac railways.

The likelihood of interference increases with rising operational and short-circuit currents in the high-voltage installation and with increasing coating resistance of the pipeline. Voltages are induced in nearby metallic conductors by magnetic coupling with the high-voltage lines, which results in currents flowing in a conducting pipeline and the existence of voltages between it and the surrounding soil. Operational currents of high-voltage overhead power lines play an increasing role in interference, independent of the type of grounding for the neutral point of the high-voltage grid.

Frequently, close proximities between a pipeline and a high-voltage line are not only parallel but also consist of oblique approaches and crossings. With variable distances, oblique approaches have to be transformed into parallel corresponding sections to calculate interference. In such a parallel section, the longitudinal field strength induced is almost constant. Slightly curved conductor tracks and tracks with few changes of direction can be represented by an average straight-line run.

A complete solution for calculating the inductive coupling between a high-voltage line and metal conductors (e.g., pipelines) is possible with the help of series expansion [10,11] and computers. The following section provides advice on these calculations.

23.3.2 Calculation of Pipeline Potentials in the Case of Parallel Routing of a High-Voltage Transmission Line and a Pipeline

To calculate the voltage between the pipeline suffering from induced interference and the reference ground, the induced field strength E in the axial direction of the pipeline and the characteristic parameters of the pipeline, the impedance load Z' and the admittance load Y' are used. These quantities are analogous to the quantities R' and G' in Section 24.4.2 [see Eqs. (24-56) and (24-57)]. By introducing the capacity related to the length of the pipeline (capacity load C') and the inductivity load L', the following relations between these complex parameters can be given as follows:

$$Z' = R' + i\omega L' = |Z'| \exp(i\alpha) \tag{23-1}$$

with

$$|Z'|^2 = R'^2 + (\omega L')^2 \text{ and } \tan \alpha = \frac{\omega L'}{R'} \tag{23-2}$$

and

$$Y' = G' + i\omega C' = |Y'| \exp(i\beta) \tag{23-3}$$

with

$$|Y'|^2 = G'^2 + (\omega C')^2 \text{ and } \tan \beta = \frac{\omega C'}{G'} \tag{23-4}$$

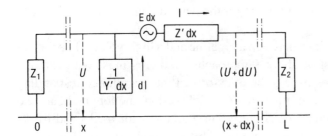

Fig. 23-4 Equivalent circuit diagram for a pipeline subjected to interference.

where $\omega = 2\pi f$, f is the frequency and $i = \sqrt{-1}$. The following approximations are made in the calculation using transmission theory [12]:

- The pipeline runs parallel to the interfering power line between points $x = 0$ and $x = L$ (see Fig. 23-4). The field strength is constant over this length.
- The pipeline has constant parameters (R', G', Z', Y') and therefore also the same r_u values in the close proximity region.
- The specific electrical soil resistivity is constant in the region under consideration. With short-term interference, the ground short circuit occurs outside the region under consideration.

The equivalent circuit diagram in Fig. 23-4 shows a differential fundamental section and the closed-circuit impedance at the end of the close proximity region.

It follows from the equivalent circuit diagram for a loop pipe/soil of the differential fundamental section:

$$E \, dx = -U + I\,(Z'\,dx) + (U + dU) \quad I\ Z'\,dx + dU$$

$$E - I\,Z' = \frac{dU}{dx} \tag{23-5}$$

$$dI\left(\frac{I}{Y'\,dx}\right) = -U; \quad \frac{dI}{dx} - Y'\,U \tag{23-6}$$

Here I is induced current, U is pipeline potential, and x is the position coordinate.
With the locally constant field strength, E, it follows from Eqs. (23-5) and (23-6):

$$\frac{d^2U}{(dx)^2} = Z'\,Y'\,U = \gamma^2\,U \tag{23-7}$$

Interference Effects of High-Voltage Transmission Lines on Pipelines

where

$$\gamma = \sqrt{Z' Y'} \tag{23-8}$$

Equation (23-7) corresponds to Eq. (24-58) in Section 24.4.2. There are identical relations for $E = 0$, $Z' = R'$ and $Y' = G'$ as in Eqs. (23-7) to (23-11). Equation (23-7) has solutions analogous to those of Eqs. (24-59) and (24-60):

$$U = -Z \left[A \, \exp(\gamma \, x) - B \, \exp(-\gamma \, x) \right] \tag{23-9}$$

$$I = A \, \exp(\gamma \, x) + B \, \exp(-\gamma \, x) + \frac{E}{Z'} \tag{23-10}$$

where

$$Z = \frac{Z'}{\gamma} = \sqrt{Z'/Y'} \tag{23-11}$$

The transfer coefficient γ and the characteristic impedance Z are important parameters of the pipeline and consist of complex quantities, in contrast to the analogous data in Section 24.4. Substituting Z' and Y' from Eqs. (23-1) to (23-4) leads to:

$$\gamma = |\gamma| \, \exp(i \, \phi_\gamma) \tag{23-12}$$

$$|\gamma|^4 = |Z'|^2 \, |Y'|^2 = \left[R'^2 + (\omega L')^2 \right] \left[G'^2 + (\omega C')^2 \right] \tag{23-13}$$

and

$$\phi_\gamma = \frac{1}{2} (\alpha + \beta) = \frac{1}{2} \left[\arctan \frac{\omega L'}{R'} + \arctan \frac{\omega C'}{G'} \right] \tag{23-14}$$

and

$$Z = |Z| \, \exp(i \, \phi_z) \tag{23-15}$$

$$|Z|^4 = \frac{R'^2 + (\omega L')^2}{G'^2 + (\omega C')^2} \tag{23-16}$$

$$\phi_z = \frac{1}{2} (\alpha - \beta) = \frac{1}{2} \left[\arctan \frac{\omega L'}{R'} - \arctan \frac{\omega C'}{G'} \right] \tag{23-17}$$

The constants A and B in Eqs. (23-9) and (23-10) result from the terminal conditions at the end of the parallel run [12]:

$$A = \frac{E}{2Z'} \frac{(1+r_1)r_2 - (1+r_2)\exp(\gamma L)}{\exp(2\gamma L) - r_1 r_2} \qquad (23\text{-}18a)$$

$$B = \frac{E}{2Z'} \frac{(1+r_2)r_1 - (1+r_1)\exp(\gamma L)}{\exp(2\gamma L) - r_1 r_2} \times \exp(\gamma L) \qquad (23\text{-}18b)$$

The reflection factors at the beginning and the end of the parallel stretch follow from Ref. 12:

$$r_{1,2} = \frac{Z_{1,2} - Z}{Z_{1,2} + Z} \qquad (23\text{-}19)$$

The quantities $Z_{1,2}$ are given in Fig. 23-4. A few examples are now given:

(a) The pipeline continues for several kilometers after the end of the parallel routing with the power line. For this case $Z_1 = Z_2 = Z$ and according to Eq. (23-19), $r_1 = r_2 = 0$. After substituting in Eq. (23-18a, b) it follows from Eqs. (23-9) and (23-10):

$$U = \frac{E}{2\gamma} \{\exp[\gamma(x-L)] - \exp(-\gamma x)\} \qquad (23\text{-}20a)$$

$$I = \frac{E}{Z'} \left\{1 - \frac{\exp[\gamma(x-L)] - \exp(-\gamma x)}{2}\right\} \qquad (23\text{-}20b)$$

According to Eq. (23-20a), the greatest pipeline potential is at the ends ($x = 0$ and $x = L$)

$$U_{max} = \frac{E}{2\gamma} [1 - \exp(-\gamma L)] \qquad (23\text{-}21)$$

In the middle ($x = L/2$), $U = 0$.

Outside the parallel stretch, the pipeline potential decreases with increasing distance according to the following relation [12]:

$$U(y) = U_{max} \exp(-y/l_k) \qquad (23\text{-}22)$$

Here the characteristic length, l_k, the reciprocal real part of the transfer constant according to Eq. (23-12), is analogous to the data in Section 24.4.2:

$$l_k = \frac{1}{|\gamma| \cos \phi_\gamma} \quad (23\text{-}23)$$

(b) The pipeline continues at $x \leq 0$ and is isolated at $x = L$ with an insulating flange. For this case $Z_1 = Z$, $Z_2 = \infty$ and according to Eq. (23-19), $r_1 = 0$, $r_2 = 1$. After substitution in Eq. (23-18a, b) it follows from Eq. (23-9):

$$U = \frac{E}{2\gamma}\left\{\exp(\gamma\, x)\left[2\exp(-\gamma\, L) - \exp(-2\gamma\, L)\right] - \exp(-\gamma\, x)\right\} \quad (23\text{-}24)$$

According to Eq. (23-24), the greatest pipeline potential is at the insulating flange:

$$U_{max} = \frac{E}{\gamma}\left[1 - \exp(-\gamma\, L)\right] \quad (23\text{-}25)$$

(c) The pipeline is isolated at both ends of the close proximity region with insulating flanges. For this case $Z_1 = Z_2 = \infty$ and according to Eq. (23-19), $r_1 = r_2 = 1$. After substitution in Eq. (23-18a, b) it follows from Eq. (23-9):

$$U = \frac{E}{\gamma}\frac{\exp(\gamma\, x) - \exp[\gamma(L - x)]}{\exp(\gamma\, L) + 1} \quad (23\text{-}26)$$

According to Eq. (23-26), the greatest pipeline potential is at the ends ($x = 0$ and $x = L$):

$$U_{max} = \frac{E}{\gamma}\frac{\exp(\gamma\, L) - 1}{\exp(\gamma\, L) + 1} \quad (23\text{-}27)$$

In the middle ($x = L/2$), $U = 0$.

(d) The pipeline is grounded at $x = 0$ and continues beyond $x \geq L$. For this case $Z_1 = 0$, $Z_2 = Z$, and according to Eq. (23-19) $r_1 = -1$, $r_2 = 0$. After substitution in Eq. (23-18a, b), it follows from Eq. (23-9):

$$U = \frac{E}{2\gamma}\left[\exp(\gamma\, x) - \exp(-\gamma\, x)\right]\exp(-\gamma\, L) \quad (23\text{-}28)$$

According to Eq. (23-28) at the point $x = 0$, $U = 0$, and the highest pipeline potential is at the point $x = L$:

$$U_{max} = \frac{E}{2\gamma}\left[1 - \exp(-2\gamma L)\right] \tag{23-29}$$

Equation (23-29) corresponds to Eq. (23-21) with the length $2L$ for the case (a), since in the middle $(x = L)$ $U = 0$ because grounding may be assumed. Correspondingly, case (c) can be applied to the following situation:

(e) The pipeline is grounded at the point $x = 0$ and isolated with an insulating flange at $x = L$. It follows from Eq. (23-19) $r_1 = -1$ and $r_2 = 1$. Substituting in Eqs. (23-18a, b) and (23-9) leads for the maximum voltage at the insulating flange to Eq. (23-27) with the length $2L$.

(f) The pipeline is grounded at both ends of the region of close proximity $(x = 0$ and $x = L)$. In this case with $Z_1 \approx Z_2 \ll Z$ and from Eq. (23-19) $r_1 \approx r_2 \approx -1$, from Eq. (23-18a, b), $A \approx B \approx 0$ and finally from Eq. (23-9), $U \approx 0$.

The parameters necessary for calculating pipeline potentials are given by Eqs. (23-1) to (23-4) and (23-12) to (23-17). The absolute values of the terms in brackets in Eqs. (23-21), (23-25) and (23-29) are still of interest. These follow from Eqs. (23-12) to (23-14):

$$\left|1 - \exp(-\gamma L)\right|^2 = \left(1 - \frac{\cos b}{\exp a}\right)^2 + \left(\frac{\sin b}{\exp a}\right)^2 \tag{23-30a}$$

where

$$a = |\gamma| L \cos\phi_\gamma \quad \text{and} \quad b = |\gamma| L \sin\phi_\gamma \tag{23-30b}$$

23.3.3 Obliquely Routed Sections of the Lines

For assessing a close proximity situation with oblique sections, a map drawn to scale is necessary that shows the tracks of the interfering high-voltage power line or the stretch of electric railway line and the pipeline that is interfered with.

Slightly curved line routes and tracks with many but small changes in direction can be approximated by the average straight line. Redrawing them in a simplified diagram with a straight-line high-voltage overhead power line or railway line may be an advantage.

Usually in calculating ac interference it is assumed that $f = 50$ Hz, soil resistivity $\rho = 50 \, \Omega$ m and with $f = 16⅔$ Hz due to the greater depth of penetration, $\rho = 30 \, \Omega$ m. The following distances from the pipeline on both sides of the center of the right-of-way rail are those in which the extent of the interference should be determined:

Interference Effects of High-Voltage Transmission Lines on Pipelines

- 1000 m for interference from grounding short-circuit currents in three-phase overhead power lines or operational and short-circuit currents in rail power and supply lines,
- 400 m for interference from operational currents in three-phase overhead power lines.

The pipeline is projected onto the interfering high-voltage overhead power line, which gives a subdivision from limiting distances, parallelism, and angle of approach of the high-voltage overhead power line and the pipeline, as well as from their points of intersection. Further investigations may be necessary with changes in the induced current, the reduction factors, the soil resistivity and the parameters of the pipeline (changes in diameter or specific coating resistance). Subdivisions are particularly necessary with oblique approaches in sections, for which a uniform field strength can be given with sufficient accuracy. The finer the subdivision, the greater the accuracy of the calculation. The geometric mean $(a_1 a_2)^{1/2}$ is used in the calculation for the average, a, from the distances a_1 and a_2 at the ends of the subdivided section, where a_1/a_2 must be less than 3 [13].

23.3.4 Simplified Calculation Methods

23.3.4.1 Interference by Fault Currents and by Railway Operating Currents

The field strength, E, induced by grounding short-circuit currents in high-voltage overhead power lines or railway power lines, is basic in calculating the pipeline potentials (see Section 23.3.2). The field strength, E, follows from Refs. 2 and 13:

$$E = 2\pi f M' \, Irw \tag{23-31}$$

where f is the frequency of the interfering current, M' is the mutual inductivity per unit length between the current-carrying lines and the line interfered with, I is the induction current, r is the reduction factor (general), and ω is the probability factor. The quantity M' can be assumed from Fig. 23-5 (see also plan 3 in Ref. 13).

The values in Fig. 23-5 have to be corrected in the case of differing resistivity values because of the dependence of mutual inductance on the soil resistivity. This is achieved by substituting an equivalent distance, a', instead of the real distance, a, giving:

$$a' = a\sqrt{\frac{50}{\rho(\Omega \times m)}} \quad \text{for} \quad 50 \text{ Hz} \tag{23-32a}$$

$$a' = a\sqrt{\frac{30}{\rho(\Omega \times m)}} \quad \text{for} \quad 16\frac{2}{3} \text{ Hz} \tag{23-32b}$$

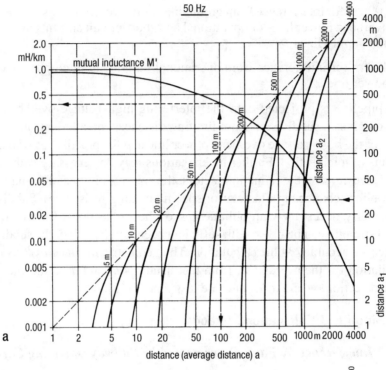

a

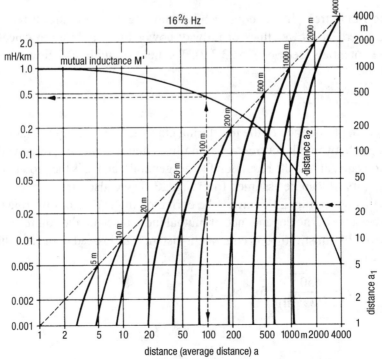

b

Interference Effects of High-Voltage Transmission Lines on Pipelines

Fig. 23-5 Mutual inductivity load M' as a function of the average spacing between two individual lines. Average distance, a, with equivalent counter-inductivity for an oblique approach with distances a_1 and a_2 at the end of the close proximity ($a_1 < a_2$). (a) $f = 50$ Hz and $\rho = 50\ \Omega$ m, (b) $f = 16⅔$ Hz and $\rho = 30\ \Omega$ m.
◂

The pipeline potential has to be estimated for every subdivided section using the equations in Section 23.3.2 both inside and outside the region of close proximity. The complex individual potentials must be superimposed to give the resulting pipeline potential using the sequence of the subdivided sections.

23.3.4.2 Interference from a Three-Phase High-Voltage Transmission Line in Normal Operation

Long-term interference arises in normal operation as a result of the geometry of the high-voltage pylon, and the magnetic fields arising from the operating currents at the location of a secondary conductor (cable, pipeline), because the two do not completely add up to zero. In addition, the grounding wire of the interfering power line interferes with the secondary conductor.

Calculation of long-term interference voltages is involved with a multiconductor problem which, in contrast to the short-term interference that derives from a one-pole grounding short circuit, in this case is related to the superposition of alternating magnetic fields of all the conductors of one or several three-phase systems as well as the ground wire.

Equations (23-33) to (23-36) are used to determine the extent of the induced field strengths in an ideally insulated conductor [14,15]. The positive terms give the induced field strengths from the main conductors, while the negative terms represent the field strengths induced from the loop ground wire/ground return line. The field strength, E, is related to the main conductor current, I, because of the proportionality according to Eq. (23-31).

$$\frac{|E|(V \times \mathrm{km}^{-1})}{|I|(\mathrm{kA})} = C\sqrt{A^2 + B^2 - AB} \tag{23-33}$$

$$A = -\ln\frac{\delta^2}{y_0^2 + a^2} \times \ln\frac{(y_0 - y_2)^2 + x_2^2}{(y_0 - y_1)^2 + x_1^2}$$

$$+ 22 \times \ln\frac{y_2^2 + (a - x_2)^2}{y_1^2 + (a - x_1)^2} \tag{23-34}$$

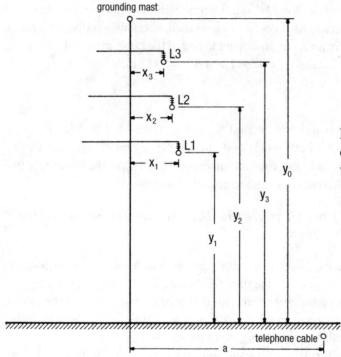

Fig. 23-6 Definition of distances at a high-voltage pylon.

$$B = -\ln\frac{\delta^2}{y_0^2 + a^2} \times \ln\frac{(y_0 - y_3)^2 + x_3^2}{(y_0 - y_1)^2 + x_1^2}$$

$$+ 22 \times \ln\frac{y_3^2 + (a - x_3)^2}{y_1^2 + (a_1 - x_1)^2} \qquad (23\text{-}35)$$

$C = 1.4$ for 50 Hz $C = 1.7$ for 60 Hz

Referring to Fig. 23-6, y_0 through y_3 are hereby the average distances of the conductor from the ground; x_1 through x_3 are the average distances of the conductor from the pylon axis; a is the distance of the affected conductor from the pylon axis, and δ is the penetration depth in the soil from:

$$\delta(\text{m}) = 658\sqrt{\frac{\rho(\Omega \times \text{m})}{f(\text{Hz})}} \qquad (23\text{-}36)$$

With long-term interference by several current circuits, the corresponding partial field strengths must be determined. The resultant longitudinal field strength can be assumed to be the square root of the sum of the squares of the partial field strengths.

Interference Effects of High-Voltage Transmission Lines on Pipelines

If, however, long-term interference results from current circuits that are connected in parallel (e.g., double conductors) or from current circuits that generally exhibit the same load flow direction (e.g., power station connection conductors), then the partial field strengths have to be added.

To determine the pipeline potentials, the resultant induced field strengths have to be included in the equations in Section 23.3.2. Such calculations can be carried out with computers that allow detailed subdivision of the sections subject to interference. A high degree of accuracy is thus achieved because in the calculation with complex numbers, the phase angle will be exactly allowed for. Such calculations usually lead to lower field strengths than simplified calculations. Computer programs for these calculations are to be found in Ref. 16.

23.3.5 Representation of the Characteristics of a Pipeline

Important parameters of a pipeline for calculating pipeline potentials are dealt with in Section 23.3.2. These quantities depend on the frequency, f, of the interfering current and on the material data of the pipeline. These are as follows:

$$R' = \frac{\sqrt{\rho_{st}\mu_0\mu_r\omega}}{\pi d\sqrt{2}} + \frac{\mu_0\omega}{8} \tag{23-37}$$

$$\omega L' + \frac{\mu_0\omega}{2\pi} \times \ln\left[\frac{3.7}{d}\sqrt{\frac{\rho}{\omega \times \mu_0}}\right] + \frac{\sqrt{\rho_{st}\mu_0\mu_r\omega}}{\pi d\sqrt{2}} \tag{23-38}$$

$$G' = \frac{\pi d}{r_u} = \frac{\pi d J_s}{0.3 \text{ V}} \tag{23-39}$$

$$\omega C' = \frac{\omega \pi d \varepsilon_0 \varepsilon_r}{s} \tag{23-40}$$

The following data are the basis of the diagrams in Figs. 23-7 through 23-11:

(a) Variables: d = pipe diameter and r_u = specific coating resistance [whose relation with the protection current density J_s follows from Eq. (5-11)];
(b) Constants: $\rho = 100 \, \Omega \, \text{m}$ for the soil; $\rho_{st} = 1.6 \times 10^{-5} \, \Omega \, \text{m}$ and $\mu_r = 200$ for steel; $\varepsilon_r = 5$ and the thickness of the coating $s = 3$ mm.

Some of the data assumed to be constant (ρ_{st}, μ_r, ε_r) can deviate more or less but this affects the results only slightly. The parameters provided in the following

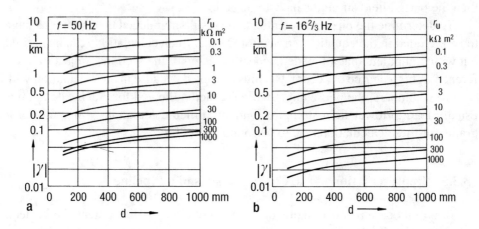

Fig. 23-7 Transfer coefficient $|\gamma|$ for $\rho = 100\ \Omega$ m. (a) $f = 50$ Hz and (b) $f = 16^2/_3$ Hz.

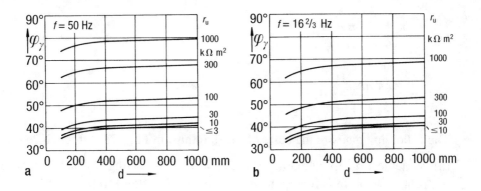

Fig. 23-8 Phase angle ϕ_γ of the transfer coefficient for $\rho = 100\ \Omega$ m. (a) $f = 50$ Hz and (b) $f = 16^2/_3$ Hz.

Interference Effects of High-Voltage Transmission Lines on Pipelines

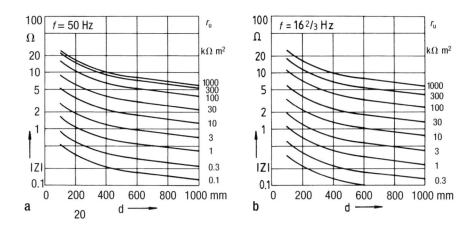

Fig. 23-9 Characteristic impedance |Z| for $\rho = 100\ \Omega$ m. (a) $f = 50$ Hz and (b) $f = 16\frac{2}{3}$ Hz.

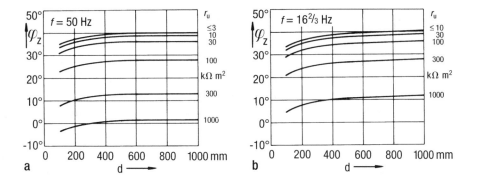

Fig. 23-10 Phase angle ϕ_z of the characteristic impedance for $\rho = 100\ \Omega$ m. (a) $f = 50$ Hz and (b) $f = 16\frac{2}{3}$ Hz.

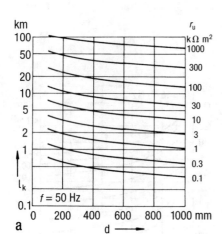 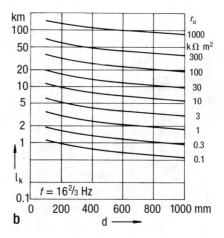

Fig. 23-11 Characteristic length l_k for $\rho = 100\ \Omega$ m. (a) $f = 50$ Hz and (b) $f = 16\frac{2}{3}$ Hz.

diagrams are within the scope of the otherwise stated error limits [17]. Figures 23-7 to 23-11 show the values for $|\gamma|$, ϕ_γ, $|Z|$, ϕ_Z and l_k from Eqs. (23-13), (23-14), (23-16), (23-17) and (23-23) as a function of pipe diameter and coating resistance for 50 Hz (Figs. 23-7a to 23-11a) and $16\frac{2}{3}$ Hz (Figs. 23-7b to 23-11b).

23.4 Limiting Lengths and Limiting Distances

Figure 23-12 shows the relation between limiting lengths and the distance between parallel-routed high-voltage power lines and pipelines [2]. For all pairs of values (L,a) which lie over the limit L_{lim}, testing of the pipeline potentials is necessary according to Section 23.3. Testing can be waived in the following cases [2]:

- Proximity of unlimited length to a high-voltage overhead powerline with a distance greater than 1000 m for interference from grounding short-circuit currents and currents in railway power and supply lines and greater than 400 m from interference by operating currents;
- Crossing with a high-voltage overhead power line at an angle greater than 55°;
- Distance from power stations, switching and transformer installations greater than 300 m.

Interference Effects of High-Voltage Transmission Lines on Pipelines 525

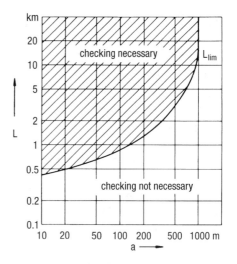

Fig. 23-12 Limiting lengths L_{lim} for parallel run of high-voltage overhead power line and pipeline as a function of distance, *a*.

23.4.1 Allowable Contact Voltages

23.4.1.1 Short-Term Interference

The leakage load of a pipeline with a bitumen coating is very dependent on the pipe/soil voltage [17]. Above a few hundred volts, glow discharge occurs at defects or pores, which raises the leakage load. Glow discharges occur from about 1 kV and finally arc between the pipe and soil, raising the leakage load by orders of magnitude. PE coating behaves similarly at defects.

With bitumen coating, there is a natural limit to the voltage so that, even under the most unfavorable conditions, pipeline potentials scarcely ever exceed 1.5 kV. With PE coating, a voltage limit is to be expected only at higher pipeline potentials of a few kilovolts.

With short-term interference due to faults in the high-voltage grid, no measures are necessary even with PE coating, if the calculated pipeline potentials do not exceed 1 kV. The reduction effect of grounding wires, railway tracks, and other pipelines and cables has to be taken into account. Furthermore, a probability factor $\omega = 0.7$ can be included in Eq. (23-31) for power lines with operational voltages >110 kV [13]. If the calculated pipeline potentials lie between 1 and 2 kV, constructive measures are necessary on the pipeline and on valves as well as measures to protect personnel working on the pipeline [2]. Grounds can be connected to the pipeline (see Section 25.5.3) to lower the contact voltage to values <1 kV. If the pipeline potential is over 2 kV, connection with grounds is always necessary.

23.4.1.2 *Long-Term Interference*

With long-term interference from operational currents of high-voltage power lines or railway power and supply lines, the contact voltage must not exceed 65 V [9]. The relevant operational current has to be stated by the operator responsible for the basic calculation. If the calculated pipeline potentials are above 65 V, connection with grounds is always necessary.

23.4.2 Determination of Pipeline Potentials

If pipeline potentials have to be calculated according to Fig. 23-12, then all the necessary data must be requested from the energy supply companies or the railway companies (e.g., the Bundesbahn Central Office in Munich). This includes, among other items, route plans, short-circuit diagrams, operational currents (thermal limiting currents), pylon diagrams, and grounding wire information (see also the data in Section 23.3.4 in this connection). The following information is required on the pipeline that is interfered with: route plans, position of insulating units, location of objects acting as grounds and cathodic protection stations, and parallel-routed pipelines and cables (see also the data in Section 23.3.5 in this connection). The method of calculation is given in Section 23.3.

23.5 Protection Measures against Unallowably High Pipeline Potentials

23.5.1 Short-Term Interference

For working at the pipeline, standing surface insulation or equipotential grounding mats at valves [2] must be provided for voltages of 1 to 2 kV. Rubber boots and water-repelling protective clothing have to be used in wet pits. According to DIN 4843 part 1, high rubber boots are sufficient. In addition, commercial insulating padding of 2.5 mm minimum thickness (e.g., safety mats for welding or insulating rubber or plastic mats) must be used in the case of working in a lying or sitting position. Generally, the pipeline potentials can be reduced by decreasing the leakage load through grounding.

23.5.2 Long-Term Interference

Installation of grounds, standing surface insulation, or equipotential grounding mats are necessary in the case of pipeline potentials higher than 65 V [2]. In the case of long sections of parallelism for an overhead power line and a pipeline, continuous grounding with respect to the distances and the resistances of the grounds

Interference Effects of High-Voltage Transmission Lines on Pipelines

are favored to obtain an allowable voltage distribution within the close proximity for a short circuit to ground.

23.5.3 Protective Measures by Grounding

The leakage load, G', as well as the transfer coefficient in Eq. (23-13) is raised and the pipeline potential according to Eq. (23-20a) is lowered by grounding. It has to be deduced, for a given interference, what the minimum value of G'_m is which must be achieved to keep the pipeline potential sufficiently low. For a given value of G', the leakage G for a section of the pipeline of length, L, with n grounds and with grounding resistance, R_G, is given by:

$$G = G'_m L = G' L + \frac{n}{R_G} \tag{23-41}$$

With $n' = n/L$ (number of grounds per unit length), it follows from Eq. (23-41):

$$n'/R_G = G'_m - G' \tag{23-42}$$

The minimum value of G'_m corresponds to the highest value of $r_{u,m}$ of the coating resistance according to Eq. (23-39):

$$r_{u,m} = \frac{\pi d}{G'_m} \tag{23-39'}$$

Hot-dipped galvanized steel with a coating of at least 550 g m^{-2} (about 70 μm) of zinc is usually employed for grounds. The usual material for horizontal grounds that can be laid in the pipeline trench is a galvanized steel strip with a cross section of 30 × 3.5 mm. Deep grounds mostly consist of assembled galvanized rods with a diameter of at least 20 mm. Grounds with these cross sections always have dimensions more than sufficient for their current loading if they are distributed along the trajectory subject to interference.

Protective measures using grounds can, according to Eq. (23-42), be achieved by choosing values of n' and R_G suitable from an economic point of view. It is often convenient to lay strip grounds in the pipe trench at the time of construction. Deep grounds are advisable if the soil resistivity decreases with depth. They have the additional advantage of not taking up much space and are particularly suitable for postconstruction installation. With deep grounds, a derived grounding resistance can be achieved by twisting together corresponding lengths of the grounding material. In this case, particularly with inhomogeneous soils, it is advisable to measure the grounding resistance when installing the ground (e.g., after each additional rod length).

The connection of grounds to a pipeline must not only be within the close proximity of length l_R, but also on both sides of the close proximity region over a length l_k, on account of the cutoff of the pipeline by the characteristic impedance, Z (see Fig. 23-4). The length l_k is obtained from Eq. (23-23) or Fig. 23-11, at which, after connection of the grounds, valid values of G'_m (i.e., the values of $r_{u,m}$ calculated from Eq. (23-39′) have to be applied). The number of grounds required is given by:

$$n = n' \left(l_R + 2 l_k \right) \tag{23-43}$$

If the pipeline ends within a close proximity region or if the pipeline is electrically cut off by an insulating unit, a ground must be connected at this point, whose grounding resistance corresponds roughly to the characteristic impedance, Z, according to Eq. (23-16) or with $r_{u,m}$ in Fig. 23-9.

Instead of installing a large number of grounding rods distributed along the pipeline, each with the characteristic length of l_0, a concentrated ground rod may be connected at each end of the area of close proximity. The resistance to the ground of these concentrated rods must be selected to provide a termination with the characteristic impedance Z. In addition, the details outlined in Section 23.5.4 must be taken into account. The different resistances to ground of various grounding rods are given in Table 24-1.

23.5.4 Grounding Electrodes and Cathodic Protection

The protection current for cathodic protection of a pipeline is raised by grounds that are directly connected to the pipeline. The additional current requirement of the ground is small, however, due to the relatively negative potential of galvanized steel. The usual grounds, with a circumference of about 70 mm, have a current requirement of about 0.5 to 1 mA per meter. If, however, concentrated grounds are connected at the ends of the close proximity region or if station grounds, concrete foundations or sheet piling walls are required as protective measures against high interference, the grounding for the ac must be dc-decoupled from the cathodically protected object (i.e., in the range of the pipe/soil potentials of the cathodic protection there must be sufficiently large Y' values and sufficiently low G' values). Overvoltage arresters described in Sections 14.2.2.1 to 14.2.2.4 have this property. There are various types of construction and methods available for its implementation.

23.5.4.1 Overvoltage Arresters for Short-Term Interference

This section deals with overvoltage spark gaps (gas discharge spark gaps) which produce a connection between pipeline and ground only when a trigger voltage (about 250 V) is exceeded. Care has to be taken in selecting the location for an installation so that the trigger voltage is actually reached in the case of a fault.

23.5.4.2 *Voltage Arresters for Long-Term and Short-Term Interference*

Polarization cells and diode voltage arresters can be used for this purpose.

23.5.4.2.1 *Polarization Cells*

The construction and action of polarization cells are described in Section 14.2.2.4. Difficulties in measuring off potentials can occur when polarization cells are connected in the circuit. The voltage of the polarization cell is the difference

$$U_{PC} = U_{on} - U_G \tag{23-44}$$

between the on potential and the potential of the ground, U_G, and charges up the polarization capacitance. On switching off the protection current, the capacity is discharged via ground/soil/pipeline, with the time constant determined by the resistance of this circuit (see Section 3.3.1). With very high coating resistances, a slow discharge is to be expected, so that too-negative off potentials are measured. By disconnecting the polarization cell from the pipeline, the correct off potential can be determined. This error in the measurement of the off potential is represented in Fig. 23-13 for a well-coated pipeline [18]. With low coating resistances and slow electrochemical depolarization, the errors become negligible after 1 s.

23.5.4.2.2 *Diodes*

The construction and action of overvoltage arresters with silicon diodes are described in Section 14.2.2.5 and Fig. 14-8. Smaller diodes can be used for this

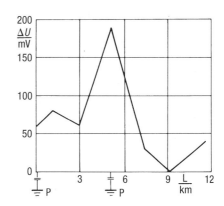

Fig. 23-13 Difference in off potentials with and without connected polarization cells (P = location of polarization cells).

530   Handbook of Cathodic Corrosion Protection

application than for power cables. The diodes must be sufficiently stable against the ac in normal operation and during short-term interference. In addition, they must have sufficient surge current resistance so that possible lightning currents do not destroy them.

As described in Section 14.2.2.5, a nonsymmetric diode overvoltage arrester creates a dc voltage in operation

$$U_{DA} = \frac{1-n}{2} U_{TV} \qquad (23\text{-}45)$$

which determines the on potential of the pipeline corresponding to Eq. (23-44):

$$U_{DA} = U_{on} - U_G \qquad (23\text{-}46)$$

where n is the number of diodes in the right-hand branch in Fig. 14-8 ($n = 4$), and $U_{TV} = 0.7$ V is the threshold voltage.

According to the details of Eq. (23-44), off potentials cannot be measured if the diode overvoltage arrester cannot be switched off from the power line momentarily (e.g., by using a time switch).

For measuring off potentials, thyristors can also be connected between pipeline and overvoltage arrester; these operate only if inadmissibly high pipeline potentials arise by making the connection to the arrester. During the rest of the time (low load time), no grounding occurs and therefore there is no problem in making *IR*-free measurements by the switching method [19]. The number of diodes follows approximately from Eq. (23-46) since U_G must not become too positive due to the anodic danger.

23.6   Measurement of Pipeline Potentials

Pipeline potentials should be measured to control interference calculations and for the determination of environmental reduction factors and ohmic interference (see Section 23.2). With inductive interference, a ground rod is used as a reference electrode in the region of the remote ground (i.e., outside the voltage cone of the pipeline or of other grounds). This is usually the case at a distance of 20 m from the pipeline. The measuring conductor must be perpendicular to the interfering high-voltage line to avoid induced voltages in the measuring conductor. The reference electrode is brought over the pipeline in the region of the voltage cone of high-voltage installations to measure the contact voltages due to ohmic interference. According to the type of interference, corresponding preparations should be made by the operator of the high-voltage installation and/or recordings carried out.

23.6.1 Measurement of Short-Term Interference

A measuring current is injected into the nonoperating high-voltage overhead power line. Pipeline potentials can be calculated for short-circuit currents because of the generally valid proportionality between current and pipeline potential as indicated in Eqs. (23-9), (23-18a, b) and (23-31). The following methods are used to avoid errors from foreign voltages [9].

23.6.1.1 Beat Method (by Superposition)

This method is used where there are interfering foreign voltages with a constant frequency (e.g., $f = 50$ Hz). In other cases a suitable filter must be connected in series. The measuring current must have a slight difference from 50 Hz (e.g., $f = 49.5$ Hz). The superposition of the disturbing voltage, U_{DV}, on the pipeline potential, U, which is being interfered with, leads to a beat with $f = (50 \text{ Hz} - 49.5 \text{ Hz})/2 = 0.25$ Hz. Given a knowledge of the extreme values of the beat (U_{max} and U_{min}) and disturbing voltage U_{DV} determined by periodically switching the measuring current on and off (25 s on and 5 s off), the pipeline potential is given by the equations:

$$\text{for } 2U_{DA} > U_{max} : U = \frac{U_{max} - U_{min}}{2} \tag{23-47}$$

$$\text{for } 2U_{DA} = U_{max} : U = \frac{U_{max}}{2} \tag{23-48}$$

$$\text{for } 2U_{DA} < U_{max} : U = \frac{U_{max} + U_{min}}{2} \tag{23-49}$$

23.6.1.2 Reverse Polarity Method

In this method a voltage source synchronized with the grid is used whose phase is reversed through 180° after a pause with no current. From vector analysis, the pipeline potential follows from the equation:

$$U = \sqrt{\frac{U_a^2 + U_b^2}{2} - U_{DV}^2} \tag{23-50}$$

where U_a and U_b = voltage with flowing measuring current (a) before and (b) after reversing the polarity; U_{DV} = disturbing voltage measured during the pause in the measuring current. Disturbing voltages of 16⅔ Hz can also be eliminated by this method provided they are constant during the period of the measurement. In other cases they must be removed by use of a filter.

23.6.2 Measurement of Long-Term Interference

To obtain accurate values, the operators of the high-voltage installations must provide recordings, during the period of the measurement, of the operating currents and the phase angles of all high-voltage lines contributing to the interference. The pipeline potentials should be recorded over a period of about 15 min synchro-

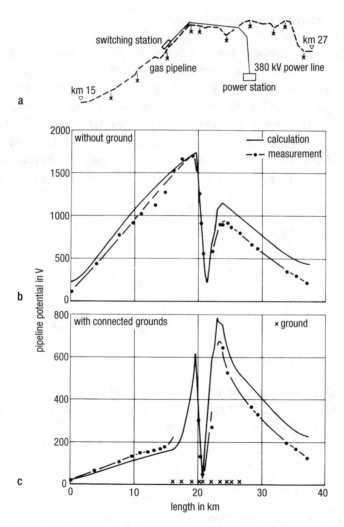

Fig. 23-14 Close proximity of a PE-coated DN 600 gas pipeline and a 380-kV high-voltage line. (a) Plan view, (b) comparison between measured and calculated pipeline potentials without connected grounds, and (c) comparison between measured and calculated pipeline potentials with connected grounds.

nized to the interfering operational current at every measuring point. Only by this means can correlations be obtained between the pipeline potentials and the data of the interfering high-voltage lines.

Where there is no pipeline measuring point in a section of the line (e.g., in the vicinity of a transposition of pylon lines), the pipeline can be simulated by a specially designed measuring line [20]. The measuring line is electrically connected to the nearest measuring points on either side of the pylon. The pipeline potentials can then be measured along the pipeline by connecting grounding rods to the measuring line.

23.6.3 Results of Pipeline Potential Measurement

Figure 23-14a shows a sketch of a region of close proximity between a PE-coated DN 600 gas pipeline and a 380-kV high-voltage overhead power line. Figure 23-14b shows the measured and calculated pipeline potentials before the connection of grounds. The pipeline is electrically connected to two other pipelines at 38 km and grounded at 0 with 1 Ω. The measuring current amounted to 256 A and the represented measured values relate to a given short circuit current of 12.2 kA. Figure 23-14c shows the comparison between calculated and measured pipeline potentials after connection to the grounds shown in Fig. 23-14a with $R_G = 1$ to 2 Ω.

23.7 References

[1] H.-U. Paul, Elektrizitätswirtschaft *85*, 98 (1986).
[2] AfK-Empfehlung Nr. 3, ZfGW-Verlag, Frankfurt 1982; Techn. Empflg. Nr. 7, VWEW-Verlag, Frankfurt 19.
[3] H.-U. Paul, Elektrizitätswirtschaft *86*, 389 (1987).
[4] DIN VDE 0210, Beuth-Verlag, Berlin 1985.
[5] H.-J. Sowade, Elektrizitätswirtschaft *75*, 603 (1976).
[6] AfK-Empfehlung Nr. 2, ZfGW-Verlag, Frankfurt 1985.
[7] H. Spickmann, ETZ-A *11*, 261 (1969).
[8] DIN VDE 0115, Beuth-Verlag, Berlin 1982.
[9] DIN VDE 0141, Beuth-Verlag, Berlin 1976.
[10] J. R. Carsson, Bell System Techn. J. S. 539 (1926).
[11] F. Pollaczek, E. N. T. *3*, 339 (1926).
[12] G. Röhrl, Elektrische Bahnen *38*, 19 u. 38 (1967).
[13] Techn. Empflg. Nr. 1, VWEW-Verlag, Frankfurt 1987.
[14] W. v. Baeckmann, H.-U. Paul u. K.-H. Feist, CIGRE-Bericht Nr. 36-02, Paris 1982.
[15] DIN VDE 0228 Teil 1, Beuth-Verlag, Berlin 1987.
[16] J. Pestka u. B. Knoche, Elektrizitätswirtschaft *81*, 518 (1982).
[17] J. Pohl, CIGRE-Konf., Bericht Nr. 326, Paris 1966.
[18] W. v. Baeckmann u. D. Weßling, 3R intern. *23*, 343 (1984).
[19] H. Rosenberg, Balslev Cons Eng. (DK), pers. Mitteilung 1987.
[20] H.-U. Paul, ÖZE *35*, 245 (1982).

Table 24-1 Calculation formulas for simple anodes (anode voltage $U_0 = IR$) (Continued)

Line	Anode shape	Anode arrangement	Grounding resistance	Remarks	Voltage cone
6	Ring-shaped ground, band width b, radius r_0		$R = \dfrac{\rho}{2\pi^2 r_0} \ln\left(\dfrac{16 r_0}{d}\right)$	$d = \dfrac{b}{2}$	$U_r = \dfrac{I\rho}{\pi^2(r_0+r)} F\left(\dfrac{2\sqrt{r_0 r}}{r_0+r}\right)^{\text{a}}$
7	Vertical anode, length l, diameter d, depth below surface t		$R = \dfrac{\rho}{2\pi l}\ln\left(\dfrac{2l}{d}\sqrt{\dfrac{4t+3l}{4t+l}}\right)$	$t \gg d$ $d \ll l$	$U_r = \dfrac{I\rho}{2\pi l}\ln\left(\dfrac{t+l+\sqrt{r^2+(t+l)^2}}{t+\sqrt{r^2+t^2}}\right)$
8	Vertical anode		$R = \dfrac{\rho}{2\pi l}\ln\left(\dfrac{2l}{d}\right)$	$t \gg l$	$U_r = \dfrac{I\rho}{2\pi l}\ln\dfrac{\sqrt{t^2+r^2+\left(\dfrac{l}{2}\right)^2}+\dfrac{l}{2}}{\sqrt{t^2+r^2+\left(\dfrac{l}{2}\right)^2}-\dfrac{l}{2}}$
9	Horizontal anode, length l, diameter d, depth below surface t		$R = \dfrac{\rho}{2\pi l}\ln\left(\dfrac{l^2}{td}\right)$	$d \ll l$ $t \ll l$	
10	Horizontal anode		$R = \dfrac{\rho}{2\pi l}\ln\left(\dfrac{2l}{d}\right)$	$t \gg l$	$U_x = \dfrac{I\rho}{2\pi l}\ln\dfrac{\sqrt{t^2+\left(x+\dfrac{l}{2}\right)^2}+x+\dfrac{l}{2}}{\sqrt{t^2+\left(x-\dfrac{l}{2}\right)^2}+x-\dfrac{l}{2}}$

[a] F is an elliptical integral and is given in Ref. 3.

The grounding resistance of an ellipsoid of rotation in infinite space is [4]:

$$R = \frac{\rho}{4\pi b} \frac{\alpha+1}{\alpha-1} \ln \alpha \qquad (24\text{-}14\text{a})$$

where

$$\alpha = 2\frac{b}{a}\left[\frac{b}{a} + \sqrt{\left(\frac{b}{a}\right)^2 - 1}\right] - 1 \qquad (24\text{-}14\text{b})$$

Here a and b are the two axes of the ellipsoid which is rotated about the b axis. On rearranging with $x = b/a$:

$$R = \frac{\rho}{2\pi a} \frac{\ln\left(x + \sqrt{x^2-1}\right)}{\sqrt{x^2-1}} = \frac{\rho}{2\pi a} \frac{\arccos x}{\sqrt{1-x^2}} \qquad (24\text{-}15)$$

For $a = b = d$ this gives the grounding resistance of the spherical anode according to Eq. (24-8) and for $a = d$ and $b \to 0$ it gives the formula for the grounding resistance of a circular plate ground in full space:

$$R = \frac{\rho}{4d} \qquad (24\text{-}16)$$

The formula for the grounding resistance of a circular plate in half space [2] which can be used as an approximation for a defect in the pipe coating gives double the resistance

$$R = \frac{\rho}{2d} \qquad (24\text{-}17)$$

For an elliptical plate in half space with the axes a and b, the resistance is approximately

$$R = \frac{\rho}{a+b} \qquad (24\text{-}18)$$

For a rectangular plate with sides a and b the resistance is somewhat lower [6]:

$$R = \frac{\rho}{1.1(a+b)} \qquad (24\text{-}19)$$

For these formulas the grounding resistance on the water surface ($\rho = 54\ \Omega$ cm) of an iron plate was measured with a thin plate with dimensions $a \times 1$ in the electrolytic trough (Fig. 24-1) and compared with the calculated values from Eq. (24-19). The agreement is sufficiently good to give the grounding resistance of a plate anode in an insulated coated surface. In full space (i.e., with sufficient distance between the anode and the surface to be cathodically protected) the resistance would have been half.

The grounding resistance of a rod anode with $b = l$, $a = 2r$ and $a \ll b$, is from Eq. (24-15) in full space:

$$R = \frac{\rho}{2\pi l} \ln \frac{l}{r} \tag{24-20}$$

The derivation for the general case of the horizontal rod anode in half space with the notation in Table 24-1, line 9 [7] gives:

$$R = \frac{\rho}{4\pi l} \left[\ln \frac{+l + \sqrt{d^2 + l^2}}{-l + \sqrt{d^2 + l^2}} \right.$$

$$\left. + \ln \frac{+l + \sqrt{(4t - d)^2 + l^2}}{-l + \sqrt{(4t - d)^2 + l^2}} \right] \tag{24-21}$$

Usually $d \ll 4t$ and $d \ll l$ so that:

$$R = \frac{\rho}{4\pi l} \left[2 \ln \frac{2l}{d} + \ln \frac{+l + \sqrt{(4t)^2 + l^2}}{-l + \sqrt{(4t)^2 + l^2}} \right] \tag{24-22}$$

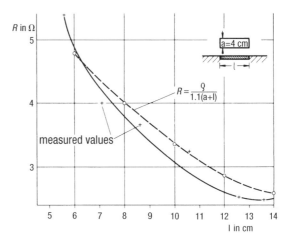

Fig. 24-1 Grounding resistance R of an iron plate at the water surface ($\rho = 54\ \Omega\ \text{cm}$) as a function of the plate length l.

For extended anodes or strip grounds, $4t \ll l$ so that the formula in line 9 in Table 24-1 becomes:

$$R = \frac{\rho}{2\pi l} \ln \frac{l^2}{td} \tag{24-23}$$

For $t \to \infty$, Eq. (24-22) for full space becomes:

$$R = \frac{\rho}{2\pi l} \ln \frac{2l}{d} \tag{24-24}$$

A formula frequently seen in western literature which is derived by integrating the potential along the rod anode is:

$$R = \frac{\rho}{2\pi l} \left(\ln \frac{2l}{r} - 1 \right) \tag{24-25}$$

which approximately agrees with Eq. (24-24) for a long, thin anode. Inserting $1 = \ln 2.72$ in Eq. (24-25) gives:

$$R = \frac{\rho}{2\pi l} \ln \frac{2l}{2.72 r} = \frac{\rho}{2\pi l} \ln \frac{1.5 l}{d} \tag{24-26}$$

For a strip ground it follows from Eq. (24-21) with $t = 0$ and $d \ll l$ in half space:

$$R = \frac{\rho}{\pi l} \ln \frac{2l}{d} \tag{24-27}$$

The grounding resistances of short rod anodes are also used to determine anode grounding resistances in seawater. Care has to be taken here that the anodes are sufficiently deep below the water surface and at least a few anode lengths away from uncoated or coated steel surfaces. Figure 24-2 shows the effect of distance x on the grounding resistance R of a rod anode with $t = 0$ near to a metal plate (curve 1) and to a plastic plate (curve 2). Furthermore, in the absence of the plate, the effect of immersion depth t on the grounding resistance of the rod is given by curve 3. Curve 3 shows that the resistance at a depth corresponding to the length of the anode is equal to the grounding resistance in full space. The measured value is, however, greater than half the value for the immersion depth at $t = 0$.

For a vertical rod anode, the general formula comparable with Eq. (24-21) is given by:

Distribution of Current and Potential in a Stationary Electric Field

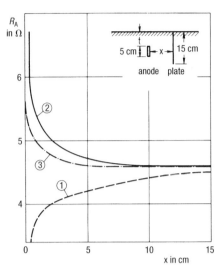

Fig. 24-2 Grounding resistance R_A of a rod anode ($d = 0.43$ cm, $L = 5$ cm) in water with resistivity $\rho = 46\ \Omega$ cm:
(1) With the approach of a metal plate as a function of the distance x,
(2) with the approach of a plastic plate as a function of the distance x, and
(3) as a function of immersion depth, t.

$$R = \frac{\rho}{4\pi l}\left[\ln\left(\frac{\sqrt{d^2+l^2}+l}{\sqrt{d^2+l^2}-l}\right)\right.$$

$$\left.+\ln\left(\frac{3l+4t+\sqrt{d^2+(3l+4t)^2}}{l+4t+\sqrt{d^2+(l+4t)^2}}\right)\right] \tag{24-28}$$

To a first approximation for $d \ll l$:

$$R = \frac{\rho}{4\pi l}\left[2\ln\left(\frac{2l}{d}\right)+\ln\left(\frac{3l+4t}{l+4t}\right)\right]$$

$$= \frac{\rho}{2\pi l}\ln\left(\frac{2l}{d}\sqrt{\frac{4t+3l}{4t+l}}\right) \tag{24-29}$$

For a full space ($t \to \infty$), Eq. (24-29) becomes the same as Eq. (24-24) (i.e., there is no difference between a horizontal and a vertical anode). For $t = -l/2$ and $d \ll l$, the grounding resistance of a rod anode of length $l/2 = l'$ at the soil surface in half space is given by:

$$R = \frac{\rho}{4\pi l}\left[2\ln\left(\frac{2l}{d}\right)+\ln\left(\frac{l+\sqrt{d^2+l^2}}{-l+\sqrt{d^2+l^2}}\right)\right]$$

$$\approx \frac{\rho}{\pi l}\ln\left(\frac{2l}{d}\right) = \frac{\rho}{2\pi l'}\ln\left(\frac{4l'}{d}\right) \tag{24-30}$$

The most important equations are collected together in Table 24-1 [2,3].

24.2 Interference Factor with Several Anodes

If anode installations for cathodic protection consist of several individual anodes of length, l, spacing, s, and grounding resistance, R, these are usually spaced far enough from each other ($s > l$) that their mutual interference can be treated by the potential distribution of a spherical anode. In practice, the anodes are covered to a certain depth, but the formula for the grounding resistance at the soil surface is mostly used (i.e., in half space) because the soil resistance is frequently greater in the upper layers of the soil, and in winter, as the result of freezing, the conduction of current is not possible. The total resistance of n anodes is given by:

$$R_n = \frac{R}{n} F \tag{24-31}$$

The interference factor, F, lies between 1.2 and 1.4 depending on the anode spacing [8,9]. The total grounding resistance of a group of anodes is obtained by summing the resistance contribution of individual anodes spaced at vs and dividing by the number of anodes [4]. This average resistance is added to each anode that physically corresponds to a uniform current loading of all the anodes. Naturally this average value ascribes too low a current to the outer anodes and too high a current to the inner anodes. An approximate average value is given by:

$$R_n = \frac{1}{n}\left[R + \frac{2}{n}\sum_{v=1}^{n-1}(n-v)\ R\ (vs)\right] \tag{24-32}$$

Because $s > l$ it follows from Eq. (24-11) that $R(vs) = \rho/(2\pi vs)$ so that the total grounding resistance of the group of anodes is:

$$R_n = \frac{1}{n}\left[R + \frac{\rho}{\pi s}\sum_{v=2}^{n}\frac{1}{v}\right] = \frac{R}{n} F \tag{24-33}$$

The following relation exists between the harmonic series of the above equation and the Euler constant ($\gamma = 0.5772$) for large values of n

$$\sum_{v=2}^{n}\frac{1}{v} \approx \gamma + \ln n - 1 = \ln(0.66\,n) \tag{24-34}$$

This gives the interference factor, F

$$F \approx 1 + \frac{\rho}{\pi s R}\ln(0.66\,n) \tag{24-35}$$

# Distribution of Current and Potential in a Stationary Electric Field

The interference factor therefore depends on the spacing s_v and the quotient R/ρ, which is determined by the anode dimensions according to Table 24-1. Figure 9-8 contains examples of various practical cases.

In Ref. 2 an accurate calculation of the potential distribution is derived for rod anodes (line 3 in Table 24-1) in a uniform, circular arrangement with a radius, r. For rod anodes with grounding resistance, R, the total resistance is given by:

$$R_n = \frac{1}{n}\left[R + \frac{\rho}{4\pi l}\sum_{v=2}^{n}\ln\frac{\sqrt{s_v^2+l^2}+l}{\sqrt{s_v^2+l^2}-l}\right] \tag{24-36}$$

where $s_v = 2r\sin[(\pi/n\,(v-1)]$ is the distance between the anodes. The interference factor from Eq. (24-36) is given by:

$$F = 1 + \frac{\rho}{4\pi l\,R}\sum_{v=2}^{n}\ln\frac{\sqrt{s_v^2+l^2}+l}{\sqrt{s_v^2+l^2}-l} \tag{24-37}$$

24.3 Potential Distribution at Ground Level

24.3.1 Soil Resistance Formulas

With the arrangement of four electrodes for measuring soil resistance according to Section 3.5.1, Fig. 3-14, an alternating current is conducted via the two outer electrodes A and B into the soil, producing a dipole field in the soil and at the earth's surface. A corresponding voltage drop, U, is experienced at electrodes C and D. To calculate the soil resistivity, ρ, from the resistance determined by $R = U/I$, the potential field arising from currents $+I$ and $-I$ must be considered at a point. Using Eq. (24-6) the potential field produced by current $+I$ at point A in the half space is given by:

$$\phi(r) = \frac{\rho I}{2\pi r} \tag{24-38}$$

Taking into consideration the negative sign of the current at point B, the potential at points which are a distance r from A and a distance R from B is

$$\phi(r) = \frac{\rho I}{2\pi}\left(\frac{1}{r}-\frac{1}{R}\right) \tag{24-39}$$

and the potential difference U between points C and D is

546 Handbook of Cathodic Corrosion Protection

$$U = \phi_C - \phi_D = \frac{\rho I}{2\pi}\left[\left(\frac{1}{b} - \frac{1}{a+b}\right) - \left(\frac{1}{a+b} - \frac{1}{b}\right)\right]$$

$$= \frac{\rho I}{\pi}\frac{a}{b(a+b)} \tag{24-40}$$

Solving for the specific soil resistivity leads to [10]:

$$\rho = \pi a \frac{U}{I}\left(\frac{b}{a} + \frac{b^2}{a^2}\right) = \pi R\left(b + \frac{b^2}{a}\right) = F(a,b)R \tag{24-41}$$

$F(a,b)$ is shown in Fig. 24-3 for various spacings, a, of the inner electrodes as a function of the spacing, b, of the outer electrodes.

24.3.2 Anodic Voltage Cone

The potential distribution around the anodes is the important factor for interference of foreign pipelines or cables in the region of the voltage cone from anodes in cathodic impressed current installations. This is obtained from the derivation of the corresponding grounding resistances and is given in Table 24-1 for various anode shapes. The hemispherical ground is considered in order to recognize the important parameters. The voltage causing interference U_r at a distance, r, is according to Eq. (24-11):

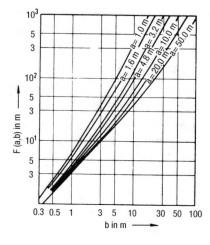

Fig. 24-3 Function $F(a,b)$ in Eq. (24-41).

Distribution of Current and Potential in a Stationary Electric Field

$$U_r = U_0 \frac{r_0}{r} = \frac{I\rho}{2\pi r} \qquad (24\text{-}42)$$

The voltage U_r increases linearly with increasing anode voltage (i.e., with increasing soil resistivity, ρ) for a given current, I. If these interfering voltages are, for example, below 0.5 V, they follow the observed spacings, r. Other voltage cones are described in Chapter 9. Figures 9-5 and 9-6 show the relative anode voltages $U_z/U_A = U_r/U_0$ or U_x/U_A for various distances and shapes of anode.

24.3.3 Cathodic Voltage Cone in a Cylindrical Field

With an uncoated pipe or one with very poor coating and many defects close together, uniform current distribution for the pipeline can be assumed in the soil even at quite short distances from it (see Section 3.6.2.2, case b).

Without taking into consideration the soil surface, the current density $J(r)$ at a distance r from the pipe axis for a pipeline (pipe diameter, d) with a protection current density J expressed in the terminology of Fig. 24-4 and corresponding to Eq. (24-4), is given by:

$$2rJ(r) = dJ \qquad (24\text{-}43)$$

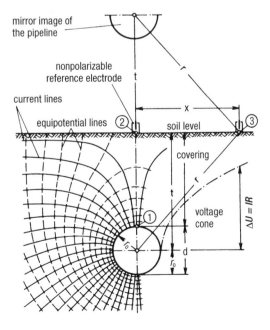

Fig. 24-4 Cylindrical field around an uncoated pipeline in soil.

548 Handbook of Cathodic Corrosion Protection

From Eqs. (24-1) and (24-43):

$$E = J(r)\rho = J\rho \frac{d}{2r} \tag{24-44}$$

After substitution in Eq. (24-3) and integrating the cylindrical potential field with integration constant, C:

$$\phi(r) = J\rho \frac{d}{2} \ln r + C \tag{24-45}$$

The electrical potential field in the soil, taking into account the soil surface, is obtained by the image method, usually applied in potential theory, for the pipeline at the soil surface and superposition of the two fields.

From this the following potentials at points (1), (2) and (3) in Fig. 24-4 are found:

$$\phi_1 = J\rho \frac{d}{2} \ln\left(dt - \frac{d^2}{4}\right) + 2C \approx J\rho \frac{d}{2} \ln(dt) + 2C \tag{24-46}$$

$$\phi_2 = J\rho \frac{d}{2} \ln t^2 + 2C = J\rho\, d\, \ln t + 2C \tag{24-47}$$

$$\phi_3 = J\rho \frac{d}{2} \ln\left(x^2 + t^2\right) + 2C \tag{24-48}$$

The important potential differences for measuring cylindrical fields are given in Section 3.6.2.2.

24.3.4 Interference from the Cathodic Voltage Cone

A foreign pipeline that is not cathodically protected can suffer anodic interference if it is within the voltage cone of defects in the coating of a cathodically protected pipeline [11]. To calculate the anodic current density J_a at the affected pipeline, the following data are assumed: the radius of the defect on the cathodically protected pipeline is r_1, the soil potential at this point with respect to the remote ground is U_0 and at a distance, a, there is a defect of radius, r_2, on the affected pipeline. The potential of the cathodic voltage cone against the remote ground at this point is calculated from Table 24-1, line 2, as:

$$\phi_1 = \frac{2}{\pi} U_0 \arctan\left(\frac{r_1}{a}\right) \tag{24-49}$$

Distribution of Current and Potential in a Stationary Electric Field

On the other hand, the anodic current, I_a, leaking via the grounding potential ($\rho/4r_2$), produces according to Eq. (24-17) a potential against the remote ground of:

$$\phi_2 = I_a \frac{\rho}{4r_2} = J_a \frac{\pi\rho}{4} r_2 \tag{24-50}$$

With $\phi_1 = \phi_2$ it follows that:

$$J_a = \frac{8 U_0}{\pi^2 \rho r_2} \arctan\left(\frac{r_1}{a}\right) \tag{24-51}$$

and, because $a \gg r_1$, by expansion of the series:

$$J_a = \frac{8 U_0 r_1}{\pi^2 a \rho r_2} \tag{24-51'}$$

24.4 Calculation of Current and Potential Distribution

The generally applicable relations for a two-conductor model are derived in the following section. For simplicity, local potential uniformity is assumed for one of the two conductor phases. Relationships for the potential and current distributions, depending on assumed current density-potential functions, are derived for various applications.

24.4.1 General Relationships for a Two-Conductor Model

Figure 24-5 contains the general data for the two-conductor model [12]. The conductor phase II has a locally constant potential based on a very high conductiv-

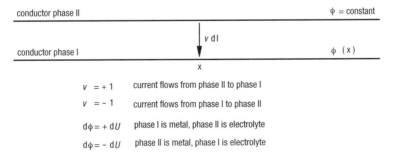

Fig. 24-5 General terms for a two-conductor model.

550   Handbook of Cathodic Corrosion Protection

ity. At point x a current exchange, $v\mathrm{d}I$, occurs between the two conductors in which the factor v indicates the sign. Inside phase I, there is a potential $\phi(x)$ dependent on position. Since this phase can be a metal or an electrolyte, there are different relationships between $\mathrm{d}\phi$ and $\mathrm{d}U$. U is the metal/electrolyte potential.

The current in phase I is given by Ohm's Law:

$$I(x) = -S\varkappa \frac{\mathrm{d}\phi}{\mathrm{d}x} \tag{24-52}$$

The relationship for the current interchange at x is:

$$v\frac{\mathrm{d}I}{\mathrm{d}x} = lJ(x) \tag{24-53}$$

where S = cross-sectional area of conductor phase I, $\varkappa$ = electrical conductivity of conductor phase I, l = contact line between the two conductor phases in a cross-sectional plane where the current, $v\mathrm{d}I$, transfers, and J = current density between the two conductor phases at point x.

The Laplace equation follows from Eqs. (24-52) and (24-53):

$$\frac{\mathrm{d}^2\phi}{(\mathrm{d}x)^2} = -\frac{vl}{S\varkappa} J(x) \tag{24-54}$$

24.4.2   Calculation of Ground Electrodes Having a Longitudinal Resistance

A ground with locally constant values of S and l in full space is regarded as conductor phase II. Therefore $\mathrm{d}\phi = \mathrm{d}U$. A linear current density-potential function is assumed for the current transfer:

$$J = -v\frac{U}{r_\mathrm{p}} \tag{24-55}$$

where r_p is the polarization resistance. According to Fig. 24-5, $v = -1$ for anodic polarization, and $v = +1$ for cathodic polarization. The resistance load R' and the leakage load of the ground for the parameters S and l are given by:

$$R' = \frac{\mathrm{d}R}{\mathrm{d}x} = \frac{1}{\varkappa S} \tag{24-56}$$

and

Distribution of Current and Potential in a Stationary Electric Field

$$G' = \frac{dG}{dx} = \frac{l}{r_p} \tag{24-57}$$

Substitution of Eqs. (24-55) into (24-57) in Eq. (24-54) gives:

$$\frac{d^2 U}{(dx)^2} = \alpha^2 U \quad \text{where} \quad \alpha = \sqrt{R'G'} \tag{24-58}$$

and after integration:

$$U = A \cosh(\alpha x) + B \sinh(\alpha x) \tag{24-59}$$

Finally from Eqs. (24-52), (24-56) and (24-59) it follows that:

$$I = -\frac{1}{Z}\left[A \sinh(\alpha x) + B \cosh(\alpha x)\right] \quad \text{where} \quad Z = \sqrt{R'/G'} \tag{24-60}$$

The constants A and B are determined by the boundary conditions.

The boundary conditions for a ground of length, L, are: $U = U_0$ and $I = I_0$ for $x = 0$ as well as $I = 0$ for $x = L$. It follows from Eqs. (24-59) and (24-60): $A = U_0 = ZI_0 \coth(\alpha L)$ and $B = -I_0 Z$ and after substituting:

$$U = U_0 \frac{\cosh[\alpha(L-x)]}{\cosh(\alpha L)} \tag{24-61}$$

$$I = I_0 \frac{\sinh[\alpha(L-x)]}{\sinh(\alpha L)} \tag{24-62}$$

The effective grounding resistance, R_w, is given by:

$$R_w = \frac{U_0}{I_0} = Z \coth(\alpha L) \tag{24-63}$$

For a better overview, the hyperbolic functions are shown in Fig. 24-6.

For an infinitely long ground, it follows from Eqs. (24-61) to (24-63) for $L \to \infty$:

$$U = U_0 \exp(-\alpha x) \tag{24-64}$$

$$I = I_0 \exp(-\alpha x) \tag{24-65}$$

$$R_w = \frac{U_0}{I_0} = Z = \sqrt{\frac{R'}{G'}} \tag{24-66}$$

552 Handbook of Cathodic Corrosion Protection

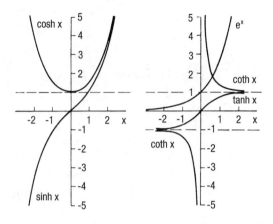

Fig. 24-6 Graphs of hyperbolic and exponential functions: $2 \sinh x = e^x - e^{-x}$; $2 \cosh x = e^x + e^{-x}$.

The characteristic resistance Z is therefore the lowest value of the grounding resistance. The path constant $1/\alpha = 1/\sqrt{R'G'} = l_k$ is the characteristic length of the infinite ground.

24.4.3 Range of Cathodic Protection and Current Requirement

A linear function corresponding to Eq. (24-55) is assumed for cathodic polarization [12-14]:

$$\eta = r_p J \tag{24-67}$$

where $\eta = -(U - U_R)$, the cathodic overvoltage. The following are taken as the limits for the protected region:

upper potential limit: $\eta_L = -(U_s - U_R)$ (24-68a)

lower potential limit: $\eta_0 = -(U_h - U_R)$ (24-68b)

Equation (24-68a) means that the protection potential, U_s, must be reached at the end of the pipeline ($x = L$), and that at the drainage point ($x = 0$), the potential must not fall below U_h because due to hydrogen evolution, according to Eq. (2-19), Eq. (24-67) is no longer valid. From experience, the following approximate values apply:

$$\eta_0 - \eta_L = \Delta U \approx \eta_L \approx 0.3 \text{ V} \tag{24-69}$$

For the lowest current density, J_s, it follows from Eq. (24-67)

$$J_s = \frac{\eta_L}{r_p} \tag{24-67'}$$

Distribution of Current and Potential in a Stationary Electric Field

It follows from Eqs. (24-56), (24-57) and (24-67′) for a pipeline of radius, r, and wall thickness, s:

$$R' = \frac{1}{2\pi rs\varkappa} \tag{24-70}$$

$$G' = \frac{2\pi r}{r_p} = \frac{2\pi r J_s}{\eta_L} \tag{24-71}$$

Substituting in Eqs. (24-58) to (24-61) with $U_0 = \eta_0$ leads to:

$$\alpha^2 = \frac{J_s}{\varkappa \eta_L s} \tag{24-72}$$

$$Z = \frac{1}{2\pi r}\sqrt{\frac{\eta_L}{\varkappa s J_s}} \tag{24-73}$$

$$\cosh(\alpha L) - 1 = \frac{\Delta U}{\eta_L} \approx \frac{(\alpha L)^2}{2} \tag{24-74}$$

The approximation after expansion of the series is permissible because of Eq. (24-69). The protected length, $2L$, is given by Eq. (24-74) with drainage at the midpoint of the pipeline:

$$(2L)^2 = \frac{8\Delta U}{\alpha^2 \eta_L} = \frac{4\Delta U}{\pi r J_s R'} = \frac{8\Delta U \varkappa s}{J_s} \tag{24-75}$$

The protection current requirement $I_s = 2I_0$ follows from Eq. (24-63) with Eq. (24-69):

$$I_s = 4\pi r \sqrt{\frac{\Delta U}{\eta_L}(\Delta U + 2\eta_L)\varkappa s J_s} \approx 4\pi r \sqrt{3\Delta U \varkappa s J_s} \tag{24-76}$$

The protected length $2L^*$ for an infinitely long pipeline corresponding to Eq. (24-68a) can be derived with the aid of Eqs. (24-64) to (24-66) for $U_0 = \eta_0$.

In contrast to the limited pipe length, the pipe current at the point $x = L^*$ is, however, not zero. From Eqs. (24-64), (24-68a) and (24-68b):

$$2L^* = \frac{2}{\alpha}\ln\left(1 + \frac{\Delta U}{\eta_L}\right) \tag{24-77}$$

554 Handbook of Cathodic Corrosion Protection

Comparison with Eq. (24-74) and taking account of Eq. (24-69) gives:

$$\frac{L^*}{L} = \sqrt{\frac{\eta_L}{2\Delta U}} \ln\left(1 + \frac{\Delta U}{\eta_L}\right) \approx \frac{\ln 2}{\sqrt{2}} = 0.49 \qquad (24\text{-}78)$$

The protection current requirement $I_s^* = 2 I_0$ follows from Eqs. (24-66) and (24-73):

$$I_s^* = 4\pi r \left(\Delta U + \eta_L\right) \sqrt{\frac{J_s \varkappa s}{\eta_L}} \qquad (24\text{-}79)$$

Comparison with Eq. (24-76), taking into account Eq. (24-69), gives:

$$\frac{I_s^*}{I_s} = \frac{\Delta U + \eta_L}{\sqrt{\Delta U(\Delta U + 2\eta_L)}} \approx \frac{2}{\sqrt{3}} = 1.16 \qquad (24\text{-}80)$$

With an infinitely long pipeline, the current requirement is raised by 16% and the protected length is reduced by 51% as a consequence of omitting the insulating unit at the point $x = L$.

In Eq. (24-67) it is assumed that r_p is constant in the potential range $U_h < U < U_R$. This could be true for resistance polarization at defects in the coating. In this case, however, a change in the IR-free potential is only possible if electrochemical polarization is also occurring. Linear polarization is, however, very unlikely in the given potential range.

On the other hand, it can be assumed for the oxygen corrosion of steel in aqueous solutions and soils that there is a constant minimum protection current density, J_s, in the protective range, $U_h \le U \le U_s$, which corresponds to the limiting current density for oxygen reduction according to Eq. (4-5) (see Section 2.2.3.2). Then it follows, with $v = +1$, $l = 2\pi r$, $S = 2\pi rs$ and $d\phi = dU$ from Eq. (24-54), instead of Eq. (24-58) [12-14]:

$$\frac{d^2 U}{(dx)^2} = -\frac{J_s}{s\varkappa} \qquad (24\text{-}81)$$

Integration with the limits $U = U_s$ and $dU/dx = 0$ for $x = 0$ as well as $U = U_h$ at $x = L$ leads to Eq. (24-75), which can therefore be regarded as correct without the approximation according to Eq. (24-74), where J_s applies for every point and not only for $x = L$. Correspondingly, the protection current requirement, I_s, of the pipeline surface area $(2\pi r)(2L)$ is, instead of Eq. (24-76), given by $4\pi r(2\Delta U \varkappa s\, J_s)^{1/2}$. Equation (24-81) and its solutions are also valid if there is an ohmic leakage load. With the constant current density J_s, this leads to a constant voltage drop at the pipeline surface, which need not be taken into account if only IR-free potentials are used in Eq. (24-81).

24.4.4 Potential Distribution in the Case of Overprotection

There is a constant protection current density in the protection region $U_h < U < U_s$. At potentials $U < U_h$, overprotection occurs so that hydrogen evolution takes place according to Eq. (2-19). The current density is clearly dependent on the potential, so that usually activation polarization is assumed to occur (see Section 2.2.3.2). Approximately from Eq. (2-35):

$$J = J_s \exp z \text{ where } z = -\frac{U - U_h}{\beta} \tag{24-82}$$

Substitution in Eq. (24-54) leads to, instead of Eq. (24-81) [12,13]:

$$\frac{d^2 z}{(dx)^2} = \frac{J_s}{s \varkappa \beta} \exp z \tag{24-83}$$

Integrating with the boundary condition $z = 0$ and $(dz/dx) = 2\Delta U/L\beta$ gives for $x = L$:

$$\frac{x - L}{L} = \frac{1}{2} \sqrt{\frac{\beta q}{\Delta U}} \ln \frac{\left(\sqrt{1 + q e^z} - 1\right)\left(\sqrt{1 + q} + 1\right)}{\left(\sqrt{1 + q e^z} + 1\right)\left(\sqrt{1 + q} - 1\right)} \tag{24-84}$$

where $q = \beta/(\Delta U - \beta)$. L corresponds to Eq. (24-75) and $(x - L)$ is the extension of the protected length due to overprotection. For unlimited overprotection ($z \to \infty$), Eq. (24-84) gives a limiting value:

$$\frac{L_{max} - L}{L} = \frac{1}{2} \sqrt{\frac{\beta q}{\Delta U}} \ln \frac{\sqrt{1 + q} + 1}{\sqrt{1 + q} - 1} \tag{24-85}$$

Using practical numbers shows that the extension of the protected length can only be about 10%.

With very high overvoltages, Eq. (24-82) is no longer applicable because of resistance polarization. It is assumed approximately that:

$$J = J_s (1 + z) \text{ where } z = -\frac{U - U_h}{b} \tag{24-86}$$

and b/J_s corresponds to the ohmic polarization resistance in the range of overprotection. Substitution in Eq. (24-54) leads in place of Eq. (24-81) to [12,13]:

$$\frac{d^2 z}{(dx)^2} = \frac{J_s (1 + z)}{s \varkappa b} \tag{24-87}$$

Integrating with limiting condition $z = 0$ and $dz/dx = 2\Delta U/Lb$ for $x = L$ gives:

$$\frac{x - L}{L} = \frac{1}{p} \ln \frac{1 + z + \sqrt{z^2 + 2z + p^2}}{1 + p} \tag{24-88}$$

where $p = \sqrt{\frac{2\Delta U}{b}}$

In contrast to Eq. (24-84), the protected length increases logarithmically with increasing overprotection. This effect is, however, small and can be neglected. The protected length can thus only be increased insignificantly.

24.4.5 Cathodic Protection in Narrow Gaps

With nonadherent coatings (e.g., as a result of disbonding; see Section 5.2.1.5), there is the question of possible corrosion damage. It was shown in Sections 2.2.5 and 5.2.4 that there is apparently cathodic protection in the gap between the coating and a pipe surface with thick coatings. Measurements indicate this [15-17] (see Fig. 5-12). Referring to Fig. 24-5, the object to be protected is phase II and the water film in the gap, phase I. The gap has breadth, w, and height, t. With cathodic polarization in the gap, the following relationships apply: $v = -1$, $d\phi = -\Delta U$, $A = tw$, $l = w$ as well as $U = U_0$ and $J = J_0$ for $x = 0$ at the mouth of the gap.

Substituting in Eq. (24-54) gives [12,17]:

$$\frac{d^2 U}{(dx)^2} = -\frac{J}{\varkappa t} \tag{24-89}$$

Two different functions $J(U)$ are introduced into Eq. (24-89). After integration the following equations concerning current and potential distribution result:

(a) linear $J(U)$ function (with resistance polarization and small overvoltages)

$$U = U_R - r_p J \tag{24-90}$$

$$J(x) = J_0 \exp(-x/a) \tag{24-91}$$

$$U(x) = U_R + (U_0 - U_R) \exp(-x/a) \tag{24-92}$$

where

$$a^2 = r_p \varkappa t \tag{24-93}$$

(b) exponential $J(U)$ function (activation polarization with large overvoltage)

Distribution of Current and Potential in a Stationary Electric Field

$$U = U_R - \beta \ln(J/J_c) \tag{24-94}$$

(J_c is the corrosion current density in free corrosion)

$$J(x) = J_0(1 + x/b)^{-2} \tag{24-95}$$

$$U(x) = U_0 + 2\beta \ln(1 + x/b) \tag{24-96}$$

where $\quad b^2 = \dfrac{2\varkappa\beta t}{J_o} \tag{24-97}$

Cathodic protection can be observed beneath nonadherent PE coating as in Eq. (24-96) [16].

24.4.6 Distribution of Current and Potential Inside a Pipe at Insulating Units

Insulating units are installed in pipelines to limit cathodic protection or to separate different materials in a mixed installation. If the pipelines are transporting electrolytes, anodic interference can occur on the pipe interior if a dc voltage, ΔU, exists at the insulator of length, L. The current flowing through the insulating unit by anodic polarization of the internal wall of the pipe comes to:

$$I_0 = -\dfrac{\Delta U \pi r^2 \varkappa}{L} \tag{24-98}$$

In the terminology of Fig. 24-5, the metal pipe is phase II and the electrolyte in the pipe is phase I. With anodic polarization of the pipe, the relationships are: $v = +1$, $d\phi = -dU$, $l = 2\pi r$, $S = \pi r^2$, as well as $I = I_0$ and $dU/dx = -\Delta U/L$ for $x = 0$ at the pipe and insulating unit interface. Substituting in Eq. (24-54) leads to [12,18]:

$$\dfrac{d^2 U}{(dx)^2} = \dfrac{2J}{r\varkappa} \tag{24-99}$$

Two different functions $J(U)$ are used in Eq. (24-99) so that on integrating, equations for the current and potential distribution are obtained:

(a) linear $J(U)$ function (with resistance polarization and small overvoltages)

$$U = U_R + r_p J \tag{24-100}$$

$J(x)$ and $U(x)$ corresponding to Eqs. (24-91) and (24-92) with

$$a^2 = \dfrac{r r_p \varkappa}{2} \tag{24-101}$$

$$J_0 = \frac{\Delta U}{L}\sqrt{\frac{r\varkappa}{2r_p}} \qquad (24\text{-}102)$$

(b) logarithmic $J(U)$ function (with activation polarization and large overvoltage)

$$U = U_R + \beta \ln(J/J_c) \qquad (24\text{-}103)$$

J_c has the same definition as in Eq. (24-94); $J(x)$ and $U(x)$ correspond to Eqs. (24-95) and (24-96) with

$$`b = \frac{2L\beta}{\Delta U} \qquad (24\text{-}104)$$

$$J_0 = \left(\frac{\Delta U}{2L}\right)^2 \frac{r\varkappa}{\beta} \qquad (24\text{-}105)$$

J_0 alone serves for evaluating the corrosion danger (see Section 20.1.4). It is of interest to note that with protective measures to reduce J_0 there is no change, with the quotients $\Delta U/L$ and r/L^2 remaining the same. Dimensionless relationships are very complicated [12]. With the linear function, the polarization parameter $k = \varkappa r_p$ only occurs in the path constants in Eq. (24-101) and not in Eq. (24-102).

24.5 General Comments on Current Distribution

The primary distribution of protection current density (see Section 2.2.5) for a given geometry and driving voltage, U_T, can be seen as follows:

$$J(x) = \frac{U_T}{f(x)}\varkappa \qquad (24\text{-}106)$$

$f(x)$ has its lowest value at the point nearest the anode with coordinate $x = 0$. The secondary current distribution can be obtained approximately by addition of the resistances:

$$J(x) = \frac{U_T}{\dfrac{f(x)}{\varkappa} + r_p} \qquad (24\text{-}107)$$

Here r_p is the specific cathodic polarization resistance which is assumed to be a constant in what follows. The cathodic $J(U)$ curve is therefore given by:

Distribution of Current and Potential in a Stationary Electric Field

$$U = U_R + r_p J(x) \tag{24-108}$$

From Eqs. (24-106), (24-107) and (2-44), it follows:

$$U = U_R + \frac{U_T}{1 + \frac{f(x)}{k}} \tag{24-109}$$

$k = r_p \varkappa$ is the polarization parameter. The width of the protection region, a, follows from Eq. (24-107) with U_1 ($x = 0$) as the lower potential limit and U_2 ($x = a$) = U_s.

$$U_1 - U_2 = \Delta U = U_T \left(C - \frac{1}{1 + \frac{f(a)}{k}} \right) \tag{24-110}$$

where $C = \dfrac{1}{1 + \dfrac{f(0)}{k}}$

The quantity, C, is dependent on k and smaller than 1; because of the high value of $J(0)$, $f(0)$ is small and therefore C is approximately equal to 1. By rearranging, it follows from Eq. (24-110):

$$f(a) = k \frac{\Delta U + U_T(1 - C)}{CU_T - \Delta U} \approx \frac{k\Delta U}{U_T - \Delta U} \tag{24-111}$$

Since $f(a)$ is a monotonically increasing function, the protection region, a, increases with the polarization parameter, k. As an example, a symmetrical coplanar electrode arrangement with equally large anodic and cathodic polarization resistances is considered. Here $f(x)$ is defined as [19]:

$$f(x) = \pi x + k \tag{24-112}$$

From this, $C = 0.5$ from Eq. (24-110) and with Eq. (24-111):

$$a = \frac{4}{\pi} \frac{k\Delta U}{U_T - 2\Delta U} \tag{24-113}$$

The protection range, a, is therefore proportional to the polarization parameter.

24.6 References

[1] K. Küpfmüller, Einführung in die theoretische Elektronik, Springer-Verlag, Berlin 1955.
[2] W. Koch, Erdungen in Wechselstromanlagen über 1 kV, 4. Auflage, Springer-Verlag 1968.
[3] F. Ollendorf, Erdströme, 2. Auflage, Birkhäuser-Verlag, Basel 1971.
[4] E. D. Sunde, Earth Conduction Effects in Transmission Systems, D. van Nostrand Company, New York 1949.
[5] AfK-Empfehlung Nr. 9, ZfGW-Verlag, Frankfurt 1979.
[6] E. R. Sheppard u. H. J. Gresser, Corrosion *6*, 362 (1950).
[7] J. Pohl, in: Europäisches Symposium, Kathodischer Korrosionsschutz, Deutsche Gesellschaft für Metallkunde, Köln 1960, S. 325.
[8] J. H. Morgan, Cathodic Protection, Leonard-Hill-Books, London 1959, S. 66.
[9] L. M. Applegate, Cathodic Protection, McGraw-Hill-Book, New York 1960, S. 129.
[10] W. v. Baeckmann u. G. Heim, gwf *101*, 942 u. 986 (1960).
[11] F. Schwarzbauer, B. Thiem u. E. Sachsenröder, gwf gas/erdgas *120*, 384 (1979).
[12] W. Schwenk, Corrosion Sci. *23*, 871 (1983).
[13] W. Schwenk, 3R international *20*, 466 (1981).
[14] W. v. Baeckmann, gwf/gas *104*, 1237 (1963).
[15] W. Schwenk, gwf gas/erdgas *123*, 158 (1958).
[16] W. Schwenk, 3R international *26*, 305 (1987).
[17] R. R. Fessler, A. J. Markworth u. R. N. Parkins, Corrosion *39*, 20 (1983).
[18] H. Hildebrand u. W. Schwenk, 3R international *21*, 367 (1982).
[19] H. Kaesche, Korrosion der Metalle, 1. Auflage, Springer-Verlag Heidelberg, New York 1966, S. 231 f.

Index

Anode bed. *See* Impressed current ground beds
Anodes. *See* Specific type of anode
Anodes and grounds, grounding resistance of, 536–44
Anodic protection
    basis, 40
    history, 14
Attached anodes, 405–6

Basket anodes, 222–23
Blistering, electrochemical, 163–66, 169, 170
Boilers. *See* Cathodic protection of water tanks and boilers, internal

Cable anodes
    advantage, 221
    reinforcing steel in concrete structures, 434–35, 436
    telephone cables, 331–32
Cables. *See* Power cables; Telephone cables
Cathodic disbonding, 156, 166–69, 172–73, 443
Cathodic protection
    ac interference, 150–51
    basis, 40
    cathodic hydrogen and, 33
    criteria, 45
    history, 9–20
    metallic coatings, 176
Cathodic protection of reinforcing steel in concrete structures
    application, 431–38
    corrosion, causes of, 428
    criteria, 429–31
    electrolytic properties of concrete, 428–29
    steel-concrete corrosion system, 427–28
    stray current effects and protective measures, 438–39
Cathodic protection of ships
    below the waterline, 397–409
    bilges, 412–13
    docks, 413
    economics, 500
    Edison, Thomas, 12
    effect of materials and coating parameters, 395–97
    heat exchangers, condensers and tubing, 412
    tanks and containers, 410–12
    water parameters, 391–94
Cathodic protection of water tanks and boilers, internal
    boilers with electrolytically treated water, 456–58
    boilers with enamel linings, 450–55
    description and function of objects to be protected, 441–47
    drinking water requirements, 462
    filter tanks, 461
    galvanic anodes, 447–48
    hydrogen evolution, danger prevention, 446–47
    impressed current, 448–50
    interference prevention, 444–46, 454–55
    preconditions, 443–44
    water storage tanks, 458–61
Cathodic protection of well casings
    causes of corrosion danger, 415–17
    commissioning, maintenance and control, 425
    description, 415
    design and construction, 421–24
    economics, 499–500, 501
    measurements, 418–21
Cathodic protection station
    measurement protocol, 238
    troubleshooting, 239
Chemical plant, anodic protection of, 474–85
Coating defects, location of, 125, 127–31
Coating resistance
    comparison of specific, 157
    corrosion rate and specific, 159
    influence of temperature on, 158

Coating resistance (*continued*)
    of long-distance pipelines, 158
    measurement, 110, 112
    protection current demand and,
        156–62
Coatings
    cement mortar, 154, 173–75
    corrosion of steel under, 172–73
    enamel, 154, 175–76
    for galvanic anode supports, 199
    and linings for water tanks and boilers, 443
    metallic, 154–55
    objectives and types, 153–55
    organic, 153–73
        See also Organic coatings
Cohen's rule, 165
Columbia rod, 117, 118
Compact anodes, 200–201
Corrosion
    aqueous electrolytes, 34–36
    in aqueous media, 148
    in aqueous solutions and soil, 139–52
    basic thermodynamics, 37–40
    defined, 27
    determining likelihood in uncoated metals, 142–48
    due to ac interference, 150–51
    due to cell formation, 148–50
    due to stray currents, 148–50
    electrical conductivity and, 34–36
    electrochemical, defined, 29
    field tests in soils, 146, 147
    flow diagram of processes, 28
    fundamentals and concepts, 27–78
    hydrogen-induced, 65–69, 70–71, 73
    metallic materials, 30–33
    mixed electrodes, 44–51
    phase boundary reactions, 36–44
    types, 27–30
        See also Specific type of corrosion
Corrosion fatigue, 29, 69–71
Corrosion protection
    fundamentals and concepts, 27–78
    importance of potential, 52–75
    overview, 30
        See also History of corrosion protection
Costs. *See* Economics

Crevice corrosion, 29
Current density measurement, 110, 112
Current distribution in a stationary electric field, 535–59
    calculation of, 549–58
    general comments, 558–59
    interference factor with several anodes, 544–45
Current measurement
    general advice, 107–8
    in pipe, 108–9
Cylindrical anodes, 219–22
Cylindrical double anodes, 220

Deep anodes
    first use, 17–18
    local cathodic protection, 311, 312–16, 320
    pipeline protection, 285
    power cables, 343
        See also Impressed current ground beds, deep anode beds
DIN numbers, explanation of DIN 50900, Part I, 1
DIN standards, development of, 23–24
Diodes, silicon
    construction and action of, 342–43
    as voltage arresters for pipelines, 529–30
Disc-shaped anodes, 221

Economics
    buried pipelines, cathodic protection of, 492–99, 501
    general comments, 491–92
    internal protection, 501–3
    seawater, corrosion protection in, 500–501
    well casings, corrosion protection of, 499–500, 501
Edison, Thomas, 12
Electrochemical kinetics, 40–44
Electrodes
    cell formation and heterogeneous mixed, 46–50
    Columbia rod, 117, 118
    homogeneous mixed, 44–46
    Shephard rod, 116–17, 118
    Wenner rod, 117, 118

Electrolytic corrosion, 30–51
Fault location, 119–31
Floating anodes, 221–22
Four-electrode process, 113–16

Galvanic anodes
    advantages and disadvantages, 204–5
    aluminum, 188–91
    apparatus for determining weight loss, 195
    backfill materials, 196–98
    current capacity, 180–83
    current discharge from, 183–85
    in drinking water, 462
    forms, 199–203
    general information, 179–80
    iron, 185
    magnesium, 191–96
    offshore, 202
    pipeline protection, 278–79
    power cables, 343
    processing equipment, vessels, and tubes, 466–67
    pure metals as, 181
    quality control and performance testing, 203–4
    rest potentials of various, 193
    special forms, 202–3
    storage tanks, 295–98
    supports, 198–99
    surface films, 179–80, 187
    for tanks, 201–2
    water tanks and boilers, 443–44, 446–48, 450–53
    zinc, 185–88
Galvanostat, 234–36
Gas production platforms. *See* Marine structures, platforms
Gathering anodes, 402

History of corrosion protection
    for buried pipelines, 1–8
    cathodic protection, 9–20
    development of DIN standards, 23–24
    evolution of ferrous metals technology, 4
    by painting, 8–9
    stray current protection, 20–23

Horizontal anodes
    grounding resistance, 538–39, 541, 543
    local cathodic protection, 311, 313
Hydrogen induced corrosion, 65–71

Impressed current anode beds. *See* Impressed current ground beds
Impressed current anodes
    cables for, 218
    data for use in soil, 209
    general comments, 207–8
    insulating materials for, 217–18
    for internal application, 222–23
    materials, 208–17
    pipeline protection, 279–80, 285
    polymer cable, 217
    processing equipment, vessels, and tubes, 467–83, 485–86
    solid, composition and properties of, 212
    storage tanks, 295
    suitable for soil, 219–21
    suitable for water, 221–22
    for telephone cables, 329–33
    water tanks and boilers, 444, 446–62
    See also Impressed current equipment; Impressed current ground beds; Local cathodic protection
Impressed current equipment
    control rectifiers, 233–36
    design and circuitry, 228–29
    protection measures, site and electrical, 226–27
    rectifier circuit, 229–30
    rectifiers resistant to high voltage, 232–33
    transformer-rectifiers, adjustable, 230–32
    transformer-rectifiers, equipment and control of, 237, 240–41
    transformer-rectifiers, without mains connections, 237
Impressed current ground beds
    anode beds, continuous horizontal, 244–48
    anode beds, deep, 250–53
    design and cost considerations, 254–56

Impressed current ground beds (*continued*)
   general information, 243
   interference with foreign pipelines and cables, 256–63
   single anode installations, 248–50
Ingot-shaped anodes, 221
Intensive measurement technique, 131–37
Interference effects of high-voltage transmission lines on pipelines
   capacitive interference, 505, 506–7
   defined, 505–6
   inductive interference, 505, 510–24
   limiting lengths and distances, 524–26
   measurement of pipeline potentials, 530–33
   ohmic interference, 505, 507–10
   protection measures against unallowably high pipeline potentials, 526–30
Interference problems
   impressed current ground beds, 256–63
   storage tank protection, 294
   telephone cables, 326, 330
   water tanks and boilers, 444–46, 454–55
   well casings, 417, 423–24, 425
   See also Stray current interference
Intergranular corrosion, 29

Kinetics, electrochemical, 40–44

Local anodes, location of, 124–25
Local cathodic protection
   applications, range of, 309
   data for examples of, 314
   installations with small steel-reinforced concrete foundations, 317
   oil refineries, 314, 315–17
   power cables, 336
   power stations, 312–15
   special features, 310–12
   tank farms, 317–21
   well casings, 422, 425
Local corrosion
   data from field experiments, 147
   described, 48–50
   metallic coatings, 176

   processing equipment, vessels, and tubes, 474
Location
   of coating defects, 125, 127–31
   of faults using alternating current, 122–23
   of faults using direct current, 120–22
   of heterogeneous surface areas, 123–24
   of local anodes, 124–25
   of pipeline with pipe locator, 122–23, 124

Marine structures
   cathodic protection measures, 367–73
   economics of protection, 500–501
   harbor structures, 376–80, 387
   offshore pipelines, 383–84, 388–89
   piling foundations, 380–83
   piling, steel sheet, 380
   platforms, 373–76, 385–86
Measurement
   coating resistance, 110, 112
   current density, 110, 112
   electrical, fundamentals and practice, 79–138
   grounding resistance, 118–19
   intensive measurement technique for pipelines, 131–37
   of pipeline potentials, 530–33
   resistivity, 112–19
   specific soil resistivity, 114–18
   survey of instruments for corrosion protection, 82–83
   See also Potential measurement
Metals
   for anode supports, 198–99
   as coatings, 154–55, 176
   corrosion in aqueous solutions and soil, 148–51
   as galvanic anodes, 180–83, 185–96
   noble, 176, 207, 213–16, 223, 448, 465
   safety and, 446, 462, 465
   substrate, 176
   valve, 208, 213–17, 221, 448, 451

Oil production platforms. *See* Marine structures, platforms

Organic coatings
    blistering, electrochemical, 163–66, 169, 170
    cathodic disbonding, 156, 166–69, 172–73, 443
    cell formation, effectiveness of cathodes and, 162–63
    properties, electrical and electrochemical, 155–69
    properties, mechanical, 170–71
    properties, physicochemical, 169–70
    stress corrosion, 172–73
    for water tanks and boilers, 443

Pearson method, 125, 128
Phase boundary reactions, 36–44
Pipelines
    cathodic protection, costs of, 492–99
    cathodic protection, design of, 276–85
    cathodic protection, monitoring and supervision of, 287–88
    commissioning the cathodic protection station, 285–87
    historical protection, 15–23
    historical use, 1–8
    interference, capacitive, 505, 506–7
    interference, inductive, 505, 510–24
    interference, ohmic, 505, 507–10
    measurement of potential difference, 124–25, 126
    measurement of potentials to control interference, 530–33
    preconditions for protection, 268–75
    steel, electrical properties of, 265–67
    See also Interference effects of high-voltage transmission lines on pipelines; Marine structures, offshore pipelines

Pitting corrosion
    described, 29
    economics of protection, 498–99
    passivated metals, 61–63
Plate anodes
    docks and, 413
    grounding resistance, 538, 540–41
    for internal application, 222–23
    supports for, 198
    uses, 200–201
    water turbines, 470–72

Platforms. *See* Marine structures, platforms
Polarization cells
    construction and action of, 341
    as voltage arresters for pipelines, 529
Potential
    conversion factors, 37
    diffusion, 85–88
    electric, 37–38
    electrochemical, 37–38
    pitting, 62–63
    ranges, determining, for effective protection, 52–78
    ranges, survey of critical, 71–75
    rest, 49
    standard, 39
Potential distribution in a stationary electric field, 535–59
    calculation of, 549–58
    at ground level, 545–49
Potential measurement
    alternating currents, influence of, 102–3
    of difference in pipelines, 124–25, 126
    electrodes with flowing current, 88–96
    instruments, measuring, and their properties, 96–97
    pipelines, to control interference calculations, 530–33
    pipelines and storage tanks, 97–100
    potential test probes, 106–7
    reference electrodes, 85–88
    stray currents, influence of, 100–102
Potential reversal, of zinc galvanic anodes, 187
Potentiostat
    described, 234–36
    processing equipment, vessels, and tubes, 477–80
Power cables
    buried, properties of, 335–36
    cathodic protection of the steel conduits, 336–45
    dc decoupling devices, 337–43
    stray current protection, 345
Power plant condenser, cooled by seawater, internal cathodic protection of, 469
Processing equipment, vessels, and tubes, internal protection of
    chemical plant, 474–85

Processing equipment, vessels, and tubes,
    internal protection of (*continued*)
        galvanic anodes, 466–67
        impressed current, 467–74
        power plant condenser cooled by
            seawater, 469
        special features, 464–66
        trends, 485–86
        water turbine, 469–74
        wet gasometer, 468–69
        wet oil tanks, 467–68
Protection criteria
    application of, 103–7
    for nonalloyed ferrous materials,
        104, 106
    for steels, plain carbon and low-alloy,
        in soil, 105

Recessed anodes, 405, 406–7
Rectifiers. *See* Impressed current equipment
Reference electrodes
    data and application range of
        important, 80
    grounding rods as, 103
    potential measurement, 85–88
    processing equipment, vessels, and
        tubes, 465–66, 468–73, 477–80
    ships, 408–9
    water tanks and boilers, 448–49,
        459–60
Reinforcing steel. *See* Cathodic protection of
    reinforcing steel in concrete structures
Resistance
    coating, 51
    measurement of grounding, 118–19
    polarization, 48–51
    three-electrode method of measuring
        grounding, 118–19
Resistance of anodes and grounds,
    grounding, 536–43
Resistivity
    four-electrode process of measuring,
        113–16
    measurement of, 112–19
    measurement of specific soil, 114–18
Rod anodes
    alloys for, 191
    docks, 413

grounding resistance, 538, 541–43,
    545
for internal application, 222–23
power plant condensers, 469
supports for, 198
uses, 200
in water, 221
water turbines, 471
Rudder anodes, 401

Sacrificial anodes
    first use, 18–19
        See also Galvanic anodes
Safety
    measures for control and maintenance,
        490–91
    pipeline failure statistics, 489–90
Shepard rod, 116–17, 118
Ships. *See* Cathodic protection of ships
Soils
    characteristics, 145
    corrosion likelihood in various,
        144–48
Spherical anodes, grounding resistance of,
    536–38, 540
Storage tanks
    coordinating protection measures, 306
    current demand, evaluation, and
        connections, 292–95
    galvanic anodes, 295–98, 305, 306
    impressed current installations, 295,
        298–99
    internal protection, 304–5
    measures for dissimilar metal
        installations, 304
    operation and maintenance of cathodic
        protection stations, 307
    preparatory measures, 290–92
    protection methods, choice of, 295
    special problems, 290
    special problems near railways,
        300–304
    tank farms and filling stations,
        299–300
Stray current interference
    causes, 347–48
    from dc railways, 348–53
    due to telluric currents, 355–58

24

Distribution of Current and Potential in a Stationary Electric Field

W. V. BAECKMANN AND W. SCHWENK

In this chapter some important equations for corrosion protection are derived which are relevant to the stationary electric fields present in electrolytically conducting media such as soil or aqueous solutions. Detailed mathematical derivations can be found in the technical literature on problems of grounding [1-5]. The equations are also applicable to low frequencies in limited areas, provided no noticeable current displacement is caused by the electromagnetic field.

The stationary electric field is described by the following equations:

1. The vector Ohm's Law

$$\mathbf{E} = \mathbf{J}\rho \tag{24-1}$$

from which because

$$\mathbf{E} = -\operatorname{grad}\phi = -\frac{d\phi}{dr} \tag{24-2}$$

the potential ϕ in polar coordinates is given by

$$\phi(r) = -\int E \, dr \tag{24-3}$$

2. The field outside a local current lead is source free. This means that the surface integral is equal to a current introduced by a ground:

$$I = \int_S J(r) dS = J(r) \, 4\pi r^2 \tag{24-4}$$

Equations (24-3) and (24-4) correspond to Kirchhoff's laws for electrical networks. Figure (24-2) gives the voltage distribution for every point in space for a given field strength.

24.1 Grounding Resistance of Anodes and Grounds

For the simplest case, the grounding resistance of a spherical anode in surrounding space is derived. The resistance between the spherical anode of radius, r, and a very distant, very large counter electrode (remote ground) is termed the grounding resistance of the anode. The major part of the resistance lies in the soil in the immediate vicinity of the anode. The total grounding resistance of the anode, that is, the resistance between its lead and the infinitely large and distant remote ground, is composed of three terms:

1. The resistance in the lead and the anode itself, which usually is so small that it can be neglected: with extended cable connections, anodes or pipelines, the voltage drop in the metal must, of course, be taken into account.
2. The transition resistance between the surface of the metal and the electrolyte: with uncoated iron anodes in coke backfill, the transition resistance is usually low. With metals in soil, it can be increased by films of grease, paint, rust or deposits. It contains in addition an electrochemical polarization resistance that depends on the current [see Eq. (2-35)].
3. The grounding resistance that is given by the current and potential distribution in the electrolyte and that will be considered in detail in what follows.

With a spherical-shaped anode of radius, r_0, which is immersed very deep in an electrolyte ($t \gg r_0$), the current radiates uniformly outward in all directions. Between this spherical anode and the infinitely large and distant remote ground, a voltage U is applied which involves a current $I = U/R$. R is the grounding resistance.

The impressed current, I, flows from the spherical anode radially in a symmetrical field (i.e., the equipotential lines represent spherical shells). It follows from Eq. (24-1)

$$E = \rho J(r) = \frac{\rho I}{4\pi r^2} \tag{24-5}$$

The potential ϕ at position r relative to the remote ground ($r \to \infty$) is given from Eq. (24-3):

$$\phi(r) = -\int_r^\infty E\, dr = -\frac{\rho I}{4\pi} \int_r^\infty r^{-2} dr = \frac{\rho I}{4\pi r} \tag{24-6}$$

From this, the voltage U_0 of the anode with $r = r_0$ is given by

$$U_0 = \phi(r_0) = \frac{\rho I}{4\pi r_0} \tag{24-7}$$

Distribution of Current and Potential in a Stationary Electric Field

and therefore the grounding resistance R of the spherical anode is

$$R = \frac{U_0}{I} = \frac{\rho}{4\pi r_0} \tag{24-8}$$

There are analogous formulas for calculating capacity. The capacity of a sphere is given by:

$$C = \frac{Q}{U} = 4\pi r_0 \varepsilon_0 \varepsilon \tag{24-9}$$

If one forms the quotients R/ρ and $\varepsilon_0\varepsilon/C$, this gives identical functions which contain only geometric parameters. This regularity is independent of the geometric shape; i.e., it follows generally from the capacity formula and also the formula for grounding resistance [1] that

$$R = \frac{\varepsilon_0 \varepsilon \rho}{C} \tag{24-10}$$

Current and voltage distribution remain completely the same if a plane cuts the sphere through its center and the upper half is removed (half space). Since only half the current flows from the hemisphere, its grounding resistance is obtained by substituting $I/2$ for I in Eq. (24-8) (see Table 24-1, column 1):

$$R = \frac{2U_0}{I} = \frac{\rho}{2\pi r_0} = \frac{\rho}{\pi d} \tag{24-11}$$

If the spherical anode is situated at a finite depth, t, the resistance is higher than for $t \to \infty$ and lower than for $t = 0$ (hemisphere at the surface of the electrolyte). Its value is obtained by the mirror image of the anode at the surface ($t = 0$), so that the sectional view gives an equipotential line distribution similar to that shown in Fig. 24-4 for the current distribution around a pipeline. This remains unchanged if the upper half is removed (i.e., only the half space is considered).

For the anode voltage, it follows from Eq. (24-7) with $2r_0 = d$:

$$U_0 = \frac{\rho I}{4\pi r_0} + \frac{\rho I}{8\pi t} = \frac{\rho I}{2\pi}\left(\frac{1}{d} + \frac{1}{4t}\right) \tag{24-12}$$

The grounding resistance is given by:

$$R = \frac{U_0}{I} = \frac{\rho}{2\pi}\left(\frac{1}{d} + \frac{1}{4t}\right) \tag{24-13}$$

Table 24-1 Calculation formulas for simple anodes (anode voltage $U_0 = IR$)

Line	Anode shape	Anode arrangement	Grounding resistance	Remarks	Voltage cone
1	Hemisphere, radius r_0, diameter d		$R = \dfrac{\rho}{\pi d}$	Spherical field	$U_r = U_0 \dfrac{r_0}{r} = \dfrac{I\rho}{2\pi r}$
2	Circular plate, diameter d, radius r_0		$R = \dfrac{\rho}{2d}$	Surface	$U_r = \dfrac{2}{\pi} U_0 \arcsin\left(\dfrac{r_0}{r}\right)$
				Depth	$U_t = \dfrac{2}{\pi} U_0 \arctan\left(\dfrac{r_0}{t}\right)$
3	Rod anode, length l, diameter d		$R = \dfrac{\rho}{2\pi l} \ln \dfrac{4l}{d}$	$l \gg d$	$U_r = \dfrac{I\rho}{2\pi l} \ln\left(\dfrac{l + \sqrt{l^2 + r^2}}{r}\right)$
4	Horizontal anode, length l, diameter d		$R = \dfrac{\rho}{\pi l} \ln \dfrac{2l}{d}$	$l \gg d$	$U_r = \dfrac{I\rho}{\pi l} \ln\left(\dfrac{l}{2r} + \sqrt{1 + \left(\dfrac{l}{2r}\right)^2}\right) \approx \dfrac{I\rho}{2\pi r}$
					$U_x = \dfrac{I\rho}{2\pi l} \ln\left(\dfrac{2x+l}{2x-l}\right) \approx \dfrac{I\rho}{2\pi x}$
					[the approximation holds for $(r, x) \gg l$]
5	Sphere, diameter d, depth below surface t		$R = \dfrac{\rho}{2\pi}\left(\dfrac{1}{d} + \dfrac{1}{4t}\right)$	$t \gg d$	$U_r = \dfrac{I\rho}{2\pi \sqrt{t^2 + r^2}}$

from high-voltage dc power lines,
353–55
protective measures, 358–65
reinforcing steel in concrete structures
and, 438–39
Stray current protection
development, 20–23
economics, 497–98
for power cables, 345
for telephone cables, 327–29
Stray currents, 34
historical reference, 7, 18–23
potential measurement, influence on,
100–102
Stress corrosion
organic coatings, 172–73
passive metals, 63–65
processing equipment, vessels, and
tubes, 481–82, 485
well casings, 417
Surface films
described, 139–42
pH value, 34–36
power plant condensers, 469
zinc galvanic anodes, 187

Tanks. *See* Storage tanks
Telephone cables
impressed current anodes for, 329–33
laying cables, 323–24
protection, cathodic, 326–33
protection, from passive corrosion,
324–26
protection, from stray currents, 327–29
rest and protection potentials in soil,
325

Thermodynamics, basic, 37–40
Three-electrode method, 118, 119
Towing anodes, 405
Transformer-rectifiers. *See* Impressed current
equipment
Tubes. *See* Processing equipment, vessels,
and tubes, internal protection of

Uniform corrosion, 29

Vessels. *See* Processing equipment, vessels,
and tubes, internal protection of

Water tanks. *See* Cathodic protection of
water tanks and boilers, internal
Water turbine, internal cathodic protection of,
469–74
Weight loss corrosion
of active metals, 53–59
of passive metals, 59–61
Well casings. *See* Cathodic protection of well
casings
Wenner method, 113, 115–17
Wenner rod, 117, 118
Wet gasometer, internal cathodic protection
of, 468–69
Wet oil tanks, internal cathodic protection of,
467–68
Wire anodes
ships and, 412–13
supports for, 198

MORE TITLES for MECHANICAL ENGINEERS from Gulf Publishing Company

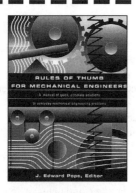

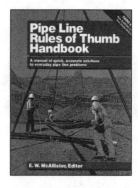

Rules of Thumb for Mechanical Engineers
J. Edward Pope, Editor

Save time with this collection of straightforward, common-sense techniques that provide quick, accurate solutions to your engineering problems. Now assembled into one convenient volume, this easy-to-read manual brings you: hundreds of shortcuts, useful calculations, practical "how-to" methods, and concise background reviews.

Experts share tips in the book's 16 chapters packed with design criteria and practical methods—all without pages of theory.

1996. 406 pages, figures, tables, charts, appendix, index,
8 3/8" x 10 7/8" paperback.
ISBN 0-88415-790-3
#5790   £42 $79

Pipeline Risk Management Manual
Second Edition
W. Kent Muhlbauer

Here's the ideal tool if you're looking for a flexible, straightforward analysis system for your everyday design and operations decisions.

1996. 438 pages, figures, tables, glossary, 6" x 9" hardcover.
ISBN 0-88415-668-0
#5668   £55 $75

http://www.gulfpub.com

Visit Your Favorite Bookstore

Or order directly from:

Gulf Publishing Company
P.O. Box 2608 • Dept. LG
Houston, Texas 77252-2608
713-520-4444
FAX: 713-525-4647

Send payment plus $9.95 ($11.55 if the order is $75 or more, $15.95 if the order is $100 or more) shipping and handling or credit card information. CA, IL, PA, TX, and WA residents must add sales tax on books and shipping total.

Prices and availability subject to change without notice

Thank you for your order!

Pipe Line Rules of Thumb Handbook
Third Edition
E. W. McAllister, Editor

This handbook provides you with quick and accurate solutions to your everyday pipeline problems. With 100 more pages than the Second Edition, it saves you valuable time and effort with useful tips on conversion factors, construction and design, gas engineering, oil products, corrosion, economics, and much more.

Significant new material on welding, electrical design, and corrosion has been added. It also covers hydrostatic testing, pipeline drying, liquids, pumps, valves, leak detection, and pipeline maintenance.

1993. 542 pages, figures, tables, graphs, photos, index, large-format flexible cover.
ISBN 0-88415-094-1
#5094   £48 $69

Rules of Thumb for Chemical Engineers
Carl Branan, Editor

This standard reference for chemical engineers helps solve field engineering problems with its hundreds of common sense techniques, shortcuts, and calculations. The safety chapter covers lower explosive limit and flash, flammability, and static charge.

1994. 368 pages, figures, tables, charts, appendixes, index, large-format flexible cover.
ISBN 0-88415-162-X
#5162   £57 $79